Ecology of the Bay of Quinte: Health, Management and Global Implications

ECOVISION WORLD MONOGRAPH SERIES

Series Editor

M. Munawar

Managing Editor

I. F. Munawar

Ecology of the Bay of Quinte: Health, Management and Global Implications

Edited by:

C.K. Minns, M. Munawar and M.A. Koops

AQUATIC ECOSYSTEM
HEALTH & MANAGEMENT SOCIETY

ECOVISION
World Monograph Series

Ecovision World Monograph Series
Aquatic Ecosystem Health and Management
Burlington, Ontario, Canada

Cover photo credits: Front cover inset, top to bottom: Satellite image of the Great Lakes, Bay of Quinte marked by a star - NASA; The Bay Bridge, Belleville - "Sunsets and Sunrises" Gary Dafoe; Fishing on the Bay at dawn – "Waiting for Sunrise" Art Saunders; Picton Harbour - R. Bonnell, Fisheries & Oceans Canada; Winter on the Bay - Eldon Burchart, Bay of Quinte Restoration Council; Picton Harbour - R. Rozon, Fisheries & Oceans Canada. Back: Picton Harbour - Fisheries & Oceans Canada.

Cover design*: J. Lorimer, M. Munawar, AEHMS*

ISBN: Book: 978-0-9921007-5-9
Electronic book: 978-0-9739184-6-9

First published 2022

Aquatic Ecosystem Health and Management Society (AEHMS),
P.O.BOX 80001 Appleby Line,
Burlington, ON. L7L 6B1, Canada

Typeset by Alpha ebook, 301 Platinum Complex, Kasturba Road, Rajkot 360001, India

Printed by Integrated Books International 22883 Quicksilver Drive, Dulles. VA 20166, USA

Distributed by Michigan State University Press,
1405 S. Harrison Rd., Ste 25, East Lansing, Mi-48823, USA.

Citation:

Minns, C.K., Munawar, M., Koops, M.A
Title: Ecology of the Bay of Quinte: Health, Management and Global Implications
Ecovision World Monograph Series,
Aquatic Ecosystem Health and Management Society,
Burlington, ON, Canada.

Ecovision Editors

Mohiuddin Munawar, Series editor
Iftekhar Fatima Munawar, Managing editor

Ecovision World Monograph Series
Aquatic Ecosystem Health & Management Society
Burlington, ON, Canada

Mohi.munawar@dfo-mpo.gc.ca
If.munawar@gmail.com
www.aehms.org

Advisory Committee

Table of Contents

Ecology of the Bay of Quinte: Health, Management and Global Implications

C.K. Minns, M. Munawar and M.A. Koops (Editors)

Dedication

Fond memories of Dr. Joe Leach: A visionary friend of the Great Lakes, 1931-2019

Obituaries usually have a sad touch and are imbued with feelings of loss and a sense of grieving. We therefore decided to devote this publication as a memorial dedication, to avoid the inevitable sense of loss and associated gloom, and instead attempt a celebration of Joe Leach's productive life and his immense contributions to Great Lakes science.

I knew Joe for more than 40 years, since the time he used to work for the Ontario Ministry of Natural Resources (OMNR) in Wheatley. We used to visit his lab during our trips to sample the algal blooms of Western Lake Erie. Lake Erie was the geographical focus of Joe's research, but his ideas about needing to understand how stressors like eutrophication affected the foodweb and ultimately fisheries, were important to the development of Project Quinte and research programs throughout the Great Lakes. These ideas permeate the research presented here and for this reason, the book is dedicated to him.

Although he retired from OMNR in early 1996, our connection was not affected. Joe continued to be in touch through various activities of the Aquatic Ecosystem Health & Management Society (AEHMS) from 1989 until 2019. He was our regular and prompt editorial advisory board member, reviewing many Great Lakes papers, especially on Lake Erie.

The AEHMS celebrated his career by convening a special symposium in collaboration with Gary Sprules during the 39th International Association for Great Lakes Research (IAGLR) conference at the University of Toronto, Canada. This symposium resulted in the publication of a book "State of Lake Erie: Past, present and future" (SOLE) (Munawar et al. (Eds.), 1999). It contained a preface by renowned scientist Jack Vallentyne (1999) and Joe's impressive biography (Nepszy, 1999), covering his educational and research training from the University of Toronto in 1954, University of Guelph in 1956 and finally his Ph.D. from the University of Aberdeen in 1969. The SOLE book also contained a well-written article by Joe on Lake Erie (Leach, 1999). We took the opportunity of organizing another celebratory party during IAGLR, 2009 in Toledo, when Joe was honoured with a Lifetime Achievement award. His later years were then engrossed in research and advisory roles on the Great Lakes, AEHMS and especially Lake Erie health.

Dr. Leach's Great Lakes contributions include a long list of topics. His pioneering work, in collaboration with Dr. Edward Mills, Hugh MacIsaac and Gary Sprules (MacIsaac, 1991, 1992; Mills et al., 1993) investigated the threats of exotic species (Zebra Mussels) in the Great Lakes and was noteworthy in this regard (Leach, 1991; Mills et al., 1994). For additional reading on Joe's pioneering publications on exotic species (Zebra Mussels) in the Great Lakes from 1991–1999, see Neary and Leach (1992), Wormington and Leach (1992), Martel et al. (1994), Fitzsimons et al. (1995), and MacIsaac et al. (1995; 1999). One of his papers discussing exotics in the Great Lakes was the most cited article in IAGLR (614 times). It is difficult to cover all the extensive activities in which Joe was involved, but a few are high-lighted here to convey the breadth of his knowledge:

- Pioneered research on exotics in the Great lakes
- Organized 21st conference and president of IAGLR in 1978
- Received Anderson Everett award, 1992
- Received Lifetime Achievement award, 2008
- Member of Great lakes Fishery Commission committees
- Member of International Joint Commission committees
- Linked fisheries and water quality management
- Promoted an ecosystem management approach
- Member of the Fishery Commission's Board of Technical Experts on the Great Lakes
- Member of several committees and boards, both scientific and community based
- Advisor to AEHMS, *AEHM* and Ecovision World Monograph Series

Joe had amazing farsightedness in supporting ecosystem-based sciences, In fact, I owe him much gratitude for his mentoring of the Society, as well as highlighting my Great Lakes research by nominating me for various IAGLR awards.

Mark Fitzpatrick and I were lucky to have the opportunity to visit Joe at his Kingsville home for the afternoon of February 19, 2018. We met with he and his wife, Mary, and he was in great spirits, entertaining us with stories about the Great Lakes, his travel, family and friends. We spent a couple of hours in his company and enjoyed going through photographs over tea and delicious cake (Figure 1).

I remember sending him these photos, for which he called to thank me. Unfortunately, my last contact with him was during a phone conversation in the summer of 2019, as he had called to congratulate me for receiving the Jack Vallentyne outreach award from IAGLR in June 2019, at their Brockport conference. His enthusiastic, congratulatory voice still echoes in my ears, loud and clear. We have indeed lost a great mentor, thinker and friend and I can honestly say, the Great Lakes will never be the same again without him.

I greatly appreciate the assistance of Drs. E. Mills, Ken Minns, John Hartig, and Mark Fitzpatrick in the preparation of this memorial dedication. Jennifer Lorimer edited the piece and prepared the attached collage of photos.

References

Fitzsimons, J.D., Leach, J.H., Nepszy, S.J., Cairns, V.W., 1995. Impacts of Zebra Mussel on Walleye (*Stizostedion vitreum*) Reproduction in Western Lake Erie. Canadian Journal of Fisheries and Aquatic Sciences 52(3), 578–586.

Leach, J.H., Lewis, C.A., 1991. Fish Introductions in Canada - Provincial Views and Regulations. Canadian Journal of Fisheries and Aquatic Sciences 48(Supp 1), 156–161. Leach, J. H., 1999. Lake Erie: passages revisited. In: M. Munawar, T. Edsall and I.F. Munawar (Eds.), *State of Lake Erie (SOLE)-Past, present and future*, pp. 5-22. Ecovision World Monograph Series, Backhuys Publishers, Leiden, The Netherlands. MacIsaac, H.J., Sprules, W.G., Leach, J.H., 1991. Ingestion of Small-Bodied Zooplankton by Zebra Mussels (*Dreissena polymorpha*) - Can Cannibalism on Larvae Influence Population-Dynamics. Canadian Journal of Fisheries and Aquatic Sciences 48(11), 2051–2060.

MacIsaac, H.J., Sprules, W.G, Johannsson, O.E., Leach, J.H., 1992. Filtering Impacts of Larval and Sessile Zebra Mussels (*Dreissena polymorpha*) in Western Lake Erie. Oecologia, 92(1), 30–39.

MacIsaac, H.J., Lonnee, C.J., Leach, J.H., 1995. Suppression of Microzooplankton By Zebra Mussels - Importance of Mussel Size. Freshwater Biology 34(2), 379–387. MacIsaac, H.J., Johannsson, O.E., Ye, J., Sprules, W.G., Leach, J.H., McCorquodale, J.A., Grigorovich, I.A., 1999. Filtering impacts of an introduced bivalve (Dreissena polymorpha) in a shallow lake: Application of a hydrodynamic model. Ecosystems 2(4), 338–350.

Martel, A, Mathieu, A.F., Findlay, C.S., Nepszy, S.J., Leach, J.H., 1994. Daily Settlement Rates of the Zebra Mussel, *Dreissena polymorpha*, on an Artificial Substrate Correlate with Veliger Abundance. Canadian Journal of Fisheries and Aquatic Sciences 51(4), 856–861.

Mills, E.L., Leach, J.H., Carlton, J.T., Secor, C.L., 1993. Exotic Species in the Great Lakes - A History of Biotic Crises and Anthropogenic Introductions. Journal of Great Lakes Research 19(1), 1–54.

Mills, E.L., Leach, J.H., Carlton, J.T., Secor, CL, 1994. Exotic Species and the Integrity of the Great-Lakes - Lessons from the Past. Bioscience 44(10), 666–676. Munawar,

M., Edsall, T., Munawar, I.F. (Eds.), 1999. *State of Lake Erie (SOLE)-Past, present and future.* Ecovision World Monograph Series, Backhuys Publishers, Leiden, The Netherlands.

Neary, B.P., Leach, J.H., 1992. Mapping the Potential Spread of the Zebra Mussel (*Dreissena polymorpha*) in Ontario. Canadian Journal of Fisheries and Aquatic Sciences. 49(2), 406–415.

Nepszy, S., 1999. Joseph Leach: a biography. In: M. Munawar, T. Edsall, I.F. Munawar (Eds.), *State of Lake Erie (SOLE)-Past, present and future*, pp. 3-4. Ecovision World Monograph Series, Backhuys Publishers, Leiden, The Netherlands.

Vallentyne, J. R., 1999. Preface. In: M. Munawar, T. Edsall and I.F. Munawar, (Eds.), *State of Lake Erie (SOLE)-Past, present and future,* pp. xv-xvi. Ecovision World Monograph Series, Backhuys Publishers, Leiden, The Netherlands.

Wormington, A., Leach, J.H., 1992. Concentrations of Migrant Diving Ducks at Point-Pelee-National-Park, Ontario, in Response to Invasion of Zebra Mussels, *Dreissena polymorpha.* Canadian Field-Naturalist 106(3), 376–380.

Mohi Munawar, President and Chief Editor

Aquatic Ecosystem Health &
Management Society, Canada

Figure 1. A collage of memories of Joe Leach. Top: Gary Sprules and Joe Leach at IAGLR 2003; Mid-Left: Mary and Joe. Leach with Mohi Munawar, 2018; Mid-Right: Joe Leach and Ed Mills in Toledo, 2009; Bottom Left: Party in Toledo, 2009 (Bob Heath, Joe Leach, Mohi Munawar, Ed Mills); Bottom Right: Mary Leach and Joe Leach with Mark Fitzpatrick, 2018.

Editorial

Any major enterprise in life has its own history. The Bay of Quinte project is no exception. It arose from a local nutrient control program and turned into a major contribution to Great Lakes science. Since the 1970s, the project has contributed a tremendous amount of information about nutrient management similar to other major initiatives such as Schindler and Vallentyne (1974) and Vollenweider et al. (1974). Major initial players in Project Quinte were Murray Johnson, (Ontario Water Resources Commission (OWRC) and DFO's Great Lakes Laboratory for Fisheries and Aquatic Sciences (GLLFAS)) as well as active researchers like Glen Owen, Michael Michalski (OWRC), Jack Christie and Don Hurley from Ontario Ministry of Natural Resources. They succeeded in persuading the government of Ontario to coordinate the timing of sewage treatment plant (STP) upgrades (Michalski et al., 1975) at sites emptying into the Bay and to establish a long-term multi-trophic and multi-disciplinary ecosystem research project in 1972—the birth of Project Quinte. During the same period, Johnson and Brinkhurst (1971) promoted an ecosystem energy flow approach to Project Quinte. This approach was supported by active scientists, such as Jack Christie and Don Hurley at the Glenora Fisheries Station, who were involved with the assessment of important fisheries in the Bay and eastern Lake Ontario. Jack Christie was instrumental in incorporating "Ecosystem Approach" into a renewed Great Lakes Water Quality Agreement (Christie et al., 1986). The core program consisted of weekly monitoring trips during open-water to sample a series of stations down the bay alongside a fish community assessment program at a similar series of sites. Assessment of benthos, temperature and oxygen conditions were also included early in the program. The assessment proceeded to establish baseline conditions prior to the winter of 1977-78, when new STPs started up. Dramatic changes were observed from the summer of 1978 onwards, showing the reduction of nutrient, chlorophyll, and phytoplankton concentrations.

In the late 1970s, the project team settled on an overall goal statement: While many elements were linked to nutrient management efforts, more were concerned with emerging theories about the structure and function of ecosystems, which lead to excellent contributions to Great Lakes science.

Goal and Sub-goals of Project Quinte

"To understand the relationship among nutrient loading, ecological characteristics and processes, and societal benefits of the Bay of Quinte ecosystem"

By the 1980s, the Project Quinte team started to consider how to organize and publish the preliminary results of this whole ecosystem "experiment". The group realized their work had originally been a limited study of the recovery of an impacted ecosystem, and that they had been laying the foundation for a long-term comprehensive study of an ecosystem experiencing ongoing succession of multiple stresses and disturbances.

The first substantial publication by Project Quinte members consisted of a set of 24 papers in the Canadian Special Publication of Fisheries and Aquatic Sciences (Minns et al., 1986). There, the authors showed how various components of the ecosystem were responding to decreased P loading. However, P loading was still decreasing, and it was apparent that insufficient time had passed for noticing the full effects of decreased eutrophication. Also, recovery of the Walleye fishery, a major goal of restoration, was a complicated response driven mainly by a climatic phenomenon and not directly linkable to reduced P levels. This complication had a positive effect stimulating further ideas about the potential for research of foodweb interactions and natural events shaping the responses of ecosystems to a range of management actions. Nicholls and Hurley (1989) showed that the crash of White Perch stock size explained more of the changes in algal biomass than P changes in the 1972-1984 period straddling the 1977-78 drop in Point-source P loading and challenged the conventional thinking about the effects of eutrophication. Another milestone was the advent of ecosystem-based modelling efforts (Koops et al., 2009) and inclusion of microbial foodweb assessment to explore the autotrophic to heterotrophic dynamics relevant to the microbial loop and algal blooms (Munawar et al., 2011).

Eventually, the Aquatic Ecosystem Health and Management Society (AEHMS) got involved with Project Quinte and played a key role in promoting its ecosystem-based research by convening a major symposium at the 53rd IAGLR conference held at the University of Toronto in 2010. The symposium was successful, well attended, and resulted in the publication of a special issue in *AEHM* (AEHMS, 2011). Due to an overwhelming response, a second issue was later published in the following year (AEHMS, 2012).

The idea of this book "Ecology of the Bay of Quinte: Health, Management and Global Implications" was then conceived to capture and synthesize the huge data base into a coherent story by republishing various printed papers from the journal, along with some new additions to promote Bay of Quinte ecosystem-based science globally. Dr. Ken Minns was invited by the AEHMS to lead the book synthesis, which he did splendidly and enthusiastically.

For the convenience of readers, the book's 26 chapters are slotted into 6 themes, followed by a collage and acknowledgements.

Introduction
Physical and chemical regime
Community structure and foodweb interactions
Ecosystem health indicators and management
Emerging themes
Synthesis
Down memory lane: Collage of project Quinte photos
Acknowledgements

The word cloud given below tries to capture the highlights of the book.

This project benefited from several colleagues through their contributions in the conception and development of Project Quinte, who are included above. In addition, hundreds of summer students and technical assistants contributed to the intensive fieldwork. Landmark work included the establishment of the Project Quinte database, which sowed the seeds of several other papers included in the book. Personally, I immensely enjoyed working with Ken Minns and Martin Koops during the long process of monograph preparation, which broadened my ecosystem understanding and horizons. Sincere thanks are also due to Edward Mills, Professor Emeritus, Cornell University, Ithaca for contributing an excellent Preface for this book. Financial support of Fisheries and Oceans Canada for the publication of this monograph is gratefully acknowledged. and we are very pleased and grateful to include the Executive Prelude by Dr. Gavin Christie, Director, for which we sincerely thank him. The efforts of Lynn Bouvier and Dorothy Bothwell are appreciated in arranging the Executive Prelude.

The AEHMS provided much needed professional and editorial support during the production of the monograph and contributions of Jennifer Lorimer and Susan Blunt are noteworthy for their patient, meticulous editing and coordination of many manuscripts. In a way, it is sad to see the end of a golden era with the project being wrapped up after almost 50 years. Surely, in these ever-shifting modern times, researchers will regard with envy, this long-term research project that resulted in a large number of excellent contributions, not only to Great Lakes ecosystem health, but also as a beacon offering guidance, lessons and inspiration to the world of aquatic sciences.

References

AEHMS (Aquatic Ecosystem Health & Management Society), 2011. Special issue: Ecosystem Health and Recovery of the Bay of Quinte, Lake Ontario. Aquat. Ecosyst. Health Mgmt. 14(1).

AEHMS (Aquatic Ecosystem Health & Management Society), 2012. Special issue: Ecosystem Health and Recovery of the Bay of Quinte, Lake Ontario: Part II. Aquat. Ecosyst. Health Mgmt. 15(2).

Christie, W.J., Becker, M., Cowden, J.W., Vallentyne, J.R., 1986. Managing the Great Lakes Basin as a home. J. Great Lakes Res. **12**, 2–17.

Johnson, M.G. and Brinkhurst, R.O., 1971. Benthic community metabolism in Bay of Quinte and Lake Ontario. J. Fish. Res. Board Can. 28, 1715–1725. Koops, M.A., Dermott, R.M., Leisti, K.L. Johannsson, O.J., Millard, S., Minns, C.K., Munawar, M., Nicholls, K.H., Hoyle, J., 2009. The Bay of Quinte: a model for large lake ecosystem management. Verh. Internat. Verein. Limnol. 30(7), 1024-1029.

Michalski, M.F.P, Nicholls, K.H., Johnson, M.G., 1975. Phosphorus Removal and Water Quality Improvements in Gravenhurst Bay, Ontario. Verh. Internat. Verein. Limnol. Vol. 19, 644-659.

Minns, C.K., Hurley, D.A., Nicholls, K.H. (Editor), 1986. Project Quinte: point-source phosphorus control and ecosystem response in the Bay of Quinte, Lake Ontario. Can. Spec. Publ. Fish. Aquat. Sci. 86.

Munawar, M, Fitzpatrick, M., Niblock, H., Lorimer, J., 2011. The relative importance of autotrophic and heterotrophic microbial communities in the planktonic foodweb of the Bay of Quinte, Lake Ontario, 2000-2007. Aquat. Ecosyst. Health Mgmt. 14(1), 21-32.

Nicholls, K.H. and Hurley, D.A., 1989. Recent Changes in the Phytoplankton of the Bay of Quinte, Lake Ontario: the Relative Importance of Fish, Nutrients, and Other Factors. Can. J. Fish. Aquat. Sci. 46(5), 770–779

Schindler, D.W., Vallentyne, J.R., 2008. *The Algal Bowl: Overfertilization of the World's Freshwaters and Estuaries.* University of Alberta Press, AB.

Vollenweider, R.A., Munawar, M., Stadelmann, P., 1974. Comparative review of phyto plankton and primary production in the Laurentian Great Lakes. J. Fish. Res. Board Can.31, 739–762.

M. Munawar, Series editor

Ecovision World Monograph Series

Aquatic Ecosystem Health & Management Society

Executive Prelude

The Bay of Quinte was wonderfully cool and strikingly clear as I swam on a recent warm summer morning. Being surrounded by the beauty of that little cove on the lower Bay, I had chance to reflect on how much good has come to this place. That good has come from thoughtful people taking on the challenge of understanding the causes of problems and how to tackle those issues. That good has come from the commitment of government to use that understanding to take action. The Bay of Quinte, like all the degraded Areas of Concern around the Great Lakes, suffered decades of assault with the impact of European settlement and development. Without the good work that has happened here, I would not be enjoying that swim. That good work came in the face of amazing changes, as I was reminded when I stepped on sharp shells of Zebra Mussels wading out of the Bay.

Project Quinte has long been part of my life. My late father, Jack Christie, a research scientist with the Ontario Ministry of Natural Resources, was among those who saw the need to understand the problems facing the Bay and who collaborated to launch Project Quinte in 1972 (Christie et al., 1986). Since its launch, until its end in 2018, Project Qunite has become the longest ecosystem-based monitoring effort and dataset in the Great Lakes. This monograph reflects that full history building on previous publications, including a special bulletin of Canadian Journal of Fisheries Science (Minns et al., 1986) and more recently two special issues of *Aquatic Ecosystem Health Management* (AEHMS, 2011, 2012). This monograph provides a broad sweep of the many facets of aquatic ecosystem structure and function which were explored in efforts to define the problems and solve the challenges to ecosystem health of the Bay of Quinte ecosystem.

The Project Quinte science documented here has been critical to decisions to take restoration actions that have done important good and will sustain the Bay far into the future. Many papers provide an extensive perspective that highlight the need for long-term ecosystem monitoring and for science to understand underlying functions. The monograph provides an excellent coverage of Great lakes science in development through 26 in-depth chapters and opens a wide horizon for exploration and application of the ecosystem approach with much remaining to be done by those that follow in these footsteps. Today, I am proud to work with the Great Lakes Laboratory for Fisheries &

Aquatic Sciences (GLLFAS DFO) of Fisheries and Oceans Canada. GLLFAS-DFO has been involved deeply in this project all along from beginning to end. Overall, about 100 co-contributors have collectively participated towards the preparation of 26 peer-reviewed chapters, out of which 18 chapters are contributed by 13 DFO scientists. This amounts to 70% share of the monograph by GLLFAS-DFO scientists. This is a tremendous contribution by a single agency, Fisheries & Oceans Canada, towards the publication of the monograph. The Department provided continued financial and logistic support throughout the life of the project, together with critical support from Environment and Climate Change Canada and its Great Lakes Action Plan.

I am also pleased that publication of this magnitude was initiated, organized and edited by senior staff of GLLFAS-DFO namely Drs. Ken Minns, Mohi Munawar & Marten Koops. The endeavor was supported by the editorial expertise of AEHMS, Canada under the umbrella of Ecovision World Monograph Series.

I am indeed excited and delighted to inaugurate the launch of this monograph by congratulating the editors and authors for generating a great contribution to Great Lakes science full of ideas, approaches, tools, and techniques for doing more good on the Great Lakes. Best wishes to Michigan State University, East Lansing MI for worldwide success in the distribution of the monograph.

References

AEHMS (Aquatic Ecosystem Health & Management Society), 2011. Special issue: Ecosystem Health and Recovery of the Bay of Quinte, Lake Ontario. Aquat. Ecosyst. Health Mgmt. 14(1).

AEHMS, 2012. Special issue: Ecosystem Health and Recovery of the Bay of Quinte, Lake Ontario: Part II. Aquat. Ecosyst. Health Mgmt. 15(2).

Christie, W.J., Becker, M., Cowden, J.W., Vallentyne, J.R., 1986. Managing the Great Lakes Basin as a home. J. Great Lakes Res.12, 2–17.

Minns, C.K., Hurley, D.A., Nicholls, K.H. (Eds), 1986. Project Quinte: point-source phosphorus control and ecosystem response in the Bay of Quinte, Lake Ontario. Can. Spec. Publ. Fish. Aquat. Sci. 86.

Fisheries and Oceans Canada Pêches et Océans Canada Canada

Gavin Christie, Division Manager

Great Lakes Laboratory for Fisheries and Aquatic Sciences

Fisheries and Oceans Canada

867 Lakeshore Rd., Burlington ON. Canada. L7R 4A6

Adolphus Reach, Lower Bay of Quinte, July 2020. G. Christie

Foreword
The Bay of Quinte: An ecological treasure

Long-term ecological studies require a dedication to scientific excellence, a passion for the ecosystem, and often begin in response to a public concern or crisis. Such studies are becoming rare, but are vital if we are to understand the interactions of human activities with natural variability.

The central theme of this book is to empower scientists and managers with both an understanding of history and the power of current knowledge to guide the future of an important ecosystem, the Bay of Quinte (BOQ).

Project Quinte began in response to excessive phosphorus loading in an eutrophic ecosystem and evolved into a comprehensive study melding science, policy, and management. Ecological studies and monitoring of the BOQ ecosystem now spans almost 50+ years and reflects the tireless efforts of many dedicated scientists, technicians, and students. These studies have also required the sustained support of policy makers and the public for nearly 50 years. Broad based ecosystem monitoring programs such as Project Quinte, have tremendous value-added benefits toward the understanding and response to environmental problems like eutrophication, invasive species, climate change, human health, and fish recruitment. It goes without saying that the detailed information and knowledge gained from the BOQ and presented in this book represent an ecological treasure that must be sustained and maintained well into the future. There are several elements that make this book so unique. The book focuses on a unique natural resource enjoyed by the public and recognized as an economic engine for the region. Furthermore, the BOQ is linked to a larger ecosystem in the Great Lakes. This book documents a case history of ecological events, including surprises that have evolved over the last 50 years. The use of state-of-the art analyses and models to synthesize complex interactions illustrates the research talent that has made the story of the Bay of Quinte so complete and so comprehensive. Thanks to Project Quinte, the lessons learned from this ecosystem and presented in this book are now available to a global community of researchers.

In closing, the Bay of Quinte book is an ecological treasure that will provide a point of reference for the understanding of nearshore-embayment ecological systems world-wide for many decades. The Bay of Quinte now has a well-documented history that will no doubt evolve and change with time. The book appropriately ends with a synthesis, "The Ecology of the Bay of Quinte: Health, Management and Global Implications" providing insights that will guide future research studies and strategies for management while the BOQ acts as a barometer of global change. Good show!!

Edward L. Mills

Professor Emeritus
Department of Natural Resources
College of Agriculture and Life Sciences
Cornell University Biological Field Station
900 Shackelton Point Road
Bridgeport, New York 13032 USA

Project Quinte 1972-2008 and Beyond: Long-term ecosystem studies in the Bay of Quinte, Lake Ontario: A prospectus

Charles K. Minns[1,2], Mohuiddin Munawar[1], Marten A. Koops[1], E. Scott Millard[1]

[1]*Fisheries and Oceans Canada, 867 Lakeshore, Rd., ON L7S 1A1*
[2]*Department of Ecology and Evolutionary Biology, University of Toronto, 25 Willcocks St., Toronto, ON M5S 3B2*

Introduction

Janzen (2009) described the sites of long-term ecological and ecosystem studies as, "*'listening places' – places where we press our ears to the earth and strain to hear its pulse*". The Bay of Quinte is one of those places, though we mainly 'listen' by dipping our nets, probes and sample bottles. This prospectus provides some historical context and an over view of two sets of papers, which document our growing, but still incomplete, understanding of the Bay of Quinte (hereafter Quinte) ecosystem.

A brief history of Project Quinte

By the late 1960s, the problem of freshwater eutrophication was evident throughout the developed world and there was a consensus that excess phosphorus in untreated sewage, human waste and laundry detergents, was the primary cause (cf Schindler and Vallentyne, 2008: a recent update of Jack Vallentyne's 1974 book telling the whole story). Efforts to upgrade sewage treatment plants (STPs) had begun around the Great Lakes. A pilot P-control project on Gravenhurst Bay in Lake Muskoka, Ontario (Michalski et al., 1975) likely provided the early thinking for Project Quinte. Murray Johnson, then at the Ontario Water Resources Commission (OWRC) and later with DFO's Great Lakes Laboratory for Fisheries and Aquatic Sciences (GLLFAS), and colleagues Glen Owen, Michael Michalski at OWRC, along with Jack Christie and Don Hurley at Ontario Ministry of Natural Resources, saw an opportunity in Quinte to study the whole ecosystem response to reduced phosphorus loading. They persuaded Ontario to coordinate the timing of STP

Ecology of the Bay of Quinte: Health, Management and Global Implications
Editors: C.K. Minns, M. Munawar and M.A. Koops

Aquatic Ecosystem Health & Management Society, Canada

upgrades at sites emptying into Quinte and to establish a long-term multi-trophic ecosystem project beginning in 1972; this became Project Quinte. Substantial upgrades to STPs at Trenton, Belleville, Deseronto, Napanee and Picton came on-stream in the winter of 1977-78.

Johnson had done his Ph.D. thesis on energy flow in the Bay of Quinte (Johnson and Brinkhurst, 1971) and promoted an ecosystem energy flow approach to Project Quinte.

This was supported by fishery scientists, Jack Christie and Don Hurley, at the Glenora Fisheries Station established to assess the important fisheries present in Quinte and eastern Lake Ontario. Christie participated in the discussions that eventually led to an "Ecosystem Approach" being incorporated into a renewed Great Lakes Water Quality Agreement (Christie et al., 1986).

Project Quinte was comprised of a variety of research and monitoring activities; however, the core activity took the form of weekly monitoring cruises visiting a series of stations down the productivity gradient that exists through the bay, alongside a fish community assessment program at a similar series of sites. Assessments of benthos, temperature and oxygen conditions were also key monitoring activities established early in the program. The assessment proceeded to establish baseline conditions prior to the winter of 1977-78. Project Quinte saw dramatic changes from the summer of 1978 onwards. Nutrient, chlorophyll, and algal metrics declined. Assessing the response of the fish community was complicated by the occurrence of a substantial concurrent winterkill in the winter of 1977-78. This winterkill adversely affected the then-dominant Alewife, *Alosa pseudoharengus*, and White Perch, *Morone americana*, stocks while releasing percid recruitment notably through a dramatic rise in Walleye, *Sander vitreus*, abundance in subsequent years. As the 1980s began, Project Quinte started to consider how to organize and publish the preliminary results of this whole ecosystem "experiment".

1980 goals statement

Project Quinte participants recognized, as they started to assemble data and begin synthesis for a major publication of results (Minns et al., 1986), that they were not participating in a fixed-term study of the recovery of an ecosystem from the impact of severe eutrophication. Instead, they were laying the foundation for a long-term study of an ecosystem as it experiences an ongoing succession of changes and disturbances. In the late 1970s, this realization led project members to spend much time and debate on defining an overall goal: "To understand the relationship among nutrient loading, ecological characteristics and processes, and societal benefits of the Bay of Quinte ecosystem". The statement appeared in the 1980 Annual Report of Project Quinte (see Supplementary Materials). The overall goal was supported by eight (A through H) sub-goals, each with up to seven objectives. While many elements were linked to nutrient management efforts, many were concerned with emerging theories about the structure and function of ecosystems.

These discussions of goals helped focus thinking, provided strategic direction for Project Quinte members and led to the publication of a set of 24 papers in a Canadian Special Publication of Fisheries and Aquatic Sciences (Minns et al., 1986). In this publication, authors showed how various components of the ecosystem were responding to decreased P loading. However, P loading was still decreasing and it was apparent that insufficient time had passed for the full effect of less eutrophic conditions to be seen. Also, recovery of the Walleye fishery, a major goal of restoration, was a complicated response as it was driven mainly by a climatic phenomenon and not directly linkable to reduced P levels. This complication had a positive effect in that it further stimulated ideas about the potential for foodweb interactions and natural events to shape the responses of ecosystems to a range of management actions. Nicholls and Hurley (1989) showed that the crash of White Perch stock size explained more of the changes in algal biomass than [P] changes in the 1972-1984 period straddling the 1977-78 drop in point source P loading and challenged the conventional thinking about the effects of eutrophication.

Project Quinte prepared to settle in for the long haul taking appropriate actions to streamline monitoring through the mid-1980s, keeping costs under control and expanding research investigations as events dictated or scientific understanding grew. Project Quinte became embedded in the Lake Ontario Bioindex (LOB) program that began in 1981 (Johannsson et al., 1998) and provided fresh impetus for the study of productivity, particularly at the lower trophic levels, along a wider gradient of trophic and lake conditions. A similar program ran on Lake Erie (LEB) for several years. These intensive temporal monitoring programmes led onto synoptic spatial programs on Lake Ontario and Erie (LOTT and LETT). Jointly, these monitoring programs facilitated a wide range of ecosystem research. For example, Millard et al.

(1999) compared primary production processes at the Great Lakes' sites and in a size series of inland lakes, and Dermott (2001) was able to document the rapid demise of the benthic amphipod *Diporeia* using data from Project Quinte and LOB sampling.

Remedial action planning for the Bay of Quinte Area of Concern

In 1986, the International Joint Commission (IJC) re-designated Quinte and 42 other degraded sites with continuing water quality and a range of environmental problems around the Great Lakes as Areas of Concern (AOCs). AOCs were areas with persistent problems identified via beneficial use impairments (BUIs). Quinte had eleven of the fourteen possible BUIs. Coupled with the AOC designation were demands that the responsible government agencies at all levels develop and implement Remedial Action Plans (RAPs). This was followed in 1987 with the signing of the renewal of the Great Lakes Water Quality Agreement (GLWQA) between Canada and the United States. As Project Quinte had been active since 1972, the Quinte RAP process was able to get to work very quickly with a strong science-based agenda with many Project Quinte members participating in the RAP's Technical Advisory Committee. The RAP adopted the "Ecosystem Approach" that was central to the 1987 GLWQA. The RAP Steering Committee developed a program of public engagement with much support from Project Quinte (Stride et al., 1992; see www.BQRAP.ca). An intensive program of activities was undertaken drawing on all aspects of Project Quinte science. Many of the Project Quinte team served on the RAP Technical Advisory Committee. After much investigation, analysis and modelling, the RAP produced the Stage 1 report, Time To Decide, in 1990 wherein remedial targets and options were examined, and later the Stage 2 report, Time To Act, in 1993 wherein a program of remediation was laid out. Stage 3, implementation, has been actively under way ever since. The RAP Steering Committee was succeeded by the RAP Implementation Council, which oversees all aspects of implementation, monitoring, and assessment. This transition also saw a gradual shift from central to local engagement with Conservation Authorities rightly assuming leadership. Project Quinte then became the principal source of monitoring and assessment of Quinte as implementation proceeded.

The final stage of the RAP process will be delisting, which depends on evidence being presented that designated beneficial uses have been restored. Once delisted AOCs may be re-designated Areas of Recovery if all restorative actions have been implemented but impaired beneficial uses remain. If all beneficial uses are deemed unimpaired, the area would cease to have any special status. The scientific understanding necessary to guarantee beneficial uses have been restored and will be sustainable into the future was incomplete in the 1980s; as a result many of the early BUI definitions were incomplete or erroneous. Thus, the ongoing monitoring and assessment work has included efforts to develop better delisting criteria. The delisting process has been made harder by the combination of the limited, largely non-quantitative definitions of beneficial use impairment laid out when AOCs were first designated (George and Boyd 2007) and the large-scale ecosystem engineering induced by invasive species like Driessenid Mussels. Also, growing understanding in ecosystem science has made it ever clearer that the real challenges of ecosystem management are not delisting criteria but the enduring task of maintaining ecologically sustainable ecosystems beyond recovery, preferably as "Areas of Sustainability".

Invasions of driessenids and other species

As often occurs in a natural 'experiment' unexpected phenomena intervene. Until the late 1980s Project Quinte was primarily seen as a study of the response of an ecosystem to decreased nutrient loading after a lengthy period of excess loading. Through the 1980s, P loadings from STPs continued to decline as facilities expanded and operating performance was enhanced. Then a series of species invasions began, due largely to the continued release of ballast water from transoceanic ships entering the Great Lakes basin (Ricciardi, 2006). The post-P control observations were confounded by changes wrought by the invaders.

First, the Zebra Mussel, *Dreissena polymorpha*, arrived in Quinte (1992) and expanded rapidly as observed in other parts of the lower Great Lakes. Other species had invaded much earlier, such as White Perch, which arrived in the 1950s establishing large populations in Lake Ontario. White Perch were particularly abundant in Quinte and became a large part of the commercial harvest up until the early 1970s. Much earlier, Alewife had invaded and become a major food source for top predators in Quinte and Lake Ontario. The invaders arriving prior to P control had been considered naturalized and hence not a concern for Project Quinte.

However, the Zebra Mussel and soon after its cousin the Quagga Mussel, *D. rostriformis bugensis*, achieved immense abundances and produced major changes in the structure and material flows in ecosystems like Quinte. In particular, Dreissenid Mussels altered the relationship between P levels and phytoplankton biomass as their filtering diverted energy away from pelagic foodweb pathways. During this period, two predatory zooplankters, *Bythotrephes longimanus* (1982) and *Cercopagis pengoi* (1998), invaded the Great Lakes causing widespread changes in zooplankton production. Fortunately, *Bythotrephes* has not invaded Quinte and *Cercopagis* has only been observed in the lower bay although remaining a threat. More recently, another invader, the Round Goby has become established since 1998 with high abundances in Quinte and exerts considerable pressure on energy flow (Taraborelli et al., 2010). These invasions drew a lot of attention given concerns for Great Lakes fishery resources although the on-going declines in P loadings and the eventual RAP goal of delisting continued to be assessed.

Species invasions significantly shifted baselines with regard to earlier delisting criteria identified when P loading was the main issue. For example, extensive filtering by Dreissenid Mussels increased water clarity, particularly in the upper and middle sections of the bay, thereby allowing the submerged macrophyte communities to expand to levels not seen since the early 1950s. Macrophytes by their presence generated a physical restructuring of large areas of the upper bay, providing surfaces for epiphytes which are grazed by a greater variety of biota than are present when bare substrates dominante. The macrophyte abundance has allowed an expansion of centrarchid fish populations. When setting P loading targets for Quinte there had been an expectation that macrophytes would recover once external P loading was reduced and in situ P regeneration rates declined as surface sediments were renewed over time. However, before Dreissenid Mussels arrived there had been little evidence of a resurgence in macrophytes. The pivotal change due to the dreissenids was the change in the Phosphorus: Chlorophyll relationship which provided the basis for managing eutrophication via P loading reductions.

Current status

In the current phase of Project Quinte, three linked topics have been the major focus: a) preparing for delisting targetted for 2013, b) reporting on the data and science, and c) generational turnover in Project Quinte membership. Many Project Quinte members are actively engaged in updating and assessing the quantitative criteria that will provide the basis for delisting. Project Quinte decided to strengthen engagement in the delisting assessment process through the generation of a new set of peer-reviewed publications to present new results and understanding of the Quinte ecosystem to the broader scientific community. Hence, Project Quinte undertook the presentation of a symposium at the 53rd Annual Meeting of the International Association of Great Lakes Research held in Toronto, 17-21 May 2010, along with a commitment to produce two issues of this journal containing 20 papers covering all aspects of Quinte science. Many of the papers include analysis and discussion of BUIs and the potential for delisting. These combined efforts also addressed the third topic.

In recent years, many of the original Project Quinte members have begun to retire posing challenges of generational turnover and the need to find successors to carry on the analysis and interpretation of the accumulating data series. The loss of institutional knowledge and history is a problem and strengthens the case for up-to-date reporting and documentation of Project Quinte studies in these special issues. Fortunately many of the retirees have maintained some continuing involvement in Project Quinte completing and publishing new analyses. The symposium and publications also provided the means for advertising the character and availability of a rich dataset accumulated since 1972 on many aspects of the Quinte ecosystem.

Future of Project Quinte

Project Quinte reaches its 40th field season in the spring of 2011. As has been the case throughout its history the array of issues driving the science is continually changing. The delisting milestone will not be the end of the Quinte story. Project Quinte has outgrown its origins in eutrophication, AOCs, and RAPs, and established a continuing role in the study and management of aquatic ecosystems. The length and breadth of the time-series on Quinte makes Project Quinte almost unique in the Great Lakes. As Janzen (2009) noted the longer the record the greater intrinsic value and relevance of the careful scientific observations to future decision-making. Studies like Project Quinte can be hard to sustain using current institutional funding models for science that tend to emphasize short-term needs for science advice.

Long-term studies create "just in case" repositories of data and understanding that can be applied to new questions that might arise. The long history of ecosystem management issues in the Great Lakes has provided ample evidence that such questions, like the consequences of new species invasions and global climate change, are inevitable and that long-term ecosystem monitoring like that being conducted in Project Quinte should be an essential element of responsible ecosystem management.

The AOC-RAP processes have been largely concerned with solving specific problems and might be equated with the traditional emphasis in medicine on illness and the reduction of undesirable symptoms (Minns, 1999). In recent decades, the focus has shifted to health both in humans, with efforts to reduce behaviours that promote illness, and ecosystems as with the Great Lakes' Ecosystem Approach. An ecosystem health approach recognizes the cumulative effects of human development activities on the ability of ecosystems to sustain healthy structures and processes, i.e. ecological goods and services. Many of the BUIs in AOCs were symptoms and have not been completely eliminated. For instance, as human populations around Quinte continue to grow, potential phosphorus loading will increase requiring further improvements at STPs and the imposition of absolute loading caps. The continuing population growth and linked cumulative land use transformations and waste outputs have now been augmented with the emerging threats of climate change.

These continuing pressures will require constant attention if the beneficial uses of Quinte and other aquatic ecosystems in the Great Lakes and beyond are to be sustained. Thus, beyond RAP delisting there is a perpetual need to ensure Quinte and similar areas become "Areas of Sustainability" and that current non-AOC areas around the Great Lakes do not become new AOCs. Completing the restoration of Quinte, a biodiversity and productivity hotspot, and ensuring it is sustainable must be augmented with complete coastal zone ecosystem management plans for each of the Great Lakes.

Project Quinte has been carried along in the growing stream of ecosystem knowledge and understanding. Sampling technologies have changed with the shift from water bottles, to dangled probes, to fully-computerized geo-referenced profilers. Statistical analysis tools have expanded as computer hardware has become smaller. The role of modelling has grown substantially, providing both the means to integrate immense amounts of data and knowledge as well as the ability to predict potential consequences of alternative management actions. For example in Quinte, Diamond et al (1996) provided a basis for remedial actions on a range of chemical contaminants; Minns and Moore (2004) developed a simple model of phosphorus which provided the foundation for a long-term P management strategy; and Koops et al (2006) used ecosystem modelling in a comparative study to help unravel the changes in foodweb dynamics due to eutrophication and the dreissends invasion. Yet, the core elements of the Project Quinte monitoring program remain essentially the same thereby providing the means to detect trends, patterns and especially regime changes.

Summary

Project Quinte can best be described as a long-term ecosystem study of the Bay of Quinte, Lake Ontario. Starting in 1972, Project Quinte was initially established to study whole ecosystem effects of controlling phosphorus loadings in a eutrophic ecosystem. Since then, the Bay of Quinte ecosystem has experienced reduced nutrient loads, climatic events that changed the dominance of fish species, multiple invasions by non-native species, a resurgence of macrophytes and increasing annual temperatures. Through this, the Bay of Quinte has gone from a study site, to a Great Lakes Area of Concern, to now the prospect of being delisted. The data that Project Quinte has assembled since its inception represents a unique opportunity to examine how ecosystems function; the papers presented in this special issue provide evidence of the scientific and management benefits of careful long-term monitoring of ecosystem structures and processes.

Project Quinte has and continues to serve as a model of the ecosystem approach (Koops et al., 2009) and an anchor and point of reference for a wide range of research studies. Events in Quinte since 1972 have demonstrated the importance of long-term monitoring and research. Regime shifts can be detected only through the deliberate long-term maintenance of well-defined sampling programs. All-too-often society chooses a self-imposed “environmental myopia” (Silvertown et al., 2010), seeming to prefer reliance on “shifting baselines” (Pinnegar and Engelhard, 2008) based on subjective anecdotal evidence. The set of Project Quinte papers in this book and in other AEHMS publications provide ample evidence of the continuing scientific and management benefits of careful long-term monitoring of ecosystem structures and processes.

References

Christie, W.J., Becker, M., Cowden, J.W., Vallentyne, J.R., 1986. Managing the Great Lakes Basin as a home. J. Great Lakes Res. 12, 2–17.

Dermott, R., 2001. Sudden disappearance of the Amphipod Diporeia from Eastern Lake Ontario. 1993-1995. J. Great Lakes Res. 27(4), 423-432.

Diamond, M.L., Mackay, D., Poulton, D.J. and Stride, F.A., 1996. Assessing chemical behaviour and developing remedial actions using a mass balance model of chemical fate in the Bay of Quinte. Water Res. 30(2), 405-421.

George, T.K., and Boyd, D., 2007. Limitations on the development of quantitative monitoring plans to track progress of Beneficial Use Impairment restoration at Great Lakes Areas of Concern. J. Great Lakes Res. 33, 686-692.

Janzen, H.H., 2009. Long-term ecological sites: musings on the future, as seen (dimly) from the past. Global Change Biology 15, 2770-2778.

Johannsson, O.E., Millard, E.S., Ralph, K.M., Myles, D.D., Graham, D.M., Taylor, W.D., Giles, B.G., and Allen, R.E., 1998. The changing pelagia of Lake Ontario (1981 to 1995): A report of the DFO Long-Term Biomonitoring (Bioindex) Program. Can. Tech. Rep. Fish. Aquat. Sci. 2243.

Johnson, M.G. and Brinkhurst, R.O., 1971. Benthic community metabolism in Bay of Quinte and Lake Ontario. J. Fish. Res. Board Can. 28, 1715–1725.

Koops, M.A., B.J. Irwin, J.E. MacNeil, E.S. Millard and E.L. Mills. 2006. Comparative Modelling of the Ecosystem Impacts of Exotic Invertebrates and Productivity Changes on Fisheries in the Bay of Quinte and Oneida Lake. Completion Report, Fisheries Research Program, Great Lakes Fishery Commission (Available at www.glfc.org).

Koops, M.A., Dermott, R.M., Leisti, K.E., Johannsson, O.E., Millard, E.S., Minns, C.K., Munawar, M., Nicholls, K.H., Hoyle, J.A., 2009. The Bay of Quinte: a model for large lake ecosystem management. Verh. Internat. Verein. Limnol. 30(7), 1024-1029.

Michalski, M.F.P, Nicholls, K.H., Johnson, M.G., 1975. Phosphorus Removal and Water Quality Improvements in Gravenhurst Bay, Ontario. Verh. Internat. Verein. Limnol. Vol. 19, 644-659.

Millard, E.S., Fee, E.J., Myles, D.D., and Dahl, J.A., 1999. Comparison of phytoplankton photosynthesis methodology in Lakes Erie, Ontario, the Bay of Quinte and the Northwest Ontario Lake Size Series. In: M. Munawar, T. Edsall, I.F. Munawar (Eds.), *State of Lake Erie (SOLE)—Past, Present and Future*, pp. 441–468, Ecovision World Monograph Series, Backhuys Publishers, Leiden, The Netherlands.

Minns, C.K., 1999. The ecosystem pyramid and the means for attaining ecological sustainability: an essay in honour of Jack Christie. Aquatic Ecosystem Health and Management 2(3),197-208.

Minns, C.K. and Moore, J.E., 2004. Modelling phosphorus management in the Bay of Quinte, Lake Ontario in the past, 1972 to 2001, and in the future. Can. Manuscr. Rep. Fish. Aquat. Sci. 2695.

Minns, C.K., Hurley, D.A. and Nicholls, K.H. (Editor), 1986. Project Quinte: point source phosphorus control and ecosystem response in the Bay of Quinte, Lake Ontario. Can. Spec. Publ. Fish. Aquat. Sci. 86.

Nicholls, K.H. and Hurley, D.A., 1989. Recent Changes in the Phytoplankton of the Bay of Quinte, Lake Ontario: the Relative Importance of Fish, Nutrients, and Other Factors. Can. J. Fish. Aquat. Sci. 46(5), 770–779.

Pinnegar, J.K. and Engelhard, G.H., 2008. The 'shifting baseline' phenomenon: a global perspective. Revs. Fish Biol. Fisheries 18(1), 1-16.

Ricciardi, A., 2006. Patterns of invasion in the Laurentian Great Lakes in relation to changes in vector activity. Diversity Distrib. 12, 425-433.

Schindler, D.W. and Vallentyne, J.R., 2008. *The Algal Bowl: Overfertilization of the World's Freshwaters and Estuaries.* University of Alberta Press, AB.

Silvertown, J., Tallowin, J., Stevens, C., Power, S.A., Morgan, V., Emmett, B., Hester, A., Grime, P.J., Morecroft, M., Buxton, R., Poulton, P., Jinks, R., Bardgett, R., 2010. Environmental myopia: a diagnosis and a remedy. Trends in Ecology and Evolution 25(10), 556-561.

Stride, F., German, M., Hurley, D.A., Millard, E.S., Minns, C.K., Nicholls, K.H., Owen, G.E., Poulton, D.A., and de Geus, N., 1992. An overview of the modelling and public consultation processes used to develop the bay of Quinte Remedial Action Plan. In: J.H. Hartig and M.A. Zarull (Eds.), *Under RAPs: toward grassroots ecological democracy in the Great Lakes Basin*, pp. 161-183. University of Michigan Press, Ann Arbor, Michigan.

Taraborelli, A.C., Fox, M.G., Johnson, T.B., Schaner. T., 2010. Round Goby (*Neogobius melanostomus*) Population Structure, Biomass, Prey Consumption and Mortality from Predation in the Bay of Quinte, Lake Ontario. Journal of Great Lakes Research 36 4, 625-63.

The above article is a reprinted version of the following previously published article:

C.K. Minns, Munawar, M., Koops, M.A., Millard, E.S., 2011.
Project Quinte 1972-2008: *Long-term Ecosystem Studies in the Bay of Quinte*, Lake Ontario, a Prospectus. Aquat. Ecosyst. Health Mgmt., 14(1), 3-8 https://doi.org/10.1080/14634988.2011.551750

Surface meteorology, hydrology and physical limnology features of the Bay of Quinte

Yerubandi R. Rao[1,*], Leon Boegman[2], Weitao Zhang[3], Ali Oveisy[1,2], Hadiseh Bolkhari[2] and Jun Zhao[1]

[1]Environment and Climate Change Canada, 867 Lakeshore Rd. Burlington, ON L7S 1A1
[2]Environmental Fluid Dynamics Laboratory, Department of Civil Engineering, Queen's University, Kingston, ON, K7L 3N6, Canada [3]Department of Physical & Environmental Sciences, University of Toronto, Toronto, ON, M1C 1A4
**Corresponding author: Ram.Yerubandi@canada.ca*

Keywords: water retention time, modeling

Introduction

The Bay of Quinte (BoQ) is a significant freshwater source for drinking, recreation and industry. This narrow and sheltered bay also receives excessive pollutant and nutrient loads (Minns et al., 1986) leading to persistent toxic and bacteriological contamination, undesirable algae growth, fish toxicity, and taste and odour problems, which are exacerbated near Belleville and Hay Bay (Fig. 1). In 1986, the Bay of Quinte was classified as an Area of Concern in the Great Lakes basin by the International Joint Commission and a Remedial Action Plan (RAP) was initiated.

The Bay of Quinte is a narrow, Z-shaped inlet, 70 km in length with an 18000 km^2 watershed which is located on the northeastern shore of Lake Ontario (44.15° N, 77.25° W). The upper bay (from the Murray Canal to the Napanee River is 35 km long with several sub-basins approximately 4-8 m deep. The middle bay is 13 km long from the north end to south with depths increasing in the range of 6 to 17 m. The lower bay is 16 km long, and the depth increases from 17 m to the maximum of 52 m toward the bay mouth. Two outlets at each end join the bay and Lake Ontario. The western passage, the Upper Gap, is 13 m deep connected to Lake Ontario by the Murray Canal for navigation, and the eastern passage, the Lower Gap, with twice the depth is connected.

Ecology of the Bay of Quinte: Health, Management and Global Implications
Editors: C.K. Minns, M. Munawar and M.A. Koops

Aquatic Ecosystem Health & Management Society, Canada

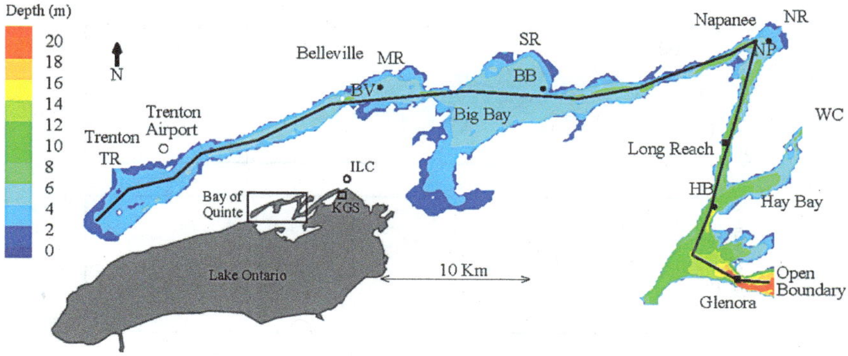

Figure 1. Location of the Bay of Quinte in Lake Ontario (inset) indicating Kingston Gage Station (KGS) and Integrated Learning Center (ILC). Bathymetric map of the model domain for the Bay of Quinte showing locations of tributaries, meteorological stations (open circle), field observation stations (solid circles), and curtain plot (black line).

to the Kingston Basin of Lake Ontario. The flow through the Murray Canal is neglected due to its weak oscillatory flow with a negligible influence on the nutrient dynamics in bay (Johnson and Owen 1971).

The Upper Bay is supplied by the Trent River, Moira River, and Salmon River and is the most sensitive region for nutrient-induced plankton growth (Robinson, 1986). The Napanee River and Wilton Creek have watersheds approximately one twelfth that of Trent River and they are also closer to the Lake Ontario open boundary; therefore, these rivers will not have a significant effect on nutrient dynamics (Oveisy et al., 2014). However, high concentrations can persist near the mouths of these rivers in side bays, away from the main canal, where low currents are expected.

Retention, mixing and flushing of nutrients from the eutrophic BoQ to the oligotrophic waters of Lake Ontario influences water quality in the lake. The dynamics are a function of the surface meteorological forcing conditions and the hydraulic flow through the system. To guide management of BoQ waters, this paper provides a brief review of these processes and predictions on how they might be influenced by climate change.

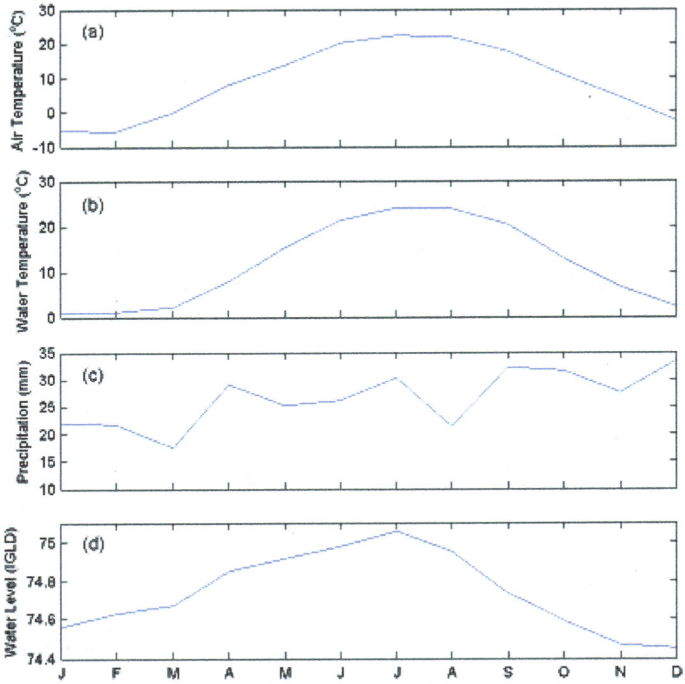

Figure 2. Monthly means of air temperature (a), precipitation at Trenton A; (c) water temperature (b), and water levels at Belleville (d)

Meteorological observations

Te surface weather stations located at Belleville and Trenton Airports provide meteorological and limnological observations throughout the year. Long-term monthly means (2002-2012) of selected meteorological and limnological parameters at Belleville are shown in Figure 2. The mean annual air temperature is around 7 °C. The mean monthly air temperatures rise above the freezing mark by March and reach an average high temperature of 20.5°C in July. The temperature then falls below 0 °C again in late November and reaches a low of -7.5 °C in January. Unlike in large lakes, the BoQ due to its shallowness and relatively smaller size does not exhibit a lag between changes in air and water temperatures. In general, the Bay freezes by December and is covered with ice until March.

Although some variation exists between Trenton and the eastern end of the Bay, in general mean annual precipitation for the

Bay of Quinte ranges from 850-890 mm. Most of the annual precipitation falls as rain. During the winter, between December and February, most precipitation falls as snow. In November and April, precipitation is mixed with a trend toward greater rainfall. Generally, March and August appear to be drier months.

The thermal structure of the Bay is closely tied to the surface forcing represented by the heat flux components and there is strong correspondence between the surface heat fluxes and the observed water temperature and heat storage cycles. Changes in the radiative and turbulent exchanges, as a consequence of regional or global scale weather patterns and/or climate warming, may impact the surface energy exchange and resulting thermal structure. In addition, changes in the lake clarity, e.g. an increase in Secchi depth from 3 m to 6 m resulting from filter feeding by invasive Dreissenid Mussels (Lesti et al., 2012), has influenced the transmission of photosynthetic light through the shallow water column and, therefore, affects both the thermodynamic and biological components of the ecosystem.

The surface heat flux (Q_T) is calculated, from buoy (44°9'30", 76°34') data close to Bay of Quinte in Lake Ontario in 2006, as a balance between the heat sources and sinks,

$$Q_T = Q^- + Q - Q_h - Q_e \tag{1}$$

The total incoming radiation ($Q^- = K^- + L^-$) is obtained from the instrument output of the radiometer (Q) and the temperature of a blackbody (T_{cav}) by $Q^- = Q + s\, T^4_{cav}$, where s is Stephen-Boltzmann constant. Here, K^- and L^- are incoming solar radiation and incoming longwave radiation (W m^{-2}), respectively. Since K^- is available from the pyranometer, we can obtain the incoming longwave radiation by the difference (i.e. $L^- = Q^- - K^-$). The average incoming longwave radiation increases during the spring period, and reaches a maximum value during the summer. Further, the reflected solar radiation is obtained as K= a K^-, where surface albedo (a) is assumed to be 0.08 based on previous experience on Lake Ontario. The reflected longwave radiation is a small component of the radiation balance, and L is obtained as $(1\text{-}e)\, L^-$, where e (=0.97) is the emissivity of the water surface (Schertzer 1987). Finally, the total outgoing radiation is obtained as ($Q = K + L$). The net radiative exchange at the surface is calculated as QSTAR (Q*) = $Q^- + Q$. Turbulent sensible (Q_h) and latent heat fluxes (Q_e) are calculated as described in Schertzer (1987). Figure 3a shows the daily averaged radiative and turbulent heat flux during the April to October, 2006. The surface heat flux clearly indicates periods of heat gain and loss for eastern Lake Ontario. Since the buoy is located 20 km from the

BoQ, some nearshore-offshore gradients in the net heat flux are expected, particularly during the spring because of intense nearshore warming, and wind-induced upwelling and downwelling events in the main lake basin. However, these variations are not important for a qualitative understanding of the expected heat balance for the Bay, as it is relatively sheltered.

Available energy, represented by the net radiative flux (Q^*), is positive (BoQ is gaining heat) during most of the year; however, the flux decreases to smaller values and is expected to become negative during November through March. The energy fluxes cause lake heating/cooling and drive evaporation from the waterbody. The losses of heat from the lake surface through latent heat flux (heat transfer through evaporation) increases are episodic in nature, and in general begin increasing from late August through the fall. In general, sensible heat flux (heat transfer through conduction) is smaller than the latent heat transfer. The daily

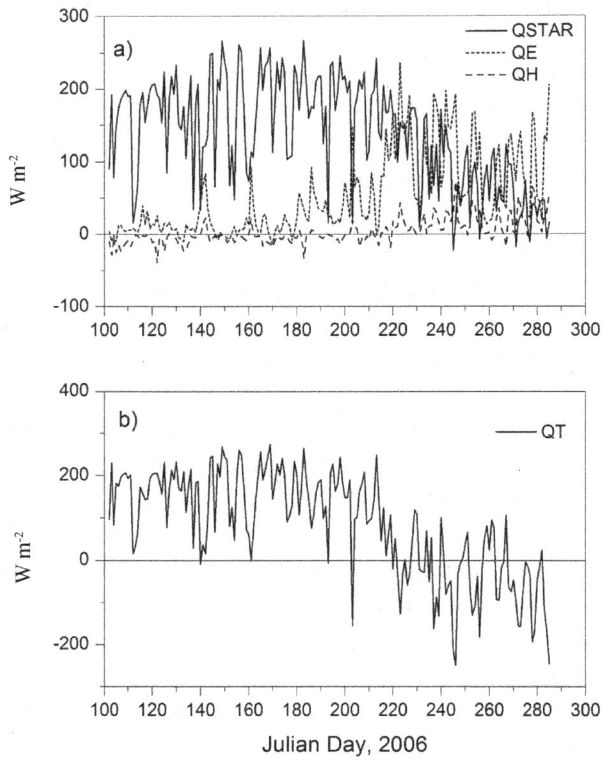

Figure 3. An example of computed surface heat fluxes from the Kingston basin meteorological buoy during 2006.

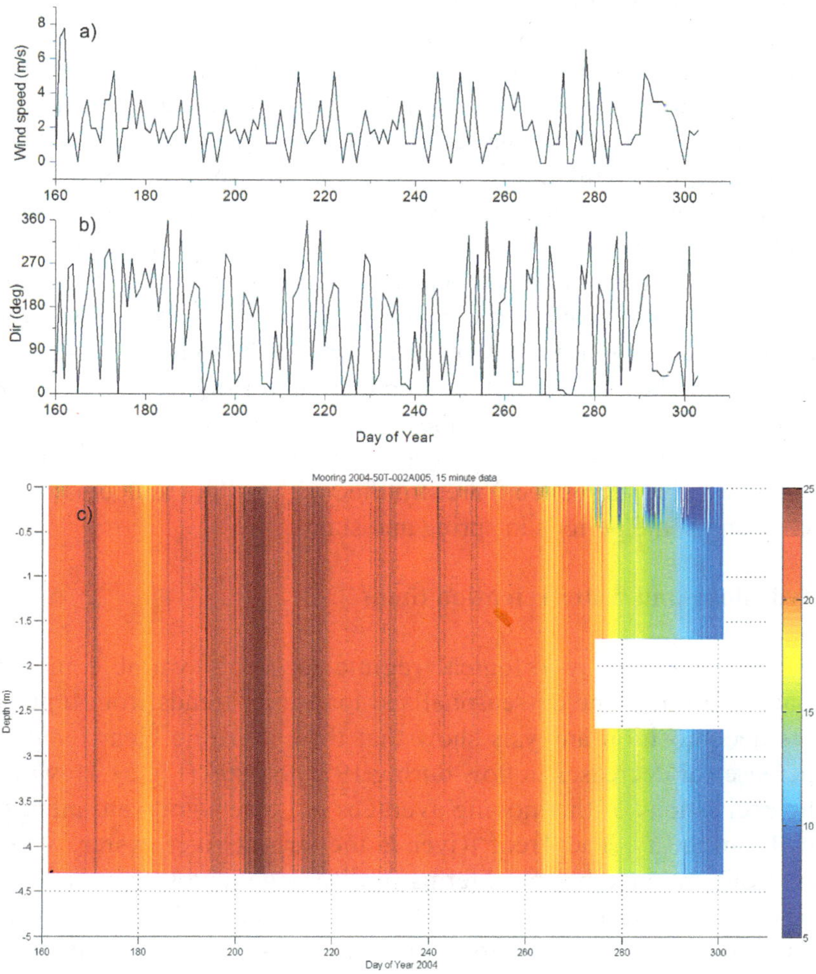

Figure 4. Observed wind speed (a) and direction (b) at Trenton A and water temperature profile from a mooring in the Big Bay during 2004.

averaged heat flux ranged from 200 W m^{-2} to -200 W m^{-2} during this period (Figure 3b). Combining the energy budget information with wind data (Figure 4a) showed that the major wind events cause intense turbulent heat and mass exchanges at the air-water interface. Heat storage occurs until an August maximum (Figure 3b), and heat losses occur from September onwards.

A thermistor chain with Tidbit (accuracy of ±0. 2°C) and RBR loggers (accuracy of ±0. 01°C) was deployed by Environment Canada in Big Bay (BB) at a ~4.7 m deep location from June 09 to October 26, 2004. Temperature was recorded at 15 minutes intervals. Figure 4 shows wind speed and direction at Belleville and the time series of vertical temperature profiles in the Big Bay. Significant variations of temperature between the surface and bottom are not expected due to the shallow depth (~5 m) at this location as the entire profile is epilimnetic water. Although the water column is mixed, we can still observe slight stratification (0.6°C $<\Delta T<$1.5°C) on several occasions during the calm conditions and warm air temperatures. During the rest of the observations, the stratification is very weak (ΔT <0.6 °C), where ΔT is the temperature difference between surface and the bottom. The Z-shape of the bay and the changes in topography around BoQ impose a marked variability of wind stress over the surface. Consequently, Oveisy et al. (2014) in their simulations using a hydrodynamic model, showed that deeper waters (~12 m deep) are stratified with bottom temperatures ~8 °C during some periods in spring and summer.

Hydrology and water retention times

In the BoQ, the hydrological regime is highly variable in both space and time. The de-seasonalized long-term trends from time-series decomosition analysis show that there were no clear temporal increases or decreases in flow during 1972 to 2009 (Fig 5). However, the oscillations of the moving averages indicate significant inter-annual variability. The Trent River is the largest river system within the BoQ drainage basin, in terms of discharge, and moving average flows vary from 60 m^3 s^{-1} to 220 m^3 s^{-1}. Direct discharges, from the rest of the four gauged rivers are much less than the Trent River. The long-term component of the Moira River flow varies between 20 m^3 s^{-1} to 50 m^3 s^{-1}. The long-term flows of the Salmon and Napanee Rivers are approximately 5 m^3 s^{-1} to 20 m^3s^{-1}, and the average flows for Wilton Creek are rarely over 2 m^3 s^{-1} due to the small size of its watershed. The seasonal pattern of flows is repeated annually with inter-annual details determined by changing weather conditions. For example, April is usually characterized by the spring runoff peak, whose magnitude depends on both the amount of snow accumulation over the previous winter and the weather conditions during the snowmelt period itself. The summer low-flow period is usually from July to August.

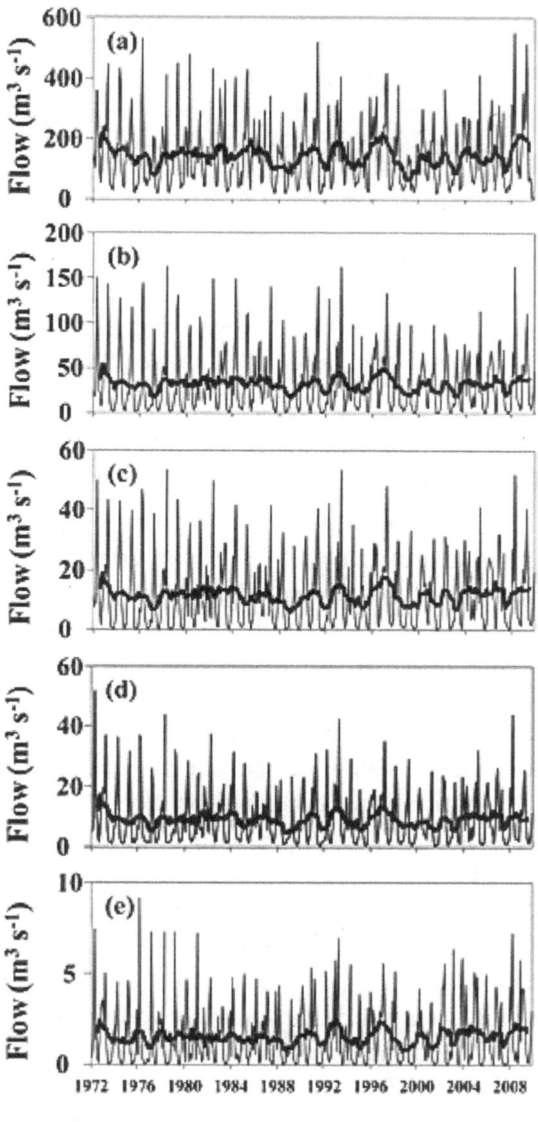

Figure 5. Time series of tributary flows of the Bay of Quinte (a) Trent River; (b) Moira River; (c) Salmon River; (d) Napanee River; and (e) Wilton Creek. Thin grey lines represent monthly average of flow measurements. Thick black lines represent de-seasonalized long-term components derived from Ratio-to-Moving Averages Classical Seasonal Decomposition method. The daily discharge data obtained from Environment Canada (Water Survey of Canada).

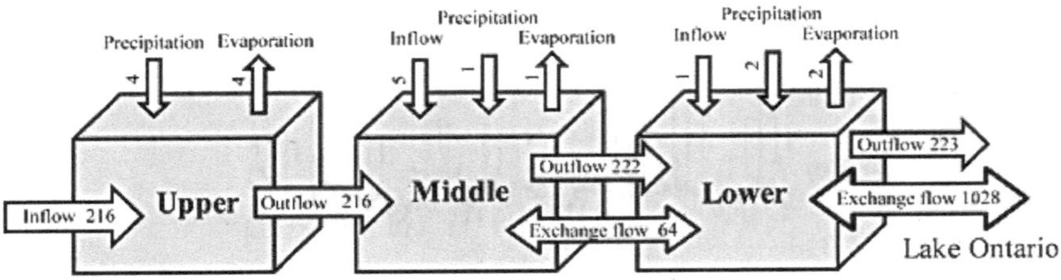

Figure 6. Schematic representation of the water balance of the Bay of Quinte. The arrows indicate water flows through the system and the numbers are fluxes ($m^3\ s^{-1}$) averaged over the 1972–2009 period. The light grey boxes represent the water column Upper, Middle and Lower bays.

Water balance calculations were performed by including the water entering (e.g. river inflow, exchange flow with Lake Ontario and precipitation) and water leaving (outflow and evaporation) the BoQ segments (Fig. 6). The water fluxes associated with individual processes were averaged over the 1972–2009 period. The exogenous inputs from metered and un-metered rivers approximately contribute 216, 5, and 15 $m^3\ s^{-1}$ into the upper, middle and lower segments, respectively. The precipitation inputs are responsible for a very small portion of the total sources. Moreover, 216, 222 and 223 $m^3\ s^{-1}$ of water are lost from the water column of the three segments. The exchange of water masses with Lake Ontario is an important cause of mixing and material transport. Over 1000 $m^3\ s^{-1}$ is exchanged between Lake Ontario and the lower segment, and subsequently ~60 $m^3\ s^{-1}$ is exchanged with the middle segment of the bay.

The annual water retention time (WRT) of the Bay of Quinte can be calculated by dividing the segment annual averaged volume by annual averaged total inflows (the rate of total amount water flow it the lake system, including inflow through river/adjacent segments and precipitation) (Fig. 7). Estimates for individual segments showed large differences, with longer WRTs in the upper and middle segments and larger variation (in the range from 8 days to 16 days for last 40 years) than the lower segment (about 9 to 10 days). The large inter-annual variability of WRTs in the upper and middle segments is driven by variations in the river inflows. The relatively stable pattern of WRT in the lower segment is caused by a large amount of backflow exchange with Lake Ontario. Time-series of the WRTs do not show any long-term trends during the period from 1972 to 2009 (Fig 7).

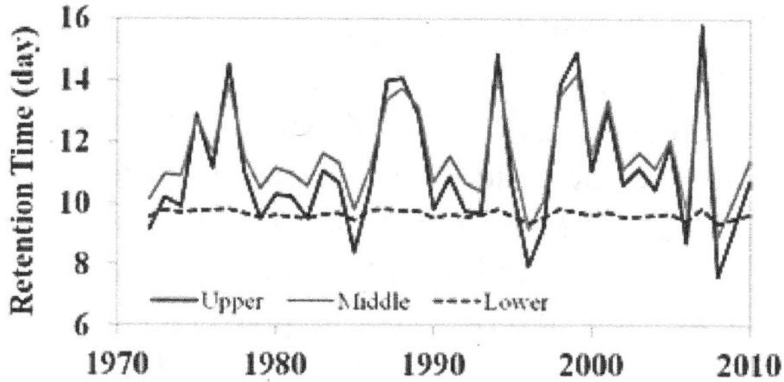

Figure 7. Annual-averaged water retention time of three spatial sections of the Bay of Quinte.

Hydrodynamics and flushing in Bay of Quinte

The long complex Z-shape geometry of the Bay of Quinte increases nutrient residence times in the warm shallow embayments providing favourable conditions for algae growth. Consequently, Freeman and Prinsenberg (1986) deployed current meters during the summers of 1976 and 1978 in the interconnecting gaps of the Adolphus Reach/ North channel basin to measure the exchange flows between Lake Ontario and upper BoQ. During the summer a two-layer baroclinic exchange flow was observed through Glenora whereas in fall the currents were generally barotropic. The bottom flow at Glenora during the summer is towards upper Bay of Quinte. The most energetic oscillations coincide with Lake Ontario surface seiches at a period close to 5.2 h. Although the long-term flushing rate is 40-50 days, they observed that a wind episode of 2-3 days might completely flush the lower part of the bay. The current meter records also show baroclinic oscillations around 2.7 h and 1.1 h. They also observed a long-term clockwise circulation around Amherst Island due to seasonal lowering of Lake Ontario water levels. The direct wind forced exchange, at a period of 3-5 days, is observed to be a dominant mechanism for mass transport between Lake Ontario and the Bay of Quinte.

Several studies using box models were conducted to simulate nutrient balances (Minns et al., 1986; Kim et al., 2013), but better estimates can be achieved by applying spatially explicit transport models that can accommodate both spatial variability and exchange flows from the lake. In one such study Oveisy et al (2014) investigated flushing of the Bay and horizontal transport (advection and mixing timescales) using a three-dimensional (3D) hydrodynamic model Estuary and Lake Computer Model (ELCOM). They divide the bay into two regions: the main channel and side bays that are connected to the main channel. Several methods can be used to investigate the effects of advection and mixing on the transport and dilution of nutrients. In this study the flushing time is calculated from the e-folding timescale associated with the rate at which a 30 day initial pulse of river inflow was flushed from the bay; the bulk residence time is computed from the volume of the bay divided by the river inflows, and the water age (i.e. age of each fluid parcel) as tracked during the simulations are computed. In the main channel, flushing is governed by high advective transport (seasonal averaged RMS velocity ~ 3 cm s^{-1}), while in the side bays the velocities are much smaller (RMS velocity < 1 cm s^{-1}) and diffusion predominates (Oveisy et al., 2014).

In agreement with Freeman and Prinsenberg (1986), the main channel of the bay is modelled to flush ~5 times per year and isolated embayments, such as southern Big Bay and Hay Bay, flush 3 times per year consistent with the estimate by Minns et al. (1986) for the entire bay (including side embayments). Flushing is predominantly driven by riverine inputs, which have high nutrient and potentially high algae loads, as opposed to intrusions from Lake Ontario, which were simulated to propagate as far as Hay Bay. The embayments have a water age of approximately 4-5 months (Figure 8), much longer than the segment averaged values (Figure 7), showing that quiescent warm water may trap nutrients in nearshore regions with sufficient time and

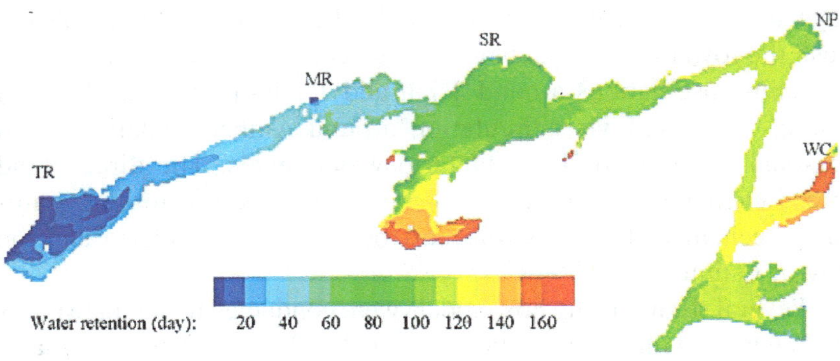

Figure 8. Water age plot at the end of a 170-day computational simulation of the Bay of Quinte hydrodynamics with ELCOM. Reproduced from Oveisy et al (2014).

conditions for algae blooms to occur. The upper reaches of the Bay contains 'young' water from the Trent River, which provides 72% of the mean annual flow to BoQ. The majority of inflow from the two main rivers (Trent and Moira) shortcuts through the northern part of Big Bay without mixing throughout Big Bay, resulting in significant amount of Trent River water in the lower reaches of the Bay. The combination of these circulation dynamics with the significant nutrient loading from the Trent River, suggests that this to be the most critical river for nutrient control (Zhang et al., 2013; Kim et al., 2013; Oveisy et al., 2014).

Effects of climate change on Bay of Quinte temperatures and biogeochemistry

There have been theoretical and modelling studies that have considered potential responses of large lakes to climate variability and change, however, there is a paucity of direct observations from which to understand the actual responses. (Schertzer and Rao, 2009). Climate (weather) has a significant role in the physical processes occurring in aquatic systems. It is anticipated that future atmospheric changes could have significant effects on the physical limnological processes and the aquatic ecosystem components of the Great Lakes. As demonstrated in the previous discussion, major forcing occurs through the action of heat exchanges and wind. Huang et al. (2012) showed that mean annual air and surface water temperatures have increased by 1.43 °C and 1.26 °C, respectively, over 1970–2009 in Lake Ontario, whereas Bay of Quinte's maximum summer surface waters have warmed almost 1 °C from 1972 to 2008, with some increase in water column stability consistent with Lake Ontario (Minns et al., 2011). The consequences of climate warming will have a significant influence on the stratification season, thus influencing the water quality in the BoQ. Several investigations have been conducted on the Great Lakes using a 1D modelling approach to simulate thermal responses using various climate scenarios (Boyce et al., 1993; Blumberg and DiToro, 1990; McCormick, 1990). The general approach adopted is to simulate the vertical temperature structure for a lake or basin under current climate conditions and then to impose a climate change scenario on the key meteorological inputs and compare the resulting temperature fields for the base case and climate change case to infer potential changes resulting from the climate scenario.

In a recent study, Trumpickas et al. (2009) employed empirical relationships between surface water temperatures and air temperatures for estimating the future temperatures in the Great Lakes. In their study, using future air temperatures predicted from General Circulation Models (GCMs), they calculated that BoQ temperatures could rise by 3.3 to 6.7 °C under the A2 scenario (high emission) and by 2.4 to 4.6 °C under the B2 scenario (moderate increases in emissions). Recently, the effects of climate change on hydrodynamics and biogeochemistry in the BoQ have been investigated by combining the ELCOM with the Computational Aquatic Ecosystem Dynamics Model (CAEDYM; Bolkhari 2014). ELCOM is an advanced 3D numerical model, which simulates the hydro dynamics and the thermodynamics of inland and coastal waters, and was already calibrated for BoQ (Oveisy et al., 2014). ELCOM is coupled with CAEDYM to study the spatial and temporal distributions and interactions of physical, biological, and chemical variables in different lakes from small to large size (Leon et al., 2011). CAEDYM simulates the carbon (C), nitrogen (N), phosphorous (P), DO, and silicon (Si) cycles, along with inorganic suspended solids and phytoplankton.

The models were forced with 3-hr meteorological predictions from the North American Regional Climate Change Adaptation Program (NARCCAP) data for the Special Report on Emission Scenario (SRES) A2 climate scenario during 2000-2004 (present) and 2066-2070 (future). The hydrodynamic model was validated against observed data from 2004 (Oveisy et al., 2014) with ice cover parameters as in Oveisy et al. (2012). Stream flow was predicted from precipitation using the rational method and inflow and initial condition nutrient concentrations were maintained at 2004 levels. Their simulations show that the future changes in air temperature (under the A2 scenario) will increase the surface water temperature by 2-3.5°C in BoQ, which is in the range of the empirical model study (Trumpickas et al., 2009).

The Bay of Quinte Remedial Action Plan (RAP) has target total phosphorus (TP) and total chlorophyll-*a* (chla) concentrations of 0.03 mg l^{-1} and 4-5 μg l^{-1}, respectively (BoQ RAP 1995). The ELCOM CAEDYM simulations (Figures 9a-d) show average May-October TP concentrations are in the range of 0.1 mg l^{-1} in present and future simulations to be higher than the RAP target (0.03 mg l^{-1}), from loads to the upper Bay (Bolkhari, 2014). With current loads chla biomass is predicted below the target range (<4 μg l^{-1}) in the middle and lower sections of the Bay and above the target (10-15 μg l^{-1}) in the upper reaches during 2000 to 2004. In future climate simulations, the average chla biomass will increase (to 6-8 μg l^{-1}), being mostly high in the upper Bay (14-15 μg l^{-1}). A sensitivity analysis (Bolkhari, 2014) shows future changes in surface water temperature by 2-3.5 °C to have minimal simulated effects on increased chla and cyanobacteria biomass. Rather, enhanced primary

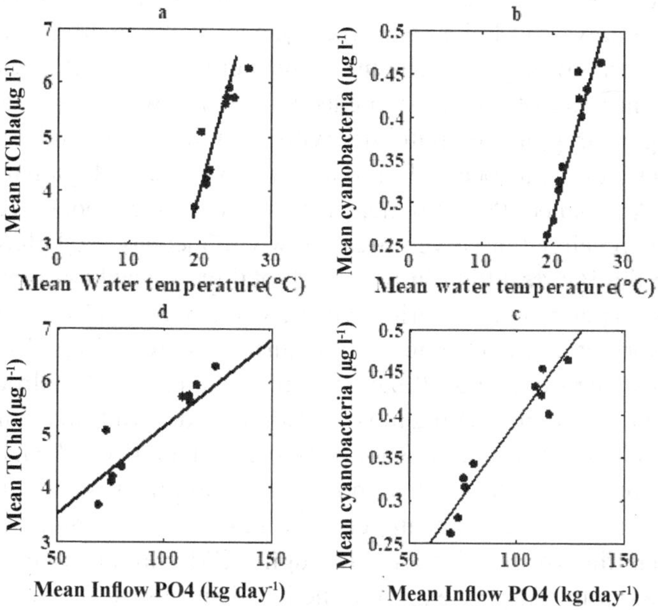

Figure 9. a, b) Mean surface water temperature (°C) versus mean TChla (a-d l^{-1}) and c, d) mean inflow PO4 (kg d^{-1}) versus mean TChla (a-d l^{-1}) aver aged over the bay surface in simulation period (day 133 to 275) during 2000-2004 and 2066-2070. Reproduced from Bolkhari (2014).

production results from increased phosphorus loads, from increases in precipitation and stream flow through the watershed, acted to increase chla and preferential growth of cyanobacteria. The predicted changes in hydrology will increase phosphate loads from (60-70 kg d^{-1}) to (100-130 kg d^{-1}), which will increase the chla biomass concentrations to between 5.5 and 6.5 μg l^{-1}, higher than the RAP targets. If the predicted changes in streamflow are realized due to climate change, the phosphate inputs to the Bay would need to be decreased by ~50%, compared to present values, to maintain recommended RAP phytoplankton concentrations.

Summary

The Bay of Quinte receives excessive nutrient loads from point and non-point sources and suffers from poor water quality due to limited flushing with Lake Ontario. In this study, a brief description of meteorological, hydrological and physical limnological characteristics of Bay of Quinte are presented. Using a three-dimensional model the effects of river inflows and Lake Ontario exchange on the hydrodynamics and mixing were investigated. The main channel of the Bay flushes 5 times a year, whereas isolated embayments have higher water residence times, resulting in trapping of nutrients with sufficient time and conditions for algae blooms to occur. Using a North American Regional Climate Change Assessment Program meteorological data, the model predicts that increased precipitation driven streamflow will increase total phosphorous loads to the Bay resulting in increased total phytoplankton biomass.

This paper provides a brief overview of current knowledge regarding important characteristics of the surface meteorology, hydrology, thermal structure, energy fluxes, water exchanges and hydrodynamic and water quality modelling in BoQ. Relevant descriptions of measurements and model simulations are presented. In the BoQ, the hydrological regime is highly variable in both space and time resulting in longer water retention times in the upper and middle segments and larger variation than the lower segment. Although GCM climate change scenarios have limitations, we explored the implications of climate warming by considering the most recent results available from the NARCCAP scenarios. Changes in water temperature have profound effects on ecosystem components in Bay of Quinte. These investigations indicate that under climatologically changed conditions the thermal structure and water quality of Bay of Quinte may be significantly altered.

References

Blumberg, A.F., and DiToro, D.M., 1990. Effects of climate warming on dissolved oxygen concentrations in Lake Erie. Trans. Amer. Fish. Soc. 119, 210-223.

Bolkhari, H., 2014. Modelling the impact of climate change on the hydrodynamics and biogeochemistry of Lake Simcoe, Hamilton Harbour and the Bay of Quinte Ph.D thesis, Queen's University, Kingston, ON http://qspace.library.queensu. ca/handle/1974/765/browse?type=author&order=ASC&rpp=20&value=Bolkhari%2C+HADISEH

Boyce, F.M. Hamblin, P.F., Schertzer, W.M., and McCrimmon, R.C., 1993. Impact of heat load on Lake Ontario to deep cooling water withdrawals and to global warming. J. Great Lakes Res. 19(3), 603-616.

Freeman, N.G. and Prinsenberg. S.J., 1986. Exchange flows in the Adolphus Reach/ North Channel, pp 27-39. In: C.K. Minns, D.A. Hurley, and K.H. Nicholls (Eds.), Project Quinte: point-source phosphorus control and ecosystem response in the Bay of Quinte, Lake Ontario. Can. Spec. Publ. Fish. Aquat. Sci. 86.

Huang, A., Rao, Y.R., and Zhang, W., 2012. On recent trends in atmosphere and lake variables in Lake Ontario, J. Climate, 25, 5807-5816.

Johnson, M. G., and Owen, G. E., 1971. Nutrients and Nutrient Budgets in the Bay of Quinte, Lake Ontario. Journal Water Pollution Control Federation 43(5), 836-853

Kim D.-K., Zhang W., Rao Y., Watson S., Mugalingam S., Labencki T., Dittrich M., Morley A., Arhonditsis G.B., 2013. Improving the representation of internal nutrient recycling with phosphorus mass balance models: A case study in the Bay of Quinte, Ontario, Canada. Ecol. Model. 256, 53-68.

Leon, L.F., Smith, R.E.H., Hipsey, M.R., Bocaniov, S.A., Higgins, S.N., Hecky, R.E., Antenucci, J.P., Imberger, J.A., and Guildford, S.J., 2011. Application of a 3D hydrodynamic–biological model for seasonal and spatial dynamics of water quality and phytoplankton in Lake Erie. Journal of Great Lakes Research 37, 41–53.

McCormick, M.J., 1990. Potential changes in thermal structure and cycle of Lake Michigan due to global warming. Trans. Amer. Fish. Soc. 119, 183-194. Minns, C. K., Owen, G. E., and Johnson, M. G., 1986, Nutrient Loads and Budgets in the Bay of Quinte, Lake Ontario. In: C.K. Minns, D.A. Hurley, and K.H. Nicholls (Eds.), *Project Quinte: point-source phosphorus control and ecosystem response in the Bay of Quinte, Lake Ontario.* Fish. Aquat. Sci. Can. Spec. Publ. Fish. Aqua. Sci. 86, 59-76.

Minns, C.K., Moore, J.E., Doka, S.E., St. John, M.A., 2011, Temporal trends and spatial patterns in the temperature and oxygen regimes in the Bay of Quinte, Lake Ontario, 1972–2008, Aquatic Ecosystem Health Management, 14(1), 9-20, DOI: 10.1080/14634988.2011.547327

Oveisy, A., Boegman, L., and Imberger, J., 2012. Three-dimensional simulation of lake and ice dynamics during winter. Limnology and Oceanography 57, 43–57. doi: 10.4319/lo.2012.57.1.0043.

Oveisy, A., Boegman, L., Rao, Y.R., 2014. A model of the three-dimensional hydrodynamics, transport and flushing in the Bay of Quinte. J. Great Lakes Res. 41(2), 536-548.

Robinson, G. W., 1986. Water quality of the Bay of Quinte, Lake Ontario, before and after reduction in phosphorus loading, Lake Ontario. In: C.K. Minns, D.A. Hurley, and K.H. Nicholls (Eds.), *Project Quinte: point-source phosphorus control and ecosystem response in the Bay of Quinte, Lake Ontario.* Fish. Aquat. Sci. Can. Spec. Publ. Fish. Aqua. Sci.86, 50-58.

Schertzer, W.M., 1987. Heat balance and heat storage estimates for Lake Erie, 1967 to 1982. J. Great Lakes Res. 13(4), 454-467.

Schertzer, W.M, and Rao, Y.R., 2009. An overview of the characteristics of Lake Superior Meteorology, Hydrology and Physical Limnology. In: M. Munawar, I.F. Munawar (Eds.), *State of Lake Superior* pp. 3-50., Aquatic Health and Management Society, Canada.

Trumpickas, J., Shuter, B.J., Minns, C.K., 2009. Forecasting impacts of climate change on Great Lakes surface water temperatures. J. Great Lakes Res. 35, 454-463.

Zhang W., Kim D.-K., Rao Y., Watson S., Mugalingam S., Labencki T., Dittrich M., Morley A., Arhonditsis G.B., 2013. Can simple phosphorus mass balance models guide management decision?: A case study in the Bay of Quinte, Ontario, Canada. Ecol. Model. 257, 66-79.

Long-term biogeochemical monitoring in the Bay of Quinte: Four decades of ecological impacts

Warren J.S. Currie* and Marten A. Koops

Fisheries and Oceans Canada,
867 Lakeshore Road, Burlington ON L7S 1A1
**Corresponding author: Warren.Currie@dfo-mpo.gc.ca*

Keywords: time-series, seasonality, ions, nutrients, chemistry, ecosystem, Long Term Ecological Research

Introduction

Project Quinte is a long-term, collaborative study investigating biological and physico-chemical changes to the Bay of Quinte (BoQ) since 1972. The impetus for the study was the increased concern over highly eutrophic conditions within the bay during the late 1960s and early 1970s. In the early years, Project Quinte primarily monitored the ecosystem changes associated with phosphorus controls. A special publication was devoted to the early assessment of improvements due to phosphorus management (Minns et al., 1986). In a recent special issue on the nearshore in Lake Ontario, changes in water chemistry and major ions has been analysed for coastal areas around the lake (Chapra et al., 2012, Makarewicz et al., 2012, Winter et al., 2012), but little attention has been given to the extensive physical and chemical water properties that were measured in embayments such as the BoQ beyond phosphorus, clarity (Secchi and light attenuation) or chlorophyll-*a* (Minns et al., 2004; Leisti et al., 2012; Nicholls, 2012).

Here we emphasize analysis of biologically relevant geochemical measures, specifically inter-annual changes over the entire series, climatic forcing and regime shifts associated with invasive species and expansion of macrophytes. Identification of the primary physical forcings and strong correlations between measures can also provide insight into which measures can be best used to characterize the biogeochemistry. To date, there has been no comprehensive publication which catalogues the extent of the biological (including under-reported seston data), physical and water chemistry data which were collected as part of the station sampling in the Bay of Quinte since 1972. In light of this, our objective

Ecology of the Bay of Quinte: Health, Management and Global Implications
Editors: C.K. Minns, M. Munawar and M.A. Koops

Aquatic Ecosystem Health & Management Society, Canada

is to provide an overview of the underutilized biogeochemical dataset that has been collected over the 40+ years of Project Quinte, and extensively documented in Currie and Frank (2015).

Methodology

Some methodological changes have occurred since sampling started in 1972 (noted when possible), but biological (e.g. phyto- and zooplankton, benthos), physical (e.g. light attenuation, mixed depth) and water chemistry (e.g. ions, pH) data were collected annually from spring to fall (May-Oct). Approximately 40 biogeochemical measures were collected from three primary sites: Belleville (B), Hay Bay (HB), and Conway (C) (see Fig. 1), two often sampled sites, Napanee (N) and Glenora (GL), and three occasional locations, Trenton (T), Picton (P) and Lennox (L) (Fig. 2). Only Belleville and Hay Bay have been sampled every year while Conway was sampled continuously until 2009 when sampling was reduced due to budgetary and time constraints. Early sampling from 1972 – 1981 was weekly (~23 sampling trips/yr^{-1}), but after 1982, sampling was changed to biweekly (13 sampling trips/yr^{-1}) due to budget constraints.

The Bay of Quinte is a long (90km) Z-shaped embayment separated from the eastern (Kingston) basin of Lake Ontario by a steep bluff barrier that forms Prince Edward County (Fig. 1). The shape of the BoQ is the result of multiple glacial scouring and outflows through shale limestone deposits (Sly, 1986). Water depth increases along the axis of the bay from very shallow conditions of 2 m near Trenton to 60 m near the outlet by Amherst Island. The Belleville site was chosen to represent the eutrophic conditions of the upper bay and is most afflicted by algal blooms including cyanobacterial blooms (see Nicholls et al., rg 2002). The shallow site (5 m) is well-mixed and influenced by flows and loadings of the Moira River (2.5 km) which passes through the City of Belleville and the much larger Trent River which flows through the City of Trenton (19 km upstream). The two tributaries account for 60-75% of the external phosphorous loadings to the BoQ (Kim et al., 2013). The City of Belleville had a stable population from 1971-1991 (~35 000) but increased by 30% (~46 000) by 2001 and has stabilized since, while Trenton (now Quinte West) and other settlements in the BoQ have not changed in population significantly (Statistics Canada, 2011). Hay Bay (HB) station is the middle bay transitional site located at the southern end of Long Reach and the outlet of Hay Bay. The site has complicated

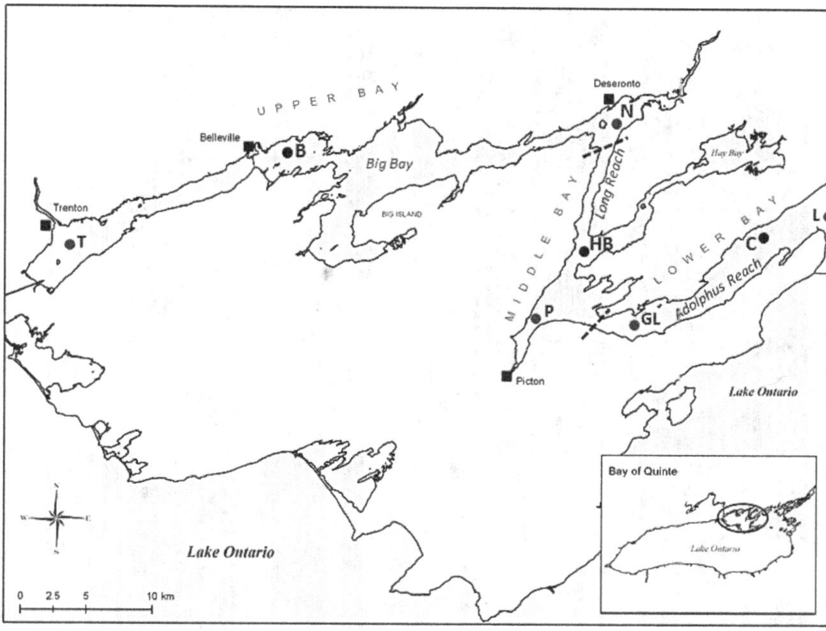

Figure 1. The Bay of Quinte and long-term sampling locations for Project Quinte. The stations with solid circles Belleville, Hay Bay and Conway (B, HB, C) are sites with complete time-series for upper, middle, and lower bay.

hydrology with water flows coming in from the upper bay, Wilton Creek which flows into Hay Bay and periodic intrusions of deep water from the lower bay (pers comm. Ram Yerabundi, Environment and Climate Change Canada). Due to its greater depth (11-12 m), HB normally stratifies during the summer (Minns et al., 2011) and as a result can occasionally have higher biomass of phytoplankton and zooplankton than occurs in the eutrophic upper bay (B) location. While many measures are very similar to the Belleville location, some measures are more similar to that found in the outer bay station Conway. Conway (C) was used as the lower bay station and reference condition for the AOC because it shows little effects of eutrophication. The station is however profoundly influenced by layered exchange flows of Lake Ontario water driven by the lake's seiche, circulation and wind-forcing (Freeman and Prinsenberg, 1986). The station at Conway has historically been located off the mid-channel on a wide-shelf of 35 m depth even though the maximum depth is closer to 50 m. Conway was sampled continuously until 2009

Figure 2. Biogeochemical properties measured for Project Quinte for 1972-2012. Parameters are organized into 3 groups: carbon and particulates, water chemistry and physical measures. Solid bars link years where samples were analysed for that parameter. Locations sampled by year are indicated at the bottom.

when it was discontinued, but sampling for some parameters has been reinstated on a monthly basis since 2013.

Two other stations have been sampled more than half of the years of Project Quinte. Sampling at Napanee (N) has a long gap in the mid-80s to 90s. Glenora (GL) is situated across from the OMNR Glenora Fisheries Station and except for a gap in the mid-80s, was sampled until 2008. This station exhibits layered flows, with the hypolimnion being quite distinct from the epilimnion, amplified by the rapid change in bathymetry (Freeman and Prinsenberg, 1986). It was felt that this station reflected either HB or C conditions depending on the nature of the flows on that day, so was deemed redundant. The uppermost bay location, Trenton (T) was sampled until 1980 and again in the early 90s but was discontinued as a station because it was felt it was more reflective of the dynamics of the Trent River than the BoQ itself, and adding the station back during the 1980s was opposed due to increased macrophyte growth. The other stations, Picton (P) and Lennox (L) were only sampled until 1978.

For standard station sampling, water is transported from the field site to the Canada Center for Inland Waters (CCIW) in Burlington, Ontario and dropped at the Ontario Ministry of the Environment (OMOE, since 2018 called Ministry of Environment, Conservation and Parks) for water chemistry testing (later being shipped with a delay of approximately a day), though a number of measures have also been conducted at the National Laboratory for Environmental Testing (NLET) at CCIW since 1988, so there is duplication for some parameters. Some analytical measurements are done at the Great Lakes Laboratory for Fisheries and Aquatic Sciences (GLLFAS) at CCIW including chlorophyll-*a* extraction. Additional details of these procedures can be found in (Graham et al., 1996). Seston procedures involve filtration of 0.2-2.0 L (depending on station or season) of epilimnetic water through ashed, pre-weighed Whatman GF/C filters, which were dried overnight at 60°C, weighed, ashed again at 480°C and reweighed to calculate ash-free dry weight.

Analysis of this dataset is complicated by having analytical laboratory analyses done by different agencies (Ontario Ministry of the Environment, Environment Canada, Fisheries and Oceans) with different methodologies for sample storage, time to analysis and pre-filtering. For instance, the MOE samples are always unfiltered whereas the NLET samples are only filtered when testing for dissolved ions. Whether a sample is filtered is noted in Fig. 2. Also, some analytical procedures only take place after the late 1980s (NLET – Environment Canada) and some are no longer measured due to past budget constraints (e.g. total nitrogen).

As a result, only some parameters and stations have been sampled for the entire period whereas many have gaps where they were not measured or dropped entirely (Fig 2).

Most values measured by both OMOE and NLET or GLLFAS were consistent given differences in methodology, but some values were commonly different (see Currie and Frank, 2015). While TP values were reliable between OMOE and NLET showing no significant difference, SRP values measured at OMOE were consistently about 1.3-1.8x higher than measured at NLET (paired-means, $p<0.01$). Likewise, ammonia values measured by the OMOE lab were generally higher except for some high spikes ($p<0.01$). This may well be due to the time delay associated with analysis of the OMOE samples which can allow for additional SRP and ammonia conversion in the bottle. Chlorophyll-*a* (uncorrected) values were consistently lower in the OMOE samples than measured at GLLFAS (~69%, 8 $\mu g\ l^{-1}$) until 1996 when they were much more comparable (1 $\mu g\ l^{-1}$) (paired means regression, $p<0.01$). The OMOE chl-*a* methods changed in 1985 from the use of cellulose-nitrate filters to nylon filters which was reported to increase values by 35% (See Nichols and Hopkins, 1993) but is still lower than found at GLLFAS until 1996. Methods at GLLFAS have been consistent since 1977 (ground GF/C filters in 90% acetone) so this change is most likely due to another change in methodology at the OMOE laboratory.

When multiple series were available from different agencies, the longest continuous time-series was selected, though some series are compared for bias in Currie and Frank (2015). The data were initially examined to determine if any outliers occurred beyond those found during standard QA/QC. In all, 8 values were removed as a result of implausibility (>5 SD units) or deviation from a concurrent measure. Correlation between measures was done on monthly averages. Strongly non-normal data were transformed using log-transformation or reciprocal (noted in text). All statistical analyses were completed using JMP® v10 (SAS Institute Inc. 2012) or R v3.0.1 (R Core Team 2013).

Results and discussion

The data have been assembled in 3 major groupings: 1) physical and climate measures, 2) carbon and particulates and 3) chemical measures including nutrients (Fig. 2), but for this summary, time-series are grouped into biologically or chemically-driven groupings.

Water clarity has undergone drastic changes in the BoQ at all sites (Fig. 3). The Secchi depth has increased at all sites from the low in the 1970s prior to phosphorus controls, to slightly higher values in the 80s to the mid-90s. In 1994-1995 the Secchi depth begins to increase sharply over several years first and strongest at Conway, with Belleville the last station to increase. Prior to 1995 (pre-1995 "turbid stanza"), the B and HB sites Secchi values were not significantly different. Secchi depth data was strongly nonlinear, so was transformed by reciprocal (power -1). This is due to the fact that light forms an exponential relationship with depth, so inversion yielded a strongly linear relationship with light attenuation. The Secchi depth values have declined slightly since 1995 (post-1995 "clear stanza"), but remains deeper than earlier periods at all sites (two-way(stanza, month) ANOVA, B ($F(1,5,239) = 38.2$), HB ($F(1,5,240) = 47.8$), C ($F(1,5,222) = 27.6$), $p<0.001$). The PAR light attenuation (k_d) (% light absorbed or scattered per meter) data has shown a steady decline since the early 70s to a minimum in the early 2000s indicating significantly increased water transparency at all sites >1995 (two-way(stanza, month) ANOVA, B($F(1,5,239) = 33.7$), HB($F(1,5,240) = 52.5$), C($F(1,5,222) = 18.4$), $p<0.0001$). There is some indication that k_d has increased since 2004, but this difference is not significant.

The lower trophic level production measures have changed considerably since the start of Project Quinte (Fig. 3). Seston and chlorophyll-*a* values are examined here but other organic carbon measurements: ash free dry weight (AFDW), total phytoplankton biomass, and particulate organic carbon (and particulate organic nitrogen) all indicate similar patterns over time (further detail found in Currie and Frank, 2015 and Millard, 1986). Conway's biomass measurements are consistently lower than upper and middle bay sites and have shown a slower decline over time. Except for the early years (72-77) when Belleville had consistently higher values, generally, the middle bay (HB) and upper bay (B) sites pre-1995 values were not significantly different in chl-*a* (two-way(stanza, site) ANOVA(1,2,680), $t=1.36$, $p=0.17$), with one or the other having a slightly higher value on any given year though B has more annual variability. Conway always had significantly lower chlorophyll-*a* values (two-way(stanza, site) ANOVA(1,2,680), $t=-13.9$, $p<0.0001$). Though patterns are similar, the seston and chl-*a* time-series differ in their breakpoints, with seston showing a much stronger change in 1994. For chl-*a*, partition analysis (JMP) using AIC (Akaike Information Criteria) indicates that the best model to separate years is 1978 for B and 1993 for

HB and C. For seston the best AIC model uses 1995 then 1978 for B and

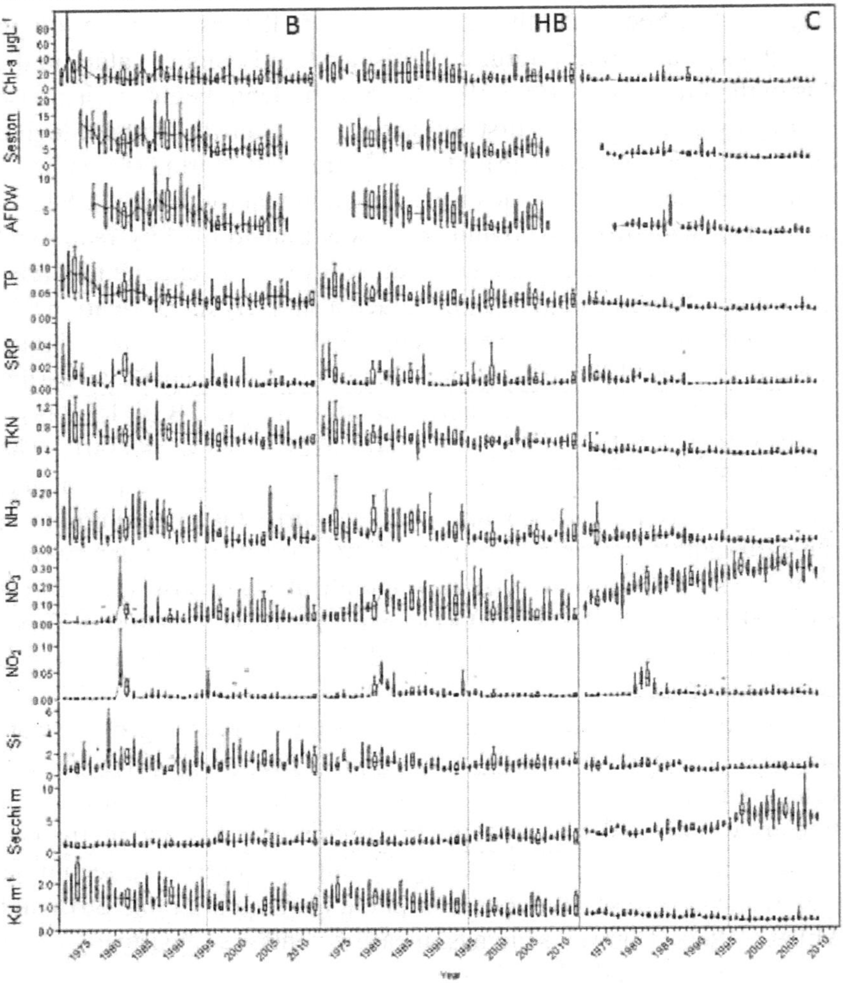

Figure 3. Annual box plots of some monthly biological, nutrient and clarity data from Project Quinte. Boxes representing 25th and 75th quartiles with whiskers indicating 1.5 of quartile range. The line in the box notes the median and outliers are shown as points. Chlorophyll-*a*, seston, ash-free dry-weight, total phosphorus, soluble reactive phosphorus, total Kjeldahl nitrogen, ammonia, nitrate, nitrite, silica, Secchi depth and light attenuation (k_d) for the 3 major sites are shown for the upper bay Belleville (B), middle bay Hay Bay (HB) and lower bay Conway (C). Measures without units are in mg l^{-1}. Solid line connects the annual means. Vertical fine-dotted lines separate time groupings of pre-1995 and post-1995.

1995 for HB and C. The pre-1995 period (see Fig. 3) exhibits the highest levels of biomass with a precipitous drop in 1995 following the record cold winter of 1994 with 1996 showing the lowest annual measurements. A similar drop in chl-*a* and seston is seen following the cold winters of 1978 and 1986 (See table 1 and Fig 5). The period from 1997 exhibits a rebound to conditions approaching the 1980s and has been slightly elevated since (Fig. 3). The post-1995 period is significantly different from the pre-1995 period for both seston and chl-*a* (two-way (stanza, site) ANOVA chl-*a*(1,2,680) F=73.3, seston F(1,2,568)=206.7), $p<0.0001$), but while the recent rebound period post-2000 is slightly elevated it is not significantly different than the 1995-2000 period.

Nutrients measured for Project Quinte are broken into phosphorus and nitrogen groups, but for summary here, we focus on the total fractions only: total phosphorus (TP) and total Kjeldahl nitrogen (TKN) (see Currie and Frank (2015) for details on other nutrient chemistry series). Silicate limitation has been implicated as a factor during the extremely high productivity years in the early 1970s (Nicholls et al., 2002) but has been invariant for most of the series. Both nitrogen and phosphorus are found in much higher concentrations at HB and B than occur at C (Fig. 3). The BoQ is like a river for transport of substances, but the sediment water interface can also increase persistence of biologically relevant compounds (Diamond et al., 1994, Minns et al., 2011). Most of the total phosphorus loadings to the BoQ is sourced from 3 sewage treatment plants and 2 rivers upstream of the upper bay Belleville site (Nicholls, 1999). TP had shown a significant decline since the high in the 1970s but has been steady since 1990 near 0.04 mg l^{-1} at B and HB and 0.01 mg l^{-1} at C (two-way(pre-90,site) ANOVA, F(1,2,703)=135.4, $p<0.0001$). Soluble Reactive Phosphorus (SRP) has similarly declined (Currie and Frank, 2015) but showed strong increases in 1978 and 1994, both which had extremely cold winters, low precipitation and reduced algal biomass that persisted for 5-8 years suggesting a change in system function, likely due to changes in sediment remineralisation since a time lag is seen between HB and C.

The role of nitrogen in the BoQ has received much less attention than that of phosphorus (though see Liao, 1977), but there are strong patterns in a variety of nitrogen forms. The nitrogen cycle is more complex than phosphorus, including an atmospheric component and nitrogen has more forms. Total Kjeldahl Nitrogen is the sum of all organic nitrogen forms, not including inorganic nitrate (NO_3^-) and nitrite (NO_2^-). TKN has shown the same steady decline as TP since the early 70s for all sites in the BoQ

(Fig. 3) and is similarly significantly lower post-1990 (two-way (pre-90, site) ANOVA, $F(1,2,703)=210.9$, $p<0.0001$). Nitrite was at very low levels in the BoQ in the 1970s, suggestive of nitrogen limitation (Liao 1977) but has been reasonably stable except for a period of exceptionally high values from 1981-1983 with multiple peaks in July 1981 of greater than 0.25 mg l^{-1}. This value is very high and is at the toxicity level for salmonids (Lewis and Morris, 1986). The source of this nitrite spike is unknown, but it occurred at all the BoQ sites. The largest fraction of Total Nitrogen, Nitrate has increased steadily in the BoQ but most strongly at the lower bay C site (where it is greatest) as influenced by the continued buildup of nitrate in the Great Lakes (Winter et al., 2012), but also showed a large spikes following years of high nitrite in the upper and middle bay. In recent years, nitrate is depleted during the summer in the upper and middle bay which strongly suggests nitrogen limitation of phytoplankton growth during the late summer (July-Sept) post-1995 (Currie and Frank, 2015).

The dissolved to total fraction of phosphorus (SRP:TP) ratio shows wide swings over time but are similar across the BoQ sites with an initial decline in the 70s though greater SRP persists until 1988 when it declines again (Currie and Frank, 2015). Years of increased SRP seem to follow years of very cold winters, with a sharp increase 1996 onwards to about 12% of TP since at the upper and middle bay locations. If biological activity is reduced following a cold winter, remineralized SRP from the sediments could increase during the summer leading to peaks that occur for that year (Nicholls 1999). SRP recycling peaks during the warmer summer when regenerated from the sediments (Nürnberg and Peters, 1984; Song and Burgin, 2017), so is expected to change over the season because of seasonal variability in TP load from tributary inflows from spring highs to lows during summer months (Nicholls, 1999, Minns et al., 2004). The ratio of total phosphorus to combined nitrate and nitrite ($TP:NO_{2+3}$) can indicate if nitrogen is a limiting nutrient and whether nitrogen-fixing algal groups may have an advantage (Liao, 1977). In the 1970s phosphorus was in excess, and it was not until the large increase in nitrogen in 1980 that the ratio was below 2 (Robinson, 1986). Generally, seston in freshwater require N:P of 20 (Sterner, 2011), so much of this phosphorus should be unutilized. After a slight increase again in 1984, the P:N ratio has stayed near 1-2 and post-1995, is similar for both B and HB sites. While phosphorus is not in excess as it was in the 1970s, the ratio P:N is still higher than would be expected for biological uptake (Redfield, 1958).

Table 1. Anomalies and Forcings. Meteorological measurements downloaded from National Centers for Environmental Information (NCEI) for Trenton Airforce Base (AFB). Other sources for biological and climatic forcings are noted.

Date	Meteorological / Climatic	Biological or Societal Changes
1972	High precip (>85mm) & snow (>200mm)	
1973-1974	La Niña	
1976	Cold summer (<20°C mean July & Aug) and winter, high snowfall (>200mm)	
1977	High snowfall (>200mm)	Phosphate controls ↓TP (28-38%), ↓TN (13-20%) (Minns et al., 2004)
1978-1979	Very cold winter	↓ Phytoplankton biomass ↑P/B ↓ Predaceous zooplankton biomass
1980	Mt. Saint Helen's eruption	↓ Herbivorous zooplankton, copepod biomass
1980-1982	Cold winter, ↑ ice winters	↑ Cormorants, ↑ Bosmina (1981)
1982-1983	El Niño, very mild winter (≤5°C Jan), Hot summer	↑↑ Herbivorous zooplankton biomass
1984-1985 1985-1986	Cold winter, ↑ ice winters, Cold summer (<20° mean July & Aug)	Record high Walleye YOY recruitment (Ridgeway et al., 1990)
1987		BoQ listed as Area of Concern (AOC) under the Great Lakes Water Quality Agreement
1988-1989	La Niña, Hot summer (>22°C), very low precipitation (<60mm)	↑ Gizzard Shad and Piscivore biomass
1991	Mt. Pinatubo eruption (June 15), second largest terrestrial eruption of the 20th Century, ↓global sunlight by 10%, ↓temp 0.5°C in northern hemisphere	Driessenid Mussels (DM) first recorded BoQ on hard substrate
1991-1992	El Niño, Cold summer (<20°C July & Aug)	↓↓Herbivorous zooplankton and copepod biomass, ↓↓ Gizzard shad and Alewife
1993	↓global temp 0.73°C, wet cold summer in US and flooding	
1993-1994	Extreme cold winter (25 days mean temp below -15°C), record ice cover on Great Lakes (>90%)	

Table 1 Contd...

Date	Meteorological / Climatic	Biological or Societal Changes
1996	Very high precipitation (>85mm), but low runoff and BoQ outflow	Begin ↑ DM at Conway, first adult DM in benthic surveys in Big Bay, ↓ Phytoplankton biomass but no P/B change, ↓ Alewife
1997	High snowfall (>200mm)	↑ DM veligers in plankton
1997-1998	Strong El Niño, very mild winter (≤5°C)	
1998		Bythothrephes reported in BoQ (lower Bay)
1999	Hot summer (>22°C)	Cercopagis and Round Goby reported in BoQ, begin ↑↑ DM biomass adults
2000	Cold summer (>20°C mean July & Aug), very high precipitation (>85mm)	↓ Phytoplankton biomass ↑P/B, DM adult biomass peak
2001	Very low precipitation (<60mm)	
2002	Very mild winter (≤5°C Jan), hot summer (>22°C)	
2003	Very high precipitation (>85mm)	↑ Round Gobies
2004	Very high precipitation (>85mm)	
2005	Hot summer (>22°C)	
2006	Very mild winter (≤5°C Jan), hot summer (>22°C), very high precipitation (>85mm)	
2007	Very low precipitation (<60mm)	
2008	Very high precipitation (>85mm), high snowfall (>200mm)	↑ Zebra Mussel veligers 25% of zooplankton biomass for all sites
2009	El Niño, very mild winter (≤5°C Jan), very high precipitation (>85mm)	
2010	Hot summer (>22°C)	

Climatic summary from Table 1:

1976-1982 regular cold winters and no hot summers, several high snowfall years (72,76,77)

1983-1989 mild winters, warm summers

1990-1994 cold winters including coldest (94) and coldest summer (92)

1995-2001 mild conditions

2002-2012 hot summers, interspersed with cold summers (05, 09) and high precipitation (03, 04, 06, 08, 09), no high snowfall years except 2008

Water chemistry reporting contains a range of measurements that include physical water characteristics, but are generally dissolved ions, including biologically important ones such as micro-nutrients. Many important water chemistry values have only been measured since 1995 including pH, total hardness, total calcium, whereas chloride and sodium, alkalinity and conductivity have been measured since the 1970s (Fig. 4). Other than nutrient ions, the lower bay site C has higher values of dissolved fractions than occurs in the upper and middle bay sites. This is due to evaporation which concentrates substances and additive loadings of conservative constituents from rivers in all the Great Lakes (Chapra et al., 2012). Iron, potassium, magnesium or sulphates show no discernable pattern over time though these have only been measured since the major shift in the BoQ which occurred post-1995 (Currie and Frank, 2015). Both chloride and sodium have significantly increased in the upper bay at B steadily, doubling from values of 3.6 to 8.7 (Na^+, $r^2=0.68$, $p<0.0001$) and 6.9 to 15.3 (Cl^-, $r^2=0.64$, $p<0.0001$) mg l^{-1}, but not as clearly at HB for Na^+ ($r^2=0.10$, $p<0.0001$) and not for Cl^-. The relationships in the lower bay are not significant for Na^+, though Cl^- decreased at C until 1995, and have since steadily increased to concentrations nearer to those found in the 1980s, likely due to recent increased loadings from salting activities as have been seen in other parts of Lake Ontario (Winter et al., 2012). Alkalinity and conductivity were relatively stable for much of the study but have recently exhibited abrupt increases in the 2000s at B and HB (Fig. 4). The BoQ acts as an elongated freshwater estuary with layered flow in the middle and lower bay, so reduced flows and increased evaporation are more likely to concentrate substances rather than to reduce them. A drop in alkalinity, total hardness, calcium, and pH was seen in 1999 during the very hot dry summer (Fig 5). There was also an increase in chl-*a* and seston values in 1999, so part of the chemical change could be due to increased algal production (Brewer, 1976). Since 1999, there have been warmer conditions, and the total hardness and sodium values of the upper and middle bay are now comparable to those found in the lower bay even though values in Lake Ontario have been dropping over this period (Chapra et al., 2012). For dissolved carbon fractions (Fig. 4), the time-series is much shorter (1988-2012) but there is little change beyond a slight increase in inorganic carbon (DIC) in 1995 and an increase in 1996 of organic carbon (DOC). This occurs concurrently with the decline in particulate carbon so may be related to remineralisation effects in the years following the dry and cold winter in 1994.

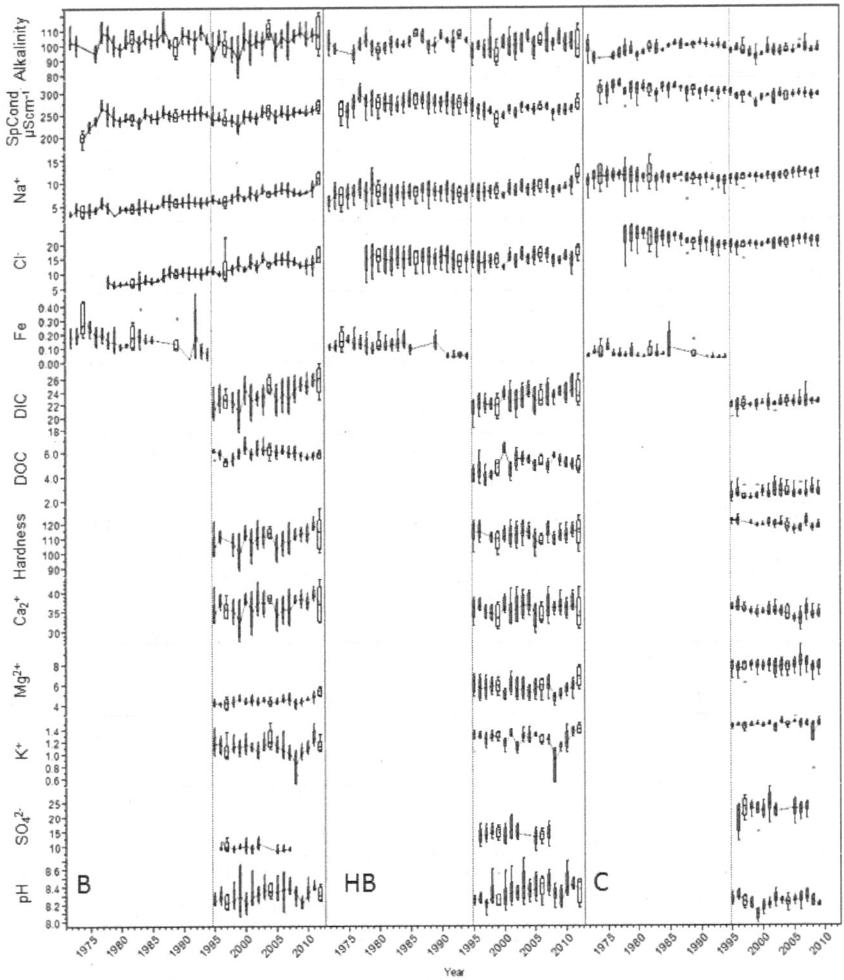

Figure 4. Annual box plots as per Fig 3 of the following dissolved measures: alkalinity, specific conductivity, sodium, chloride, iron, dissolved inorganic carbon, dissolved organic carbon, total hardness, total calcium, magnesium, potassium, sulphate and pH for the upper bay Belleville (B), middle bay Hay Bay (HB) and lower bay Conway (C). Measures without units are in mg l^{-1}. Solid line connects the annual means. Vertical fine-dotted lines separate time groupings of pre-1995 and post-1995.

While the collection of meteorological data is not specifically part of Project Quinte, it helps to explain some of the time series in this section. Data were compiled from the National Climatic Data Center (NCDC) for Trenton Air Force Base (CYTR). In Figure 5, mean, maximum and minimum monthly air temperatures were plotted with total precipitation and snow accumulation. Climatic extremes are summarized in Table 1. Extremely cold winters were experienced in 1976-1977, 1979, 1987, and 1994. The winter of 1994 was by far the coldest in the time-series with 25 days below a mean temperature of -15°C. The coldest winters often did not have increased snowfall, but the colder winter period in the 70s to early 80s had more high snowfall years. 1988 was the driest year in the series though the 2000s had many low precipitation years (2001, 2007, 2010, and 2012) along with many years with elevated summer temperatures. The period post-1999 exhibits much greater variability year to year though summers were typically hot and wet but interspersed with years of cooler temperatures and low precipitation. Since 1999, the summers have been hotter, but have been interspersed with cold wet summers. The period from 1999-2012 is 1.2 °C warmer than 1972-1998 (ANOVA $F(1,241)=26.9$, $p<0.0001$) and a linear regression with date gives a significantly positive slope with $r^2=0.31$, $F=17.3$, $p<0.001$). This is in part due to the variability in temperature across the year. The months of June (+1.3 °C), September (+1.7 °C) and October (+1.2 °C) were significantly warmer 1999-2012 (ANOVA (pre-99, month) $F(1,5,241)=236$, $p<0.0001$), and though July and August were warmer, the values were not significant. It has already been determined that the stability (stratification) of the water column has increased (Minns et al., 2011) in part due to an epilimnetic temperature increase of 1.3°C, and other physical environmental factors might also be affected. The periodicity in the precipitation output is likely to change the flow rates in the major tributaries, thus changing the loadings of nutrients to the bay (Minns and Johnson 1979). For the 1999-2012 period, the March precipitation is less (75% of 1972-1998) while the mid-summer July precipitation is significantly more (170%, mean > 90mm) (ANOVA, $F(1,241)=2.38$, $p=0.008$). This recent increase in summer precipitation could change the flow and exchanges within the bay since flushing rate is highly dependent on river inputs, and rates of remineralisation of nutrients might be affected since this increased flow occurs during the hottest period of the year.

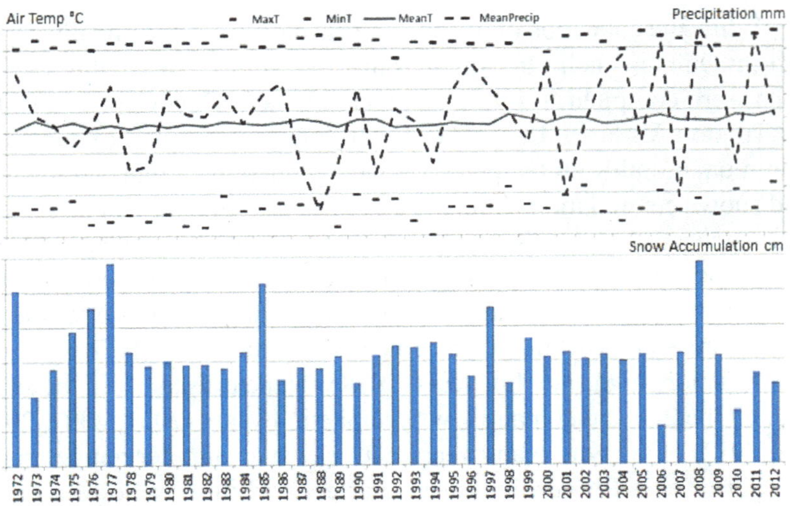

Figure 5. Climate measurements downloaded from National Centers for Environmental Information (NCEI) for Trenton Air Force Base of air temperatures, annual mean (solid line) with bars for min and max, annual precipitation (dashed line) in mm (a) and annual snow accumulation in cm (b). The shaded area represents the period of greater climactic variability post-1999 noted in the text.

Correlations

A cautious approach was taken when interpreting the Pearson Correlation Coefficients (Prairie 1996) so only values of $r>|0.7|$ (very strong) were considered relevant for monthly averages (Fig. 6). Some correlations consistently clustered as expected: the major total nutrients TP & TKN, the major organic particulate carbon measures (seston, chlorophyll-*a*), and the water column light characteristics (Secchi^{-1}, k_d). It is worth noting that without taking the reciprocal, Secchi depth had a much poorer relationship to other measures (reduced ~10%), sometimes dropping below $r=0.7$. In the upper bay (B) TP was strongly correlated with SRP, TKN, chl-*a*, seston, Secchi and k_d, although TKN had even stronger correlations with these values (Fig 6). This suggests that nitrogen plays an important role as a co-limiting nutrient in the upper bay. The light measurements were strongly correlated to major nutrients (TP, TKN) and organic particulates (seston, chl-*a*, $r>0.8$). Within the dissolved fraction (Fig. 6), DIC, Alkalinity, conductivity, and total hardness and total calcium were also strongly correlated. Total hardness was exceptionally correlated to calcium as expected since this is the largest component to hardness. Chl-*a* and seston were also negatively correlated with total hardness and calcium (r=-0.7) likely due to the correlation between hardness and TKN and TP (r<-0.6). Chloride, sodium (r>0.9) and magnesium (r>0.8) were very strongly correlated, which is expected since the primary source of chloride in the bay is expected to be from road salting. Unlike Winter et al., (2012) we did not see significant correlations with chloride and precipitation or snow accumulation, however, the highest values for these measurements are in months that are not sampled during Project Quinte (Dec-Apr).

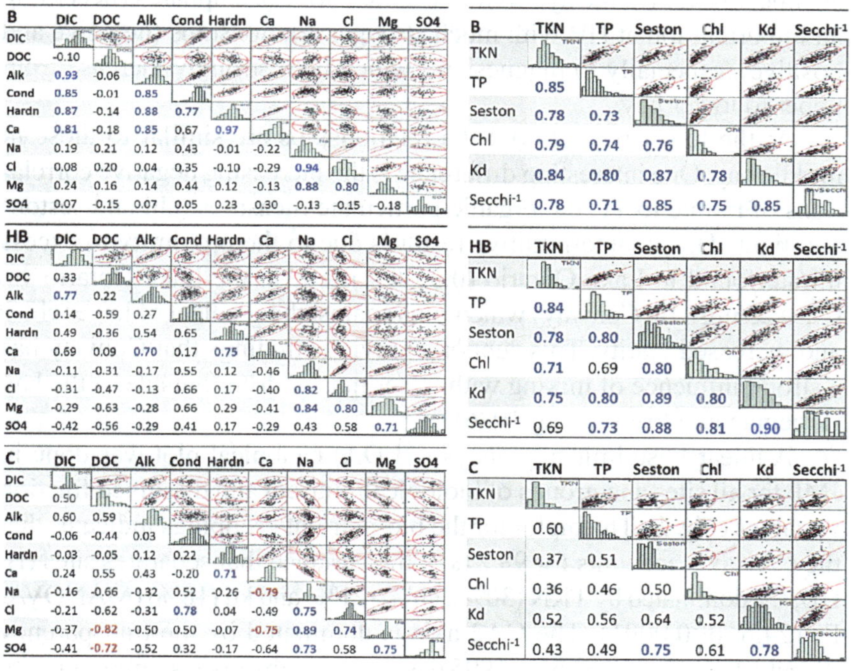

Figure 6. Correlation coefficient matrix (below), regression plots with lines of best fit and bar chart indicating distribution (above) using monthly data for the dissolved and the bio/nutrient/clarity groupings at the upper bay Belleville (B), middle bay Hay Bay (HB) and lower bay Conway (C). Values in bold represent strong positive or negative correlation coefficients with values >|0.7|. Only years where all measurements were taken (see Fig. 2) were used in the matrix.

The middle bay (HB) has similar relationships to those found in the upper bay, though a couple are weaker (Chl-TKN, Chl-TP and TKN-Secchi^{-1}). The Secchi^{-1} correlations with seston, chl and attenuation are higher likely due to reduced turbidity and resuspension at this site which is often stratified during the summer (Fig. 6). Some of the dissolved water chemistry relationships are no longer correlated or much weakened (total hardness, calcium, alkalinity), particularly to DIC and are no longer related to chl-*a* or light as they were in the shallower upper bay. This may be due to the effect of stratification at Hay Bay concentrating primary production in the epilimnion and separating some of the chemical factors to the hypolimnion. Alternatively, sulphate is becoming an important ion at HB with much stronger relationships (negative and positive), especially to magnesium, which both strongly increase from upper to lower bay.

At the lower bay site (C) the relationships are similar to those in middle bay. One interesting difference is the increasing negative correlations of nitrate to TP, the organic particulate measures (chl and seston) and light. This counterintuitive result is due to the proximity to excess nitrate found in Lake Ontario (0.35-0.4 mg·l^{-1} in the Lake Ontario vs. 0.2-0.25 mg·l^{-1} at Conway) which runs against the gradient of lower chl-*a* and increased clarity. This supports previous findings that Conway has a strong influence of mixing with water masses from outside of the bay (Freeman and Prinsenberg 1986).

A linear Discriminant Analysis (LDA) canonical plot was done in JMP for all sites and groups of biological/nutrient and dissolved fraction variables (Fig 7). The biplot for the bio-nutrient groupings indicates that the CA1 axis accounts for 98.2% of the variance that separates sites (B, HB, C), dominated by TKN (30%), chl-*a* (21%) and kd (18%) (MANOVA, $F=124.6$, $p<0.0001$). The CA2 axis is dominated by the transparency (Secchi, k_d) values. The post-1995 data indicated by open symbols on Fig. 7 are shifted to the left on CA1 for B and HB. The biplot for dissolved fractions indicates that the CA1 axis accounts for 95.2% of the variance to separate sites, dominated equivalently by Ca^{2+} (23%) and Mg^{2+} (22%) and by hardness (14%) (MANOVA, $F=196$, $p<0.001$). The CA2 axis has similar dominants with hardness and Ca^{2+} and the HB sites partially separates from B and C on the CA2 axis.

Seasonality

To illustrate the seasonality associated with some measures, values from May to Oct at B, HB and C are plotted in Fig. 8 for the time stanzas of 1972-1994 (pre-1995) and1995-2012 (post-1995). Seasonally, seston weight and chl-*a* peak in August post-1995 whereas the peak was in September pre-1995 and are now significantly reduced with the seston peak of 12 down to 7 mg l^{-1}. Summer chl-*a* is down 8 $\mu g \cdot l^{-1}$, which is 70% of pre-1995 value and seston, POC and AFDW are at 55-60% of their previous values (Currie and Frank, 2015). This differs from the findings of Winters et al., (2012) for coastal Lake Ontario which showed reduced fall and spring blooms with a slight increase in summer. The Secchi depth has deepened by 40-50% at all sites indicating considerably clearer conditions (Fig. 8). The seasonal pattern has not changed for Secchi depth though there is considerably more variability in the values from B in the upper bay, which is expected since there is more likelihood in this shallower section for sediment resuspension and river-based turbidity input. The difference in Secchi depth is most apparent in the spring and early summer since there was much reduced seasonality pre-1995.

The seasonal pattern of phosphorus closely follows that of the TKN with a significant reduction to 70% for peak TP of pre-1995 values. The impact of phosphorus on the BoQ ecosystem has been extensively discussed in several previous publications (Millard and Sager, 1994; Minns et al., 1986; Nicholls, 2012), so will not be repeated here other than some comments on seasonality. While Belleville (B) shows the highest TP values, the values in the middle bay (HB) are equal to or exceed those of B for Sept-Oct, though the values at B show more variability so the values are not statistically different. The maximum measured (peak) TP value at B has decreased since the early 1970s to 1986 and has remained at the lower value since (Currie and Frank, 2015). Belleville also experienced two abrupt drops in peak TP (77-89 and 84-86) which correspond to years with very cold winters and summers. Post-1995 the Julian day of peak TP occurs approximately 2-3 weeks earlier (in August rather than September) than occurred in the 1970s and 80s (Fig 8, see Currie and Frank, 2015 for more details).

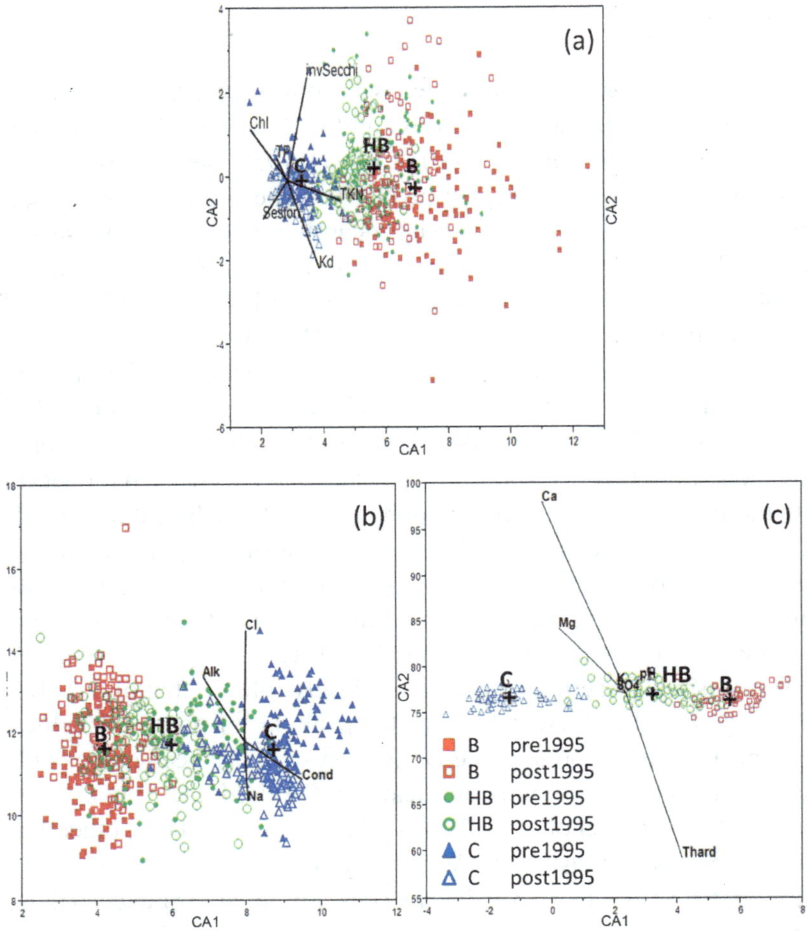

Figure 7. Linear Discriminant Analysis (LDA) canonical plots showing biplot rays and sites for **a**) biological/nutrient/physical measures: seston, chlorophyll-*a* (Chl), total Kjeldahl nitrogen (TKN), total phosphorus (TP), secchi-1 (inv Secchi) and light attenuation (kd); **b**) dissolved constituents having a long time-series: alkalinity (alk), specific conductivity (Cond), sodium (Na) and chloride (Cl); and **c**) for dissolved series with data limited to post-1995 only: total hardness (THard), calcium (Ca), magnesium (Mg), potassium (K), sulphate (SO4) and pH. All graphs use the same site symbols Belleville (B, squares), Hay Bay (HB, circles) and Conway (C, triangles) with solid symbols indicating pre-1995, open symbols post-1995 values. Each site centroid is labeled using a **+**. The rays on the biplots give the Mahalanobis distance of the observation from the site centroid and direction in canonical space.

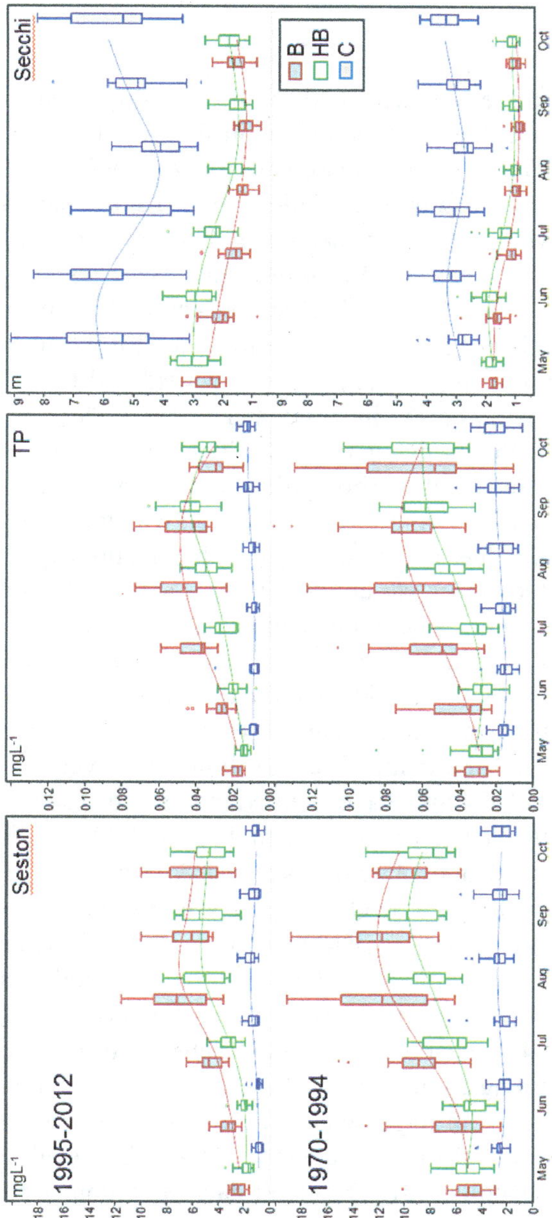

Figure 8. Seasonality illustrated by monthly box plots as in Fig. 3 for the period of May–Oct for seston, total phosphorus and Secchi depth using the pre-1995 time stanza (bottom) and post-1995 (top) for Belleville (B dark shading), Hay Bay (HB clear bars) and Conway (C light shading). Line is a spline fit though the annual mean.

Summary

The Bay of Quinte was described as impaired due to eutrophication in the 1970s. Since then, phosphorus controls and changes to ecosystem structure have resulted in increased clarity as recorded by Secchi depth and light attenuation measures. Chlorophyll-*a* and seston declined to a low in 1995, when the system changed from a turbid phase to clear phase, just prior to the increase in Driessenid Mussels (*Driessena polymorpha* and *bugensis*) in the Bay. Total phosphorus and nitrogen have declined, yet the proportion of soluble reactive to total phosphorus has remained high. Though phosphorus is generally considered to be the limiting nutrient to growth of algae in freshwater systems, there is indication of co-limitation with nitrogen in the Bay of Quinte with higher correlations ($p>0.75$) between total Kjeldahl nitrogen and productivity measures (seston, chlorophyll-*a*) and water clarity than with total phosphorus. The middle bay site values are often similar to the impaired upper bay though those sites exhibit more variability. The timing of peak productivity has resulted in a greater reduction in summer chlorophyll and seston values than during the spring. While no relationship existed with precipitation, the values of sodium, chloride and conductivity have increased within the Bay of Quinte, most likely sourced from winter road salting. There is some indication that climatic forcing following the very cold winters of 1977-1978 and 1994 may have resulted in reduced system productivity. Increased mean temperature and variability post-1999 have likely changed the timing of nutrient cycling peaks to earlier in the year.

Studies on the Bay of Quinte ecosystem date back to Pritchard (1931) detailing Cisco spawning and Tucker (1948) documenting the phytoplankton community temperature and wind. Unfortunately, a considerable gap occurs until the documentation of deteriorated conditions in the late 1960s. Project Quinte has produced a dataset of biweekly sampling from May-Oct where we have observed significant long-term changes in biogeochemical measures in the Bay of Quinte since 1972 (also see Millard and Sager, 1994; Minns et al., 1986; Minns et al., 2011, Nicholls, 2012). Though data often comes from different analytical labs, the data is quite comparable, even due to methodological differences, which helps to extend the data series because some measures were only collected by an agency for short periods of time. While there was often a difference in absolute value, many of the measures show remarkable synchrony across the stations from the upper to lower bay. In particular, there has been a distinct increase in clarity, resulting in deeper euphotic zones throughout the bay. This is due to decreased algal biomass (chlorophyll-*a*) and seston in turn driven by lower nutrient values of phosphorus and nitrogen as suggested by high correlation values with TKN and TP. A return to co-limitation by nitrogen is suggested post-1995, particularly in late summer when the phytoplankton biomass peak occurs, which could promote nitrogen-fixing cyanobacteria (Munawar et al., 2012). Light-related measures (Secchi depth, light attenuation (k_d)) were highly correlated, though Secchi depth was much less correlated to productivity measures than k_d, particularly at the turbid upper bay site. The sharp change in clarity around 1994 was likely triggered by the extremely cold winter and the accelerating macrophyte expansion in the upper bay (Leisti et al., 2006) shifting the system from a turbid to a clear phase state (Scheffer and Carpenter, 2003). This in turn helped to foster the invasion of Zebra Mussels into the upper bay with the sharp increase in biomass occurring in 1999 in the lower and middle bay (Dermott and Bonnell, 2011). While total organic nitrogen (TKN) has decreased over time, nitrite has occasionally experienced peaks, including a huge upsurge in 1981 that persisted for some years following, the source of which we could not explain. Nitrate has increased over time within the BoQ, though the largest increase at Conway in the lower bay is likely driven by the much larger increases seen in Lake Ontario over the same period (Winters et al., 2012).

The major ions were highly correlated with each other, with alkalinity, conductivity, total hardness and calcium having high correspondence. The minor ions magnesium and potassium were correlated with one another at the lower bay site, but magnesium was related to sodium at the upper bay site suggesting probable loading from urban sources (Minns et al., 2004). While we did not see the relationships of chloride with precipitation or snow accumulation seen in other studies (Winter et al., 2012), this is likely in part due to the fact that our measurements begin in May after the freshet and ice-off. Temperature and precipitation were tied with changes to productivity, particularly following very cold winters (e.g. 1978-79 and 1994) but these relationships were not significant in the annual means correlation matrix and will likely require an analysis of higher temporal resolution data to resolve them, including lag times and the temperature extremes which drive these relationships. Likewise, a further analysis of the changes in climatic forcing would be helpful, including the most recent years since 2000 which exhibit higher summer temperatures and high precipitation, interspersed with occasional colder or dry summers.

The Project Quinte dataset shows the importance seasonality, though it is limited to the months of May-Oct. The very early spring and late fall unstratified periods are not represented. In the upper bay, while the overall pattern has been a decrease over time, seasonally the major nutrients, seston and chlorophyll-*a* peak a month earlier in August post-1995. There has been no major change to the timing of the transparency maximum despite the overall change in clarity. The chlorophyll-*a* value decreased post-1995, while similar for the May-June period, is much lower in value from July-October resulting in reduced summer-fall period phytoplankton. An increase in DIC and DOC during most of the year post-1995 may be due to recirculation behaviour of Driessenid Mussels (Heath et al., 1995; Barnard et al., 2006; Miehls et al., 2009). The timing of phosphorus peak might also be mediated by Zebra Mussels, with the timing of the phosphorus peak occurring approximately 20 days earlier in recent years, but this pattern is strongest post-1999, suggestive that it may be due to the increase in summer temperatures in the upper bay during this period. During peak productivity in the summer, nitrogen limitation is implicated as water column values drop to detection limit, likely due to the release of high levels of internally loaded sediment phosphorus (Doan et al., 2018). This relationship between nitrogen and phosphorus may help to inform nutrient management within the Bay of Quinte.

The Project Quinte Long Term Ecological Research (LTER) has provided a longstanding, continuous source of biological, chemical and physical data for an aquatic ecosystem undergoing a series of stressors ranging from nutrient controls to a series of invasive species. Ecosystem studies are often hampered by gaps in data or termination of data collection before associations can be ascertained (Hofmann et al., 2013). Long term whole-ecosystem datasets like Project Quinte are extraordinarily rare and invaluable for the study of climate change or invasive species impacts, which can take many years to be expressed. Future work linking findings from the Bay of Quinte to other locations will illustrate how valuable this extensive and detailed dataset can be to determine relationships not only for a single ecosystem, but in the application of these findings to similarly impacted ecosystems around the world.

Acknowledgements

Thanks to a number of the current and former members of GLLFAS who answered questions on the history and decisions associated with Project Quinte. As a relative newcomer to the Quinte research community, it would not have been possible to knit together this manuscript without the phone calls and emails returned by those people who actually collected the data for decades. Thanks go to several anonymous reviewers and funding of Project Quinte from DFO and ECCC Great Lakes Action Plan.

References

Brewer, P.G., and Goldman, J.C., 1976. Alkalinity changes generated by phytoplankton growth. Limnology and Oceanography 21, 108-117.

Chapra, S. C., Dove, A., and Warren, G.J., 2012. Long-term trends of Great Lakes major ion chemistry. Journal of Great Lakes Research 38(3), 550-560.

Currie, W.J.S. and Frank, M.M., 2015. Multivariate analysis of the 40 year Project Quinte biogeochemistry dataset: water chemistry, physical characteristics, seston and climate. Can. Tech. Rep. Fish. Aquat. Sci. 3125.

Dermott, R., Bonnell., R., 2011. Benthic Fauna in the Bay of Quinte 2009. p. 51-71. In: Project Quinte Members. Project Quinte Annual Report. RAP Monitoring Report #20.

Diamond, M. L., Poulton, D.J., Mackay, D., Stride, F.A., 1994. Development of a mass-balance model of the fate of 17 chemicals in the Bay of Quinte. Journal of Great Lakes Research 20(4), 643-666.

Doan, P.T., Watson, S.B., Markovic, S., Liang, A., Guo, J., Mugalingam, S., Stokes, J., Morley, A., Zhang, W., Arhonditsis, G.B. and Dittrich, M., 2018. Phosphorus retention and internal loading in the Bay of Quinte, Lake Ontario, using diagenetic modelling. Sci. Total Environ., 636, 39-51.

Freeman, N.G., Prinsenberg, S.J., 1986. Exchange flows in the Adolphus Reach/North Channel, p27-39. In C.K. Minns, D.A. Hurley, and K.H. Nicholls (Eds.), Project Quinte: point-source phosphorus control and ecosystem response in the Bay of Quinte, Lake Ontario. Can. Spec. Publ. Fish. Aquat. Sci. 86.

Graham, D.M., Dahl, J.A., Millard, E.S., Johannsson, O.E., White, L.L.,1996. Assessment of abundance, biomass and production of the lower trophic levels in the eastern basin of Lake Erie, 1994. Can. Tech. Rep. Fish. Aquat Sci., 2110.

Heath, R. T., Fahnenstiel, G.L., Gardner, W.S., Cavaletto, J.F., and S. J. Hwang, S.J., 1995. Ecosystem-level effects of zebra mussels (Driessena polymorpha): An enclosure experiment in Saginaw Bay, Lake Huron. Journal of Great Lakes Research 21(4), 501-516.

Hofmann, G.E., Blanchette, C.A., Rivest, E.B., Kapsenberg, L., 2013. Taking the pulse of marine ecosystems: The importance of coupling long-term physical and biological observations in the context of global change biology. Oceanography, 26:140-148. JMP®, Version 10. SAS Institute Inc., Cary, NC, 1989-2007.

Kim, D.K., Zhang, W., Rao, Y.R., Watson, S., Magalingam, S., Labencki, T., Dittrich, M., Morley, A., Arhonditsis, G.B., 2013. Improving the representation of internal nutrient recycling with phosphorus mass balance models: A case study in the Bay of Quinte, Ontario, Canada. Ecological Modelling 256, 53-68.

Leisti, K. E, Millard, E.S., and Minns, C.K., 2006. Assessment of submergent macrophytes in the Bay of Quinte, Lake Ontario, August 2004, Including historical context. Can. Man. Rep. Fish. Aquat. Sci., 2762.

Leisti, K.E., Doka, S.E. and Minns, C.K., 2012. Submerged aquatic vegetation in the Bay of Quinte: Response to decreased phosphorous loading and Zebra Mussel invasion. Aquatic Ecosystem Health & Management, 15(4), pp.442-452.

Lewis Jr, W. M., and Morris, D.P., 1986. Toxicity of nitrite to fish: a review. Transactions of the American fisheries society 115.2 (1986), 183-195.

Liao, C. F. H., 1977. Effect of nutrient enrichment on nitrogen-fixation activity in Bay of Quinte, Lake Ontario. Hydrobiologia 56(3), 273-279.

Makarewicz, J. C., Lewis, T.W., Boyer, G.L., and Edwards, W.J., 2012. The influence of streams on nearshore water chemistry, Lake Ontario. Journal of Great Lakes Research 38, 62-71.

Miehls, A. L. J., Mason, D.M., Frank, K.A., Krause, A.E., Peacor, S.D., and Taylor., W.W., 2009. Invasive species impacts on ecosystem structure and function: A comparison of the Bay of Quinte, Canada, and Oneida Lake, USA, before and after zebra mussel invasion. Ecological Modelling 220(22), 3182-3193.

Millard, E.S., 1986. Effect of light climate, photosynthetic capacity, and respiration on gross integral primary production and net primary production potential in the Bay of Quinte, Lake Ontario. in Project Quinte: point-source phosphorus control and ecosystem response in the Bay of Quinte, Lake Ontario. Edited by CK Minns, DA Hurley, and KH Nichols. Can. Spec. Publ. Fish. Aquat. Sci, 86, 128-138.

Millard, E.S. and Sager, P.E., 1994. Comparison of phosphorus, light climate, and photosynthesis between two culturally eutrophied bays: Green Bay, Lake Michigan, and the Bay of Quinte Ontario. Can. J. Fish. Aquat. Sci. 51, 2579-2590.

Minns, C.K., and Johnson, M.G., 1979. Temporal variation in the nutrient export of rivers draining into the Bay of Quinte, Ontario, Canada. 1. JAWRA Journal of the American Water Resources Association, 15, 1061-1072.

Minns, C.K, D.A. Hurley, D.A., and Nicholls, K.H. (Eds.), 2004. Project Quinte: point-source phosphorus control and ecosystem response in the Bay of Quinte, Lake Ontario. Can. Spec. Publ. Fish. Aquat. Sci. 86.

Minns, C.K., Moore, J.E., and Seifried, K.E., 2004. Nutrient loads and budgets in the Bay of Quinte, Lake Ontario, 1972 to 2001. Canadian Manuscript Report of Fisheries and Aquatic Sciences, 2694.

Minns, C. K., Moore, J.E., Doka, S.E., and St John, M.A., 2011. Temporal trends and spatial patterns in the temperature and oxygen regimes in the Bay of Quinte, Lake Ontario, 1972-2008. Aquatic Ecosystem Health & Management 14(1), 9-20.

Munawar, M., Fitzpatrick, M., Munawar, I.F., Niblock, H., and Kane, D., 2012. Assessing ecosystem health impairments using a battery of ecological indicators: Bay of Quinte, Lake Ontario example. Aquat. Ecosyst. Health Mgmt. 15(4), 430-441.

Nicholls, K. H., 1999. Effects of temperature and other factors on summer phosphorus in the inner Bay of Quinte, Lake Ontario: implications for climate warming. Journal of Great Lakes Research 25(2), 250-262.

Nicholls, K. H., 2012. Phosphorus and chlorophyll in the Bay of Quinte: A time-series/intervention analysis of 1972-2008 data. Aquatic Ecosystem Health & Management 15(4), 421-429.

Nicholls, K. H., Heintsch, L., and Carney, E., 2002. Univariate step-trend and Multivariate assessments of the apparent effects of P loading reductions and zebra mussels on the phytoplankton of the Bay of Quinte, Lake Ontario. Journal of Great Lakes Research 28(1), 15-31.

Nicholls, K. H., Hoyle, J.A., Johannsson, O.E., and Dermott, R., 2011. A biological regime shift in the Bay of Quinte ecosystem (Lake Ontario) associated with the establishment of invasive Driessenid mussels. Journal of Great Lakes Research 37(2), 310-317.

Nürnberg, G. and Peters, R.H., 1984. Biological availability of soluble reactive phosphorus in anoxic and oxic freshwaters. Can. J. Fish. Aquat. Sci. 41, 757-765.

Prairie, Y.T., 1996. Evaluating the predictive power of regression models. Can. J. Fish. Aquat. Sci. 53, 490-492.

Pritchard, A. L., 1931. Spawning habits and fry of the cisco (*Leucichthys artedi*) in Lake Ontario. Contrib. Can. Biol. Fish. 9, 227-240.

R Core Team, 2013. R: A language and environment for statistical computing. Vienna, Austria: R Foundation for Statistical Computing. URL http://www.R-project.org/.

Redfield, A.C., 1958. The biological control of chemical factors in the environment. Amer. Sci. 46, 205-221.

Ridgway, M. S., Hurley, D. A., and Scott, K.A., 1990. Effects of winter temperature and predation on the abundance of Alewife (Alosa pseudoharengus) in the Bay of Quinte, Lake-Ontario. J. Gr. Lakes. Res. 16(1) 11, 20.

Robinson, G.W., 1986. Water quality of the Bay of Quinte, Lake Ontario, before and after reductions in phosphorus loading, p. 50-58. In: C.K. Minns, D.A. Hurley, and K.H. Nicholls (Eds.), Project Quinte: point-source phosphorus control and ecosystem response in the Bay of Quinte, Lake Ontario. Can. Spec. Publ. Fish. Aquat. Sci. 86.

Scheffer, M and S. R. Carpenter, S.R., 2003. Catastrophic regime shifts in ecosystems: linking theory to observation. Trends Ecol. Evol. 18, 648-656.

Sly, P.G., 1986. Review of postglacial environmental changes and cultural impacts in the bay of quinte. p. 7-26. In: C.K. Minns, D.A. Hurley, and K.H. Nicholls (Eds.), Project Quinte: point-source phosphorus control and ecosystem response in the Bay of Quinte, Lake Ontario. Can. Spec. Publ. Fish. Aquat. Sci. 86.

Song, K. and Burgin, A.J., 2017. Perpetual Phosphorus Cycling: Eutrophication Amplifies Biological Control on Internal Phosphorus Loading in Agricultural Reservoirs. Ecosystems, 20(8), pp.1483-1493.Statistics Canada. 2012. Focus on Geography Series, 2011 Census. Statistics Canada Catalogue no. 98-310-XWE2011004. Ottawa, Ontario. Analytical products, 2011 Census.

Sterner, R. W., 2011. C:N:P stoichiometry in Lake Superior: freshwater sea as end member. Inland Waters 1(1): 29-46. Tucker, A. 1948. The phytoplankton of the Bay of Quinte. Trans. Amer. Microscopical. Soc. 67, 365-383.

Winter, J. G., Howell, E. T., and Nakamoto, L. K., 2012. Trends in nutrients, phytoplankton, and chloride in nearshore waters of Lake Ontario: Synchrony and relationships with physical conditions. Journal of Great Lakes Research, 38, 124-132.

Temporal trends and spatial patterns in the temperature and oxygen regimes in the Bay of Quinte, Lake Ontario, 1972–2008

Charles K. Minns,[1,2,*] James E. Moore,[3] Susan E. Doka,[2] and Margaretha A. St. John[2]

[1]*Department of Ecology and Evolutionary Biology, University of Toronto, Toronto, ON M5S 3B2*

[2]*Fisheries and Oceans Canada 867 Lakeshore Road, Burlington, ON L7S 1A1*

[3]*JEMSys Software Systems Inc., 22 Marion Crescent, Dundas, ON L9H 1J1*

**Corresponding author: ken@minns.ca*

Keywords: oxygen depletion, long-term trends, spatial variation

Introduction

Temperature and oxygen conditions are fundamental limnological variables (Hutchinson, 1957). Weekly or biweekly monitoring has been conducted annually (May–October) in the Bay of Quinte, Lake Ontario, since 1972 as part of the Project Quinte long-term ecosystem studies. Since 2001, the temperature monitoring has been augmented with temperature-loggers at various inshore and offshore sites. Trends and patterns in the temperature and oxygen regimes in the Bay of Quinte were last assessed by Minns and Johnson (1986).

Many studies have shown climate impacts on the thermal regime in lakes (see Adrian et al., 2009 for review) with higher summer surface temperature, longer periods of stratification, changes in thermocline depth, and indirectly the potential of anoxia in the hypolimnion. Declining point source phosphorus (P) loadings from 1978 and the invasion of Dreissenid Mussels from 1995 (Minns et al., 2004) increased light penetration particularly after 1995 and might be expected to increase thermocline depth in parts of the Bay that stratify in summer (Mazumder and Taylor, 1994). Decreased P loadings were expected to decrease hypolimnetic oxygen depletion with some lag time due to turnover of surface sediments (Minns and Johnson, 1986; Minns et al., 2004). This study had several objectives: (1) to assess long-term trends and spatial patterns in seasonal surface temperatures and vertical thermal structure at index sites, (2) to assess spatial and temporal patterns in hypolimnetic warming and

Ecology of the Bay of Quinte: Health, Management and Global Implications
Editors: C.K. Minns, M. Munawar and M.A. Koops

Aquatic Ecosystem Health & Management Society, Canada

apparent oxygen depletion, and (3) to assess the potential influence of water quality, climatic and hydrologic conditions on the temporal trends.

Materials and Methods

Data collection

Long-term temperature and oxygen monitoring data at index stations
Temperature and oxygen measurement profiles have been collected at the main Project Quinte index stations since 1972. Monitoring cruises typically ran from April/May to October/November. In the early years the cruises were weekly and after the mid-1980s they were reduced to bi-weekly. A number of index stations were visited in various phases of the Project. Three stations have been monitored throughout (1 Belleville/B, 3 Hay Bay/HB, and 5 Conway/C) and two others for most years (2 Napanee/N and 4 Glenora/GL) (Figure 1). The measurement protocol has changed over time with three phases: (1) samples were drawn by water bottle from selected depths, temperatures measured with a thermometer, and oxygen levels determined by Winkler's titration; (2) temperature and oxygen levels measured in situ at selected depth using a calibrated YSI probe; and (3) levels measured in situ using a calibrated Hydro-Lab continuous profile measurement system. Using the depth algorithm developed by Minns and Johnson (1986) for the Bay of Quinte index stations, measurements by depth and station were classified into three layers: surface, middle, and bottom.

Spatio-temporal temperature data collection using loggers
Temperature data used in spatial analyses were taken from three near shore sites (6 Napanee/NANO, 7 Hay Bay North/HBWO2, and 8 Carnachan Bay/CBNO) deployed adjacent to the shoreline at 1m to 1.5m depths and four offshore sites (1 Belleville/BVO, 3 Hay Bay/HBO, 4 Glenora/GLO, and 5 Conway/CO) deployed at 1m depth on spar buoys at the offshore sites which marked the index sampling sites in the Bay of Quinte (Figure 1). The nearshore sites cover some of the period 2002–2007, while offshore sites cover April to November of each year during 2001–2008. Temperatures were logged at half hour intervals using Onset Tidbit loggers for both the offshore and the near-shore sites. To obtain data comparable to the index site data, daily mean temperatures were computed from the logger temperature record. At the

nearshore sites, dates with temperatures less than 7 °C were removed to produce datasets comparable to offshore and index sites (Trumpickas et al., 2009). The offshore loggers span the period of the index monitoring cruises.

Analyses

Summary statistics for long-term index sites

Summary statistics were computed for each index site and depth layer without regard to temporal or seasonal patterns to provide an overview of the database. Sample profiles were sorted into three layers (S-surface, M-middle, and B-bottom) using the formula devised by Minns and Johnson (1986) for Quinte index sites.

Surface water temperature patterns at all sites

The seasonal patterns in surface layer temperatures at index monitoring sites, and offshore and nearshore logger sites were examined using three alternate models as a starting point:

i) $T = a + b*JD + c*JD^2$; Quadratic model
ii) $T = a + b*(J_{MAX} - JD)*(JD <= J_{MAX}) + c*(JD - J_{MAX})*(JD > J_{MAX})$; Inverted-V model
iii) $T = a + b*(J_{XLO} - JD)*(JD <= J_{XLO}) + c*(JD - J_{XHI})*(JD > J_{XHI})$; Plateau model

Where T is the temperature in the surface layer, JD is Julian day of year, J_{MAX} is the day of maximum temperature in model ii, J_{XLO} and J_{XHI} are the days for the start and end of maximum temperature in model (iii). Model (i) has often been used to describe seasonal surface temperature patterns, e.g. Sharma et al., 2007, but appears to underestimate peak values. Models (ii) and (iii) are variants on an approach developed by Trumpickas et al., 2009. Model (i) was fit by ordinary linear least squares and models (ii) and (iii) were fit by non-linear least squares. Variants of the models involved adding variables to capture the effects due to sample depth (Zs), station depth (Zstn), and year (as a factor not a linear covariate). For each site's data, the best model was selected to represent the observed water temperature changes. Where year effects were significant, estimates of summer maximum temperature were derived from the models and correlations with climatic and hydrologic indicators computed.

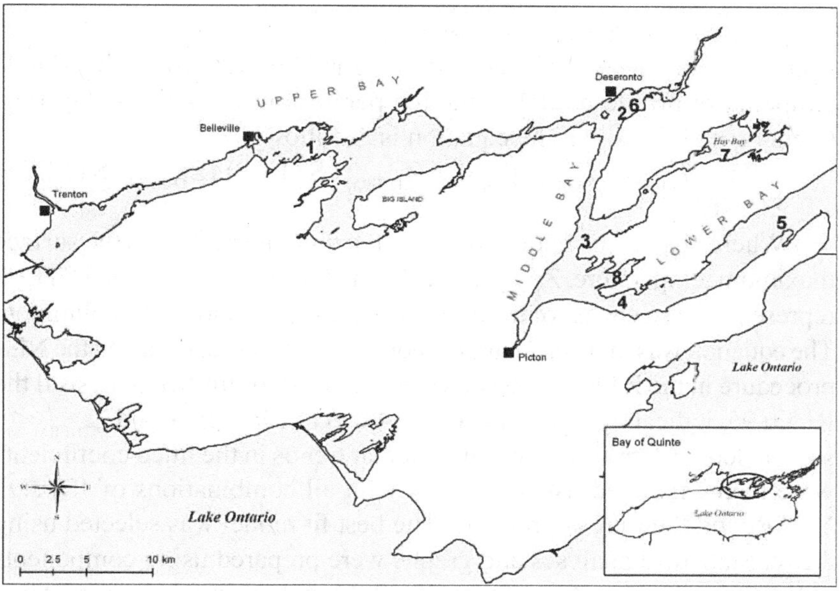

Figure 1. Limnological monitoring sites in the Bay of Quinte: long-term index sites (1-Belleville B, 2-Napanee N, 3-Hay Bay HB, 4-Glenora GL, and 5-Conway); offshore logger sites (1-BVO, 3-HBO, 4-GLO, and 5-CO); and nearshore logger sites (6-NANO, 7-HBWO2, and 8-CBNO).

Vertical thermal stability

The strength of vertical temperature stratification was assessed at the main index sites by calculating the Schmidt's stability, the wind work needed to render the instantaneous vertical density distribution uniform (Idso, 1973). Since stability was being calculated for profiles at stations rather than whole basin, we adopted the oceanic practice of integrating to a fixed depth for each station (Millard, 1990), here the 1972–2008 mean station depth, and using the hypsometric curves for the appropriate Bay section drawn from Chu et al. (2004) (Upper – B and N, Middle – HB, and Lower – GL and C). Water densities were determined from temperatures using an interpolation model of the tables in Hutchinson (1957) and the density profiles were integrated with a 0.1 m interval. Subsequently the stability values at index sites HB, GL and C were modelled using linear and non-linear regression models to examine seasonal and secular patterns. As year was included as a significant factor in the best fit model, post-hoc analyses examined Pearson correlations with annual or seasonal indicators of climate and hydrology.

Temperature profile characteristics

Using the approach of Mackenzie-Grieve and Post (2006) we fit a simple temperature profile model to the temperature profiles from Hay Bay, Glenora, and Conway. The equation is as follows:

$$T(Z) = T_{MIN} + (T_{EPI} - T_{MIN})*(Z_{THERM}\text{^STEEP})/(Z_{THERM}\text{^STEEP} + Z\text{^STEEP})$$

Where, T_{MIN} is the bottom minimum temperature, T_{MAX} is the surface maximum temperature, Z_{THERM} is the depth of the thermocline, and STEEP represents the steepness of the transition from epilimnion to hypolimnion. The equation was fit using Gaussian non-linear least squares using the NLS procedure in the R library. Results were selected for further analysis if the $R^2 > 0.95$, indicating a good fit, $T_{MIN} > 4.0$, STEEP < 25.0, and $Z_{THERM} <=$ station depth (Zstn). Seasonal and secular trends in the fitted coefficients were then examined using ANOVA with all combinations of JD, JD^2, YEAR, and their cross-products. The best fit model was selected using AICc. Statistical analyses and graphs were prepared using components of R (R Development Core Team, 2010). Model selection was performed using the Akaike Information Criterion as adjusted for sample size and the number of fitted parameters (AICc) (Burnham and Anderson, 2004).

Hypolimnetic warming and oxygen depletion in the middle and lower Bay of Quinte:

At HB, GL, and C index sites, bottom layer temperature and oxygen values were selected. Integrated calculations of areal or volumetric oxygen demand (AHOD and VHOD) such as used by Burns et al. (2005) were not attempted as the index sites do not provide complete assessment of the Bay by definable bathymetric basins. Some analyses of oxygen depletion have concentrated in the minimum late summer bottom oxygen concentration as an indicator; such analyses were not attempted here given the weekly or biweekly observation interval in different years. Where sample depths exceeded the station depth, the data points were excluded; sometimes the boat drifted to deeper water whilst profiles were being measured. Station depths varied considerably particularly before global positioning systems were used regularly. As a result data selected for analyses was restricted to the defined depth ranges at each station: 8.5 to 13.0 m at HB, 19 to 25 m at GL, and 25 to 35 m at C. After inspection of plots of bottom temperature and oxygen values at each station, data from selected day of the year ranges were analyzed using ANOVAs with all combinations of day of the year, sample depth, station depth, and year (as a factor) and the best model selected using AICc. For each best fit regression, the sum of all fitted regression coefficients with Julian day as an input variables, having set fixed choices for other variables (Zstn, Zsam, year, temperature) as required yields standardized estimates of the hypolimnetic bottom warming and apparent, or net, oxygen depletion rates

Results

Summary statistics for long-term index sites

The summary statistics showed the extent and general character of the sample database (Table S1). Station depth varied little at the upper Bay sites (B-Belleville, 4.0–7.0 m; N-Napanee, 4.3–8.8 m) but varied more at deeper sites (HB-Hay Bay, 8.0–17.0 m; GL-Glenora, 16.2–28.6 m; and C-Conway, 21.5–40.0 m). Mean temperatures (°C) ranged from 19.5 (B-middle) to 9.7 (C-bottom) while maximum temperatures ranged from 28.2 (N-surface) to 19.1 (C-bottom) in line with patterns of station depth and the potential for stratification. Mean oxygen levels (mg. L^{-1}) ranged from 10.1 (C-surface) to 8.4 (HB-bottom) with minimum values ranging from 5.5 (HB-surface) to 0.2 (HB-bottom). Lower oxygen levels occurred more in bottom waters except at GL where upwelling sometimes brought deoxygenated water into surface waters and seiche-induced mixing into the middle waters at GL and C. Percent oxygen saturation averaged close to 100% in surface waters at all sites with means decreasing into the middle and bottom layers particularly at HB, GL and C. Super-saturation was encountered at all sites/layers, reflecting the high levels of primary production in summer and the daylight sampling times.

Surface water temperature patterns at all sites

All sites showed similar dome-shaped seasonal patterns of surface temperature, with a peak in mid to late summer (Figure S1, top panel).

Table 1. Best fit models of summer surface temperature regimes for the long-term index monitoring stations, and the short-term spatial logger sites, offshore and nearshore, in the Bay of Quinte. (Details of all analyses and model selection using AICc are given in supplementary materials Table S1.)

Site	# Samples	# Years	R^2	J_{MAX}	Best model
Index sites					
Belleville (B)	2325	37	0.887	207	Quadratic + Years + StnZ
Napanee (N)	1977	30	0.864	208	Quadratic + Years + StnZ
Hay Bay (HB)	2478	37	0.861	211	Quadratic + Years + StnZ
Glenora (GL)	1900	27	0.808	230	Inverted-V + Years + StnZ
Conway (C)	3030	37	0.888	211/241	Plateau + Years + StnZ
Offshore loggers					
BVO	1635	8	0.924	180/245	Plateau + Years
HBO	1635	8	0.912	184/251	Plateau + Years
GLO	1635	8	0.888	211/248	Plateau + Years
CO	1635	8	0.918	207/248	Plateau + Years
Nearshore loggers					
NANO	1243	6	0.880	179/241	Plateau + Years
HBWO2	821	4	0.877	184/231	Plateau + Years
CBNO	1023	5	0.858	195/236	Plateau + Years

Peak values were higher in the upper Bay (B and N) and lowest at C and peaks were attained at later dates going from B down to C along the Bay's axis. The regression models selected for each site accounted for much of the variance at all sites (Table 1). Overall R^2 values ranged from 0.808 for GL to 0.924 for BVO. Generally, the index sites had lower values than the logger sites. Among the index sites, the best model for Belleville, Napanee and Hay Bay was Quadratic + factor(years) + Zstn, for Glenora was Inverted-V + factors(years) + Zstn, and for Conway was Plateau + factor(years) + Zstn. At all logger sites the Plateau + factor(years) was the model selected (Table S2). Among the logger sites, the spring warming rates (-b in model iii) were consistently smaller (0.139 at GLO to 0.198 at HBWO2) than the fall cooling rates (-c in model iii; 0.177 at CO to 0.241 at HBO and BVO). The fitted peak dates, or date ranges, at all sites followed the trend from earlier in the upper Bay to later in the lower Bay (Table 1). The estimated maximum temperatures at the offshore logger sites were only slightly higher than the corresponding values for the index

Table 2. Differences (°C) between model-estimate maximum summer surface temperature for pairs of index vs. offshore logger sites and pairs of nearshore vs. offshore logger sites

Comparison	# Years	Mean	Minimum	Maximum
Index vs. offshore				
B-BVO	8	0.3	−0.2	0.9
N-BVO	8	0.4	−0.2	1.5
HB-HBO	8	0.2	−0.1	0.7
GL-GLO	5	1.0	−0.1	1.7
C-CO	8	0.0	−0.8	0.7
Nearshore vs. offshore				
NANO-BVO	6	0.4	0.2	0.5
HBWO2-HBO	4	2.2	1.3	3.0
CBNO-GLO	5	2.1	1.6	2.6

sites (Table 2). The nearshore and offshore logger sites were similar in the upper Bay (the difference between NANO with BVO) while nearshore sites were higher in the middle and lower Bay (Table 2).

The year coefficients provided an estimated fitted maximum summer temperature (T_{MAX}) for each year at the index sites. These values were significantly correlated with both year and mean summer air temperature at Trenton A (MSAT; Figures 2A and B). The T_{MAX} values were highly correlated (range 0.577 N-GL to 0.840 HB-C) among the five sites and with year (range 0.217 B to 0.523 GL) (Table 3). T_{MAX} values were highly correlated with MSAT (range 0.369 GL to 0.774 N). Correlations with other climatic/hydrologic indicators were weaker. Multiple regressions of T_{MAX} with stations as factors and either year or MSAT as a covariate were highly significant ($R^2 = 0.648$ and 0.739, respectively). B and N had similar temperatures, while HB and GL were 1°C cooler than B and C 2.2 °C cooler. The annual warming rate was 0.026°C. yr^{-1}, or almost 1 degree over 37 years, with a one degree rise in MSAT causing a 0.485 °C rise in surface water temperature.

Vertical thermal stability

The Schmidt's stability values showed consistent patterns across years and season at each index site (Figure S1, bottom panel). In the upper

Table 3. Pearson correlations1 among fitted summer maximum surface temperatures at five index monitoring stations and with annual climatic and hydrological metrics.

Identifier	Variable	Year	B	N	HB	GL	C
B	Belleville	0.217	—				
N	Napanee	0.356	0.812	—			
HB	Hay Bay	0.416	0.783	0.834	—		
GL	Glenora	0.523	0.635	0.577	0.835	—	
C	Conway	0.409	0.768	0.779	0.840	0.806	—
MSAT[2]	Air temperature	0.318	0.633	0.774	0.579	0.369	0.689
MSWL[3]	Water level	-0.380	-0.118	-0.317	-0.283	-0.181	-0.351
MSRF[4]	River flow	0.195	-0.140	-0.175	0.046	0.153	-0.233
NAO[5]	Winter NAO	0.176	-0.112	0.042	0.140	0.293	0.027
SOI[5]	Winter SOI	0.032	-0.114	-0.254	-0.241	-0.134	-0.274

1. Sample size is 37 at B, HB, and C, 30 at N, and 27 at GL.
2. MSAT—Mean June-July-August (summer) air temperature at Trenton A weather station in the upper Bay of Quinte (Environment Canada, www.climate.weatheroffice.gc.ca).
3. MSWL—Mean summer water level on Lake Ontario (Canadian Hydrographic Service, http://www.waterlevels. gc.ca/C&A/historical_e.html).
4. MSRF—Mean June to August Trent River flow based on scaling flows at stations HK001 Trent R. at Healey Falls (9090 km^2) and HK003 Crowe R. at Marmora (1090 km^2) to the total drainage (12548.7 km^2) (Water Survey of Canada, http://www.wsc.ec.gc.ca).
5. NAO—Mean winter December to March North Atlantic Oscillation index (NOAA Earth System Res. Lab., http://www.esrl.noaa.gov/psd/data/climateindices/list). 6. SOI—Mean winter December to March Southern Oscillation Index (indicator of El Nino/La Nina events, same as item 5).

Bay, values were low throughout the summer at B reflecting complete mixing while at N values were generally low with some higher values encountered occasionally in early summer. In the middle and lower Bay, marked seasonal patterns were evident at HB, GL, and C. Significant regression models of seasonal stability were obtained at HB, GL, and C, with year as a significant factor in the Inverted-V model examined with surface water temperatures (Table 4A). At each site, peak stability varied considerably with significant temporal trends at GL and C alongside higher correlations with maximum summer surface and mean summer air temperatures (Table 4B). Correlations with other climatic/hydrologic indicators were low.

Table 4. Summary of regression results for Schmidt's stability values at Hay Bay, Glenora and Conway using the equation: (A) Non-linear least square regressions of Stability = a[Year]*(1 + b*(J_{MAX} – JD)*(J_{MAX} > JD) + c* (JD-J_{MAX})*(JD >= J_{MAX})) where JD is Julian day of year and J_{MAX} is the day of peak stability; (B) Correlations of maximum stability values by year, a[Year], with the fitted maximum surface temperature (T_{MAX}) and various climate/hydrologic indicators.

(A) Regression results for stability as a function of year and day							
Station	N	A[Min.]	A[Max.]	b	c	J_{MAX}	R^2
Hay Bay	580	63.3	123.0	–0.0166	–0.0098	176.8	0.623
Glenora	451	310.0	551.2	–0.0085	–0.0147	219.6	0.549
Conway	558	766.9	1763.0	–0.0106	–0.0150	223.7	0.822

(B) Correlations (as R^2) for annual Stability maxima with climatic and hydrologic indicators							
Station	Year	T_{MAX}	MSAT	MAWL	MSRF	NAO	SOI
Hay Bay	0.002	0.190	0.048	0.000	0.000	0.000	0.047
Glenora	0.299	0.561	0.254	0.043	0.073	0.097	0.014
Conway	0.399	0.593	0.656	0.082	0.002	0.049	0.053

Temperature profiles

At the three index sites the subsets of profiles with good fits ($R^2 > 0.95$) to the profile model were primarily associated with higher Schmidt's stability values present when stratification was present. Seasonal patterns were discernible for most parameters at HB, GL and C, though the goodness of fit varied considerably, and year to year differences were significant in the case of T_{EPI} (Table 5; See the supplementary materials for plots of seasonal patterns for each index site and parameter – Figures S3, S4, and S5). The parameter means for T_{MIN} and Z_{THERM} followed trends of station depth while STEEP values were similar among index sites. T_{MAX} values followed the patterns described above for surface temperatures. As with the surface temperature analyses above, T_{EPI} models had the highest R^2 (0.612–0.811) with years having significant effects at all three sites. The STEEP models had the lowest R^2 at each station with HB the lowest (0.014) and C the highest (0.395). Year effects were not significant for STEEP, T_{MIN}, and Z_{THERM}, and in some instances, only one of the terms in the quadratic seasonal component was significant. The R^2 values ranged from 0.489 to 0.585 for T_{MIN}. Z_{THERM} was

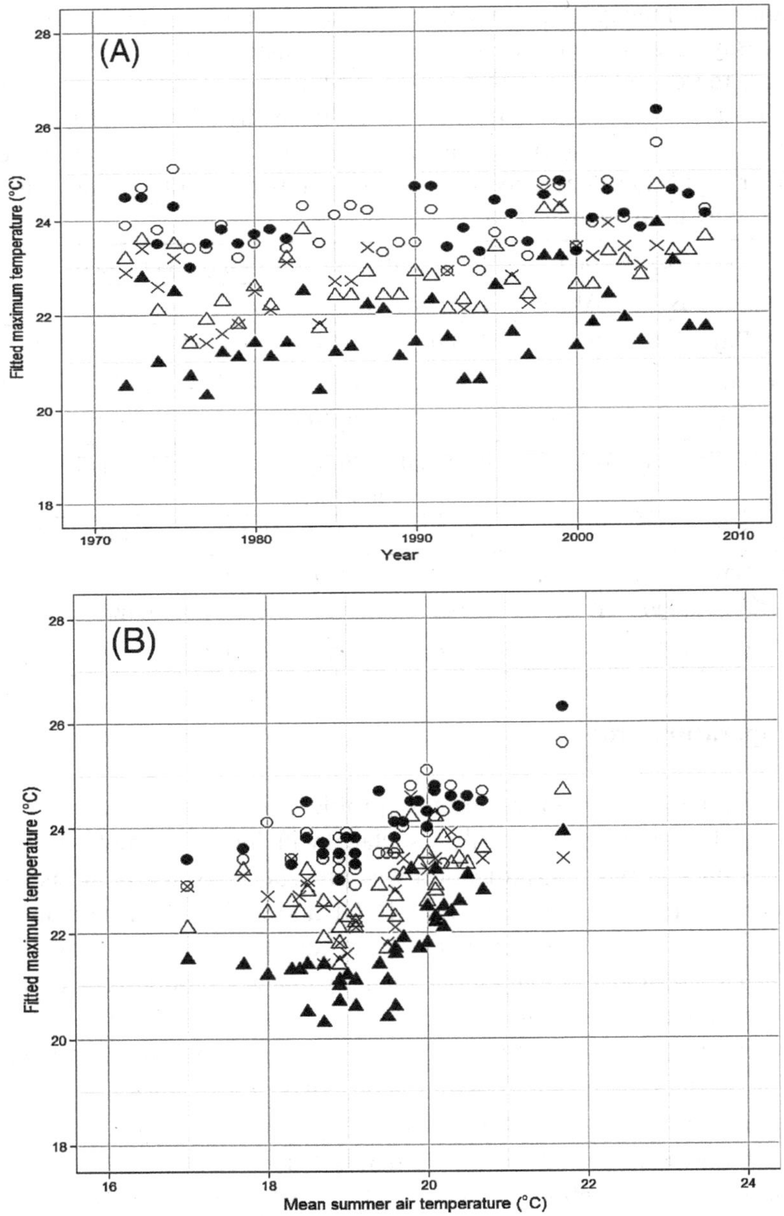

Figure 2. The relationships of the fitted maximum summer temperature values with A) Year and B) Mean summer air temperature at Trenton A at five index sites in the Bay of Quinte: Belleville (○ open circle), Napanee (● solid circle), Hay Bay (× cross), Glenora (D open triangle), and Conway (▲ solid triangle).

Table 5. Sample size, mean and standard deviation (SD), and selected regression coefficients for Julian day and the square of Julian day, and R^2 for the best fit models of the fitted parameters of the vertical temperature profile model at index sites HB, GL, and C.

Station	Parameter	N	Mean	SD	JD	JD^2	Year[1]	R^2
Hay Bay	T_{MIN}	394	12.19	4.17	+0.24	−0.00042	—	0.583
	T_{EPI}		19.73	3.46	+0.60	−0.0014	*	0.758
	Z_{THERM}		6.59	2.05	−0.064	+0.00018	—	0.071
	STEEP[2]		7.98	5.28	–	−0.0000047	—	0.014
Glenora	T_{MIN}	267	9.13	3.28	+0.056	–	—	0.585
	T_{EPI}		16.69	4.08	+0.48	−0.0011	*	0.612
	Z_{THERM}		10.79	5.51		+0.00018	—	0.361
	STEEP[2]		4.80	4.53	−0.035	+0.00010	—	0.224
Conway	T_{MIN}	408	8.24	2.46	−0.028	+0.00016	—	0.489
	T_{EPI}		17.25	4.29	+0.62	−0.00014	*	0.811
	Z_{THERM}		17.62	6.46	+0.098	–	—	0.441
	STEEP[2]		5.28	4.11	−0.0085	+0.000044	—	0.395

poorly predicted at HB (0.071) but higher values were obtained at GL and C. In the fitted parameter models, the magnitude of the coefficients for JD and/or JD^2 as fitted inputs and their ratio when both are present indicate the extent of seasonal variation. Analysis of variance (ANOVA) of Z_{THERM} with index sites (HB, GL, and C) and time stanzas (Pre-P control 1972–1977, Post-P control 1978–1994, and Post driessenid invasion 1995–2008) revealed no significant time stanza effects.

Hypolimnetic warming and oxygen depletion in the middle and lower Bay

Substantial bottom warming occurred at index sites HB, GL, and C in all years (Figure S2) with strong evidence of the intermixing of surface and bottom waters later in the summer at HB and GL as bottom temperatures approached surface values. After an initial period when oxygen concentrations were level, linear depletion was observed at all three stations after day 160. At the end of summer bottom oxygen concentrations increased again first at HB and last at C. Subsequent statistical analyses of hypolimnetic warming and apparent oxygen depletion (AOD) were restricted to Julian days 160–230 at HB, 160–240 at GL, and 160–250 at C. Best fit regression models for temperature and oxygen explained much of the variance (Table 6) with similar sets of variables included. Temperature was a significant co-factor in the apparent depletion of oxygen. Applying mean station depths (Table S1; Zstn), a reference sample depth (Zsam) one metre above bottom, and the long-term mean summer bottom temperature to the best fit regression equations obtained by analysis, the sum of all regression terms having Julian date as a variable yields standardized estimates of the yearly warming and apparent oxygen depletion rates (Table 6). The temporal trends in the warming rates showed much variability and no significant long-term trends (Figure 3A). AOD rates showed some decrease (Figure 3B) which

Table 6. Best fit regression models of hypolimnetic warming and oxygen depletion at index sites Hay Bay (HB), Glenora (GL), and Conway (C); all non-year coefficients were significant to $P < 0.01$ and temperature was included as a potential covariate in the depletion modelling.

Parameter	Hypolimnetic warming			Apparent oxygen depletion		
	HB	GL	C	HB	GL	C
N	879	640	1188	879	640	1188
# Years	34	27	34	34	27	34
R^2	0.886	0.764	0.887	0.824	0.847	0.931
Coefficients:						
Intercept	-10.7	-27.9	+8.69	+0.0532	-2.37	+30.48
JD	+0.152	+0.253	+0.0440	+0.0818	0.0524	-0.0905
Factor (Year)[1]	*	*	*	*	*	*
JD*(Year)[1]	*	*	*	*	*	*
Zsam	-0.472	+0.963	+0.814	+1.95	-0.00212	-0.312
JD*Zsam		*	-0.00523	-0.00128	-0.00653	+0.00126
Zstn		-0.00791	-0.929			
JD*Zstn			+0.00446			
Temperature	*na*	*na*	*na*	-0.335	+1.33	+0.027
JD*Temp.	*na*	*na*	*na*			-0.00140

[1] Year as a factor with fitted values (*) for each year.

was significant ($P = 0.03$; rate= $-0.00055\ y^{-1}$ or, using the overall mean depletion rate, $1\%\ y^{-1}$; Model $R^2 = 0.501$) in a regression model of depletion rates versus year with sites as factors; warming rate and mean summer Trent River flow did not contribute significantly as covariates. ANOVA of apparent depletion with index sites and time stanzas as treatments did not reveal any significant effects due to the arrival of dreissenids. Overall apparent oxygen depletion rates decreased with higher warming rates (Figure 3C), though there was no relationship when considered site by site. If GL and C sites are considered together, AOD rates increase with warming rate (Figure 3C). Warming rates decreased with increasing mean summer Trent R. flow (Figure 3D); in a regression model with warming rates versus flow and sites as factors, flow was significant ($P = 0.04$; rate −0.00027; Model $R^2 = 0.609$). The warming rates at HB and GL were positively correlated with MSAT while at C there were negative correlations with MRSF, NAO, and MSWL (Table 7). Apparent oxygen depletion rates had no strong correlations with climatic/hydrologic metrics (Table 7).

Table 7. Pearson correlations of estimated annual bottom hypolimnetic warming and apparent oxygen depletion rates with climatic and hydrologic metrics.

	Year	MSAT	MSWL	MSRF	NAO	SOI
Warming						
Hay Bay	0.177	0.452	−0.124	−0.159	−0.111	−0.214
Glenora	−0.166	0.400	0.171	−0.160	−0.009	−0.004
Conway	−0.180	−0.025	−0.348	−0.546	−0.428	−0.124
Depletion						
Hay Bay	−0.273	−0.294	0.118	−0.151	0.043	−0.031
Glenora	−0.352	0.268	0.152	−0.127	0.164	−0.256
Conway	−0.087	0.083	0.225	−0.037	0.152	−0.118

Discussion

The Bay of Quinte is a freshwater, non-tidal estuary, a "freshtuary," with flushing dominated by runoff from the rivers coming from the large drainage area in the November-May period of the year and by the two-layer flows that dominate the middle and lower Bay area in the June-October period when thermal stratification occurs in deeper areas (Minns et al., 2004; Freeman and Prinsenberg, 1986). The results reported here provide a more complete appraisal of the thermal and oxygen regimes in the Bay of Quinte than was possible for Minns and Johnson (1986). The long-term data provided a basis for assessing the contributions of changing eutrophication, species invasion and climate warming to the patterns observed.

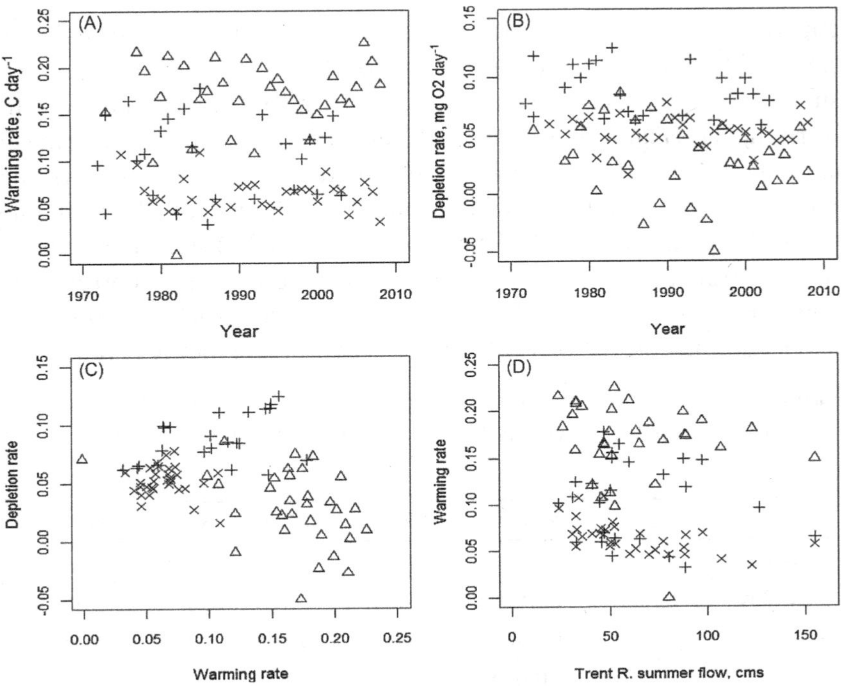

Figure 3. Plots of hypolimnetic warming (°C.d^{-1}) and apparent oxygen depletion (O_2 mg. L^{-1}.d^{-1}) rates versus year (A and B), of depletion versus warming rates (C), of warming versus means summer Trent River flow for the bottom waters at index sites HB (D), GL (+), and C (×).

Thermal regime

Surface water temperature increases, observed at the index sites in response to higher air temperatures (0.485:1°C or 0.026°C. yr^{-1}), are consistent with rates observed for Great Lakes surface temperatures (Trumpickas et al., 2009) and in lakes elsewhere (Adrian et al., 2009). The stability analyses confirmed the lack of evidence of any sustained thermal stratification at the shallow index sites B and N in the upper Bay. At N there were occasional early summer incursions of cooler water as thermal stratification and the two-layer flows were established and deeper, lower Bay waters were pushed upstream by wind-driven processes and seiche activity (Freeman and Prinsenberg, 1986). Stability values climbed rapidly from HB though GL to C, as station depth increased from roughly 10 through 20 to 30 metres, with peak values occurring later in the season going from HB to C. The strength of the stability varied in line with the annual warming of the surface waters. At HB, GL, and C, given the constant spring and fall slopes (coefficients b and c in section A, Table 4), higher peak stability values, alongside higher surface temperatures, also indicate longer stratification periods. Unlike surface temperatures, other features of the summer temperature profiles (Z_{THERM}, T_{MIN} and STEEP), while variable, showed no significant year to year variation.

Oxygen regime

Apparent oxygen depletion varied considerably from year to year but showed an overall long-term downward trend once Point-source P loading was decreased significantly. Similar responses to decreased P loading have been seen in other lake ecosystems, e.g. the central basin of Lake Erie (CB-LE; Burns et al., 2005) and Onondaga Lake, New York (Matthews and Effler, 2006). Much of the variation in apparent oxygen depletion rates is influence by physical processes as demonstrated by Rao et al. (2008) with detailed modelling in CB-LE. In the Bay of Quinte, the two-layer flow regime during stratification (Freeman and Prinsenberg, 1986) is accompanied by upwelling and down-welling notably in the area between the Glenora and Hay Bay sites (Figure 1) causing much of the substantial hypolimnetic warming evident in the lower Bay. These hydrodynamic effects also appeared to strongly influence apparent oxygen depletion (Figure 3C). Higher temperatures should directly increase oxygen demand (Chapra and Canale, 1991) in the water column, but, in the deeper areas of the Bay of Quinte, those higher temperatures (and higher warming rates) can also reflect the greater mixing down of oxygenated waters from the surface layers offsetting the higher demand. Taking GL and C together, the oxygen depletion rate – warming rate relationship was positive and linear indicating the direct effect of temperature on AOD rate whereas at HB warming rates were higher and down-mixing of oxygenated waters produced lower AOD rates. The inverse correlation of hypolimnetic warming with higher mean summer river flow may also reflect a direct influence of flow with warmer waters from the upper Bay, strengthening the intensity of the two-layer flows in the middle and lower Bay. Charlton (1980) showed how apparent oxygen depletion was inversely related to hypolimnion thickness in CB-LE, with the thickness in turn being inversely correlated with hypolimnion temperature. Although the importance of morphometry and hypolimnion thickness is acknowledged, the results here indicate the main linkage may be between hypolimnion temperature and depletion.

Eutrophication, species invasions, and climate change

Reviewing recent UK experiments involving climate warming and eutrophication of freshwater mesocosms, Moss (2010) described how these stresses add to on-going ecosystem disruption and pose serious threats to many freshwater ecosystems. Project Quinte began as a study of multi-trophic response to decreased nutrient loading. Subsequently the continuing species invasion in the Great Lakes brought Dreissenid Mussels to the Bay and initiated further ecosystem disruptions. Now the threat of climate change is adding further complexity to the long-term studies. The results here indicate oxygen depletion is slowly responding to the large reductions in Point source P loading, in line with predicted changes in internal P loading as active surface sediment layers are renewed (Minns et al., 2004). However, warming trends are now clear in the surface temperatures and in the increased stability values. Thermal stratification has not responded, perhaps because of the intense two-layer mixing. The dreissenids invasion has not affected the depth of the thermocline but may have indirectly affected oxygen depletion by promoting the retention of nutrients in littoral areas tied up in dreissenid and macrophyte biomass. However surface warming and increased stability will extend the stratification period and thereby increase the potential for existing oxygen depletion rates to produce oxygen concentrations that impair the performance of heterotrophs such as fish. In the Peterborough district of Ontario which includes the Bay of Quinte and much of its drainage, summer temperatures are expected to rise from 1971–2000 levels by 2–3 and 4–5 °C in the 2041–2070 in the 2071–2100 periods while precipitation declines by 0 to 10% (Colombo et al., 2007). As evapotranspiration increases with temperature, river flows can be expected to decline further, particularly in summer, thereby supporting higher hypolimnetic warming rates. If the warming trends follow their present upward path and nutrient loadings are allowed to grow again as demographic pressures overtake the control investments of the 1970–1980s (Minns et al. 2004), oxygen depletion may be added to the already complex list of challenges facing us as we try to shift human use of freshwater ecosystems like the Bay of Quinte onto a sustainable path.

Summary

The temperature and oxygen regimes of the Bay of Quinte have been monitored, weekly or biweekly, in the open water season at a series of offshore index sites since 1972. Since 2001, temperature has been monitored in situ using loggers at both offshore and nearshore sites. This study was undertaken to assess the potential role of morphometry, climate, nutrient loading, and invasion by dreissenids in temporal trends and spatial patterns of temperature and oxygen regimes. The thermal regime was assessed using summer surface temperatures, Schmidt's vertical stability values, and vertical stratification characteristics. When the Quinte studies began taking measurements in the hypolimnion, deeper stations were attaining low oxygen levels by the end of summer, raising concerns that the cumulative impacts of eutrophication might lead to anoxia. The bottom oxygen regime at sites that stratified was assessed using the hypolimnetic warming rates and the apparent oxygen depletion rates one meter off the bottom. Quinte's maximum summer surface waters have warmed almost 1°C from 1972 to 2008 in line with climate warming. Nearshore sites in the shallow upper Bay have similar temperatures to offshore index sites while those in the lower Bay are warmer. The warming of surface waters has increased the maximum Schmidt's stability values at the deeper sites but thermocline depth or bottom temperatures have not changed. Hypolimnetic warming rates showed no temporal change but were inversely correlated with mean summer flow from the main river draining into the Bay.

Oxygen depletion was variable but exhibited a downward trend in response to reduced phosphorus loading with neither climate warming nor the dreissenid invasion having a detectable effect. Projected 21st C climate warming poses a threat to the Bay of Quinte ecosystem.

The trends and patterns detected in the temperature and oxygen regimes of the Bay of Quinte demonstrate the value of long-term monitoring of ecosystems. Surface temperatures rose almost 1°C between 1972 and 2008, providing further confirmation of the ongoing climate warming. The surface warming has lengthened and strengthened stratification in deeper areas but, to date, no changes in thermocline depth or bottom temperatures were evident. The importance of upwelling/downwelling flows in the middle and lower Bay was confirmed. Apparent oxygen depletion during stratification was present at middle and lower Bay sites. The moderate depletion rates do not represent a threat of hypoxia and showed a long-term decrease in line with reduced Point-source phosphorus loading reductions. There was no evidence that climate warming or the invasion of Dreissenid Mussels had affected depletion rates.

Overall, the temperature and oxygen conditions in the Bay do not indicate any appreciable impairment with respect to the recovery from past eutrophication and also to recent species invasions. However, the magnitude of climate changes now forecast during the 21st Century may be expected to exacerbate ongoing pressures associated with excessive human use of freshwater resources in the Bay of Quinte ecosystem.

Acknowledgements

Special thanks to Scott Millard, DFO Great Lakes Laboratory for Fisheries and Aquatic Sciences (GLLFAS) for all his help with getting the database together for this work. Thanks to Kathy Leisti and Jocelyn Gerlofsma, DFO-GLLFAS, for assistance with hydrological and limno logical summary data. Thanks to Carolyn Bakelaar, DFO-GLLFAS, for assistance with the map of the Bay of Quinte area.

References

Adrian, R., O'Reilly, C. M., Zagarese, H., Baines, S. B., Hessen, D. O., Keller, W., Livingstone, D. M., Sommaruga, R., Straile, D., Van Donk, E., Weyhenmeyer, G. A., Winder, M., 2009. Lakes as sentinels of climate change. Limnol. Oceanogr. 54, 2283–2297.

Burnham, K. P., Anderson, D. R., 2004. Multi-model inference: understanding AIC and BIC in model selection. Sociological Methods and Research 33, 261–304. Burns, N. M., Rockwell, D. C., Bertram, P. E., Dolan, D. M., Ciborowski, J. J. H., 2005.

Trends in temperature, Secchi depth, and dissolved oxygen depletion rates in the central basin of Lake Erie, 1983–2002. J. Great Lakes Res. 31(S2), 35–49. Chapra, S. C., Canale, R. P., 1991. Long-term phenomenological model of phosphorus and oxygen for stratified lakes. Water Res. 25, 707–715.

Charlton, M. N., 1980. Oxygen depletion in Lake Erie: has there been any change? Can. J. Fish. Aquat. Sci. 37, 72–81.

Chu, C., Minns, C. K., Moore, J. E., Millard, E. S., 2004. Impact of oligotrophication, temperature, and water levels on walleye habitat in the Bay of Quinte, Lake Ontario. Trans. Amer. Fish. Soc. 133, 868–879.

Colombo, S. J., McKenney, D. W., Lawrence, K. M., Gray, P. A., 2007. Climate Change Projections for Ontario: A Practical Guide for Policymakers and Planners. Climate Change Research Report CCRR-05, Applied Research and Development Branch, Ontario Ministry of Natural Resources, Sault Ste. Marie, Ontario.

Freeman, N. G., Prinsenberg, S. J., 1986. Exchange flows in the Adolphus Reach/North Channel. Can. Spec. Publ. Fish. Aqua. Sci. 86, 27–39.

Hutchinson, G. E., 1957. *A Treatise on Limnology, Volume 1: Geography, Physics and Chemistry*. John Wiley and Sons, New York.

Idso, S. B., 1973. On the concept of lake stability. Limnol. Oceanogr. 18(4), 681–683. Mackenzie-Grieve, J. L., Post, J. R., 2006. Projected impacts of climate warming on production of lake trout (*Salvelinus namaycush*) in southern Yukon lakes Can. J. Fish. Aquat. Sci. 63, 788–797.

Matthews, D. A., Effler. S. W., 2006. Long-term changes in the areal hypolimnetic oxygen deficit (AHOD) of Onondaga Lake: Evidence of sediment feedback. Limnol. Oceanogr. 51, 702–714.

Mazumder, A., Taylor. W. D., 1994. Thermal structure of lakes varying in size and water clarity. Limnol. Oceanogr. 39, 968–976.

Millard, R., 1990. On the calculation of the Brunt-Vaisala frequency. Deep Sea Res. Part A 37(1), 167–181.

Minns, C. K., Johnson, M. G., 1986. Temperature and Oxygen Conditions in the Bay of Quinte, Lake Ontario. Can. Spec. Publ. Fish. Aqua. Sci. 86, 40–49. Minns, C. K., Moore, J. E., Seifried, K. E., 2004. Nutrient loads and budgets in the Bay of Quinte, Lake Ontario, 1972 to 2001. Can. Manuscr. Rep. Fish. Aquat. Sci. 2694. Moss, B., 2010. Climate change, nutrient pollution and the bargain of Dr Faustus. Freshwater Biol. 55(Suppl1), 175–187.

R Development Core Team, 2010. R: A Language and Environment for Statistical Computing. R Foundation for Statistical Computing, Vienna, Austria. Rao, Y. R., Hawley, N., Charlton, M. N., Schertzer, W. C., 2008. Physical processes and hypoxia in the central basin of Lake Erie. Limnol. Oceanogr. 51, 2007–2020. Sharma, S., Jackson, D. A., Minns, C. K., Shuter, B. J., 2007. Will northern fish populations be in hot water because of climate change? Global Change Biology 13, 2052–2064.

Trumpickas, J., Shuter, B. J., Minns, C. K, 2009. Forecasting impacts of climate change on Great Lakes surface water temperatures. J. Great Lakes Res. 35(3), 454–463.

The above article is a reprinted version of the following previously published article:
Minns; C. K., Moore, J.E., Doka, S.E., St. John, M. A., 2011.

Temporal trends and spatial patterns in the temperature and oxygen regimes in the Bay of Quinte, Lake Ontario, 1972–2008.

Aquatic Ecosystem Health & Management (2011) 14 (1), 9–20.
https://doi.org/10.1080/14634988.2011.547327

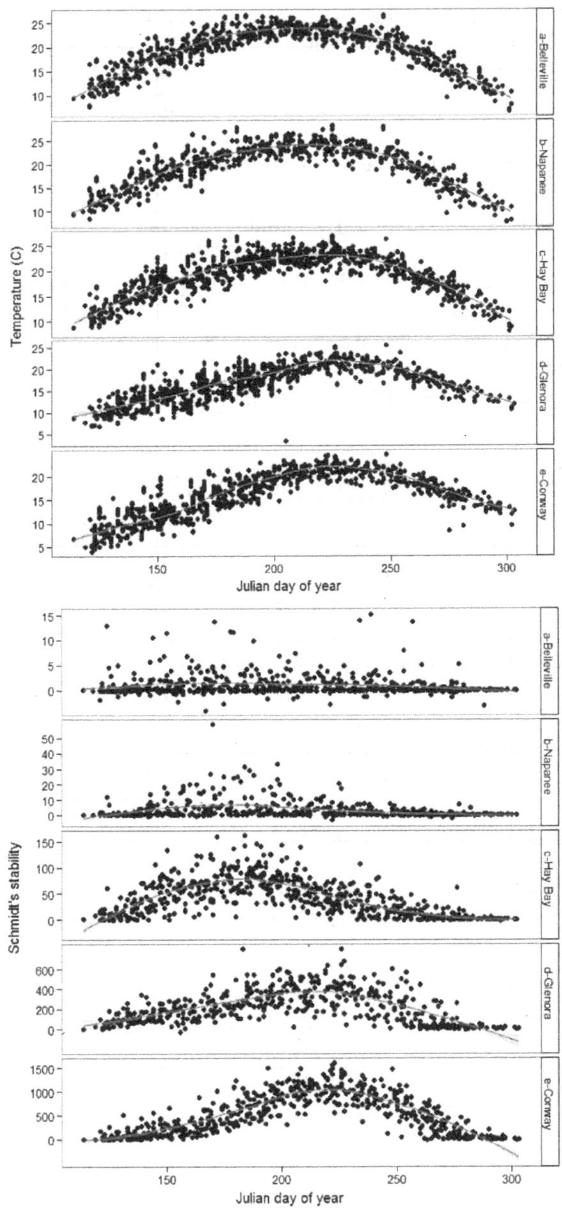

Figure S1. Pattern of surface water temperature (°C, top panel) Schmidt's stability (bottom panel) versus Julian day of the year (JD) at the five main Bay of Quinte index sites, 1972-2008: a) Belleville (mean station depth 5.1 m), b) Napanee (5.5 m), c) Hay Bay (11.6 m), d) Glenora (21.4 m), and e) Conway (32.0 m).

Table S1. Summary statistics for samples size, station depth, temperature, oxygen, and percent oxygen saturation by index site and layer, 1972-2008.

Index Site[1]	Layer[2]	N	Station depth (Zstn), m			Temperature, °C			Oxygen, mg. l^{-1}			% Oxygen Saturation		
			Mean	Min.	Max.	Mean	Min.	Max.	Mean	Min.	Max.	Mean	Min.	Max.
B	S	2548	5.1	4.0	7.0	19.4	6.5	26.7	9.2	5.3	14.9	99.9	61.0	148.4
B	M	942	5.2	4.1	7.0	19.5	7.0	26.3	9.2	2.7	16.0	99.5	32.0	147.4
B	B	2481	5.1	4.0	7.0	19.2	7.0	26.3	8.8	0.3	22.4	95.1	3.6	184.6
N	S	2190	5.4	4.3	8.8	19.5	7.6	28.2	9.4	4.7	19.7	102.6	55.6	191.9
N	M	1039	5.5	4.3	8.8	19.2	7.6	27.2	9.3	4.3	13.4	100.7	50.9	160.0
N	B	2093	5.4	4.3	7.5	18.6	7.6	26.9	9.1	1.8	14.8	96.8	20.6	177.1
HB	S	2692	11.5	8.0	17.0	19.0	8.0	26.9	9.8	5.5	16.4	105.2	49.6	194.2
HB	M	5759	11.6	8.0	17.0	16.8	6.4	25.8	9.2	1.3	13.7	95.1	13.6	153.7
HB	B	2521	11.4	8.0	14.5	14.3	2.0	23.8	8.4	0.2	13.5	81.1	2.2	128.3
GL	S	2083	21.1	16.2	28.0	17.2	3.4	25.5	9.8	0.5	13.7	101.6	64.0	154.6
GL	M	6486	21.0	16.2	28.0	14.8	4.8	24.8	9.5	0.3	14.8	93.2	2.9	154.8
GL	B	1723	21.0	16.2	28.0	12.0	1.7	21.8	8.8	1.3	16.2	81.0	13.1	128.0
C	S	3209	31.7	21.5	40.0	16.7	4.0	24.6	10.1	4.8	20.0	103.1	56.9	194.8
C	M	15374	31.6	21.5	38.0	13.9	2.5	24.4	9.7	0.5	21.6	93.1	5.3	185.2
C	B	2797	31.6	21.5	40.0	9.7	3.5	19.1	9.2	0.3	14.5	80.2	2.3	130.9

[1] Index sites: B-Belleville, N-Napanee, HB-Hay Bay, GL-Glenora, C-Conway

[2] Layers: S-Surface, M-Middle, B-Bottom; Samples were sorted by sample depth (Zs) relative to Zcrit = 5/3 + Zstn/15 with S if Zs < Zcrit or B if Zs > (Zstn – Zcrit) else M after Minns and Johnson (1986).

Table S2. Results of model fitting for surface layer temperatures at five Bay of Quinte index stations, 1972 to 2008.

	Model	**00**	**01**	**02**	**03**	**11**	**12**	**13**	**21**	**22**	**23**
	shape	-	**Quad**	**Quad**	**Quad**	**Inverse V**	**Inverse V**	**Inverse V**	**Plateau**	**Plateau**	**Plateau**
	factor	-	-	**Years**	**Years**	-	**Years**	**Years**	-	**Years**	**Years**
	covariate	-	-	-	**Stn. Z**	-	-	**Stn. Z**	-	-	**Stn. Z**
Station	**Statistic**										
Belleville	d.f.	2	4	40	41	5	41	42	6	42	43
(N=2325)	ΔAICc	4986.84	372.79	58.81	0.00	1016.90	811.62	798.29	449.86	129.51	69.30
	R^2		0.863	0.884	0.887	0.819	0.840	0.884	0.858	0.880	0.884
Napanee	d.f.	2	4	33	34	5	34	35	6	35	36
(N=1977)	ΔAICc	3879.63	277.52	25.04	0.00	755.71	570.75	566.79	329.25	47.67	38.50
	R^2		0.839	0.862	0.864	0.795	0.819	0.862	0.835	0.861	0.862
Hay Bay	d.f.	2	4	40	41	5	41	42	6	42	43
(N=2478)	ΔAICc	4802.29	430.39	3.50	0.00	905.12	594.97	591.68	425.60	12.37	12.92
	R^2		0.829	0.860	0.861	0.793	0.823	0.860	0.830	0.860	0.860
Glenora	d.f.	2	4	30	31	5	31	32	6	42	43
(N=1900)	ΔAICc	3078.09	741.47	421.25	419.27	363.07	0.00	2.05	1520.45	1200.45	1140.26
	R^2		0.708	0.760	0.761	0.761	0.808	0.814	0.774	0.809	0.814
Conway	d.f.	2	4	40	41	5	41	42	6	42	43
(N=3030)	ΔAICc	6548.48	1569.90	954.07	907.99	953.99	211.39	200.27	841.61	32.45	0.00
	R^2		0.807	0.846	0.849	0.843	0.880	0.888	0.848	0.887	0.888

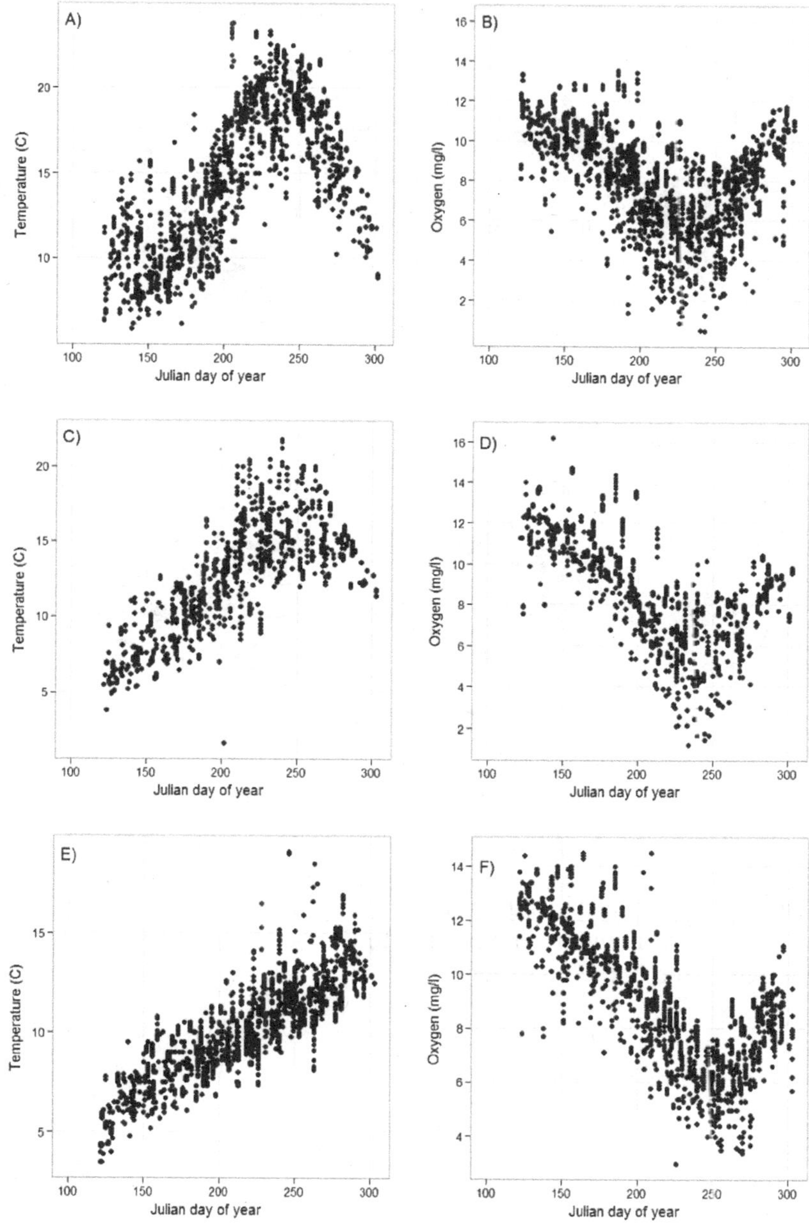

Figure S2. The overall seasonal trends in bottom water temperature °C and oxygen mg. l^{-1} observed 1972 to 2008 at the three monitoring stations Hay Bay (A, B), Glenora (C, D), and Conway (E, F).

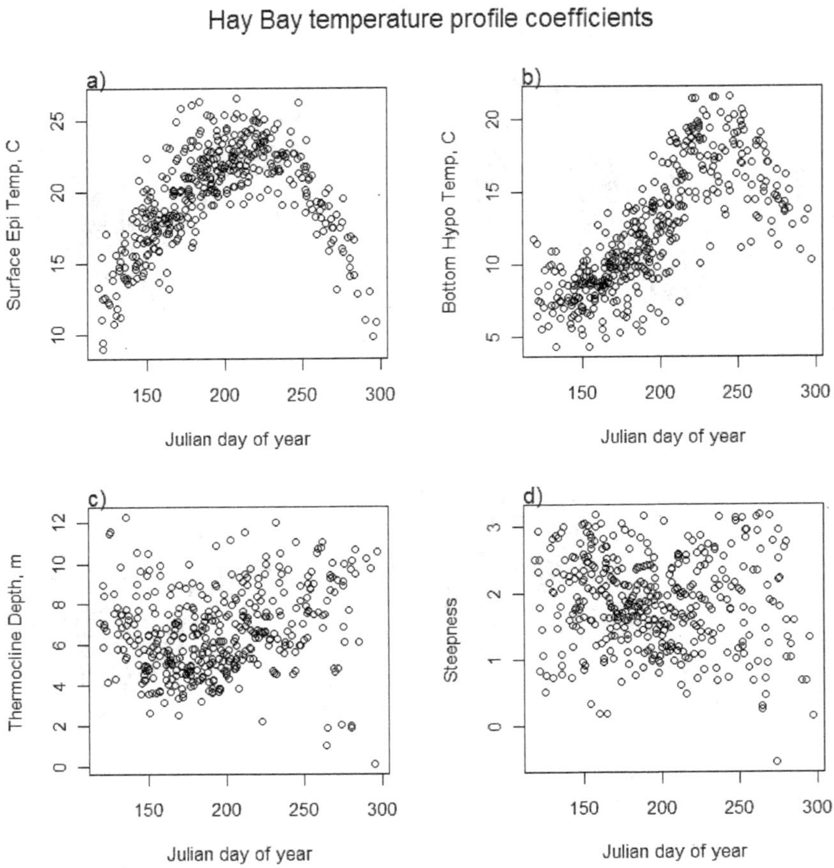

Figure S3 Seasonal temperature profile characteristics at the Hay Bay (HB) index site: a) Epilimnion temperature, b) Bottom temperature, c) Thermocline depth, and d) profile steepness.

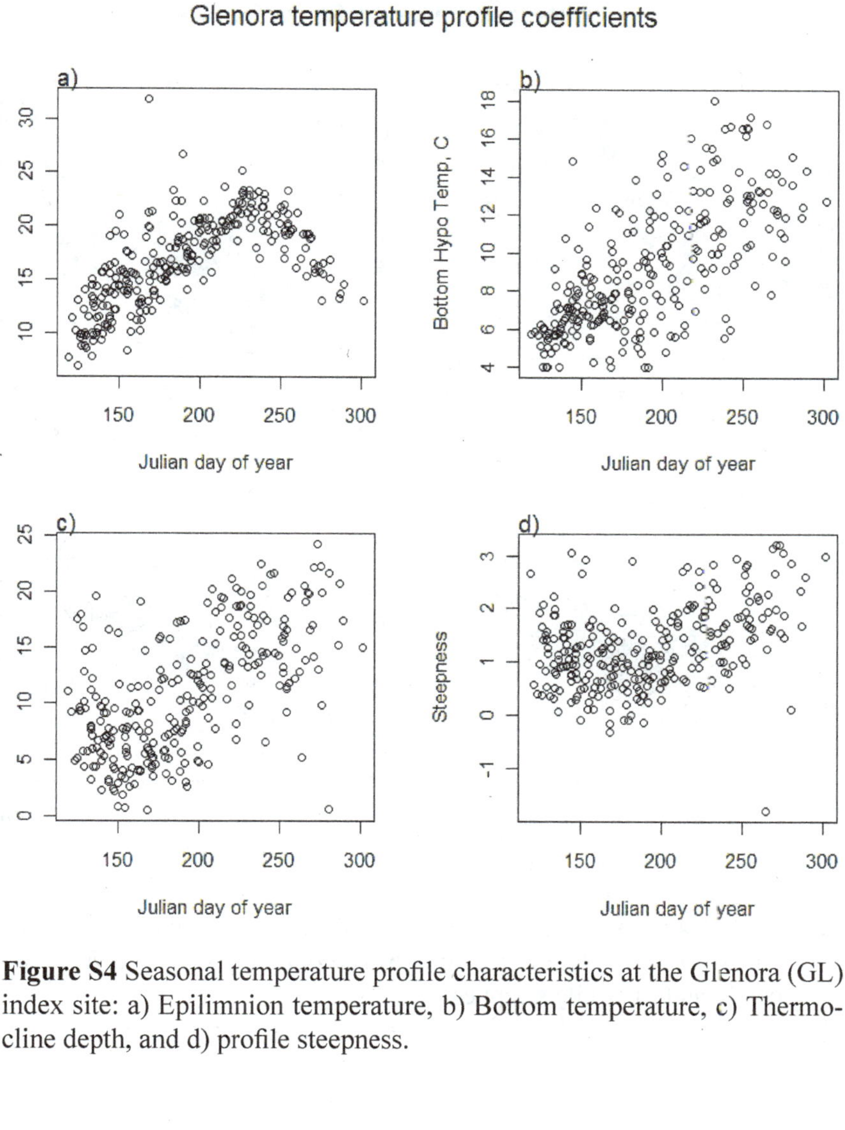

Figure S4 Seasonal temperature profile characteristics at the Glenora (GL) index site: a) Epilimnion temperature, b) Bottom temperature, c) Thermocline depth, and d) profile steepness.

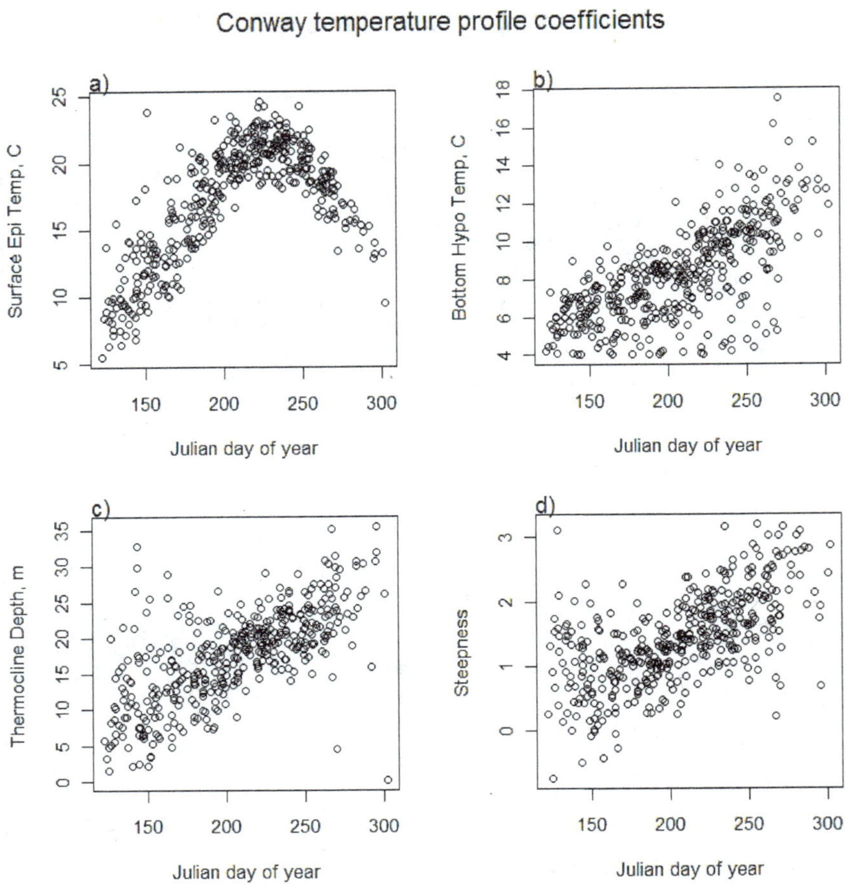

Figure S5 Seasonal temperature profile characteristics at the Conway (C) index site: a) Epilimnion temperature, b) Bottom temperature, c) Thermocline depth, and d) profile steepness.

Phosphorus and chlorophyll in the Bay of Quinte: A time-series/intervention analysis of 1972–2008 data

Kenneth H. Nicholls

S-15 Concession 1, RR #1
Sunderland, ON L0C 1H0. khnicholls@interhop.net

Keywords: time-series, Dreissenid Mussels, Point-source P-control, long-term trends

Introduction

The Bay of Quinte, a large (254 km^2) multi-basin, Z-shaped embayment at the NE end of Lake Ontario, has had a long history of eutrophication (Warwick, 1980). The first comprehensive assessment of the nutrient status of the Bay was reported by Johnson and Owen (1971). Their assessment clearly identified the important role of high-concentration/low-flow sources of phosphorus (so-called "point" sources) originating from municipal sewage treatment facilities. Also at this time, municipal, provincial, state and national governments were enacting legislation to control the phosphate content in commercial laundry detergents (Prince and Bruce, 1972) and were upgrading sewage treatment plants and processes for reduction of phosphorus in treated wastewater (Anon, 1969; Hasler, 1969; Slater and Bangay, 1980).

The database for assessing the long-term ecosystem response to P loading reductions to the Bay of Quinte began in 1972 with the launch of "Project Quinte," the preliminary results of which were published in Minns et al. (1986). That assessment continues in this article with an analysis of the apparent effects of two interventions: (1) an approximately 50% reduction in Point-source P loading to the upper bay during the winter of 1977–1978 and (2) the establishment of Zebra and Quagga Mussels (*Dreissena* spp.) in the upper bay in the mid-1990.

Environmental/biological time-series data present special challenges for detection and analysis of trends owing mainly to the characteristic serial dependencies of the data associated with seasonal, autocorrelative effects (so-called "red noise") and other issues. A major purpose of this article is to apply three largely independent methods of trend detection to the 37-year record of phosphorus and chlorophyll data collected under Project Quinte in an effort to build consensus around the significance of the observed changes in total phosphorus and chlorophyll in the Bay of Quinte.

Methods

Field and laboratory methods for collection and laboratory analyses of phosphorus and chlorophyll samples were as described in Robinson (1986). Briefly, samples were collected as composites through the euphotic zone (twice the Secchi disc depth) at the four main sampling stations (B, N, HB and C) at weekly intervals from 1972 to 1982 and at biweekly intervals, 1983–2008. Chemical analyses (automated, colourimetric) were performed by the Laboratory Services Branch of the Ontario Ministry of the Environment (OME). A correction factor (+35%) was applied to chlorophyll-*a* data generated prior to 1985 to account for a change in filtration/extraction methods (Nicholls and Hopkins, 1993). Chlorophyll-*a* data reported here represent "total" chlorophyll-*a*, uncorrected for phaeopigments. A conservative approach to "far outlier" detection (median ± 5*st. dev.) was used for the data reported here. Outliers so identified were evaluated in a case-by-case approach that took into account the concentration of other variables in the same sample (e.g. total Kjeldahl nitrogen) and the same variables on nearby dates or nearby sampling stations (within Project Quinte or the OME's water treatment plant intake monitoring program at the city of Belleville). For the most part, outliers were replaced by linear interpolation between adjacent values (before and after). All data were reduced to monthly means. For a few of the earliest years of the project, there were no October samples, in which case mean values from adjacent years were substituted.

Time-series analyses were based on three independent approaches:

(1) nonparametric statistical tests for step-trends (Cluis et al., 1989), in recognition that the upper bay theoretically is flushed by river flows several times per year and that both interventions occurred over short time intervals. Responses to the interventions were, therefore, more likely to be seen as "steps" rather than monotonic changes over several years.

(2) ARIMA/intervention (A/I) analysis (Box and Tiao, 1975) based on Box and Jenkins (1970) autoregressive integrated moving average (ARIMA) models. For each data set, an ARIMA function describing the pre-intervention data was used to forecast for

the post-intervention period. The forecast was compared with measured values (permutation test) to assess statistical difference. The PC software used (Human Edge Software, 1992) afforded full ARIMA capabilities, including inclusion of an iteratively optimized set of ARIMA parameters drawn from both seasonal and non-seasonal ARIMA models of the classic *p*, *d*, *q* type.

(3) Regime Shift Detection (RSD; Rodionov, 2004, 2006), based on sequential statistics, was applied to residual data sets after modelling the seasonal component of each data set as a peri odic function (Cohort Software, 1995).

The three procedures outlined above were applied systematically to all four stations for both total phosphorus (TP) and chlorophyll-*a* (Chla), and produced copious output that can best be presented in charts and plots, most of which could not be presented here for space reasons (see the online appendix for examples of some of this output*).

Results

Simple plots of monthly mean TP and Chla (Figure 1) suggest the initiation of important negative step-trends in the late 1970s, especially at the upper bay stations (B and N). Total P concentrations in the middle bay (Station HB) appeared to follow a negative monotonic trend until about the early 1990s (Figure 1c). The apparent response of TP to Point-source loading reductions in 1977–1978 is rendered less clear by increases in concentrations in the early 1980s at stations B and HB (also incompletely evident at Station N). Chlorophyll trends generally agreed with the TP trends; steep declines were most evident at Stations B and N beginning in 1978. Clearly, more detailed testing is required to explore the possible significance of these changes.

Nonparametric step-trend tests

The nonparametric step-trend detection tests revealed a total of 13 statistically significant negative steps associated with both the TP loading intervention and the *Dreissena* intervention out of a possible 16 cases (2 variables, 4 stations, 2 interventions; Table 1). The apparent impact of Point-source TP loading reductions on TP concentrations in the upper bay was greater than the reductions associated with the establishment of *Dreissena*. A similar conclusion can be drawn from the results of the same tests on TP data at Station N, revealing a highly significant negative step in 1978 and a nonsignificant step in 1996 (Table 1). The reverse was found for Station C in the lower bay, where a statistically significant negative step was associated with the *Dreissena* establishment but not with the upper bay TP loading reduction. Station HB in the middle bay was intermediate in its apparent response; there was a significant negative step in 1978 (though less so than at Station B), and, as at Station C, a significant negative step in 1996.

Chlorophyll *a* revealed negative steps at all four stations associated with both interventions (Table 1). The largest reductions associated with the TP loading intervention were seen in the upper bay at Stations B (45%) and N (29%), while Station N, as well as the middle and lower stations HB and C, experienced reductions of 36%, 41%, and 44%, respectively, in association with the *Dreissena* intervention. Decreases in the Chl-*a*-to-TP ratio were clearly much more strongly associated with the *Dreissena* establishment than with the TP load reductions at all stations. The statistical significance levels (*P* values in Table 1) for the inferred *Dreissena*-related 1996 negative step were 0.001, 0.003, <0.001, and <0.001 for Stations B, N, HB and C, respectively. In contrast, the corresponding *P* values for the 1978 negative steps associated with the TP load reductions were 0.026, 0.456, 0.434 and 0.150 (Table 1). The reduction in the ratio Chl-*a*/TP was clearly identified by regression analysis (Figure 5) showing significantly different relationships pre- and post-*Dreissena*. This was also demonstrated in the results from a multiple regression analysis of the entire1972–2008 data set, where the Zebra Mussel influence (ZM) was included as a dummy variable (+1 for 1995–2008 and zero for 1972–1994): LogChl-*a* = 1.127*LogTP - 0.138ZM - 0.525 (adjusted multiple $R^2 = 0.882$; $P < 0.0001$).

ARIMA/intervention analysis

ARIMA modelling was completed on all stations and variables that were included previously in the nonparametric tests (above), but to save space, only one detailed example (Figure 2) is included here. In no cases (four stations, two variables) did ARIMA modelling produce predictions for post-intervention time periods that were exclusively outside the 95% confidence band of the prediction. In many cases (e.g. Stn B, Chl-*a* [1978]; Stn B, TP [1978]; Stn N, TP [1978]; Stn N, Chl-*a* [1978]; Stn N, Chl-*a* [1996]; Stn HB, Chl-*a* [1996]; and Stn C, Chl-*a* [1996]), the majority of the measured values in the post-intervention period were well below the prediction but still within the 95%

confidence boundaries of the ARIMA prediction. In all of these cases, however, permutation tests showed that measured mean TP and Chl-*a* values were significantly below the ARIMA predictions ($P < 0.001$).

Regime shift detection (RSD)

Seasonality was a strong element of all 16 time-series investigated (4 stations, 2 variables, 2 intervention time periods), yielding periodic functions with correlation coefficients ranging from 0.750 to 0.991. In the TP time-series for Station B, for example (Figure 3a), initiation of phosphorus loading reductions in the winter of 1977–1978 was very evident in the time-series of "de-seasonalized" residuals (Figure 3b). Clearly, the strong autocorrelation effects of seasonal periodicity in the original data (Figure 3c) were removed by subtraction of the periodic function given in Figure 3a.

Regime shifts were detected over a wide range of statistical probabilities, but only those shifts that were revealed at levels of $P < 0.2$ were defined here. The most significant negative regime shifts (those with the lowest P values) included TP at Stations B and N in 1978 and Chl-*a* at Stations HB and C in 1996, all shifts being significant at $P < 0.005$. In all cases, except where no shifts were identified, detection was made with a higher level of confidence (lower P value) on model residuals (seasonality removed) than on original uncorrected series (Figure 4). In some cases this was very dramatic. At Station HB for the second (*Dreissena*) intervention, for example, the regime shift in TP beginning in 1995 was significant at $P = 0.0075$ on the residuals (seasonally "corrected" data), but on the original unadjusted data, a shift was detected only at P levels exceeding 0.18 (i.e. not statistically significant).

Rarely were any regime shifts detected that were not associated with either 1978 or 1995/1996—the dates representing the turning points for the two interventions explored here. Of those, only the positive shift

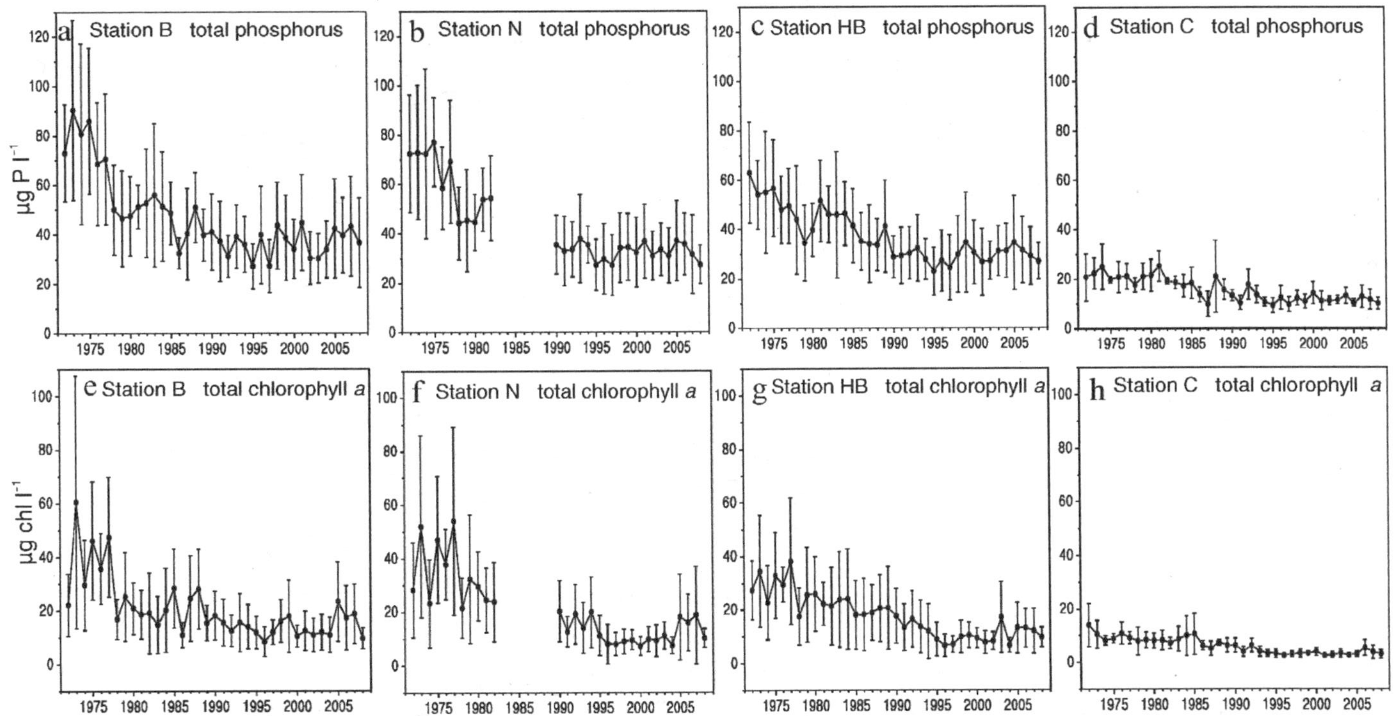

Figure 1. May to October means (±1 st. dev) of total phosphorus and total chlorophyll *a* (uncorrected for phaeopigments) based on monthly means for each of the four main sampling stations in the Bay of Quinte, B, N, HB and C, 1972–2008. Station N was not sampled during 1983–1989.

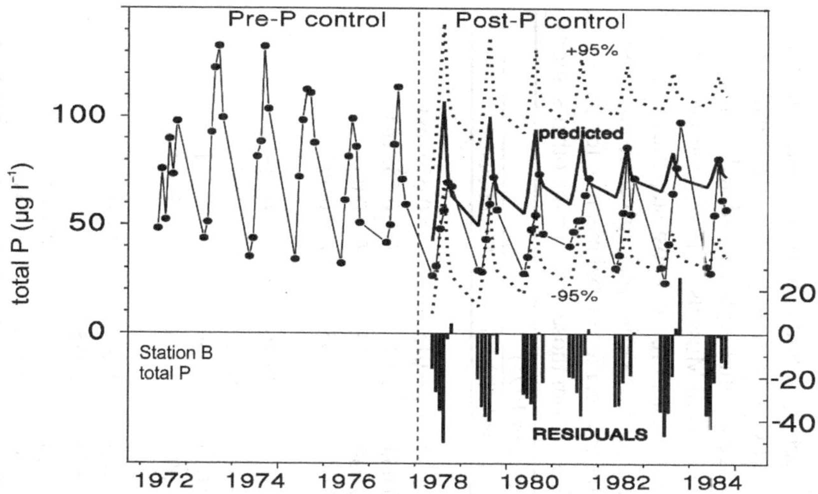

Figure 2. One example illustrating an ARIMA model prediction compared with measured data pre-and post-TP loading reductions at Station B in the Bay of Quinte.

identified in 2005 in Station B seasonally corrected Chl-*a* was statistically significant ($P < 0.05$).

Discussion

Over the past two decades, nonparametric statistical testing for long-term monotonic or abrupt step-trends has become the method of choice for detecting change in river and lake water quality data series (Cluis et al., 1989; Bodo 1989; Tsanis, 1993), including several applications to Great Lakes data series (Helsel, 1987; Tsanis and El-Shaarawi, 1992; Nicholls et al., 2001). The application of ARIMA modelling/forecasting/ intervention analysis, however, has been applied only rarely to Great Lakes data sets (Nicholls et al., 1999a) or elsewhere (Lehmann and Rode, 2001; Kurunç et al., 2005). Other newer methods, like Rodionov's (2004) RSD test, originally intended for identifying "regime shifts" in climate-related time-series data should be used more widely in other environmental applications as they become better known (Andersen et al., 2009).

Although it is very clear that the major Point-source TP reduction occurred in the winter of 1977–1978, a date for the establishment of *Dreis sena* spp. in the upper bay is less clear. Bailey et al. (1999) reported the first arrivals in 1993, but a population large enough to elicit a detectable response in limnological variables may not have been established for a

Table 1. Results of nonparametric tests (Mann-Whitney; Mann-Whitney seasonal and Mann-Whitney seasonal-Lettenmaier tests; Cluis et al. 1989) for step-trends in May to October means of total phosphorus, chlorophyll *a*, and chlorophyll *a*-to-total phosphorus ratios at each of the four main Project Quinte sampling stations, B, N, HB and C. Two time periods were analyzed to determine changes associated with two interventions: (1) phosphorus removal at sewage treatment plants initiated in the winter of 1977–1978 (step-trend tests: 1972–1977 vs. 1978–1984) and (2) the establishment of Zebra Mussels in 1995 (step-trend tests: 1985–1994 vs. 1996–2008, with 1995 data having been removed as a likely transitional year). Pre- and post-step means and the root mean square error (RMSE [in mg l^{-1} = the precision of the step model for the defined time period]) are given for each intervention. Note: Where step-trends were not significant (ns), pre-step mean = post step mean (i.e. the "global mean"); for Station N, there were no samples for 1983–1989, so the pre- and post-step time periods are different from those used for the other stations.

	1972–1984 (step in 1978)				1985–2008 (step in 1995)			
Station	Pre-step	Post-step	RMSE	Significance	Pre-step	Post-step	RMSE	Significance
a. Total P								
B	0.078	0.053	0.024	$P < 0.001$	0.038	0.038	0.015	$P = 0.059$ (ns)
N	0.071	0.050	0.020	$P = 0.001$	0.033	0.033	0.012	$P = 0.165$ (ns)
HB	0.054	0.046	0.017	$P = 0.004$	0.034	0.030	0.013	$P = 0.005$
C	0.021	0.021	0.006	$P = 0.203$ (ns)	0.015	0.012	0.005	$P = 0.004$
b. Chl-*a*								
B	0.040	0.022	0.020	$P = 0.006$	0.018	0.014	0.010	$P = 0.016$
N	0.041	0.029	0.022	$P = 0.047$	0.017	0.011	0.009	$P = 0.002$
HB	0.031	0.025	0.015	$P = 0.033$	0.017	0.010	0.009	$P < 0.001$
C	0.010	0.009	0.004	$P = 0.022$	0.006	0.003	0.003	$P < 0.001$
c. Chl-*a*-to-TP ratio								
B	0.498	0.412	0.194	$P = 0.026$	0.491	0.360	0.260	$P = 0.001$
N	0.560	0.560	0.243	$P = 0.465$ (ns)	0.480	0.330	0.185	$P = 0.003$
HB	0.560	0.560	0.230	$P = 0.434$ (ns)	0.500	0.340	0.210	$P < 0.001$
C	0.510	0.510	0.350	$P = 0.150$ (ns)	0.490	0.300	0.250	$P < 0.001$

year or two after that. Indeed, the RSD tests indicated some step changes beginning in 1995 and others in 1996, depending on the variable and sampling location. Evidence for this delayed response was also supported by earlier preliminary multivariate analysis of Bay of Quinte phyto plankton community structure (Nicholls et al., 2002), which identified 1995 as a transition year between the pre- and post-*Dreissena* ecosystem states.

The suggestion of similarities in trophic state between upper, middle, and lower Bay of Quinte and western, central, and eastern Lake Erie was first made by Johnson and Owen (1971) and has been rein forced by the clear gradients in TP and chlorophyll reported here, still existing three decades after the 1970s Point-source loading reductions. Undoubtedly, this gradient is not only the consequence of the down stream progressive dilution of the nutrient loads supplied to the upper bay but is also controlled by morphometric and hydrologic influences including the incursion of Lake Ontario water—a comparatively low-nutrient water mass that has been detected as far "upstream" in the Bay of Quinte as Station HB (Freeman and Prinsenberg, 1986).

The physical influence of Lake Ontario water on the lower bay was apparent in the phosphorus and chlorophyll concentrations at Station C. Nicholls et al. (2001) reported monotonic declines in phosphorus concentrations in the raw-water intakes for the municipalities of Cobourg and Kingston (west and east, respectively, of the lower Bay of Quinte) of 0.7–0.9 $\mu g\ l^{-1}\ yr^{-1}$ during the 1980s and 1990s. This, together with the apparent lack of a well-defined negative step-trend in TP at Station C in 1978 (when such declines were strongly manifested in the upper bay), suggests that the longer term monotonic declines in the lower bay are more strongly influenced by Lake Ontario than by the upper Bay of Quinte. There were dramatic declines in chlorophyll in north-eastern Lake Ontario (municipal water intake samples; Nicholls, 2001), but because these were mostly associated with a later time period (the 1990s), they were associated with the Zebra Mussel (*Dreissena*) establishment in Lake Ontario. Declines in chlorophyll at Bay of Quinte Station C during the 1990s were likely more related to *Dreissena* effects in northeastern Lake Ontario and in the lower bay than to lingering impacts of phosphorus loading reductions to the upper bay. In general, it can be concluded that the TP loading intervention had a greater relative impact on the upper bay than in the lower bay while the reverse is true for the *Dreissena* inter vention (greater relative impact in the lower bay than in the upper bay).

Results from the three independent analytical approaches were in very good agreement with respect to the levels of significance of the dentified steps or shifts. Those stations and variables showing no statistically significant step-trends in the nonparametric tests (i.e. global means pre- and post-interventions identical in Table 1) were also those stations/ variables that had measured data values clustered around the ARIMA prediction line for the post-intervention periods, and likewise did not reveal statistically significant RSD shifts.

In both the nonparametric step-trend and ARIMA approaches, the analyst must provide an a priori input on the date of the first sample date for the step (in the first approach) or the date of the intervention in the ARIMA modelling. Although some simple tools are available to assist in this largely subjective determination (e.g. the shape of CUSUM or doublemass curves (Nicholls, 2001), a more objective way to identify the end-date of a pre-intervention period in a data set is now available in Rodionov's (2004) RSD test. It was applied here after removal of strong seasonal

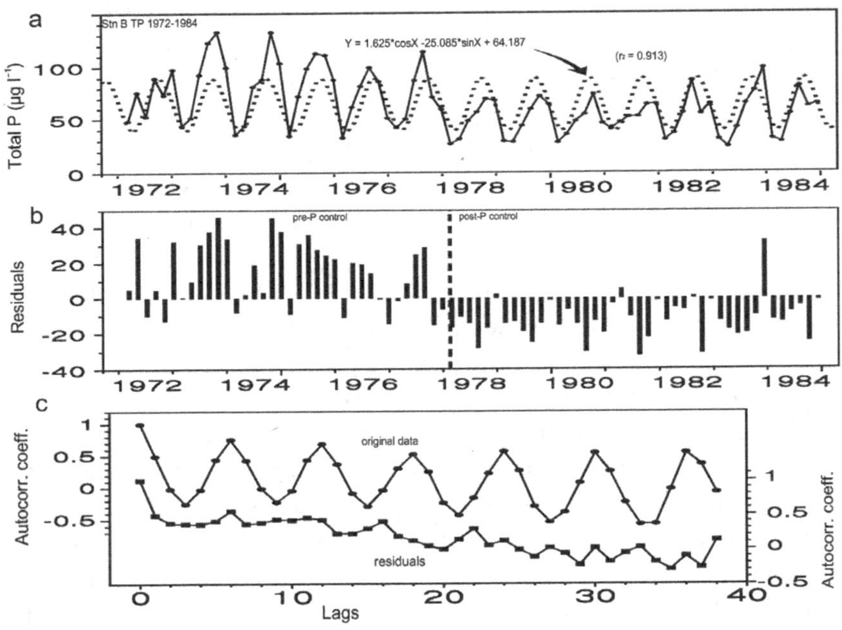

Figure 3. May to October monthly mean total phosphorus concentrations at Station B, 1972–1984 (pre- and post-TP loading intervention; evenly spaced, with the unsampled November to April periods left out): (a) monthly means with a sinusoidal function representing the monthly means, (b) residuals (monthly means minus the sinusoid in part (a)) and (c) the autocorrelation coefficients between 0 and n/2 (= 38) lags, showing strong autocorrelation in the original data and no significant autocorrelation in the residuals.

components in the data sets because the RSD procedure is most efficient for residual data sets that have been thus "pre-whitened" (Rodionov, 2006).

Regime shifts identified by RSD that coincide with those used in nonparametric tests for step-trends and for use in ARIMA/intervention modelling clearly help to build consensus about intervention effects. Conversely, regime shifts identified by the RSD approach that do not coincide with the timing of the known interventions help to provide perspective on the relative significance of factors other than those directly related to the interventions that may conspire to affect the structure of the time-series.

The declines in TP and chlorophyll reported here for the Bay of Quinte support findings from other Great Lakes locations (Nicholls, 2001; Nicholls et al., 1999a, 2001). Clearly, dreissenids are effective in removing TP from the water column during filter feeding of suspended particulate matter (containing phosphorus and chlorophyll) and delivering TP to the lake sediment layer in feces, pseudofeces and dead dreissenids. The normal transference of TP through the plankton and necton communities is thus disrupted; instead, TP is directed to the benthos (Hecky et al., 2004).

Declines in the chlorophyll-to-TP ratio associated with the establishment of *Dreissena* in the Lower Great Lakes are well documented. May to October Chl-*a*/TP ratios in northern, nearshore waters of Lake Erie were 2–6 times higher prior to *Dreissena* establishment, than after (Nicholls et al., 1999b). The likely explanation for the decline in the chlorophyll-to-TP ratio associated with *Dreissena* establishment is as follows: Essentially all chlorophyll-bearing particles contribute to the total P pool, but a portion of the total P pool is not associated with chlorophyll because it includes non-chlorophyllous particulate P and dissolved P (but, in a P-limited system, dissolved P is likely to be a minor component of the total P pool during the growing season). It follows then, that the filter-feeding habit of Zebra Mussels will remove P-bearing particles (both chlorophyllous and non-chlorophyllous particles) from the water column. Being organic, the chlorophyllous particles will largely be digested and assimilated by the mussels, but the inorganic P-bearing particles will be ejected directly or "packaged" into pseudofeces and expelled back into the environment from which they came. The net result is a reduction in both chlorophyll and total P concentrations in the water column, but because of the feedback of rejected inorganic particulate

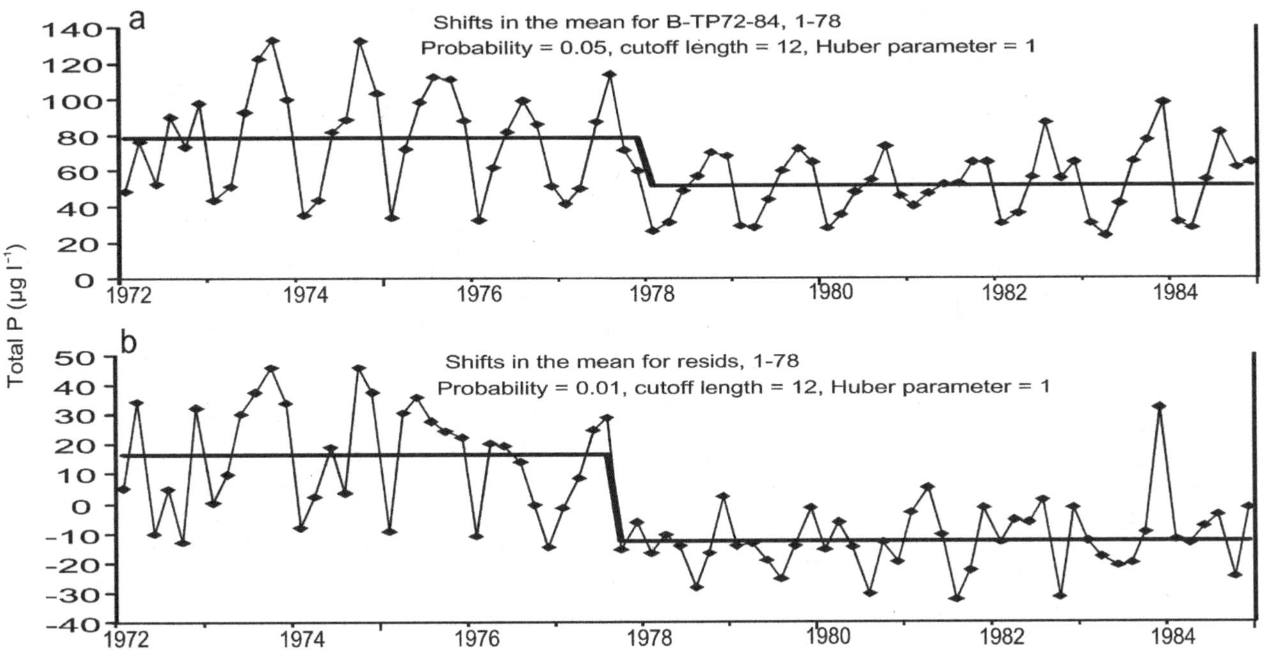

Figure 4. May to October monthly mean total phosphorus data from Station B, 1972–1984 (per Figure 3a) showing a regime shift in 1978 (a) that was significant at $P = 0.05$, but not at $P < 0.04$. In contrast, there is a very well defined regime shift in the residual data series at $P = 0.01$ (b) after removal of seasonality modelled as the periodic function given in Figure 3a.

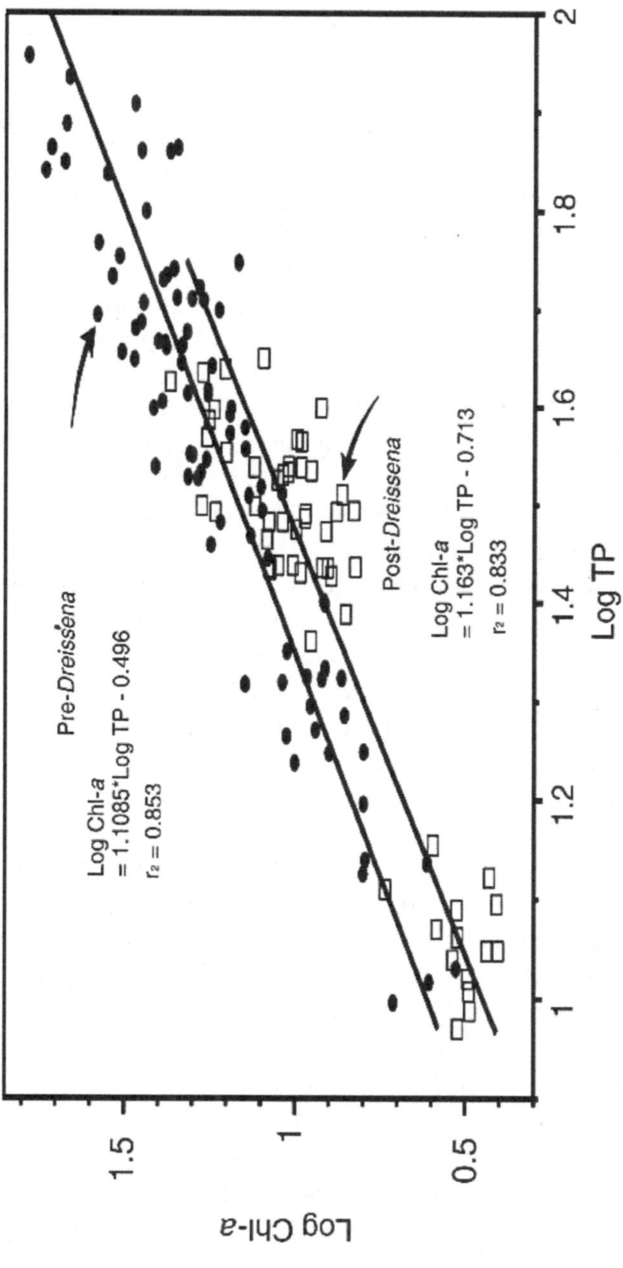

Figure 5. Shift in the relationship between May-October total phosphorus and chlorophyll-*a* at the four main Bay of Quinte stations, B, N, HB and C, pre-Dreissena and post-Dreissena (closed circles and open squares, respectively).

P, the rate of water-column chlorophyll decline exceeds the P decline, resulting in a negative step trend in the chlorophyll-to-phosphorus ratio.

Summary

Three independent methods of time-series analysis were applied to a 37-year record of total phosphorus and chlorophyll-*a* data collected at four sampling stations representing the upper, middle and lower sections of the Bay of Quinte, NE Lake Ontario, Canada. The three methods were used to build consensus around the significance of the apparent declines in total phosphorus (TP) and chlorophyll-*a* (Chl-*a*) concentrations following two interventions. (1) Point-source phosphorus loading reductions of about 50% to the upper bay in the winter of 1977–1978 and (2) the establishment of Zebra Mussels (*Dreissena* spp.) in the early to mid-1990s. The methods were applied to May to October monthly means and included: (1) nonparametric tests that accounted for persistence and seasonality and determined the statistical significance of step-trends, (2) ARIMA-Intervention modelling that produced forecasts into post-intervention time periods that were compared statistically to measured data and (3) regime shift detection for identifying the significance of persistent steps after removal of the seasonal components of the data series (modelled as periodic functions). Strong gradients in total phosphorus (TP) and Chl-*a* concentrations between the upper and lower Bay of Quinte still existed three decades after reductions in Point-source loadings of TP to the upper bay, where, during 2000–2008, May-October average TP was 3.8 times higher and Chl-*a*, 2.9 times higher than in the lower bay. The Remedial Action Plan May-October TP objective of 0.030 mg l^{-1} for the upper Bay of Quinte was not achieved consistently in recent years. Concurrence among the data analysis methods suggests that the relative decreases in the upper bay TP (31%) and Chl-*a* (37%) after phosphorus loading reductions were greater than in the lower bay (0% and 10%, respectively). The relative decreases in the upper bay TP (0%) and Chl-*a* (29%) associated with *Dreissena* establishment were less than those measured in the lower bay (20% and 50%, respectively).

The three methods of time-series analysis applied here to the 37-year record of phosphorus and chlorophyll-*a* data from the Bay of Quinte have revealed the persistence of strong gradients between the upper bay (Belleville region) to the lower bay (between Glenora and Lake Ontario). The Remedial Action Plan objective of 0.030 mg l^{-1} (May–October average) for the upper Bay of Quinte was not achieved consistently in recent years, even while under the influence of the water column clearing effects of filter-feeding Dreissenid Mussel populations established about the mid-1990s. Reductions in chlorophyll-*a* throughout the Bay of Quinte were significant after the establishment of *Dreissena* spp. Reductions in total phosphorus, however, were much more significant in the upper bay following Point-source control of phosphorus loading initiated in the winter of 1977.

Acknowledgements

I thank the many regular staff and students of the Laboratory Services Branch, Ontario Ministry of the Environment (Kingston and Toronto laboratories) for several thousand chlorophyll and phosphorus analyses of Bay of Quinte samples over the past 37 years. Students and staff of Fisheries and Oceans, Canada collected most of the water samples from the Bay of Quinte stations.

References

Andersen, T., Carstensen, J., Hernández-García, Durte, C.M., 2009. Ecological thresholds and regime shifts: approaches to identification. Trends in Ecology and Evolution 24, 49–57.

Anon. 1969. Report to the International Joint Commission on the pollution of Lake Erie, Lake Ontario and the international section of the St. Lawrence River. International Lake Erie and Lake Ontario - St. Lawrence River Water Pollution Boards. Vol. I. Queen's Printer, Ottawa.

Bailey, R.C., Grapentine, L., Stewart, T.J., Schaner, T., Chase, M.E., Mitchell, J.S., Coulas, R.A., 1999. Dreissenidae in Lake Ontario; impact assessment at the whole lake and Bay of Quinte spatial scales. J. Great Lakes Res. 25, 482–491.

Bodo, B.A., 1989. Robust graphical methods for diagnosing trend in irregularly spaced water quality time series. Environ. Monitoring and Assessment 12, 407–428.

Box, G.E.P., Jenkins, G.M., 1970. *Time Series Analysis Forecasting and Control. 2nd Ed.* Holden-Day, Oakland, CA.

Box, G.E.P., Tiao, G.C., 1975. Intervention analysis with applications to economic and environmental problems. J. Amer. Statistical Assoc. 70 (349), 70–79.

Cluis, D., Langlois, C., Van Coillie, R., Laberge, C., 1989. Development of a software package for trend detection in temporal series: application to water and industrial effluent quality data for the St. Lawrence River. Environ. Monitoring and Assessment 12, 429–441.

CoHort Software, 1995. *CoStat*. Minneapolis, Minnesota 55419, U.S.A.

Freeman, NG., Prinsenberg, S.J., 1986. Exchange flows in the Adolphus Reach/North Channel. pp 27–39. In: C.K. Minns, D.A. Hurley, K.H. Nicholls (Eds.), *Project Quinte: point source phosphorus control and ecosystem response in the Bay of Quinte, Lake Ontario*. Can. Spec. Publ. Fish. Aquat. Sci. 86.

Hasler, A.D., 1969. Cultural eutrophication is reversible. BioScience 19, 425–431.

Hecky, R.E., Smith, E.H., Barton, D.R., Taylor, W.D., Charlton, M.N., Howell, T., 2004. The nearshore phosphorus shunt: a consequence of ecosystem engineering by dreissenids in the Laurention Great Lakes. Can. J. Fish. Aquat. Sci. 61, 1285–1293. Helsel, D.R., 1987. Advantages of non-parametric procedures for analysis of water quality data. J. Hydrol. Sci. 32, 179–190.

Human Edge Software, 1992. The Forecasting Edge. Human Edge Software Corporation, San Kateo, CA.

Johnson, M.G., Owen, G.E., 1971. Nutrients and nutrient budgets in the Bay of Quinte, Lake Ontario. J. Wat. Pollut. Control Fed. 43, 836–853.

Kurunç, A., Yürekli, K., Çevik, O., 2005. Performance of two stochastic approaches for forecasting water quality and streamflow data from Yeşilırmak River, Turkey. Environmental Modelling & Software 20, 1195–1200.

Lehmann, A., Rode, M., 2001. Long-term behaviour and cross-correlation water quality analysis of the river Elbe, Germany. Water Research 35, 2153–2160.

Minns, C.K., Hurley, D.A., Nicholls, K.H. (Eds.), 1986. Project Quinte: point source phosphorus control and ecosystem response in the Bay of Quinte, Lake Ontario. Can. Spec. Publ. Fish. Aquat. Sci. 86.

Nicholls, K.H., 2001. CUSUM phytoplankton and chlorophyll functions illustrate the apparent onset of Dreissenid Mussel impacts in Lake Ontario. J. Great Lakes Res. 27, 393–401.

Nicholls, K.H., Hopkins, G.J., 1993. Recent changes in Lake Erie (north shore) phyto plankton: cumulative impacts of phosphorus loading reductions and the Zebra Mussel introduction. J. Great Lakes Res. 19, 637–647.

Nicholls, K.H., Standke, S.J., Hopkins, G.J., 1999a. Effects of Dreissenid Mussels on nitrogen and phosphorus in north shore waters of Lake Erie. In: M. Munawar, T. Edsall, I. Munawar (Eds), *State of Lake Erie (SOLE) - Past, Present and Future*, pp. 323–336. Ecovision World Monograph Series. Backhuys Publ., Leiden, The Netherlands.

Nicholls, K.H., Hopkins, G.J., Standke, S.J., 1999b. Reduced chlorophyll-to-phosphorus ratios in nearshore Great Lakes waters coincide with the establishment of Dreissenid Mussels. Can. J. Fish. Aquat. Sci. 56, 153–161.

Nicholls, K.H., Hopkins, G.J., Standke, S.J., Nakamoto, L., 2001. Trends in total phosphorus in Canadian near-shore waters of the Laurentian Great Lakes: 1976–1999. J. Great Lakes Res. 27, 402–422.

Nicholls, K.H., Heintsch, L., Carney, E.C., 2002. Univariate step-trend and multivariate assessments of the apparent relative effects of P loading reductions and Zebra Mussels on the phytoplankton of the Bay of Quinte, Lake Ontario. J. Great Lakes Res. 28, 15–31.

Prince, A.T., Bruce, J.P., 1972. Development of nutrient control policies in Canada. Tech. Bull. No. 51, Inland Waters Branch, Dept. of the Environment, Ottawa.

Robinson, G.W., 1986. Water quality of the Bay of Quinte, Lake Ontario, before and after reductions in phosphorus loading. In: C.K. Minns, D.A. Hurley, K.H. Nicholls (Eds.), *Project Quinte: point source phosphorus control and ecosystem response in the Bay of Quinte, Lake Ontario*, pp. 50–58. Can. Spec. Publ. Fish. Aquat. Sci. 86.

Rodionov, S.N., 2004. A sequential algorithm for testing climate regime shifts. Geophys. Res. Lett. 31, L09204, doi: 10.1029/2004GL019448.

Rodionov, S.N., 2006. The use of pre-whitening in climate regime shift detection. Geophys. Res. Lett. 31, L12707, doi: 10.1029/2006GL025904.

Slater, R.W., Bangay, G.E., 1980. Action taken to control phosphorus in the Great Lakes. In: R.C. Loehr, C.S. Martin, W. Rast, (Eds), Proceedings of the 1979 Conference on Phosphorus Management Strategies for Lakes. Ann Arbor Science, Ann Arbor, Michigan.

Tsanis, I.K., 1993. WQTA—a water quality trend analysis program. Water Poll. Res. J. Canada 28, 311–335.

Tsanis, I.K., El-Shaarawi, A.H., 1992. Trend evaluation of water quality parameters in the Niagara and St. Lawrence Rivers. Environ. Monit. Assessment 23, 205–218.

Warwick, W.F., 1980. Paleolimnology of the Bay of Quinte, Lake Ontario: 2800 years of cultural influence. Can. Bull. Fish. Aquat. Sci. 206.

The above article is a reprinted version of the following previously published article:

Phosphorus and chlorophyll in the Bay of Quinte: A time-series/intervention analysis of 1972–2008 data Kenneth H. Nicholls

Aquatic Ecosystem Health & Management (2012) 15 (4): 421–429.

https://doi.org/10.1080/14634988.2012.727720

Supplemental Material

Data Characterization

Preliminary explorations of the Bay of Quinte data, using the Runs test for data stationarity and autocorrelation tests, emphasized the need to assess these data for trends with methods, like those outlined in Nicholls (submitted), to which this supplemental material is attached. Such methods are specific for time series with strong seasonal and/or trend components. Virtually all Bay of Quinte chemical data sets examined revealed symptoms of non-stationarity or non-randomness. With Station B TP data, illustrated here as just one example, this was manifested strongly in both a serial dependency/seasonality component (the apparent cycling of the lagged autocorrelation coefficients) and in a trend component, inferred by increasingly negative autocorrelation coefficients with increasing lags (Figure S1). In a Runs test of the same Stn B data for stationarity, N1=129 and N2=88, with total runs = 60, contrasting with an expected runs total of 106 ($Z = -6.44$; $P<0.0001$), indicating with a very high probability that the Station B TP data were not likely taken from a normally distributed population. In contrast, Runs test results for the same random data set used in the autocorrelation comparison (per Figure S1) were: N1 = N2 = 111, with actual and expected total runs = 115 and 112, respectively ($Z = 0.404$; $P = 0.687$), clearly demonstrating stationarity, as expected.

Non-parametric step-trend tests, ARIMA/Intervention analysis and Regime Shift Detection

Of the 16 cases examined (2 variables, 4 stations, 2 interventions), nonparametric step-trend detection tests revealed a total of 13 statistically significant negative steps associated with both the TP loading intervention and the *Dreissena* intervention (per Table S1 of Nicholls). Detailed output for only TP at Station B (both interventions) is presented here to save space, and demonstrates that the apparent impact of point source TP loading reductions (Figure S2a) apparently had a much bigger impact on TP concentrations in the upper bay than did the establishment of *Dreissena* (Figure S2b). Chlorophyll *a* revealed negative steps at all four stations; only those for Station B are illustrated here to save space (Figure S3). Four examples (Figure S4) illustrate the wide differences between ARIMA forecasts and the measured values for the post-intervention time periods (all differences were statistically significant $P < 0.001$; permutation tests). Table S2 provides details of the RSD tests run on all 16 Bay of Quinte time series for TP and Chl-*a*.

Table S1. Periodic functions (sinusoids) describing May to October seasonality in chlorophyll *a* (chl) and total phosphorus (TP) data for each of the four main Bay of Quinte sampling stations determined for 1972-1984 (the Point-source P loading intervention in 1978) and for 1985-2008 (*Dreissena* establishment in the mid-1990s). In sinusoids, the independent variable (X) is an angle of arc expressed in radians where one month of one complete cycle (6 months or 360° of arc) = pi/3 radians, because 360° = 2*pi radians.

	1st intervention (1972-1984)		2nd intervention (1985-2008)	
	Function	r^2	Function	r^2
B, chl-*a*	3.741*cosX -16.30*sinX +30.32	0.914	0.217*cosX -9.474*sinX +15.86	0.958
N, chl-*a*	0.082*cosX -19.99*sinX +35.15	0.842	-0.696*cosX -8.558*sinX +12.68	0.983
HB, chl-*a*	5.234*cosX -14.41*sinX +27.29	0.856	1.373*cosX -8.037*sinX +13.05	0.939
C, chl-*a*	2.409*cosX -2.686*sinX +9.383	0.961	0.077*cosX -1.446*sinX +4.466	0.912
B, TP	1.625*cosX -25.08*sinX +64.19	0.912	-1.757*cosX -15.07*sinX +38.09	0.948
N, TP	0.818*cosX -20.97*sinX +60.27	0.902	1.164*cosX -13.38*sinX +32.90	0.96
HB, TP	11.74*cosX -14.06*sinX +49.65	0.792	6.456*cosX -12.44*sinX +31.09	0.885
C, TP	1.267*cosX -1.737*sinX +20.87	0.563	0.898*cosX -2.245*sinX +12.86	0.963
B, chl-*a*/TP	0.084*cosX -0.111*sinX +0.466	0.688	0.010*cosX -0.120*sinX +0.409	0.808
N, chl-*a*/TP	-0.011*cosX -0.172*sinX +0.575	0.572	-0.036*cosX -0.118*sinX +0.374	0.893
HB, chl-*a*/TP	-0.013*cosX -0.176*sinX +0.552	0.823	-0.047*cosX -0.127*sinX +0.410	0.82
C, chl-*a*/TP	0.141*cosX -0.109*sinX +0.497	0.867	-0.007*cosX -0.080*sinX +0.398	0.569

Table S2. Regime Shift Detector (RSD) outcomes, including probability levels (P) for statistically significant difference between pre- and post-shift means for total phosphorus (TP) and chlorophyll *a* (chl-*a*) during the Point-source phosphorus loading (1st) intervention and the *Dreissena* (2nd) intervention at the four main Bay of Quinte sampling stations, B, N, HB and C. RSD tests were run on original monthly mean data as well as monthly means that were corrected for seasonality after subtracting the appropriate sinusoid function (See table S1). Cut-off length for all RSD tests was 12, meaning that regimes shorter than two full seasons (12 May-October monthly means) were not identified in these tests. A negative or positive sign preceding the date of the shift indicates a change to a lower or higher regime, respectively. If no date is given and P > 0.20 is indicated, no shift was detected at P < 0.20.

	Date of shift: 1st intervention (1972-84)		Date of shift: 2nd intervention (1985-2008)	
	original data	seasonally corrected data	original data	seasonally corrected data
B, chl-*a*	+1974; -1978 (*P* = 0.150)	-1978 (*P* = 0.034)	-1988 (*P* = 0.075)	-1996; +2005 (*P* = 0.0375)
N, chl-*a*	+1974; -1978 (*P* = 0.20)	+1974; -1978 (*P* = 0.120)	-1996 (*P* = 0.050)	-1996 (*P* = 0.010)
HB, chl-*a*	*P* > 0.20	-1978 (*P* = 0.160)	-1996 (*P* = 0.031)	-1996 (*P* = 0.0047)
C, chl-*a*	-1973 (*P* = 0.087)	-1973 (*P* = 0.0812)	-1996 (*P* = 0.0065)	-1996 (*P* = 0.0045)
B, TP	-1978 (*P* = 0.050)	-1978 (*P* = 0.00018)	-1987; +1990 (*P* =0.090)	-1995 (*P* = 0.018)
N, TP	-1978 (*P* = 0.007)	-1978 (*P* = 0.0008)	*P* > 0.20	-1995; +1998 (*P* = 0.085)
HB, TP	-1978 (*P* = 0.077)	-1978 (*P* = 0.044)	-1995 (*P* = 0.180)	-1995 (*P* = 0.0075)
C, TP	*P* > 0.20	*P* > 0.20	-1995 (*P* = 0.067)	-1995 (*P* = 0.038)

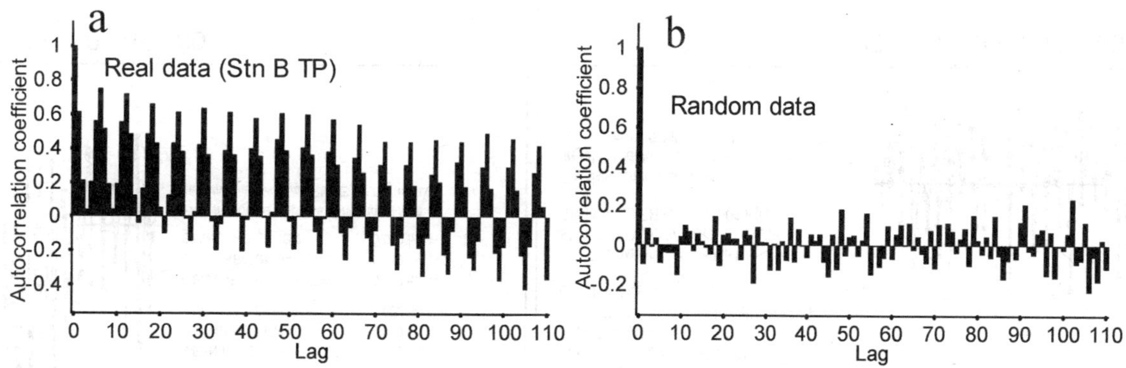

Figure S1. Autocorrelation coefficients for (a) monthly mean total phosphorus concentrations at Station B, 1972-2008, and in (b), for comparison, a random number data set with the same n = 222 (i.e. 6 months*37 years), and range (0.011<TP>0.133 mg l^{-1}). Note 1) that autocorrelations with zero lag = unity, because the correlation of any month with itself = 1, and 2), that the full range of possible lags = n/2.

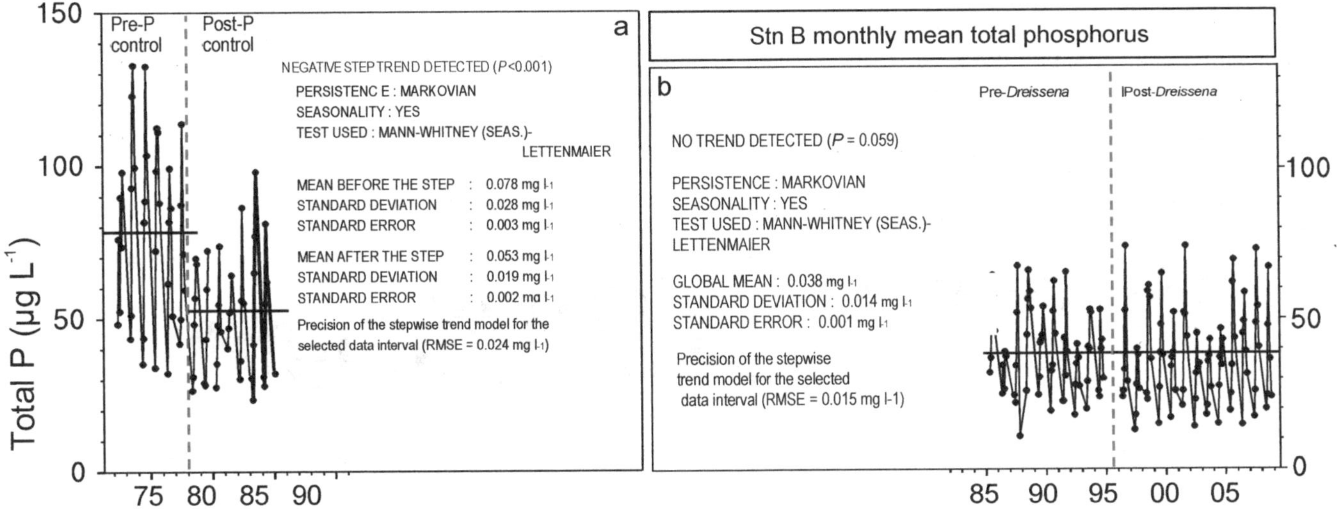

Figure S2. Non-parametric test results for step-trends in monthly mean total phosphorus at Station B centred on (a) the point-source TP load reduction in early 1978 and (b) the *Dreissena* establishment in the mid-1990s.

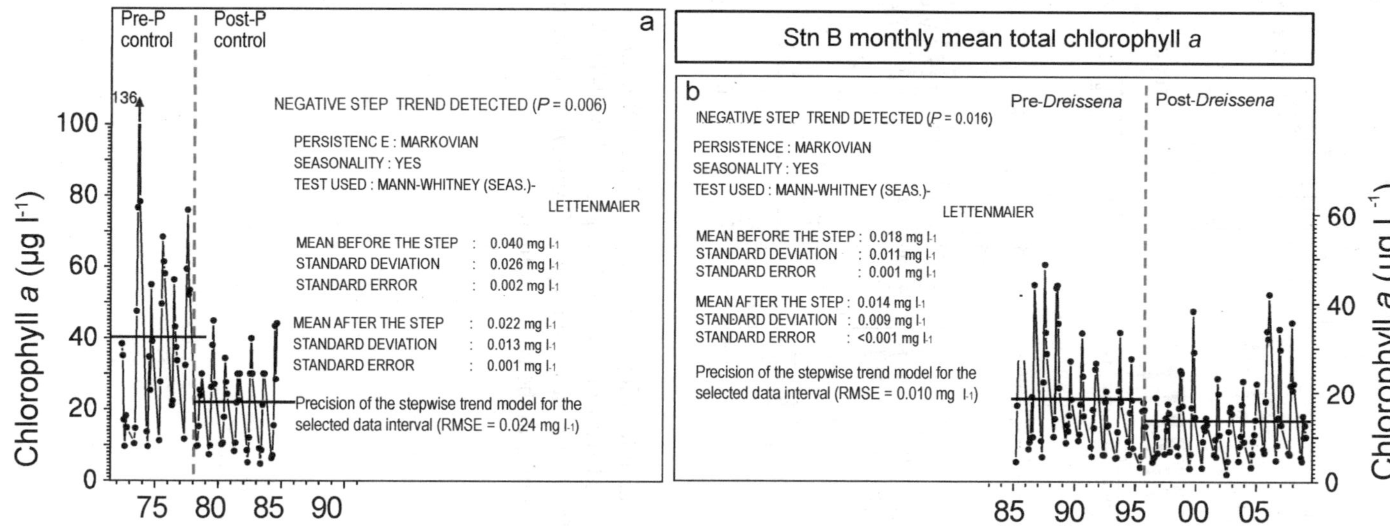

Figure S3. Non-parametric test results for step-trends in monthly mean chlorophyll *a* at Station B centred on (a) the pointsource TP load reduction in early 1978 and (b) the *Dreissena* establishment in the mid-1990s.

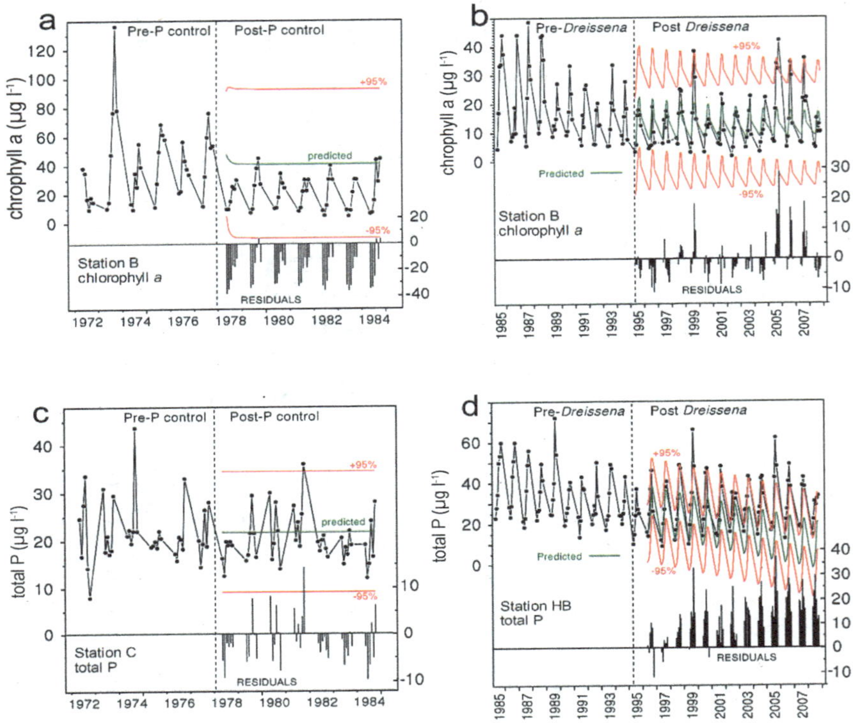

Figure S4. Four examples illustrating some of the different ARIMA model predictions compared with measured data pre-and post-TP loading reductions (winter of 1977-78) and pre-and post-*Dreissena* establishment (1995) at Stations B, HB and C in the Bay of Quinte.

The relative importance of autotrophic and heterotrophic microbial communities in the planktonic foodweb of the Bay of Quinte, Lake Ontario 2000–2007

M. Munawar,* M. Fitzpatrick, H. Niblock, and J. Lorimer

Fisheries and Oceans Canada, 867 Lakeshore Road, Burlington, ON L7S 1A1
**Corresponding author: mohiuddin.munawar@dfo-mpo.gc.ca*
Keywords: eutrophication, carbon budgets, heterotrophy, Beneficial Use Impairments

Introduction

Eutrophication has been one of the major stressors affecting the North American Great Lakes and the need to control eutrophication has driven research and management objectives since the 1970s. External phosphorus loadings, primarily from agricultural wastes and detergents, were identified as the probable cause of noxious algal blooms related to eutrophication (Vollenweider, 1971; Vallentyne, 1974; Schindler, 1978). Within the Great Lakes basin, the tightly coupled relationship between phosphorous loadings, primary production and algal standing crop (chlorophyll-*a*) was modelled by Vollenweider et al. (1974) and resulted in a phosphorus abatement program being implemented by the governments of Canada and the United States in 1978 under the terms of the Great Lakes Water Quality Agreement (GLWQA) (International Joint Commission (IJC), 1988.)

In later revisions to the GLWQA, 42 Areas of Concern (AoC) were identified as having experienced severe environmental degradation (IJC, 1989) in order to provide a focal point for remediation efforts. These areas were identified as having at least 1 of 14 Beneficial Use Impairments (BUIs) (Hartig and Zarull, 1992). The Bay of Quinte, located on the north-eastern shores of Lake Ontario, has a long history of eutrophication and was designated an AoC because it was deemed to have 10 BUIs including "Eutrophication or undesirable algae" and "Degradation of phytoplankton and zooplankton communities" (BQ RAP, 1993). There is currently some controversy regarding the application of BUIs in the

Ecology of the Bay of Quinte: Health, Management and Global Implications
Editors: C.K. Minns, M. Munawar and M.A. Koops

Aquatic Ecosystem Health & Management Society, Canada

Great Lakes basin since objective, quantitative and defensible ecological parameters were never fully defined (Krantzberg, 2004; George and Boyd, 2007).

Fisheries and Oceans Canada began a long term research and monitoring effort in the Bay of Quinte in the early 1970s in response to eutrophication which included phytoplankton, zooplankton, nutrients and chlorophyll-*a*, and these findings were essential to its designation as an AoC. Remediation efforts in the Bay of Quinte have included phosphorus load reductions, the establishment of targets for total phosphorus concentrations (30 $\mu g\ l^{-1}$), phytoplankton biomass (4–5 $g\ m^{-3}$) and chlorophyll-*a* (12–15 $\mu g\ l^{-1}$). The purpose of these targets was to provide some objective measures against which improvements in trophic state and the health of planktonic communities can be measured (BQ RAP, 1993) and, for the most part, focus on structural measurements of autotrophic organisms. Although microbial foodweb research was initiated in the Great Lakes in the late 1980s (Munawar and Weisse, 1989; Weisse and Munawar, 1989), the research and monitoring program of the Bay of Quinte was expanded in 2000 to include the microbial loop–bacteria, autotrophic picoplankton (APP), heterotrophic nanoflagellates (HNF) and ciliates–in order to assess the role of both autotrophs and heterotrophs on trophic state and ecosystem health.

A large body of research exists relating the structure and dynamics of autotrophic communities in the Great Lakes to trophic state (e.g. Munawar and Munawar, 1982; Munawar and Munawar, 1996; Millard et al., 1999; Fitzpatrick et al., 2007). Trophic state, however, is not simply a consequence of autotrophic processes. Recent work has shown that heterotrophic components of the microbial foodweb have a critical role in recycling both organic and inorganic carbon (Dodds and Cole, 2007) so a thorough examination of the microbial–planktonic foodweb is required for a holistic assessment of ecosystem health (Munawar et al., 2010). The microbial–planktonic foodweb includes bacteria, APP, phytoplankton, HNF, ciliates and zooplankton and inextricably links the primary producers with the nutrient recyclers and secondary consumers providing the energy that supports higher trophic levels (Pomeroy, 1974; Sherr et al., 1986; Sherr et al., 1988; Munawar et al., 1999). Heterotrophs, particularly bacteria, have been shown to have an increasingly important role in energy flux in oligotrophic versus eutrophic systems (del Giorgio et al., 1997; Cole et al., 2000; Biddanda et al., 2001). However, the impacts of cultural eutrophication (i.e. inputs of phosphorus and allochthonous matter, frequent algal blooms) have been shown to enhance heterotrophic microbial processes (Pace, 1993; Porter, 1996).

The aims of this paper are twofold. The first is to characterize the composition of the organic carbon pool of the Bay of Quinte into the major autotrophic and heterotrophic assemblages in order to uncover potential energy pathways and understand the linkages between organisms. The second is to comment on the health of the bay with respect to the remediation targets and beneficial use impairments. The implications for energy transfer from lower to higher trophic levels in the Bay of Quinte are discussed within the broader context of ecosystem health and potential impairments.

Materials and methods

The Bay of Quinte is a large z–shaped embayment (254 km^2) that is also relatively shallow (4–8 m), but is much deeper at the interface with Lake Ontario. One shallow station in the upper bay (Belleville–5 m) and one deep station in the lower bay (Conway–30 m) (Figure 1) were sampled bi-weekly from May 1st to October 31st for each of the years from 2000 to 2007. Integrated epilimnetic water samples were collected from each site and subsamples were drawn for total phosphorus, chlorophyll-*a*, microbial loop, ciliates and phytoplankton. Zooplankton was sampled from discrete depths. The Area of Concern is represented by the upper bay site while the lower bay site serves as a control.

Total phosphorus (TP) concentrations were determined with the automated ascorbic acid technique after persulfate digestion (Strickland and Parsons, 1968). Point source phosphorous loadings were compiled by P. Kintsler, Ontario Ministry of the Environment, Kingston, Ontario (pers. com.). Microbial loop samples, including bacteria, autotrophic picoplankton and heterotrophic nanoflagellates, were fixed with 1.6% formaldehyde and enumerated using DAPI staining (Porter and Feig, 1980) under epi-fluorescence microscopy (Munawar and Weisse, 1989). Organic carbon was estimated as 200 fg C cell^{-1} for autotrophic picoplankton (APP), 10 fg C cell^{-1} for bacteria and 14 fg C cell^{-1} for heterotrophic nanoflagellates (HNF) (Sprules et al., 1999). Ciliate samples were preserved with Lugol's iodine (Lynn and Munawar, 1999) and enumerated following the Quantitative Protargol Staining technique (Montagnes and Lynn, 1993). Ciliate organic carbon was calculated as 11% of the fresh weight biomass (Turley et al., 1986).

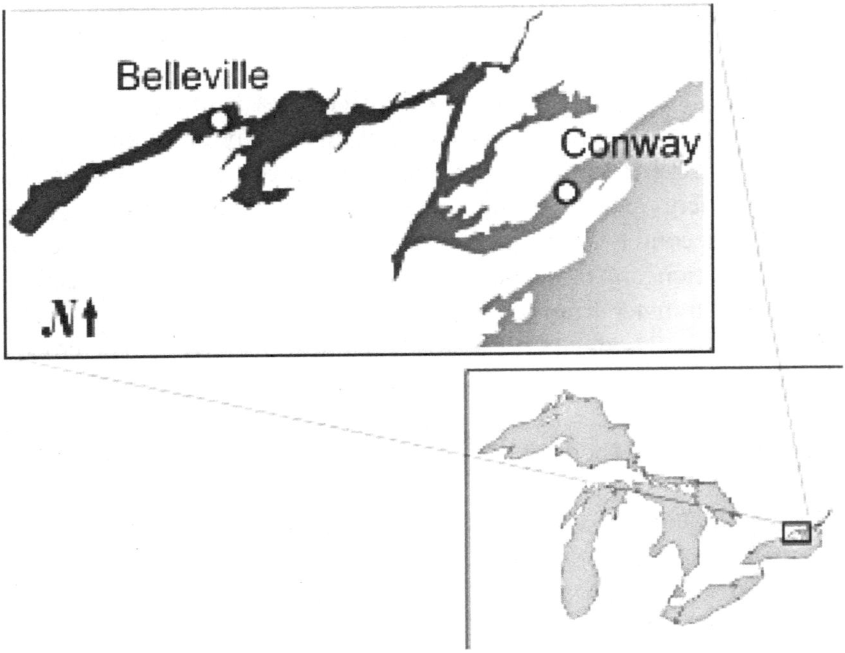

Figure 1. Map of the Bay of Quinte, Lake Ontario with Belleville and Conway stations indicated.

Chlorophyll-*a* concentrations were determined by filtering up to 1 L of water through Whatman GF/C filters followed by cold acetone pigment extraction and spectrophotometric analysis (Strickland and Parsons, 1968). Phytoplankton samples were fixed immediately with Lugol's iodine. Identification and enumeration followed the Utermöhl (1958) inverted microscope technique as described in Nicholls et al. (2002) which are broadly compatible with the established techniques of Munawar et al. (1987). Organic carbon was estimated for the various groups of phytoplankton using equations from the literature namely: Diatomeae (Strathmann, 1967), Chlorophyta, Chrysophyceae, Cryptophyceae and Dinophyceae (Verity et al., 1992), and Cyanophyta (Lee and Furman, 1987).

The zooplankton data used in this article were obtained from Bowen and Johannsson (2011). Zooplankton samples were collected using a 41 litre Schindler–Patalas trap fitted with a 64 µm mesh. Three depths were sampled at Belleville (1 m, 2 m, 3 m) and eight depths were sampled at Conway (1 m, 3 m, 5 m, 8 m, 10 m, 20 m and 25 m). Samples were preserved immediately with 4% sugared formalin. A single composite sample was constructed for each station-date by combining 50% of the sample from each depth. Biomass of each taxon was determined in each sample from length–weight regressions (Bowen and Johannsson, 2011). Organic carbon was estimated as 48% of the dry weight (Anderson and Hesson, 1991).

Annual primary production was estimated by ^{14}Carbon uptake. Samples from each cruise were spiked with $Na^{14}CO_3$ and incubated for 2–4 hours at irradiance levels ranging from 25–1300 µmol $m^{-2} s^{-1}$. Following incubation, acidification and bubbling was used to remove excess ^{14}C and radioactivity was determined by liquid scintillation counting (Millard et al., 1999). Seasonal areal primary production was then estimated according to the model of Fee (1990). These estimates were increased by 10% in order to account for winter primary production (Fitzpatrick et al., 2007; Vollenweider et al., 1974).

Results

Upper Bay (Belleville)

The biomass of a traditional planktonic foodweb (phytoplankton and zooplankton) and an integrated microbial-planktonic foodweb (phytoplankton, zooplankton and microbial loop) for Belleville is shown in Figure 2 for 2000–2007. The seasonal weighted mean fresh weight biomass was 2.4–5.5 g m^{-3} for phytoplankton and 0.5–3.2 g m^{-3} for zooplankton (Figure 2a). Phytoplankton contributed 57.5–85.7% of the biomass, on average, of a traditional planktonic foodweb. In comparison, an integrated microbial planktonic foodweb is shown in Figure 2b. The microbial loop contributes an additional 2.2–8.7 g m^{-3} of biomass and accounts for 22.6–63.5% of the combined biomass compared to 26.1–55.5% for phytoplankton and 8.1–28.6% for zooplankton. The largest single component of the microbial loop was heterotrophic nanoflagellates which ranged from 1.7–8.4 g m^{-3} followed by bacteria (0.1–0.5 g m^{-3}), autotrophic picoplankton (0.1–0.3 g m^{-3}) and ciliates (0.01–0.2 g m^{-3}).

The size of the organic pool at Belleville ranged from 546.4–1837.8 mg C m^{-3} and there was considerable inter annual variability in the relative proportions of autotrophs (APP, phytoplankton) and heterotrophs (bacteria, HNF, ciliates, zooplankton) as shown in Figure 3. During 2000 and 2001, phytoplankton contributed 50% of the organic carbon of the combined microbial and planktonic foodweb. In 2002, HNF contributed slightly more organic carbon (48%) than phytoplankton (43%), and HNF dominated the organic carbon pool (60%) in 2003 and 2004. However, during 2005–2007, phytoplankton contributed upwards to 75% of the organic carbon.

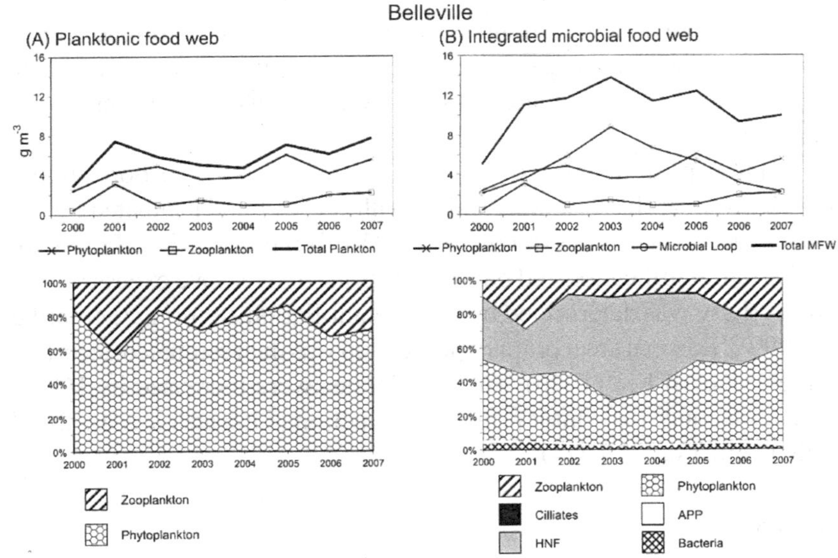

Figure 2. A comparison of traditional and integrated foodwebs in the upper Bay of Quinte (Belleville) from 2000–2007. Values are seasonal weighted means (May–October) of fresh weight biomass (mg m^{-3}).

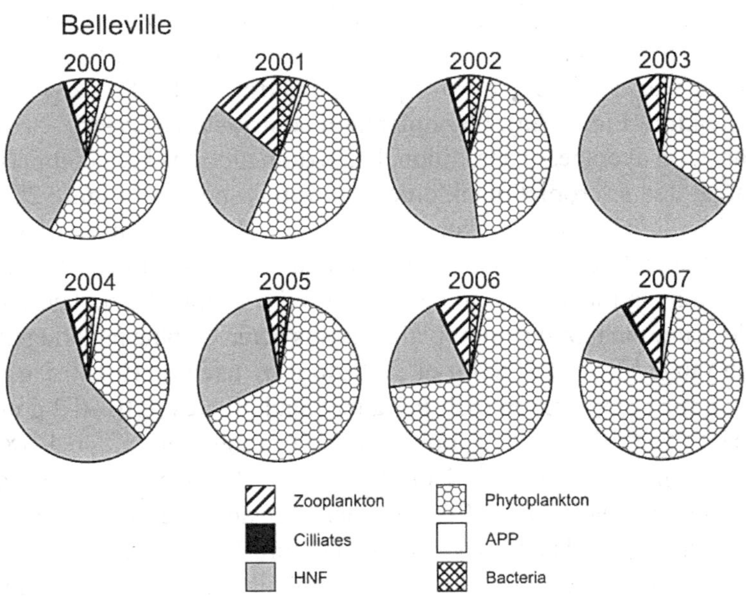

Figure 3. The percent contribution of the various components of the microbial– planktonic foodweb to the organic carbon pool in the upper bay (Belleville).

Annual estimates of primary production, (mean) chlorophyll-*a*, (mean) point source phosphorus loadings and (mean) total phosphorus concentrations are given in Table 1 for Belleville from 2000–2007. Primary production ranged from ≈300–425 g C m^{-2}, chlorophyll-*a* from 12–26 µg l^{-1}, phosphorus loadings from 8.6–22.3 kg d^{-1} and total phosphorus from 28–42 µg l^{-1}.

Table 1. Annual estimates of primary production (g C $m^{-2} y^{-1}$), mean chlorophyll-*a* (µg l^{-1}), mean point source phosphorus loadings (kg d^{-1}) and mean total phosphorus concentrations (µg l^{-1}) in the Bay of Quinte Area of Concern (Belleville). Data is compiled from Munawar et al. (2010b).

Year	Chlorophyll *a* (µg l^{-1})	Primary Production (g C $m^{-2} y^{-1}$)	Point Source Phosphorus Loadings (kg d^{-1})	Total Phosphorus (µg l^{-1})
2000	11.5	315.7	12.4	34.5
2001	15.2	297.0	10.1	41.8
2002	12.2	343.2	12.6	27.6
2003	13.0	424.6	22.3	29.8
2004	13.7	310.2	13.2	31.0
2005	25.0	365.2	11.1	41.0
2006	17.7	335.5	18.5	37.8
2007	16.2	326.7	8.6	34.0

Lower Bay (Conway)

At Conway, the combined phytoplankton and zooplankton biomass of the traditional planktonic foodweb ranged from 0.8–1.3 g m^{-3} (on a seasonal weighted mean basis) with phytoplankton accounting for 50–81% of the total (Figure 4a). The combined microbial-planktonic foodweb for Conway is depicted in Figure 4b. The microbial loop contributes an additional 1.3–6.7 g m^{-3} of biomass at Conway which accounts for 56–86% of the combined foodweb compared to 7–36% for phytoplankton and 7–17% for zooplankton. Similar to Belleville, HNF were the largest component of the microbial loop (1.1–6.3 g m^{-3}), followed by bacteria (0.09–0.4 g m^{-3}), APP (0.04–0.1 g m^{-3}) and ciliates (0.02–0.06 g m^{-3}).

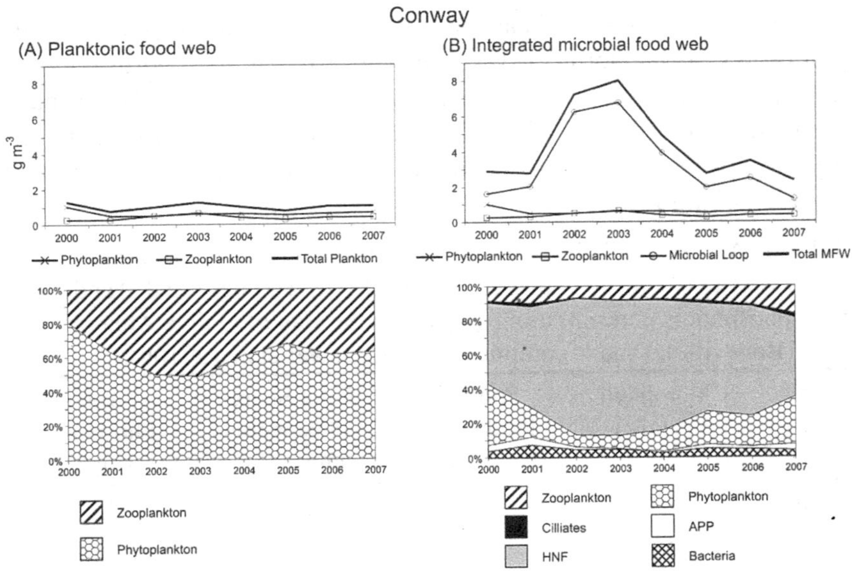

Figure 4. A comparison of traditional and integrated foodwebs in the lower

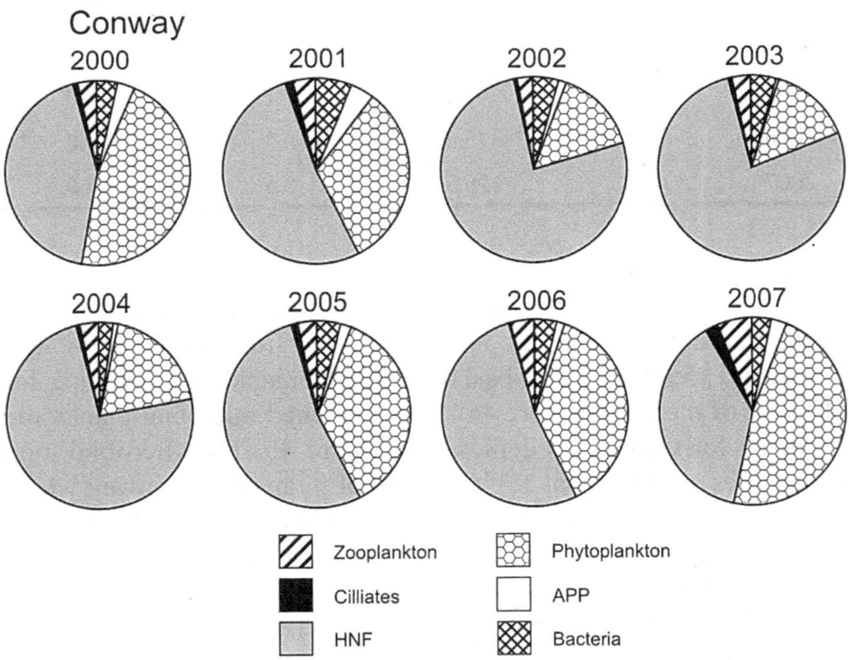

Figure 5. The percent contribution of the various components of the microbial– planktonic foodweb to the organic carbon pool in the lower bay (Conway).

Bay of Quinte (Conway) from 2000–2007. Values are seasonal weighted means (May–October) of fresh weight biomass (mg m^{-3}). The size of the organic carbon pool at Conway ranged from 303.9– 900.4 mg C m^{-3} and HNF were typically the largest component accounting for 39–77% of the organic carbon followed by phytoplankton ranging from 14–47% (Figure 5). Bacteria, APP and zooplankton accounted for less than 5% of the organic carbon and ciliates were typically <1%.

Discussion

Traditionally, studies of the lower trophic levels of aquatic ecosystems have been restricted to analyses of the phytoplankton and zooplankton communities and this was the case at the inception of Project Quinte in 1972. However, microbial loop assessments were added beginning in 2000 and continue to the present. In our analysis from 2000–2007 (Figure 2a), we observed that phytoplankton biomass at Belleville was generally greater than 3 g m^{-3} (2.4–6.0) and typical of eutrophic conditions (Munawar and Munawar, 1982). Phytoplankton biomass was 2–5× greater than zooplankton biomass which is not unexpected in eutrophic environments since it reflects a lack of top down control (e.g. Bays and Crisman, 1983; Pace, 1986; Elser and Goldman, 1991). Phytoplankton biomass at Conway fell between 0.5–1.0 g m^{-3} (Figure 4a) on an annual mean basis which is typical of oligotrophic waters. Zooplankton biomass was about half of the phytoplankton biomass, although there was some variability. In 2000 for example, zooplankton biomass was only about 25% of the phytoplankton biomass, whereas in 2002 phytoplankton and zooplankton biomass were not significantly different and in 2003 zooplankton biomass exceeded phytoplankton biomass. On the whole, the relationship between phytoplankton and zooplankton biomass appears to be more closely related at Conway than at Belleville and broadly indicative of oligotrophic conditions.

However, our results demonstrate that the traditional foodweb analyses based mainly on phytoplankton and zooplankton, give only a partial picture of the community and do not account for the significant amount of biomass which the microbial loop (bacteria, APP, HNF, ciliates) may contribute to the pelagic foodweb. Recent and historic work conducted in Lake Ontario by our lab and by others have shown that heterotrophic protists can contribute significantly to plankton abundance and biomass (Munawar and Nauwerck, 1971; Pick and Caron, 1987; Munawar et al., 2010). At Belleville, microbial biomass was equal to zooplankton biomass at its lowest point and upwards to 7× greater than zooplankton biomass (Figure 2b). In fact microbial biomass was equal to or exceeded the combined phytoplankton and zooplankton biomass in 3 of the years studied (2002, 2003, 2004). HNF dominate the microbial loop and contributed 17.4–60.9% of the total microbial and planktonic biomass.

Similarly, at Conway the microbial biomass ranged from 1.3–6.7 g m^{-3} during the study period and was greater than the combined phytoplankton and zooplankton biomass in all of years studied (Figure 4b). The largest component of the microbial loop and by extension the microbial and planktonic foodweb was HNF which contributed 45.9– 79.5% of the combined microbial and planktonic biomass. The high biomass of HNF and other microbes raises questions as to the nature of the organic carbon pool that supports the foodweb and whether or not energy is being diverted away from a traditional grazing food chain.

Organic carbon has long been considered to be a sound currency of energy exchange (Gosselain et al., 2000) so the structure of the organic carbon pool is an important consideration when determining ecosystem health because it provides an estimate of the amount of energy in the system available for transfer to higher trophic levels. Over the course of 8 years, the organic carbon pool at Belleville and the potential energy that it provides was evenly split between autotrophs and heterotrophs during the first 3 years; and then heterotrophs dominated for the next two followed by an overwhelming dominance by autotrophs once again (Figure 3). The largest contributors to the organic carbon pool in all of the years studied were HNF and phytoplankton. These two factors appear to be controlling the shift of the organic carbon pool from heterotrophic to autotrophic. Such dramatic swings in the structure of the foodweb were not observed between autotrophs and heterotrophs at Conway (Figure 5). However heterotrophs did supply 50–85% of the organic carbon (predominantly HNF). These observations are similar to our earlier studies in the offshore oligotrophic waters of Lake Ontario (Munawar et al., 2010). It is apparent that the potential dominance of HNF is not limited by the trophic state of the ecosystem.

The observed changes in the structure of the microbial–planktonic foodweb at Belleville, from being primarily heterotrophic to being primarily autotrophic, appear to be due to late summer blooms of filamentous diatoms and blue green algae such as *Melosira* spp., *Anabaena* spp. and *Lyngbya* spp. (Nicholls and Carney, 2011) which results in an overall increase in the supply of organic carbon from autotrophs.

Importantly, there is no corresponding decline in the amount of organic carbon supplied by heterotrophs so algal blooms appear to increase the size of the organic carbon pool. However, the appearance of late summer algal blooms should not be interpreted as an increase in autochthonous production. This is supported by estimates of annual primary production in the upper bay which ranged from 300–425 g C m^{-2} y^{-1} (Table 1). Despite the inter-annual variability in primary production, such changes are unlikely to be statistically significant (Porta et al., 2005; Fitzpatrick et al., 2007) indicating that autochthonous production is stable over longer time frames.

The question of whether the microbial loop is a source or sink for organic carbon has been debated in the literature for some time (Wylie and Currie, 1991; Porter, 1996). Zooplankton, particularly Daphnia, play an important role in regulating HNF biomass (Pace, 1993; Porter, 1996). While Daphnia are the most prevalent zooplankter in the Bay of Quinte (Bowen and Johannsson, 2011), the amount of zooplankton at both Belleville and Conway stations is less than expected given its trophic state (Pace, 1986). Research conducted in coastal areas of the Great Lakes has shown that both HNF and Daphnia are important bacterivores (Hwang and Heath, 1997; Hwang and Heath, 1999), so the high proportion of HNF to zooplankton would suggest that HNF are out-competing zooplankton for food resources in the Bay of Quinte. This hypothesis assumes that bacteria are the sole food resource and there is sufficient bacterial production to support such a large standing crop of HNF. However, results from stable isotope (13C) experiments suggest that zooplankton preferentially graze autochthonous matter, particularly in eutrophic environments (Pace et al., 2004; Brett et al., 2009). As we noted in a previous work, very little is known about the dynamics of HNF in the Great Lakes and production and grazing experiments are badly needed (Munawar et al., 2010). With respect to foodweb structure in the Bay of Quinte, it appears that autochthonous production can be diverted away from a traditional grazing food chain towards a microbial foodweb and support a large biomass of heterotrophs sequestering the autochthonous energy (Munawar et al., 2010).

In the first part of our discussion on microbial–planktonic webs, we treated the Bay of Quinte as a geographic entity (i.e. a shallow embayment of Lake Ontario) which includes both eutrophic and oligotrophic conditions. In the second part of the discussion, we consider the Bay of Quinte as a management entity defined by its status as an Area of Concern. The AoC includes the eutrophic waters of the bay which are represented in this study by observations at the Belleville site. As noted, the Bay of Quinte has a long history of eutrophication and Project Quinte was developed as a multi-agency effort to control eutrophication under the terms of the GLWQA. Phosphorus load reductions were implemented beginning in 1978 and point source loads were immediately reduced from more than 200 kg d^{-1} to less than 70 kg d^{-1} (Minns et al., 1986). As a consequence, from 1977 to 1978, total phosphorus concentrations declined from 70–48 μg l^{-1}, chlorophyll-*a* declined from 40 to 21 μg l^{-1} and phytoplankton biomass from 12.5 to 4.3 g m^{-3} (Nicholls, 2012; Nicholls and Carney, 2011). Despite this success, "Eutrophication or undesirable algae" and "Degradation of phytoplankton and zooplankton communities" in the upper bay (i.e. Belleville) were identified as 2 of the 10 Beneficial Use Impairments when the Bay of Quinte was designated as an AoC (BQ RAP, 1993) based on the data gathered from Project Quinte.

With respect to alleviating these Beneficial Use Impairments (BUIs), phosphorus abatement was continued and remediation targets for total phosphorus (30–35 μg l^{-1}), chlorophyll-*a* (12–15 μg l^{-1}) and phytoplankton biomass (4–5 g m^{-3}) were established to help measure the success of management efforts. A cap of 15 kg d^{-1} was also recommended for phosphorous loadings (Minns and Moore, 2004). All of these parameters were at or near target levels from 2000–2007 (Table 1), but it remains unclear whether or not these targets demonstrate enough of an improvement in the trophic status and health to conclude that the stated BUIs no longer apply. According to George and Boyd (2007), the concept of BUIs is vague and poorly defined from a policy perspective. Since precise criteria for the 42 AoCs in the Great Lakes basin were never established, it is difficult to evaluate the status of individual impairments (ibid). However, considerable progress has been made from an ecological perspective in developing integrated, sensitive and rapid indicators of ecosystem health. Consequently, a variety of multi-trophic tests and models are available which can be deployed to assess the status and recovery of BUIs.

In order to establish a scientifically sound basis for BUIs and assess progress towards recovery, our lab adopted a "battery of tests" strategy using multi-trophic indicators (Table 2). One such indicator is phytoplankton biomass and its species composition. The phytoplankton biomass at Belleville exceeded an annual mean concentration of 3 g m^{-3} establishing the upper bay as eutrophic. Species composition requires a detailed treatment of its own and hence is not included here due to lack of space. Another indicator chosen was Vollenweider's eutrophication model which classifies trophic state as a function of phosphorous loadings, primary production and chlorophyll-*a* as a proxy for algal standing crop (Vollenweider et al, 1974; Munawar et al., 2010b). Table 1 shows the values of these parameters observed during our study (2000–2007) which is summarized as follows:

Point source phosphorus loadings: >10 kg d^{-1}
Primary production: >300 g C $m^{-2} y^{-1}$
Chlorophyll-*a* concentration: >12 µg l^{-1}

Table 2. Examples of a body of evidence strategy for assessing aquatic ecosystem health.

Structural	Functional	Models
Phytoplankton Biomass	Primary Production	Vollenweider's Eutrophication Model (Vollenweider et al., 1974)
Zooplankton Biomass	Bacterial Production	Phytoplankton Biomass Trophic Ladder (Munawar and Munawar, 1982)
Biodiversity and Species Composition	P/B (Carbon Turnover)	P-IBI (Kane et al., 2009)
Edible: Inedible Phytoplankton		Linkages (*proposed*)
Plankton Size Spectra		Foodweb Dynamics and Linkages (*proposed*)
Microbial Loop		
Autotrophs: Heterotrophs		
Carbon Budgets		

Vollenweider's model defines eutrophic conditions as those where algal standing crop (chlorophyll-*a*) exceeds 9.0 µg l^{-1} and annual primary production exceeds 295.0 g C $m^{-2} y^{-1}$, which confirms that the upper bay is eutrophic as indicated by the above observations. Furthermore, the model predicts that existing remediation targets will result in the continuation of eutrophic conditions. But according to the model, the predicted loadings of phosphorous required to sustain this level of primary production is greater than our current estimates of point source loadings and suggest that non-point sources of phosphorous have a critical role in determining the Bay of Quinte's overall trophic state. These non-point sources would include: tributary loadings (estimated by Minns et al., 2004 to be >600 kg d^{-1}), storm sewer overflows and recycling by Dreissenid Mussels. Furthermore, Vollenweider's model was developed for contained lake systems and does not fully incorporate the effect of flow in riverine systems. However, the model does appear to be applicable to the Bay of Quinte although it is a flow through system.

Summary

The structure and function of the microbial and planktonic communities of the Bay of Quinte, Lake Ontario were studied for 8 years from 2000 to 2007. The bay has a long history of eutrophication and has undergone remediation efforts which included reductions in phosphorus loadings and the implementation of a long term research and monitoring program (1972–present) conducted by Fisheries and Oceans Canada along with other federal and provincial agencies. Microbial loop research was added in 2000 to the ongoing monitoring program which included nutrients, phytoplankton, zooplankton, benthos and fish so that a comprehensive picture of foodweb linkages would emerge. The structure of the microbial-planktonic foodweb was determined based on microscopic analysis of bacteria, autotrophic picoplankton (APP), heterotrophic nanoflagellates (HNF), ciliates, phytoplankton and zooplankton and was compared to a traditional grazing food chain consisting of phytoplankton and zooplankton. On a seasonal weighted mean basis, HNF biomass (fresh weight) of 1.7–8.4 g m^{-3} at Belleville and 1.1–6.3 g C m^{-3} at Conway exceeded that of zooplankton in virtually all observations and was often greater than the combination of phytoplankton and zooplankton. Furthermore, the results showed that HNF contributed upwards to 85% of the organic carbon pool on a seasonal weighted mean basis. Various parameters in the upper bay relating trophic status to autotrophic communities were measured including: point source phosphorus loadings (>10 kg d^{-1}); primary production (>300 g C m^{-2} y^{-1}); chlorophyll *a* (>12 µg l^{-1}) and phytoplankton biomass (>3 g m^{-3}) which indicated that the upper bay remained eutrophic. This was also confirmed by the "battery of tests" strategy of ecological indicators developed in our laboratory to assess trophic status, health, and potential Beneficial Use Impairments. Based on our observations spanning 8 years, it was concluded that the microbial foodweb was dominated by heterotrophic communities which are still widely ignored in Great Lakes research and monitoring efforts. Our data clearly demonstrates that future studies and management strategies should include the "microbial loop" to obtain a holistic picture of the structure, function and dynamics of the lower foodweb.

Conclusions

In conclusion, our long term study of 8 years (2000–2007) indicates that the upper Bay of Quinte (Belleville) continues to be eutrophic, while the lower bay (Conway) is oligotrophic. Given that the lower bay was intended as a control site, results from the upper bay show the Beneficial Use Impairments of "Eutrophication or undesirable algae" and "Degra dation of phytoplankton and zooplankton communities" still persist. These findings are based on a "battery of tests" of widely accepted indicators and models of trophic status which rely on measures of the autotrophic community.

Our study also shows that the microbial loop components contribute significantly to the organic carbon pool of the Bay of Quinte and that heterotrophic nanoflagellates (HNF) can contribute more organic carbon than phytoplankton and zooplankton combined. However, we observed significant shifts in the structure of the foodweb from being primarily autotrophic to being primarily heterotrophic due to the presence or absence of algal blooms. Science has developed considerably since the 1970s and 1980s and the new frontiers of knowledge include sensitive techniques and models which need to be included in ecosystem health assessments. Consequently, future research, monitoring and management strategies, including refined criteria for Beneficial Use Impairments, should not ignore the important role of the microbial loop for a holistic evaluation of the structure, function and dynamics of the lower trophic levels.

Acknowledgements

We thank Drs. C.K. Minns, J.H. Leach and M.A. Koops for their constructive reviews of the manuscript. Thanks are also due to S. Blunt and L. Elder (AEHMS) for their technical editing. Finally, we gratefully acknowledge the hard work and perseverance of the DFO staff whose dedication to Project Quinte have made this research possible, including: E.S. Millard, M. Frank (M. Burley), J. Gerlofsma, C. Timmins, R. Bonnell and A. Bedford.

References

Anderson, T. and Hessen, D.O., 1991. Carbon, nitrogen and phosphorus content of fresh water zooplankton. Limnol. Oceanogr. 36, 807–813.

Bay of Quinte Remedial Action Plan (BQ RAP), 1993. Time to Act. The Bay of Quinte Remedial Action Plan Stage 2 Report. Bay of Quinte Remedial Action Plan (RAP) Coordinating Committee, Ontario Ministry of the Environment, Kingston, ON.

Bays, J.S. and Crisman, T.L. 1983. Zooplankton and trophic state relationships in Florida lakes. Can. J. Fish. Aquat. Sci. 40, 1813–1819.

Biddanda, B., Ogdahl, M., Cotner, J, 2001. Dominance of bacterial metabolism in oligo trophic relative to eutrophic waters. Limnol. Oceanogr. 46(3), 730–739.

Bowen, K.L. and Johannsson, O.E., 2011. Changes in zooplankton biomass in the Bay of Quinte with the arrival of the mussels, *Dreissena polymorpha and D. rostiformis, and the predatory cladocerans, Cercopagis pengoi*: 1975 to 2008. Aquat. Ecosyst. Health Mgmt. 14(1), 44–55.

Brett, M.T., Kainz, M.J., Taipale, S.J., Seshan, H., 2009. Phytoplankton, not allochthonous carbon, sustains herbivorous zooplankton production. PNAS 106(50), 21197–21201.

Cole, J.J., Pace, M.L., Carpenter, S.R., Kitchell, J.F., 2000. Persistence of net heterotrophy in lakes during nutrient addition and foodweb manipulations. Limnol. Oceanogr. 45(8), 1718–1730.

del Giorgio, P.A., Cole, J.J., Cimbleris, A., 1997. Respiration rates in bacteria exceed phytoplankton production in unproductive aquatic systems. Nature 385, 148–151.

Dodds, W.K., Cole, J.J., 2007. Expanding the concept of trophic state in aquatic ecosystems: it's not just the autotrophs. Aquat. Sci. 69, 427–439.

Elser, J.J. and Goldman, C.R., 1991. Zooplankton effects on phytoplankton in lakes of contrasting trophic status. Limnol. Oceanogr. 36(1), 64–90.

Fee, E.J., 1990. Computer programs for calculating in situ phytoplankton photosynthesis. Can. Tech. Rep. Fish. Aquat. Sci. No. 1740.

Fitzpatrick, M.A.J., Munawar, M., Leach, J.H., and Haffner, G.D., 2007. Factors regulating primary production and phytoplankton dynamics in western Lake Erie. Fundam. Appl. Limnol. (Archiv. Hydrobiol.) 169, 137–152.

George, T.K., Boyd, D., 2007. Limitations on the development of quantitative monitoring plans to track progress of Beneficial Use Impairment restoration at Great Lakes Areas of Concern. J. Great Lakes Res. 33, 686–692.

Gosselain, V., Hamilton, P.B., Descy, J-P., 2000. Estimating phytoplankton carbon from microscopic counts: an application for riverine systems. Hydrobiol. 438, 75–90.

Hartig, J.H., and Zarull, M.A., 1992. Towards defining aquatic ecosystem health for the Great Lakes. J. Aquat. Ecosyst. Health 1, 97–107.

Hwang, S-J. Heath, R.T., 1997. Bacterial productivity and protistan bacterivory in coastal and offshore communities of Lake Erie. Can. J. Fish. Aquat. Sci. 54, 788–799.

Hwang, S-J. Heath, R.T., 1999. Zooplankton bacterivory at coastal and offshore sites of Lake Erie. J. Plankton Res. 21(4), 699–719.

International Joint Commission (IJC), 1988. Revised Great Lakes Water Quality Agreement of 1978. Agreement, with Annexes and Terms of Reference, between the United States and Canada signed at Ottawa, November 22, 1978 and Phosphorus Load Reduction Supplement signed October 7, 1983 as amended by protocol signed October 7, 1983, as amended by protocol signed November 18, 1987.

International Joint Commission (IJC), 1989. Revised Great Lakes Water Quality Agreement of 1978 between the United States and Canada, as amended by protocol, signed November 18, 1987.

Kane, D.D., Gordon, S.I., Munawar, M., Charlton, M.N., Culver, D.A., 2009. The Planktonic Index of Biotic Integrity (P-IBI): an approach for assessing lake ecosystem health. Ecol. Ind. 9, 1234–1247.

Krantzberg, G., 2004. Science must inform Great Lakes policy. J. Great Lakes. Res. 30, 573–574.

Lee, S., Fuhrman, J.A., 1987. Relationships between biovolume and biomass of naturally derived marine bacterioplankton. Appl. Env. Microbiol. 53, 1298–1303. Lynn, D.H., Munawar, M., 1999. Abundance, biomass, and diversity of planktonic ciliates (Ciliophora) in Lake Erie. In: M. Munawar, T. Edsall, and I.F. Munawar (Eds.), *State of Lake Erie (SOLE)–past, present and future*, pp. 155–168. Backhuys Publishers, Leiden, Netherlands.

Millard, E.S., Fee, E.J., Myles, D.D., Dahl, J.A., 1999. Comparison of phytoplankton photosynthesis methodology in Lakes Erie, Ontario, the Bay of Quinte and the Northwest Ontario Lake Size Series. In: M. Munawar, T. Edsall, I.F. Munawar (Eds.), *State of Lake Erie (SOLE)–Past, Present and Future*, pp. 441–468. Backhuys Publishers, Leiden, Netherlands.

Minns, C.K., and Moore, J.E., 2004. Modelling phosphorus management in the Bay of Quinte, Lake Ontario in the past, 1972 to 2001, and in the future. Can. Manuscr. Rep. Fish. Aquat. Sci. 2695.

Minns, C.K., Owen, G.E., Johnson, M.G., 1986. Nutrient loads and budgets in the Bay of Quinte, Lake Ontario, 1951–81. In: C.K. Minns, D.A. Hurley, K.H. Nicholls, (Eds.), Project Quinte: point-source phosphorus control and ecosystem response in the Bay of Quinte, Lake Ontario. Can. Spec. Publ. Fish. Aquat. Sci. 86.

Minns, C.K., Moore, J.E., Seifried, K.E., 2004. Nutrient loads and budgets in the Bay of Quinte, Lake Ontario, 1972 to 2001. Can Manser. Rep. Fish. Aquat. Sci. 2694: V+49p.

Montagnes, D.J.S. and Lynn, D.H., 1993. A quantitative protargol stain (QPS) for ciliates and other protists. In: P. F. Kemp, B.F. Sherr, E. B. Sherr, E. J. Cole (Eds.), *Handbook of methods in aquatic microbial ecology*, pp. 229–240. Lewis Publishers, Boca Raton, FL.

Munawar, M., Munawar, I.F., 1982. Phycological studies in Lakes Ontario, Erie, Huron and Superior. Can. J. Bot. 60, 1837–1858.

Munawar, M., Munawar, I.F., 1996. *Phytoplankton dynamics in the North American Great Lakes Vol. I: Lakes Ontario, Erie and St. Clair*. Ecovision World Monograph Series. SPB Academic Publishing, Amsterdam.

Munawar, M., Nauwerck, A., 1971. The composition and horizontal distribution of phytoplankton in Lake Ontario during the year 1970. pp. 69–78. Proc. 14th Conf. Great Lakes Res. Internat. Assoc. Great Lakes Res.

Munawar, M., Weisse, T., 1989. Is the 'microbial loop' an early indicator of anthropogenic stress? Hydrobiol. 188/189, 163–174.

Munawar, M., Munawar, I.F., McCarthy, L., 1987. Phytoplankton ecology of large eutrophic and oligotrophic lakes of North America: Lakes Ontario and Superior. Arch. Hydrobiol. Bieh. Ergebn. Limnol. 25, 51–96.

Munawar, M., Munawar, I.F., Weisse, T., van Stam, H., Fitzpatrick, M., Lorimer, J., 1999. Probing microbial food-web structure in Lake Erie. In: M. Munawar, T. Edsall, I.F. Munawar (Eds.), *State of Lake Erie (SOLE)–Past, Present and Future*, pp. 197–218. Ecovision World Monograph Series. SPB Academic Publishing, Amsterdam.

Munawar, M., Fitzpatrick, M., Munawar, I.F., Niblock, H., 2010. Checking the pulse of Lake Ontario's microbial-planktonic communities: a trophic transfer hypothesis. Aquat. Ecosyst. Heath. Mgmt. 13(4), 395–412.

Munawar, M., Fitzpatrick, M., Kane, D., Munawar, I.F., Niblock, H., and Lorimer, J., 2010b. Application of ecological indicators as a tool for assessing Beneficial Use Impairments towards delisting of the Great Lakes Areas of Concern: Bay of Quinte Example. 53rd Annual Conference on Great Lakes Research., 2010 May 17–21. International Association of Great Lakes Research (IAGLR), Toronto, ON, Canada

Nicholls, K.H. and Carney, E., 2011. The phytoplankton of the Bay of Quinte, 1972–2008: point-source phosphorus loading control, dreissenid mussel establishment, and a proposed community reference. Aquat. Ecosyst. Health Mgmt 14(1), 33–43.

Nicholls, K.H. 2012. Phosphorus and chlorophyll in the Bay of Quinte: A time-series/intervention analysis of 1972–2008 data. Aquat. Ecosyst. Health Mgmt. 15(4), 421-429.

Nicholls, K.H., Heintsch, L, Carney, E., 2002. Univariate step-trend and Multivariate assessments of the apparent effects of P loading reductions and zebra mussels on the phytoplankton of the Bay of Quinte, Lake Ontario. J. Great Lakes Res. 28(1), 15–31.

Pace, M.L., 1986. An empirical analysis of zooplankton community size structure across lake trophic gradients. Limnol. Oceanogr. 31(1), 45–55. Pace, M.L., 1993. Heterotrophic microbial processes. In: S.R. Carpenter, and J.F. Kitchell (Eds.), *The trophic cascade in lakes*, pp. 252–277. Cambridge University Press, Cambridge, UK

Pace, M.L., Cole, J.J., Carpenter, S.R., Kitchell, J.F., Hodgson, J.R., Van de Bogert, M.C., Bade, D.L., Kritzberg, E.S., Bastviken, D., 2004. Whole-lake carbon-13 additions reveal terrestrial support of aquatic foodwebs. Nature 427, 240–243.

Pick, F.R., Caron, D.A., 1987. Picoplankton and nanoplankton biomass in Lake Ontario: relative contribution of phototrophic and heterotrophic communities. Can. J. Fish. Aquat. Sci. 44(12), 2164–2172.

Pomeroy, L.R., 1974. The ocean's foodweb a changing paradigm. BioScience 24, 499–504.

Porta, D., Fitzpatrick, M.A.J. and Haffner, G.D., 2005. Annual variability of phytoplankton primary production in the Western Basin of Lake Erie (2002–2003). J. Great Lakes Res. 31 (Suppl. 2), 63–71.

Porter, K.G., 1996. Integrating the microbial loop and the classic food chanin into a planktonic foodweb. In: G.A. Polis and K.O. Winemiller (Eds). *Foodwebs: Integration of patterns and dynamics*, pp. 51–59. Kluwer Academic Publishers, Boston, MA.

Porter, K.G., Feig, Y.S., 1980. The use of DAPI for identification and enumeration of bacteria and blue-green algae. Limnol. Oceanogr. 25, 943–948.

Schindler, D.W., 1978. Factors regulating phytoplankton production and standing crop in the world's freshwaters. Limnol. Oceanogr. 23(3), 478–486.

Sherr, E.B., Sherr, B.F., Paffenhoefer, G.-A., 1986. Phagotrophic protozoa as food for metazoans: a 'missing' trophic link in marine pelagic foodwebs? Mar. Microb. Foodwebs 1, 61–80.

Sherr, E.B., Sherr, B.F., Hopkinson, C.S., 1988. Trophic interactions within pelagic microbial communities: indications of feedback regulation of carbon flow. Hydrobiologia 159, 19–26.

Sprules, W.G., Johannsson, O.E., Millard, E.S., Munawar, M., Stewart, D.S., Tyler, J., Dermott, R., Whipple, S.J., Legner, M., Morris, T.J., Ghan, D., Jech, J.M., 1999. Trophic transfer in Lake Erie: A whole foodweb modeling perspective. A

White Paper Submitted for the Great Lakes Modeling Summit. 1999 May 24–28. 42nd Conference of the International Association of Great Lakes Research. Cleveland, OH.

Strathman, R.R., 1967. Estimating the organic carbon content of phytoplankton from cell volume or plasma volume. Limnol. Oceanogr. 12, 411–418.

Strickland, J.D.H., Parsons, T.R., 1968. A practical handbook of seawater analysis. Bull. Fish. Res. Board Can. Ottawa, ON, Canada.

Turley, C.M., Newell, R.C., Robins, D.B., 1986. Survival strategies of two small marine ciliates and their role in regulating bacterial community structure under experimental conditions. Mar. Ecol. Prog. Ser. 33, 59–70.

Utermöhl, H., 1958. Zur vervolkommnung der quantitativen phytoplankton-methodik. (The improvement of quantitative phytoplankton methodology. In German.) Mitt. Internat. Verein. Limnol. 9, 1–38.

Vallentyne, J.R., 1974. the algal bowl: lakes and man. Government of Canada, Misc. Spec. Publ. 22.

Verity, P.G., Robertson, C.Y., Tronzo, C.R., Andrews, M.G., Nelson, J.R., Sieracki, M.E., 1992. Relationship between cell volume and the carbon and nitrogen content of marine photosynthetic nanoplankton. Limnol. Oceanogr. 37, 1434–1446.

Vollenweider, R.A., 1971. Scientific fundamentals of the eutrophication of lakes and flowing waters, with particular reference to nitrogen and phosphorus as factors in eutrophication. Organization for Economic Co-operation and Development, Paris, France.

Vollenweider, R.A., Munawar, M., Stadelmann, P., 1974. A comparative review of phytoplankton and primary production in the Laurentian Great Lakes. J. Fish. Res. Board Can. 31, 739–762.

Weisse, T., Munawar, M., 1989. Evaluation of the microbial loop in the North American Great Lakes. Can. Tech. Rep. Fish. Aquat. Sci. No. 1709.

Wylie, J.L. and Currie, D.J., 1991. The relative importance of bacteria and algae as food sources for crustacean zooplankton. Limnol. Oceanogr. 36(4), 708–728.

The above article is a reprinted version of the following previously published article:
The relative importance of autotrophic and heterotrophic microbial communities in the planktonic foodweb of the Bay of Quinte, Lake Ontario 2000–2007
M. Munawar; M. Fitzpatrick; H. Niblock; J. Lorimer
Aquatic Ecosystem Health & Management (2011) 14 (1): 21–32.
https://doi.org/10.1080/14634988.2011.551102

The phytoplankton of the Bay of Quinte, 1972–2008: Point-source phosphorus loading control, Dreissenid Mussel establishment, and a proposed community reference

Kenneth H. Nicholls[1]* and Elaine C. Carney[2]

[1]S-15 Concession 1, RR #1 Sunderland, ON, L0C 1H0
[2]12421 6th Line, RR #1, Limehouse, ON, L0P 1H0
**Corresponding author: khnicholls@interhop.net*

Keywords: long-term change, community similarity, ecosystem rehabilitation

Introduction

The phytoplankton of the Bay of Quinte has been investigated several times in the past 60 years (Tucker, 1948; McCombie, 1967; Nicholls and Carney, 1979, 1986; Nicholls and Heintsch, 1986; Nicholls et al., 1986). While much of this earlier work, including that reported under "Project Quinte" begun in 1972, was descriptive in scope, a preliminary attempt was made more recently using multivariate analyses (Nicholls et al., 2002), to determine the relative importance of changes in Bay of Quinte phytoplankton associated with the Point-source phosphorus loading controls of early 1978 and the mid-1990s establishment of Zebra and Quagga Mussels (*Dreissena* spp.).

Nicholls and Hurley (1989) provided evidence to suggest that the "top-down" foodweb effects of the White Perch die-off may well have had a greater impact on phytoplankton than the P loading reductions. Further testing of this hypothesis is not pursued here. Although the P loading reduction of the winter of 1977–1978 is used as a point in time for describing phytoplankton community change, the reader should recognize that a potentially major foodweb disruption occurred simultaneously with the P loading reduction, and that references herein to conditions "before and after Point-source P loading reductions" are made for convenience only and are not meant to suggest that effects on phytoplankton through

Ecology of the Bay of Quinte: Health, Management and Global Implications
Editors: C.K. Minns, M. Munawar and M.A. Koops

Aquatic Ecosystem Health & Management Society, Canada

the foodweb (so-called "top-down" effects) are necessarily less important than the P loading reductions.

This paper includes an update of the Nicholls et al. (2002) assessment for the upper bay, by including new data from nine more recent years and by extending the analysis to similar phytoplankton data sets collected from the middle and lower Bay of Quinte over the entire 37-year period of the project. Because Nicholls et al. (2002) had identified and assessed, with non-parametric statistical tests, the step-trends for several taxa associated with the P loading reductions in 1978, similar testing was not repeated here. Additionally, with the recent international emphasis on the inclusion of community structure and other biotic attributes in objective setting for aquatic ecosystem rehabilitation (Nielsen et al., 2003; Padisák et al., 2006), a proposal for defining the structure of an objective/reference/target phytoplankton community for the upper Bay of Quinte is advanced, furthering the initial work of Nicholls et al. (2004).

Methods

Sampling, laboratory and multivariate statistical methods

The phytoplankton sampling, analyses with inverted microscopes and multivariate statistical methods were as outlined in Nicholls et al. (2002, 2004). Some new information relevant to phytoplankton sample pooling or "recombining" is, however, important at this time.

Sample pooling (for purposes of analyzing a single sample representative of the entire May–October sampling period), for which 13 samples were consistently obtained for all of the "new" years reported here, 2000–2008, was done only for Station HB. Individual samples collected at biweekly intervals during 2000–2008 at Stations B and C were all analyzed as individual samples. An evaluation was done of this pooling or "recombination" procedure, which is intended to save on laboratory analytical costs, and included a total of 41 station-years representing samples from the four main Bay of Quinte stations: B and N in the upper bay, HB at the mouth of Hay Bay in the middle Bay of Quinte and C, near Conway in the lower bay. For each of these station-years, phytoplankton analyses were completed on both a single sample constituted from aliquots of all May–October samples for the given year, as well as on each of those samples, individually, and for which May– October means were calculated arithmetically from the individual sample results. The comparison of May to October means from the two methods revealed good agreement at the Division level with, for example, correlation coefficients for total phytoplankton, diatoms and blue-green algae of 0.966, 0.922 and 0.965, respectively. Agreement between the two methods was lower at the genus level (r = 0.21–0.75) and, in fact was not statistically significant for some rarely occurring genera. About one-half of all Bay of Quinte phytoplankton samples, 1972– 2008, were analyzed as pooled seasonal composites, with the exception of Station B, where 32 of the 37 years' samples were analyzed individually. For the Bay of Quinte pooled samples, enumeration errors were reduced by increasing the number of "pieces" or algal units enumerated, to totals in the 500–1000 range that were expected to yield adequate precision for the totals (Lund et al., 1958).

Phytoplankton taxonomy in Project Quinte has undergone minor revisions over the years with the splitting of several genera and species. We have, however, maintained the same major groupings in order to protect the continuity and integrity of the Project Quinte phytoplankton database, thus enabling year-to-year evaluation of changes; more details were provided in Nicholls et al. (2002). Generally, the laboratory methods have remained unchanged over the years. E.C. performed most of the analyses since the beginning of the project in 1972 and had trained others responsible for a minority of sample analyses.

For multivariate analyses, the 20 "bins" (individual genera, groups of genera or Classes/Divisions) used by Nicholls et al. (2004) were used here as follows: (i) total cryptophytes, (ii) total chlorophytes (incl. tribophytes, prasinophytes, and raphidophytes), (iii) total dinophytes, (iv) total euglenophytes, (v) total chrysophytes (incl. haptophytes), (vi) *Anabaena*, (vii) *Aphanizomenon/Cuspidothrix/Raphidiopsis*, (viii) *Gloeotrichia*, (ix) *Microcystis*, (x) *Lyngbya/Planktolyngbya/Leptolyngbya*, (xi) *Oscillatoria/Planktothrix/Limnothrix/Phormidium/Pseudanabaena*, (xii) *Gomphosphaeria/Coelomoron/Snowella*, (xiii) *Coelosphaerium/Woron ichinia*, (xiv) all remaining Cyanoprokaryota, (xv) *Stephanodiscus/Cyclostephanos*, (xvi) *Synedra/Ctenophora/Oxyneis*, (xvii) *Tabellaria*, (xviii) *Aulacoseira/Melosira*, (xix) *Cyclotella/Stephanocyclus*, and finally, (xx) all remaining diatoms.

Every phytoplankton taxon encountered during the phytoplankton enumerations completed over the 1972–2008 period was included in the multivariate analyses either in its own genus bin or a bin that grouped it with other related taxa. No taxon was double-counted (e.g. a diatom assigned to a genus bin was not included in e.g. the "remaining diatoms" bin).

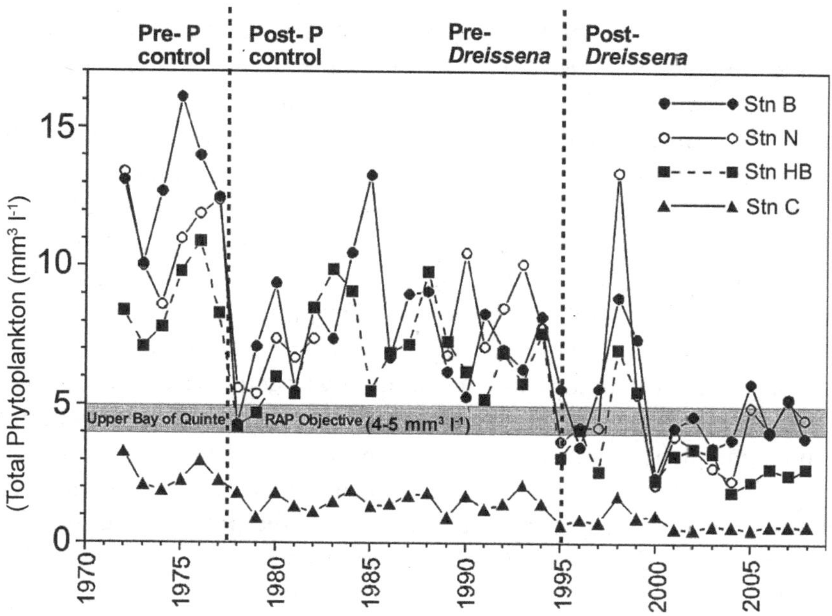

Figure 1. May to October average total phytoplankton biomass at the four main Bay of Quinte sampling stations, 1972–2008. The Remedial Action Plan (RAP) objective of 4–5 $mm^3 l^{-1}$ for the upper bay (represented by Stations B and N) is also shown. No samples were collected from Station N from 1983–1988, inclusive.

The terms "biovolume" and "biomass" are used here interchangeably, assuming unit density for planktonic algal cells; i.e. 1 mm^3 has a mass of 1 mg.

Synthesis of a reference community structure

We suggest that a useful way to evaluate long-term changes in phytoplankton community structure of the upper bay (Stations B and N) is to measure its similarity with a reference community that is considered more desirable than that which existed pre-P control (before 1978). Furthermore, this reference community likely should not be based on phytoplankton communities in existence after the arrival of Zebra Mussels owing to an associated apparent loss of certain desirable algal species and a domination by potentially toxic *Microcystis*, as seen in the Bay of Quinte and elsewhere in the Great Lakes (Vanderploeg et al., 2001; Nicholls et al., 2002). Other criteria used in constructing this reference phytoplankton community included (1) selection of nearby locations, thus ensuring local geology, basic mineral chemistry and land uses similar to those influencing the upper Bay of Quinte (Ecologistics, 1988), and (2) selection of shallow, well-mixed and highly flushed systems with hydrologic and physical characteristics comparable to those present in the upper bay between Belleville and Napanee, but with lower Point-source loads of total P. The rationale for developing a reference phytoplankton community derived from the phytoplankton communities of West Lake, the inner Bay of Quinte near Trenton (old Station T of Project Quinte), and upper Hay Bay (old station U-HB of Project Quinte; Nicholls, 1999) was based on the belief that no single location can best serve this purpose. The averaging of phytoplankton community structures at these locations during 1978–1987 has generated a synthetic phytoplankton community with some blended attributes of those existing at all three locations that are believed to typify a less degraded ecosystem than that characterizing the upper bay (Stations B and N) during the pre-1978 period. The rationale for including West Lake as a component of this reference was detailed in Nicholls et al. (2004) and will not be repeated here to save space. The inner bay near Trenton and the upper Hay Bay location meet all of the criteria outlined above for a reference site. In particular, point source loading of total P to the Trenton area in the inner bay (averaging about 10 kg d^{-1}) was considerably lower than the Point-source loading influencing phytoplankton at Stations B and N, where additional inputs from Belleville, Deseronto and Napanee totaled about 55 kg d^{-1} during the period 1978–1986 (Owen, 1989). Similarly, the upper Hay Bay location experienced insignificant Point-source loads, being located in a predominantly forested and rural agricultural watershed.

May–October means for all taxa were calculated for each of the three locations listed above, which were themselves averaged to produce a "master reference row" of the 20 column data matrix, thus constituting the synthesized reference phytoplankton community structure. (Note—the mean May–October phytoplankton biomasses at the inner bay (Station T), upper Hay Bay and West Lake were 4.03, 4.93 and 1.11, respectively. Had the mean phytoplankton biomass for any of these three locations been above 5 $mm^3 l^{-1}$ (the upper limit of the RAP upper bay phytoplankton objective), that might have rendered it inadmissible as a candidate reference location).

The SIMPER routine [SIMilarity PERcentages (PRIMER software, as described by Clarke and Warwick (1994)] was used to determine which taxa contributed most to the between-treatment dissimilarities (where "treatments," were the pre- and post-intervention community structures, with the division points being 1978 for P control and 1995 for the *Dreissena* intervention. Significant difference between treatments was determined by the ANOSIM routine (ANalysis Of SIMilarities) - run in PAST (Hammer et. al., 2001) and PRIMER (Carr, 1997; Clarke and Warwick, 1994).

In order to facilitate statistical analysis of difference between the reference community and any other multi-year group of Bay of Quinte phytoplankton communities, the "master reference" row of the data matrix needed to be replicated in a fashion that provided a level of within-group variation similar to that inherent in the treatment group to which it was intended to be compared (e.g. Station B, 1972–1978 or 1978–1994). This is because the measure of statistical difference between two groups of communities (i.e. treatment and reference groups) is determined by the ANOSIM procedure, which is based on the differences between the within-group community similarities and the between-group similarities. A reference group of community similarities that has either much greater or much lower within-group similarity values than those of the treatment group will bias the test of significant difference between groups towards either statistical non-significance (if variance is high) or significance (if variance is low). The adjustment of the within-group variance in community similarity values of the reference group was achieved with 10 sets of 20 random numbers [generated in Modstat (Knodt, 1999)] over the range 0.25 to 4.0 that were, in turn, applied to the "master reference row", thus generating 10 new rows of reference communities in the data matrix. Because the Remedial Action Plan for the Bay of Quinte had set an upper bay May–October average phytoplankton biomass rehabilitation target of 5 $mm^3 l^{-1}$, each of the 20 variables included in the now 10 rows of the reference data matrix was multiplied by a single, empirically derived factor (0.61) that ensured that the highest total phytoplankton biomass value of the 10 reference communities did not exceed 5 $mm^3 l^{-1}$. As a consequence, the 10 generated reference means ranged between 2.80 and 4.98 with an average of 4.15 $mm^3 l^{-1}$.

Results

Trends in major components

May to October average total phytoplankton biomass has fluctuated greatly over the 37-year record of Project Quinte. Dramatic declines immediately following Point-source phosphorus loading reductions in the winter of

1977–1978 at the upper and middle bay sampling stations were followed by increases again during the 1980s with total biomass values occasionally comparable to the pre-P control years (Figure 1). More recently, similar sharp declines in total phytoplankton were experienced immediately following establishment of Dreissenid Mussel species in the Bay of Quinte, but, again, there was a rapid, but brief, increase in the late 1990s to levels similar to those seen pre-*Dreissena*. Thereafter, levels were lower and more stable with May–October biomass at Stations B and N averaging at about the level of the Remedial Action Plan Objective of 4–5 $mm^3 l^{-1}$, and Station HB in the middle bay somewhat lower, in the 2–3 $mm^3 l^{-1}$ range (Figure 1). May-October mean phytoplankton biomass at Station C in the lower bay has declined nearly 80%, from a 1972–1977 (pre-phosphorus control) mean of 2.5 $mm^3 l^{-1}$ to a 2003–2008 value of 0.56 $mm^3 l^{-1}$.

Of the 6 or 7 major algal Divisions routinely encountered in the Bay of Quinte, by far the most important have been the diatoms (Bacillariophyta) and the blue-green algae (Cyanoprokaryota). The combined contribution from these two groups has routinely been between about 75 and 95% of the total phytoplankton biomass in the upper bay, and a slightly lower contribution in the middle bay, but a considerably lower contribution in the lower bay, where cryptomonads, dinoflagellates, chlorophytes, and chrysophytes and haptophytes assumed greater relative importance than in the upper and middle Bay of Quinte (as reported previously by Nicholls et al. (2002), but not illustrated here for the more recent data because of space limitations).

Of the blue-green algae, N-fixing species (predominantly of the genera *Anabaena*, *Aphanizomenon* and *Gloeotrichia*) have been an important part of the total community. Only in 1998 and 2008 was biomass of N-fixers less than 50%. During both of these years, and more generally, after the establishment of Dreissenid Mussels, the non-N-fixing genus *Microcystis* was atypically abundant, especially at the middle and lower bay locations (Figure 2f). Most other algal genera, especially in the middle and lower bay, registered well-defined reductions in biomass associated with the establishment of *Dreissena*, including bloom-forming *Anabaena*, *Aphanizomenon*, *Lyngbya*, *Oscillatoria* and *Coelosphaerium* (Figures 2a–e, respectively).

The dominant diatom genera *Aulacoseira* and *Stephanodiscus* generally declined over the long-term (Figures 2g and h), but the clearest example of "steps" associated with the interventions of 1978 and 1995 were in the upper bay where 1972–1977 May to October average biomass of *Stephanodiscus* of 1.26 $mm^3 l^{-1}$ fell to 0.46 $mm^3 l^{-1}$ for the 1978–1994 period, and 0.13 $mm^3 l^{-1}$ during 1995–2008. Biomass values for *Synedra* spp., and especially *Tabellaria* spp., have declined drastically since the establishment of *Dreissena* (Figures 2i and j).

Multivariate community structure

Based on 20 taxonomic input variables consisting of May to October averages of individual genera and groups of genera recorded in the Station B samples (see listing in Methods section), all years from 1972–2008 were correctly classified in a dendrogram into the three time periods surrounding the two interventions - Point-source P control and *Dreissena* establishment (cophenetic correlation coefficient (ccc) = 0.759). Also, not shown here to save space, were dendrograms for Station HB with 100% classification of the years into the three time blocks (ccc = 0.883), and for Station C, wherein just one year (1990) was "incorrectly" classified with the 1972–1977 group (ccc = 0.854).

Long-term change in the Bay of Quinte phytoplankton community structure was demonstrated well by a non-metric multidimensional scaling ordination of the stations representing the upper (Stn B), middle (Stn HB) and lower (Stn C) bay after summarizing as group means representing each of the time periods referred to above. Clearly, the largest changes in community structure occurred after establishment of Dreissenid Mussels (Figure 3).

The relative declines in community similarity were greater after the *Dreissena* intervention than after the P loading intervention; the opposite was true for biomass. For example, at Station B, the percentage change in phytoplankton biomass and in community similarity associated with the P loading reduction were −47% and −1.5%, respectively, but biomass declined 21% and community similarity declined 3.7% after *Dreissena* establishment.

To facilitate a more direct comparison of these "before and after" (treatment) differences, community similarity coefficients were calculated for time periods representing the two interventions, but after restricting the number of years representing each treatment to six (this number set by the available data before phosphorus load reduction, 1972–1977). Within treatments, there was a high degree of community similarity (89.3–92.8% for the P loading intervention) at all three (upper, middle, and lower) bay locations. Within treatment similarities were lower (86.1–91.8%) before and after *Dreissena* establishment (Table 1). In all cases (both interventions, all three stations), the phytoplankton community similarities were lower after the intervention, than before the intervention, as were the phytoplankton biomass values.

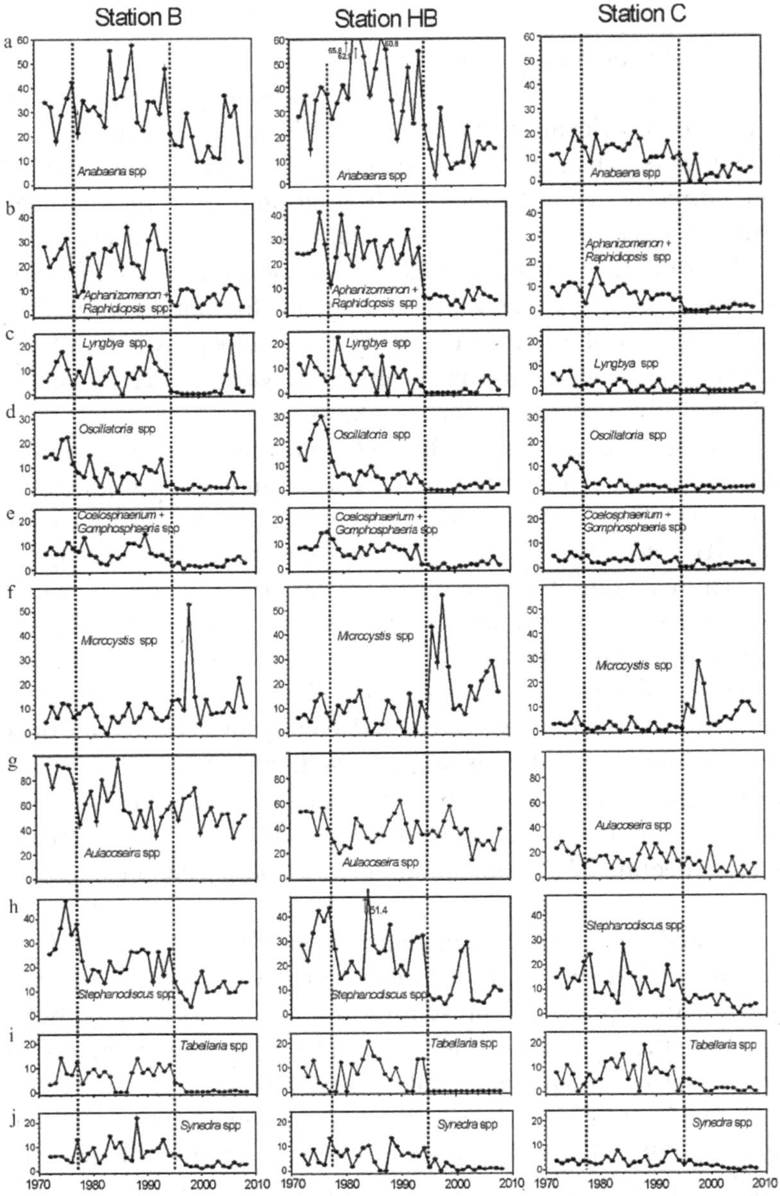

Figure 2. Changes in May–October mean biomass (square root transformed) of commonly occurring genera of blue-green algae (a-f) and diatoms (g-j) in

the upper, middle and lower Bay of Quinte (Stations B, HB and C, respectively). The units for each are $[mm^3\ m^{-3}]^{1/2}$ The vertical dashed lines coincide with the approximately 50% reduction in Point-source phosphorus loading to the upper bay in the winter of 1977-1978 and the establishment of Dreissenid Mussels in the mid-1990s. The genus names listed are representative, but not exclusive (see Methods section; for example, "Lyngbya" includes Planktolyngbya and Leptolyngbya). Similar patterns were found in the Station HB and Station C data (Table 1).

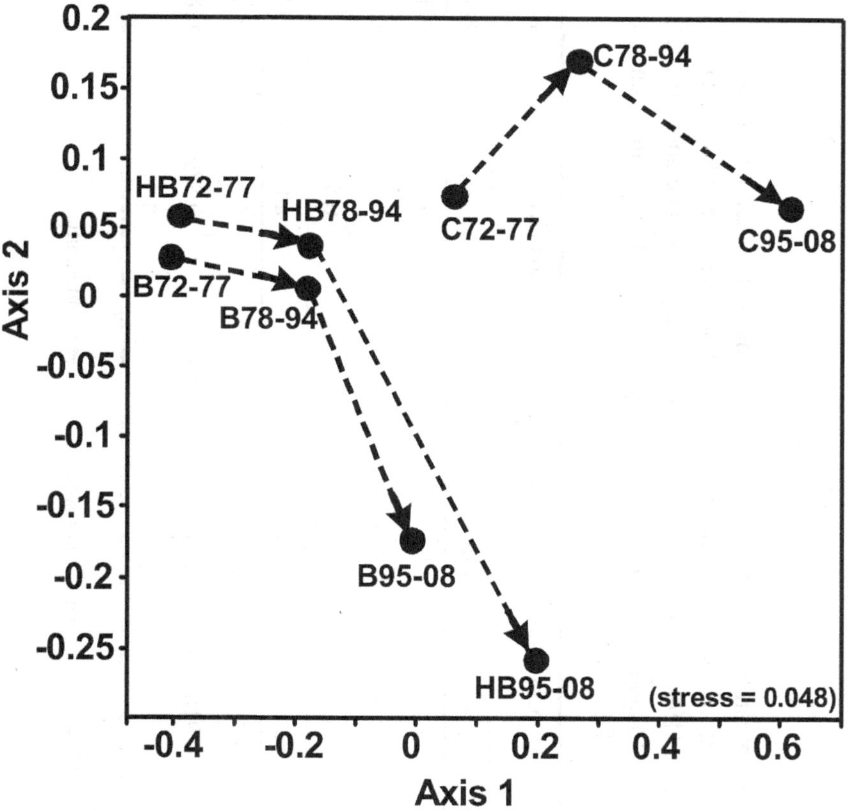

Figure 3. Non-metric multidimensional scaling ordination of Bray-Curtis phytoplankton community similarity coefficients grouped into before and after time blocks associated with the Point-source phosphorus loading reduction (winter of 1977–1978) and the Dreissenid Mussel establishment in the mid-1990s. Three regions (upper, middle and lower Bay of Quinte) are represented by Stations B, HB and C, respectively.

Table 1. Comparison of 6-year mean (within group) total phytoplankton biomass and mean community composition (Bray-Curtis similarity coefficients[1]), before and after Point-source phosphorus loading reductions (1972–1977 vs. 1978–1983) and Dreissend Mussel establishment (1989–1994 vs. 1996–2001) in the upper, middle and lower Bay of Quinte. Also shown is the average between-group phytoplankton community similarity for each of the two interventions (i.e. average of each of the "before" years paired with each of the "after" years).

	Upper (Station B)		Middle (Station HB)		Lower (Station C)	
P load reduction	before	after	before	after	before	after
Biomass (mg l^{-1})	13.1	6.91	8.93	6.3	2.47	1.42
Within-similarity	0.928	0.915	0.9143	0.905	0.92	0.893
Between-similarity		0.859		0.841		0.822
Dreissena establishment	before	after	before	after	before	after
Biomass	6.76	5.33	6.48	4.13	1.46	0.93
Within-similarity	0.918	0.884	0.8653	0.879	0.871	0.861
Between-similarity		0.79		0.747		0.762

[1] Similarity coefficients closer to 1.0 indicate a higher level of similarity between communities.

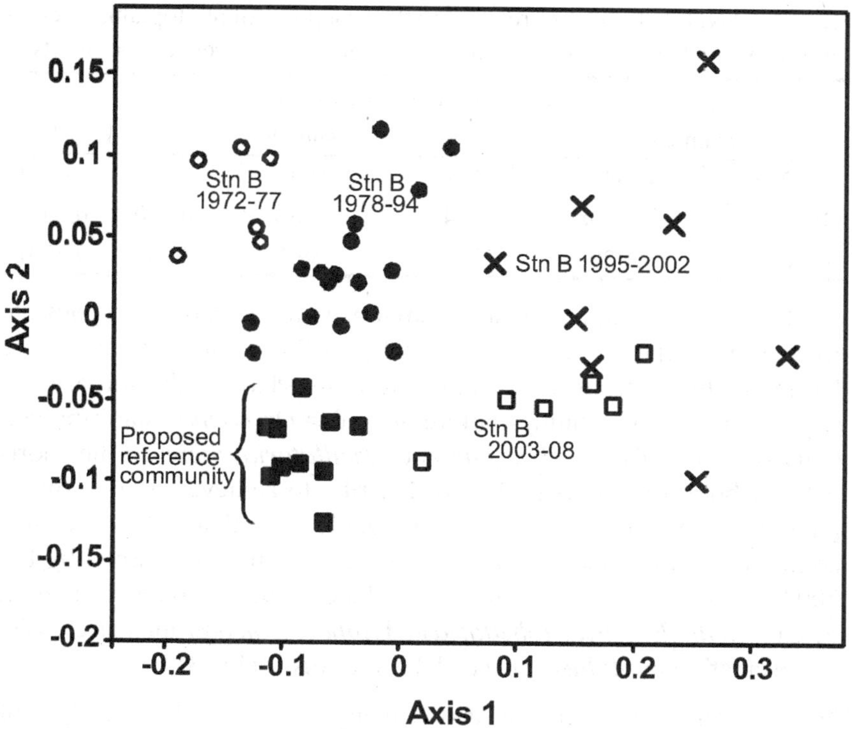

Figure 4. Non-metric multidimensional scaling ordination of Bay of Quinte Station B Bray-Curtis phytoplankton community similarity coefficients grouped as per Figure 3, and including 10 synthetically derived reference phytoplankton associations (see Methods for details).

There are three aspects of the phytoplankton structural changes over time that were well visualized in the NMDS ordination of the upper bay phytoplankton (Figure 4): (1) community structure changed suddenly in association with both the P-loading and the Dreissenid Mussel interventions, (2) within each of the three treatment periods, (i) pre-P control, (ii) post-P control/pre-*Dreissena* and (iii) post-*Dreissena*, phytoplankton structure remained within boundaries distinctly separated from other treatment boundaries, and (3) the inter-annual variability increased progressively through the treatment periods ((i), (ii) and (iii), listed above). ANOSIM results confirmed the statistical distinctiveness of each of the treatments (Table 2), and in particular that the proposed reference community structure was statistically different from the upper bay phytoplankton of recent years (2003–2008).

Table 2. ANOSIM statistics relating to the groupings of phytoplankton community-years (treatments), including the proposed Reference community.

Treatment	within-group mean rank	between-group mean rank	r	*P*
1972–77 vs. 1978–94	90.6	180.9	0.715	<0.0001
1978–94 vs. 1995–2008	135.4	326.1	0.820	<0.0001
Reference vs. 2003–2008	32.0	89.1	0.952	<0.0001

It is important to note that the phytoplankton taxa most responsible for the dissimilarity between pre- and post-P control treatments were largely different from those contributing most to the dissimilarity in phytoplankton communities before and after *Dreissena* establishment. Only two taxa (*Coelosphaerium* and *Oscillatoria*) were on the short-lists for both treatments (Table 3). In order to achieve an "improved" phytoplankton community structure (i.e. closer to that of the reference community), the contemporary upper bay structure as defined by the 2003–2008 community, will need to include an increased representation by *Cyclotella, Lyngbya, Tabellaria* and *Aphanizomenon* and a decreased representation by *Aulacoseira* and *Microcystis* (Table 3).

Table 3. The average phytoplankton biomass (Station B, upper Bay of Quinte) of the most influential taxa (ranked and combined contributions totaling at least 50% of the total between-group dissimilarity [SIMPER analyses]) and their cumulative percentage contribution to the total dissimilarity between groups (treatments).

Taxon	1972–1977 (#1) Av. Biomass ($mm^3\ m^{-3}$)	1978–1994 (#2) Av. Biomass ($mm^3\ m^{-3}$)	Cumulative percentage
Gomphosphaeria	42.9	< 0.01	11.81
Aulacoseira	7512.7	3318.7	20.93
Coelosphaerium	0.5	41.4	29.83
Oscillatoria	271.7	41.7	37.68
Stephanodiscus	1187.3	428.6	44.37
Dinophyta	410.1	120	51.02
Lyngbya	58.9	1.0	9.73
Aphanizomenon	500.5	45.7	19.43
Tabellaria	32.1	0.02	28.85

Taxon	1972–1977 (#1) Av. Biomass (mm³ m⁻³)	1978–1994 (#2) Av. Biomass (mm³ m⁻³)	Cumulative percentage
Anabaena	1171.2	317.1	36.79
Coelosphaerium	41.4	0.5	44.25
Oscillatoria	41.7	3.6	50.36
Aphanizomenon	676.5	56.4	11.09
Tabellaria	37.2	< 0.01	21.69
Aulacoseira	576.5	2138.1	30.61
Lyngbya	71.7	13.6	38.69
Microcystis	19.1	132.1	44.69
Cyclotella	124.4	22.6	50.67

Discussion

Several studies have documented declines in phytoplankton following the establishment of Dreissenid Mussels in the Great Lakes (Nicholls and Hopkins, 1993; Holland, 1993; Barbiero et al., 2006). In a preliminary assessment of the relative changes in Bay of Quinte phytoplankton associated with P load reduction and *Dreissena* establishment, Nicholls et al. (2002; their Table 4) organized the phytoplankton response in five categories. Their conclusions were based on statistical tests of differences pre- and post-intervention; those tests were not repeated here because only the data for the post-*Dreissena* period have changed (increased to 14 years from the five years available to Nicholls et al., 2002). The general conclusion from the additional nine years of phytoplankton data from Station B in the upper bay (and largely reinforced by the phytoplankton data summaries for Station HB in the middle bay) is that the five previously identified "response categories" appear still to be valid. In particular, the post-*Dreissena* increases in *Microcystis* (the sole occupant of Category 5—taxa that "did not change after P control, but increased after *Dreissena* establishment"), were dramatic and especially evident in the middle and lower bay. The taxa declining most after the arrival of Dreissenid Mussels included *Aphanizomenon, Tabellaria, Synedra* and *Oscillatoria.* Because some of these changes have been observed at other Great Lakes locations (e.g. Vanderploeg et al. (2001) re *Microcystis* increase), it suggests that at least some of the ecosystem changes associated with *Dreissena* establishment are perhaps reproducible and predictable. Declines in many phytoplankton taxa can be explained by the direct effects of the filter-feeding habit of *Dreissena* spp., but increases in other taxa offer opportunities for research on the possibly secondary ecosystem influences of *Dreissena* as modifiers of water column light climate, and of nutrient availability. Also of significance in the updated findings from the Bay of Quinte is the observation that post-*Dreissena* inter-annual phytoplankton stability (as measured by year-to-year community percentage similarities) declined to levels not seen in the pre-*Dreissena* era. This may support the idea that an increase inter-annual instability (as inferred from increased variance in community structure) might be a useful indicator of stress (e.g. Brock and Carpenter, 2006).

As part of the Remedial Action Plan (RAP) for the rehabilitation of the 42 Areas of Concern (AOC's) identified in the North American (Laurentian) Great Lakes, specific water quality and aquatic ecosystem objectives (so-call "delisting criteria") for each AOC were identified. For the Bay of Quinte, the only quantitative phytoplankton objective was a May to October mean phytoplankton biovolume of 4–5 $mm^3 l^{-1}$ for the upper bay (Bay of Quinte RAP, 1993). No quantitative objectives relating to phytoplankton community composition were identified, beyond a general statement relating to the undesirability of algal species inducing potential toxicity and undesirable tastes and odours to municipal water supplies intended for human consumption. Using the Bay of Quinte as one of two test cases, Nicholls et al. (2004) explored a multivariate statistical approach to define a target or objective phytoplankton community based on phytoplankton data from nearby West Lake. The unexpected arrival in the Great Lakes (including the Bay of Quinte) of *Dreissena* spp. (Zebra and Quagga Mussels) introduced another level of uncertainty in attempts to interpret change in the Bay of Quinte phytoplankton community (Nicholls et al., 2002).

The large changes in the Bay of Quinte phytoplankton community structure over the past many years beg the question: Is there direction to this change, and, if so, is the direction desirable (from trophic efficiency and ecosystem health perspectives)? Certainly, some of the changes associated with *Dreissena* establishment have been desirable (e.g. the decline in *Aphanizomenon*), but others such as the large increase in *Microcystis* are disturbing, as it relates to potential toxin production and the associated loss of more desirable foodweb functional species. Key elements of aquatic ecosystem rehabilitation must include objectives for biotic communities. Such objectives must be ecologically sensible and practical. Management objectives are well documented for Great Lakes fish communities (DesJardines et al., 1995), and framework proposals have been discussed for benthic communities (Reynoldson et al., 1997).

Specific objectives for Great Lakes zooplankton and phytoplankton communities, however, are in the early stages of evolution. Initiatives under the European Union's Water Framework Directive are at a similar early stage of development for these lower trophic level communities (Padisák et al., 2006).

We are reluctant to use the term "target" or "objective" for the proposed upper Bay of Quinte reference phytoplankton community because it suggests an end-point that is actively sought through "ecosystem engineering" and aggressive environmental management initiatives. While such approaches may be necessary at some future date, they have not been advocated under Project Quinte's science plan or within the Bay of Quinte RAP. Instead, Project Quinte has been more oriented towards long-term ecosystem monitoring and the resulting enhancement of understanding of aquatic ecosystem function. It is in this context that the proposed phytoplankton community reference has some value; it allows an easily visualized portrayal of community change over time that can be readily interpreted by the lay public in terms such as "improving" or "deteriorating."

Summary

The objective of this contribution is to further the understanding of long-term (37 years) changes in the composition and biomass of the phytoplankton of the Bay of Quinte (northeastern Lake Ontario), especially those changes associated with the simultaneous Point-source phosphorus loading reduction/White Perch winter kill of early 1978 and the establishment of Dreissenid Mussels in the mid-1990s. The relatively shallow and polymictic upper bay has facilitated the ice-free period domination of the phytoplankton by meroplanktonic diatoms (especially *Aulacoseira* spp.); while a more balanced representation by several algal Divisions has characterized the thermally stratified, dimictic lower bay. At all three stations (upper bay, middle bay and lower bay), phytoplankton biomass declined and community similarity decreased after both the phosphorus loading and the *Dreissena* interventions, but the biomass changes associated with the P load reduction were greater than those associated with *Dreissena* establishment. Conversely, the loss of phytoplankton community similarity after *Dreissena* establishment was greater than that associated with P loading reduction at all three stations. The Remedial Action Plan phytoplankton objective of 4–5 $mm^3\ l^{-1}$ (May-October mean) has been met inconsistently since the establishment of *Dreissena*. The post-*Dreissena* period, however, was also characterized by occasional very high biomass values for the potentially toxic cyanoprokaryote (blue-green alga) *Microcystis*, as well as by a dramatic decline in the bloom-forming blue-green *Aphanizomenon*, and the near extirpation of the diatoms, *Tabellaria* and *Synedra* spp. A partially synthetic phytoplankton community was constructed using data taken from three local aquatic systems (Trenton Bay, upper Hay Bay and West Lake). This might be used as a reference against which past and future changes in the upper Bay of Quinte phytoplankton can be compared and evaluated.

Acknowledgements

We are grateful for the helpful suggestions provided by two anonymous referees. This work was funded by the Ontario Ministry of the Environment, 1972—1999, and by Fisheries and Oceans Canada (Great Lakes Action Plan), 2000–2009.

References

Barbiero, R. P., Rockwell, D. C., Warren, G. J., Tuchman, M. C., 2006. Changes in spring phytoplankton communities and nutrient dynamics in the eastern basin of Lake Erie since the invasion of *Dreissena* spp. Can. J. Fish. Aquat. Sci. 63, 1549–1563.

Bay of Quinte RAP, 1993. Time to Act. The Bay of Quinte Remedial Action Plan Stage 2 Report. Bay of Quinte RAP Coordinating Committee, September, 1993. Ontario Ministry of the Environment, Kingston, Ontario.

Brock, W. A., Carpenter, S. R., 2006. Variance as a leading indicator of regime shift in ecosystem services. Ecology and Society 11(2), Article 9. http://www.ecologyandsociety.org/vol11/iss2/art9/

Carr, M. R., 1997. *PRIMER* User Manual (Plymouth Routines in Multivariate Ecological Research). Plymouth Marine Laboratory. Plymouth, UK. 27.

Clarke, K. R., Warwick, R. M., 1994. *Change in marine communities: an approach to statistical analysis and interpretation*. National Environment Research Council, UK.

DesJardine, R. L., Gorenflo, T. K., Payne, R. N., Schrouder, J. D., 1995. Fish community objectives for Lake Huron. Great Lakes Fishery Commission. Special Publication No. 95–1, Ann Arbor.

Ecologistics Ltd., 1988. Analysis of Agricultural Diffuse Source Loadings to the Bay of Quinte. Report to the Bay of Quinte Remedial Action Plan Coordinating Committee, September, 1988.

Hammer, Ø., Harper, D. A. T., Ryan, P. D., 2001. PAST: Paleontological Statistics Software Package for education and data analysis. Palaeontologia Electronica 4(1), 1–9.

Holland, R. E., 1993. Changes in planktonic diatoms and transparency in Hatchery Bay, Bass Island area, western Lake Erie since the establishment of the zebra mussel. J. Great Lakes Res. 19, 617–624.

Knodt, R. C., 1999. *MODSTAT - version 12. Copyright 1999.* Statware, Bellingham, WA.

Lund, J. W. G., Kipling, C., LeCren, E. D., 1958. The inverted microscope method of estimating algal numbers and the statistical basis of estimation by counting. Hydrobiologia 11, 143–170.

McCombie, A. M., 1967. A recent study of the phytoplankton of the Bay of Quinte, 1963–1964. Proc. 10th Conf. Great Lakes Res., Internat. Assoc. Great Lakes Res. 1967, 37–62.

Nicholls, K. H., 1999. Effects of temperature and other factors on summer phosphorus in the Bay of Quinte, Lake Ontario: implications for climate warming. J. Great Lakes Res. 25, 250–262.

Nicholls, K. H., Carney, E. C., 1979. The taxonomy of the Bay of Quinte phytoplankton and the relative importance of common and rare taxa. Can. J. Bot. 57, 1591–1608.

Nicholls, K. H., Carney, E. C., 1986. Nitrogen and phosphorus limitation to phytoplankton in the Bay of Quinte and implications for phosphorus loading controls. pp.139–144. In: C.K. Minns, C.K., Hurley, D.A., Nicholls, K.H. (Eds), *Project Quinte: point source phosphorus control and ecosystem response in the Bay of Quinte, Lake Ontario.* Can. Spec. Publ. Fish. Aquat. Sci. 86.

Nicholls, K. H., Heintsch, L., 1986. A comparison of the phytoplankton in the lower Bay of Quinte near Indian Point, 1945 and 1981, pp.159–165. In: C.K. Minns, C.K., Hurley, D.A., Nicholls, K.H. (Eds), *Project Quinte: point source phosphorus control and ecosystem response in the Bay of Quinte, Lake Ontario.* Can. Spec. Publ. Fish. Aquat. Sci. 86.

Nicholls, K. H., Hurley, D. A., 1989. Recent changes in the phytoplankton of the Bay of Quinte: the relative importance of fish, nutrients and other factors. Can. J. Fish. Aquat. Sci. 46, 770–779.

Nicholls, K. H., Heintsch, L., Carney, E. C., Beaver, J. L., Middleton, D., 1986. Some effects of phosphorus loading reductions on phytoplankton in the Bay of Quinte, Lake Ontario, pp.145–158. In: C.K. Minns, C.K., Hurley, D.A., Nicholls, K.H. [Eds], *Project Quinte: point source phosphorus control and ecosystem response in the Bay of Quinte, Lake Ontario.* Can. Spec. Publ. Fish. Aquat. Sci. 86.

Nicholls, K. H., Heintsch, L., Carney, E. C., 2002. Univariate step-trend and multivariate assessments of the apparent relative effects of P loading reductions and zebra mussels on the phytoplankton of the Bay of Quinte, Lake Ontario. J. Great Lakes Res. 28, 15–31.

Nicholls, K. H., Heintsch, L., Carney, E. C., 2004. A multivariate approach for evaluating progress towards phytoplankton community restoration targets: examples from eutrophication and acidification case histories. Aquatic Ecosystem Health and Management 7(1), 15–30.

Nicholls, K. H., Hopkins, G. J., 1993. Recent changes in Lake Erie (North Shore) phytoplankton: cumulative impacts of phosphorus loading reductions and the zebra mussel introduction. J. Great Lakes Res. 19, 637–647.

Nielsen, K., Sømod, B., Ellegaard, C., Krause-Jensen, D., 2003. Assessing reference conditions according to the European Water Framework Directive using modelling and analysis of historical data: an example from Randers Fjord, Denmark. Ambio 32, 287–295.

Owen, G., 1989. Point-source phosphorus loading. 1989 Project Quinte Annual Report, pp. 10–11. Bay of Quinte Remedial Action Plan, Kingston, ON, December, 1990.

Padisák, J., Borics, G., Grigorszky, I., Soróczki-Pintér, É., 2006. Use of phytoplankton assemblages for monitoring ecological status of lakes within the Water Framework Directive: the assemblage index. Hydrobiologia 553, 1–14.

Reynoldson, T. B., Norris, R. H., Resh, V. H., Day, K. E., Rosenberg, D. M., 1997. The reference condition: a comparison of multimetric and multivariate approaches to assess waterquality impairment using benthic macroinvertebrates. J. N. Am. Benthol. Soc. 16, 833–852.

Tucker, A., 1948. The phytoplankton of the Bay of Quinte. Trans. Amer. Microsc. Soc. 67, 365–383.

Vanderploeg, H. A., Liebig, J. R., Carmichael, W. W., Agy, M. A., Johengen, T. H., Fahnenstiel, G. L., Nalepa, T. F., 2001. Zebra mussel (*Dreissena polymorpha*) selective filtration promoted toxic *Microcystis* blooms in Saginaw Bay (Lake Huron) and Lake Erie. Can. J. Fish. Aquat. Sci. 58, 1208–1221.

The above article is a reprinted version of the following previously published article:

The phytoplankton of the Bay of Quinte, 1972–2008: point-source phosphorus loading control, Dreissenid Mussel establishment, and a proposed community reference
Kenneth H. Nicholls; Elaine C. Carney
Aquatic Ecosystem Health & Management (2011) 14 (1): 33–43.
https://doi.org/10.1080/14634988.2011.550847

Ciliate ecology in the eutrophic Bay of Quinte, Lake Ontario: community structure and feeding characteristics

M. Munawar,[1] H. Niblock,[1] M. Fitzpatrick[1]* and J. Lorimer[2]

[1]Fisheries and Oceans Canada, 867 Lakeshore Road, Burlington, ON L7S 1A1 [2]Aquatic Ecosystem Health and Management Society, Burlington, ON
**Corresponding author: mark.fitzpatrick@dfo-mpo.gc.ca*

Dedication

Dedicated to Denis H. Lynn, 1947 – 2018. The current manuscript began as a poster presented at a special session of the 53rd Annual Conference on Great Lakes Research titled, *Ecosystem Health and Recovery in the Bay of Quinte*. That session would inspire two special issues of Aquatic Ecosystem Health and Management, but it also coincided with Denis' retirement from the University of Guelph and subsequent move to British Columbia. A draft manuscript based on our poster was developed with input from Denis, but unfortunately never completed. When news of Denis' passing reached us, our thoughts quickly turned to how we could best honour him and completing this paper seemed obvious. We dedicate this paper to Denis in recognition of his contributions to ciliate ecology in the Great Lakes.

Keywords: Ciliophora, protozoa, microbial ecology, eutrophication, trophic interactions

Introduction

Ciliates are relatively small protozoa that occur in both terrestrial and aquatic habitats. Within aquatic ecosystems, ciliates are known to feed on bacteria and smaller phytoplankton and, in turn, may be grazed by larger zooplankton and planktivorous fish (e.g. Porter, 1979; Azam et al. 1983; Montagnes et al. 1988, Sherr and Sherr, 1994). Perhaps due to their smaller size, these micro-zooplankton tend to be overlooked in studies of the broader plankton community. With sizes typically in the range of

Ecology of the Bay of Quinte: Health, Management and Global Implications
Editors: C.K. Minns, M. Munawar and M.A. Koops

Aquatic Ecosystem Health & Management Society, Canada

10 – 80 μm, they are still smaller than mesh sizes on standard plankton nets (164 μm) and require high powered microscopy for evaluation (e.g. Montagnes et al. 1988). In the North American Great Lakes, studies began to develop in the mid 1980s which attempted to quantify the ciliate community in terms of biomass, composition, growth rates and feeding behaviour (e.g. Carrick and Fahnenstiel 1990; Taylor and Johannsson, 1991; Lynn and Munawar, 1999) and which added to the complexity of foodweb models being developed and provided an opportunity for more holistic foodweb assessments.

The ciliate community represents a fairly diverse group of microfauna. In Lake Michigan for example, 7 distinct Ciliophora taxa were identified (Carrick, 2005). On a Great Lakes wide basis, Munawar and Lynn (2002) reported that the most common genera observed throughout the (North American) Great Lakes were *Askenasia, Halteria, Balanion, Strobilidium* and *Urotricha*. Within two of the largest Great Lakes (Huron and Michigan), ciliates were estimated to have turnover rates of 1-2 d^{-1} and contributed approximately 65% of the protozoan biomass and 30% of the total zooplankton biomass (Carrick and Fahnenstiel, 1990; Carrick, 2005). Similar observations have also been reported from temperate European lakes (e.g. Weisse and Müller, 1998; Lischke et al., 2016). In contrast, our work in the offshore waters of Lake Ontario found that ciliates accounted for less than 10% of the (heterotrophic) protozoan biomass and hypothesized that ciliates were out-competed for food resources by heterotrophic nanoflagellates (hereafter HNF; Munawar et al., 2010).

Perhaps due to their high growth rates and differing feeding strategies, ciliates can exert strong top down effects and/or be subject to strong bottom up effects depending on the environment (e.g. Beaver and Crisman, 1989; Sanders and Wickham, 1993; Jürgens et al., 1999). Despite some broad similarities, the evidence suggests that the structure and function of ciliate communities are unique to their environments. This apparent contradiction in ciliate community structure and function has also been observed in eutrophic environments. In Lake Võrtsjärv (Estonia) for example, ciliates were dominated by algivores in spring and bacterivores in summer but collectively accounted for >60% of the organic carbon consumed by larval fish (Zingel et al., 2019). Research from Lake Taihu (China) found that ciliate community composition was significantly affected by Cyanobacteria (Microcystis) blooms (Li et al. 2013). The implication is that changes in ciliate community composition brought about by eutrophication can exert strong bottom up pressure on the fisheries they help support.

This paper was developed as part of a broader survey of the microbial-planktonic foodweb of the Bay of Quinte, a eutrophic embayment of Lake Ontario and will focus on the ciliate community. Geographically, the bay is at the low end of the North American Great Lakes chain near the outflow to the St. Lawrence River. A long term research and monitoring program was established in 1972 focusing on phytoplankton and zooplankton populations with assessment of ciliate and microbial populations beginning in 2000. While previous work has been published emphasizing the importance of heterotrophic protozoa within the broader microbial-planktonic foodweb (Munawar et al., 2011; Bowen and Johannsson, 2011), the contributions of ciliates were not assessed in detail. To address this deficiency, our study will:

- Consider the structure, biomass and composition of the ciliate community and relate these to the both foodweb structure and general ecological conditions
- Use Canonical Correspondence Analysis to test these relationships with particular emphasis on feeding behaviour and potential energy transfer between components of the microbial – planktonic foodweb

Materials and methods

The Bay of Quinte is a large z – shaped embayment (254km^2) that is also relatively shallow (4 – 8 m) but is much deeper (30m) at the interface with Lake Ontario. One shallow station in the upper bay (Belleville – 5m) and one deep station in the lower bay (Conway – 30m) (Figure 1) were sampled bi-weekly from May 1st to October 31st for each of the years from 2003 to 2007. Integrated epilimnetic water samples were collected from each site and subsamples were drawn for various biological and chemical parameters including: total phosphorus, soluble reactive phosphorus, nitrate+nitrite, dissolved inorganic carbon, chlorophyll-*a*, total seston, microbial loop, ciliates and phytoplankton. Temperature and dissolved oxygen were measured with a Hydrolab minisonde 4a surveyor. For the current paper, the ciliate data will be presented in its entirety. The remaining biological and chemical data were previously published in Nicholls and Carney, 2011 (phytoplankton, nutrients, water chemistry, etc.) as well as Munawar et al. 2011 (nutrients, phytoplankton, microbial loop); here they are used primarily as inputs into the Canonical Correspondence Analysis model.

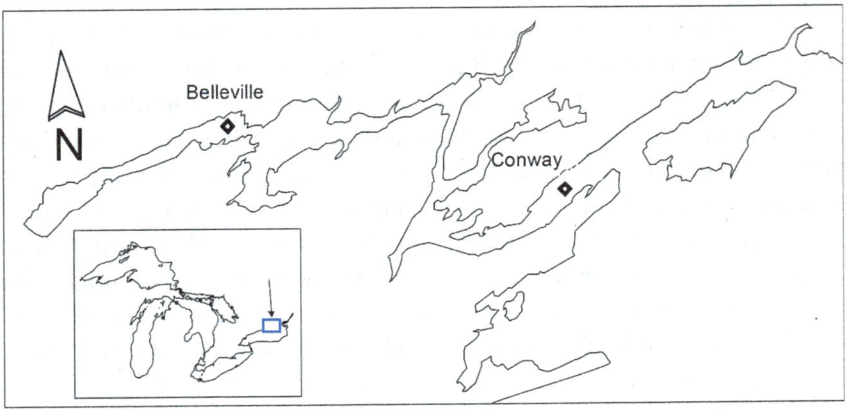

Figure 1. A detailed map of the Bay of Quinte showing the location of the Belleville site in the upper bay and Conway in the lower bay. Inset provides the location of the Bay of Quinte on Lake Ontario.

Ciliates

Whole water samples were immediately preserved in 4% acid Lugol's iodine. Within 6 months, the samples were post-fixed by adding concentrated Bouin's fluid to a final concentration of 5% and stained using the Quantitative Protargol Stain (QPS; Montagnes and Lynn, 1993). Abundance and bio-volume were calculated using standard geometric shapes and the Microbiota software developed by Roff and Hopcroft (1986). Bio-volume was converted to biomass assuming a specific gravity of 1 and 20% shrinkage of the ciliates after the QPS treatment (Jerome et al., 1993). Taxonomic groups identified under the microscope were assigned to feeding groups: fine filtrators, coarse filtrators, raptorial and encounter feeders and flagellate eaters, following the methodology of Munawar and Lynn (2002) with particular reference to: Gilbert, 1994; Foissner et al., 1999; Šimek et al., 2000 and Carrias et al., 2001. Taxonomic nomenclature followed the classifications of the World Registry of Marine Species (WoRMs, 2020). For comparative purposes, ciliates were typically grouped according to genus (*Balanion, Halteria, Strobilidium, Strombidium, Urotricha, and Vorticella*). In some cases however, sub class (Haptoria) and order (Tintinnina) were more appropriate groupings for comparison due to changes in the classification system.

Statistical analyses

Annual means along with coefficients of variation and standard errors were estimated for ciliate biomass, taxonomic composition and feeding groups using the bi-weekly monitoring data. Significant differences were assessed using a repeated measures Analysis of Variance (ANOVA). Canonical Correspondence Analysis (CCA) was run using Canoco (version 4.56) to help infer the relationships between taxonomic groups of ciliates with potential prey items and environmental variables using the biweekly monitoring data. With respect to ciliate prey, phytoplankton were classified as either "edible" or "inedible" based primarily on size and shape. CCA was designed to extract synthetic environmental gradients from ecological data-sets assuming a unimodal response curve to environmental variables. These gradients form the basis for describing and visualizing the differential habitat preferences of taxa via an ordination diagram (ter Braak and Verdonschot, 1995).

Results

Belleville

The Belleville site located in the upper Bay of Quinte was eutrophic with annual mean (May – October) total phosphorus concentrations ranging from 29.8 – 41.0 µg l^{-1} during the 2003 – 2008 period were observed in addition to chlorophyll-*a* concentrations of 12.1 – 25.0 µg l^{-1} and phytoplankton biomass of 3.5 – 5.8 g m^{-3} (Munawar et al. 2011). Ciliate biomass ranged from 1.6 – 798.0mg-m^{-3} (mean = 100.2 mg m^{-3}, CV = 71%, n = 77); annual mean biomass (Figure 2, upper panel) ranged from a low of 38.4mg m^{-3} in 2003 and increased significantly to a high of 231.3mg m^{-3} in 2008 ($F(5,71) = 5.64$; $p = 0.0002$). Average ciliate size, measured as ESD, ranged from a low of 19.5 µm in 2007 to a high of 26.0 µm in 2003.

The dominant groups observed throughout the study were: Haptoria[1] (4–50%), *Strobilidium* (9–33%), *Halteria* (7–22%), *Strombidium* (4–21%), *Vorticella* (4–14%), *Urotricha* (4–11%), *Tintinnid* (1–8%), and *Balanion* (1–2%). The wide ranges indicate considerable year to year variability in community composition. The "oligotrich" ciliates as a group (*Strobilidium* spp., *Strombidium* spp., *Halteria* spp., and Tintinnina) contributed 64% of ciliate biomass in 2003 but steadily declined to 24% by 2008.

1 Includes the taxonomic orders of: Cyclotrichiida (*Askenasia*, *Mesodinium*); Haptorida (*Dileptus*, *Lagynophrya*, *Monodinium*, Spathidiidae), and Pleurostomatida (*Litonotus*).

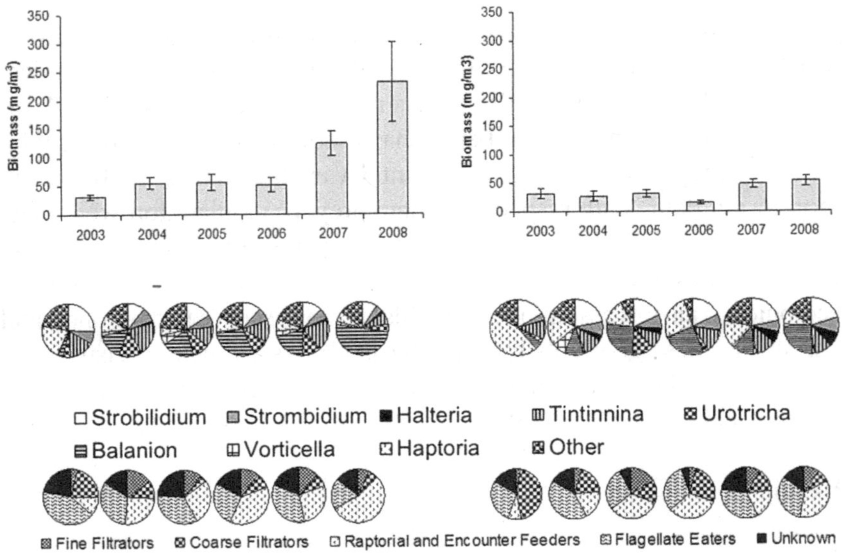

Figure 2. Top panel: Ciliate density and biomass at Belleville (a) and Conway (b) stations in the Bay of Quinte 2003 – 2008 Middle panel: May-October average ciliate biomass by taxonomic group at Belleville (a) and Conway (b) stations. 2003 to 2008. Bottom panel: May to October average Ciliate biomass by feeding group at stations Belleville (a) and Conway (b) 2003 to 2008. (Figure 2 middle panel). *Urotricha* spp. and *Vorticella* spp. were relatively stable but minor components of the ciliate community. Haptoria showed more variability in terms of relative composition but importantly accounted for most of the observed increase in ciliate biomass over the course of the study.

With regards to feeding behaviour, over 50% of the population was categorized as either "flagellate eaters" or "raptorial and encounter feeders" (Figure 2 lower middle panel). Ciliates that feed by filtration were less common; fine and coarse filtrators together comprised 25% or less of the biomass in all years of the study (Figure 2 lower middle panel).

Conway

At Conway, located near the interchange with Lake Ontario, seasonal mean total phosphorus concentrations ranged from 10.9 – 13.8 μg l^{-1}, Chlorophyll-*a* concentrations ranged from 3.1 – 4.4 μg l^{-1}, and total phytoplankton biomass ranged from 0.5 – 0.6 g m^{-3} indicating more oligotrophic conditions than Belleville (see also Munawar et al. 2011). Ciliate biomass was low compared to Belleville with individual observations ranging from 1.3–131.1mg m^{-3} (mean = 37.1mg·m^{-3}; CV = 121%;

n = 76); annual mean biomass ranged from 20.5 – 58.9mg m^{-3} with a significant increase observed between 2006 and 2008 ($F_{(5,71)}$ = 3.19; p = 0.01) (Figure 2). Average size, measured as ESD, ranged from a low of 15.9 µm in 2008 to a high of 20.2 µm during 2003.

The dominant ciliate groups observed over the course of the study were: *Haptoria* (5–27%), *Strombidium* (7–26%), *Strobilidium* (17–23%), *Vorticella* (1–19%), *Halteria* (6–13%), *Urotricha* (1–12%), and *Tintinid* (1–8%). The "oligotrich" ciliates declined over the course of the study, but not to the same extent as in Belleville, from 76% (2003) to 39% (2008) (Figure 2 middle panel). Also similar to Belleville, *Vorticella* spp. were relatively stable but minor components of the ciliate community. On the other hand, *Urotricha* spp. and Haptoria showed more variability but were typically more important components of the ciliate community (Figure 2 middle panel).

Feeding ecology was also different at Conway as compared to Belleville. In 2003, filtrators, primarily *Strombidium* spp., comprised almost 50% of the biomass while in the other years this group accounted for less than 30% (Figure 2). Both "Flagellate eaters" and "raptorial and encounter feeders" comprised over 50% of the ciliate biomass during the other years of the study.

Comparison of sites

Generally, ciliate biomass was higher at Belleville (100.2±14.7mg m^{-3}, n = 77) than Conway (37.1±3.5mg m−3, n = 76) during the 2003 – 2008 study period. The difference was found to be significant using a paired sample t test (t = -4.25421, df = 74, p<0.0001) indicating more favourable conditions for ciliates at Belleville. The results were further parsed using a repeated measures ANOVA with total ciliate biomass as the response variable along with: 1. station, 2. year and 3. sampling week as effects. Given the rigid bi-weekly sampling program, temporal effects should be apparent. The results were statistically significant (F = 3.9103, df = 152, P<0.0001) and showed that ciliate biomass during 2003 was significantly lower than the other years (t = -2.14, p<0.05). The results also show significant changes in ciliate biomass during weeks 39 (t = 2.74, p<0.01) and 41 (t = 2.53, p<0.02) which correspond to roughly September 18–24 and October 2–8.

Assessment of trophic relationships by Canonical Correspondence Analysis, 2003-2008

Canonical Correspondence Analysis (CCA) was used to relate ciliate taxonomic groups and putative prey (including autotrophic picoplankton, edible phytoplankton, bacteria, heterotrophic nanoflagellates in addition to Inedible algae) to a series of environmental parameters (surface temperature, total phosphorus, soluble reactive phosphorus, nitrate+nitrite, dissolved oxygen, dissolved inorganic carbon and total seston. Overall, the results showed the primary axis explained 90% of the variability ($p = 0.002$). The relatively low Eigen value (0.429) indicates that there is not a lot of variability in the dataset and can be attributed to the consistency of the sampling regime over the study period.

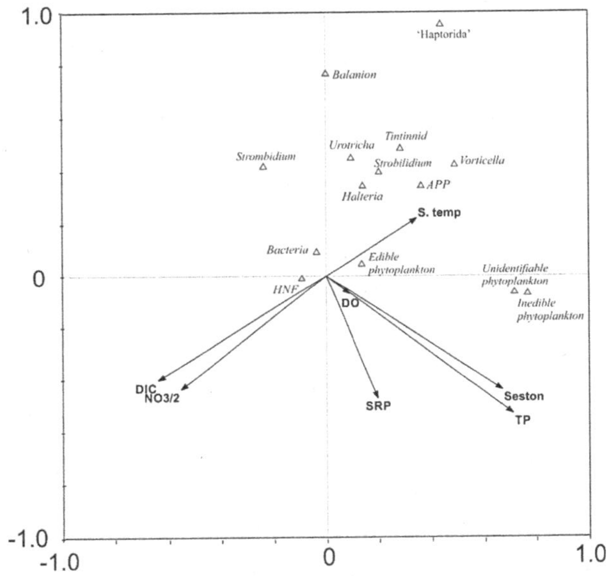

Figure 3. Canonical Correspondence Analysis of ciliates, potential prey and environmental variables in the Bay of Quinte. Approximately 12% of the variability is explained by the environmental variables; the 1st axis explains 26% of total inertia (variance) and the second axis accounts for 2.1%. Combined, both axes account for 97.2% of total inertia. Legend: S. temp = surface temperature, DO = dissolved oxygen, DIC = dissolved inorganic carbon, NO3/2 = nitrate+nitrite, SRP = soluble reactive phosphorus, TP = total phosphorus, APP = autotrophic picoplankton, HNF = heterotrophic nanoflagellates; all others are as listed.

The ordination diagram (Figure 3) shows a clustering of *Halteria* spp., *Strobilidium* spp., Tintinnina, Urotricha and Vorticella in the upper right quadrant relating to surface temperature. Edible phytoplankton and autotrophic picoplankton were also found in the same quadrant. Of the remaining ciliate groups, Haptoria and Balanion were found on the border of the upper right quadrant whereas Strombidium was found in the upper left quadrant; all three were a considerable distance from the environmental variables indicating a high degree of dissimilarity.

Discussion

The Bay of Quinte is a largely riverine embayment of Lake Ontario with a long history of algal blooms and cultural eutrophication due to phosphorus enrichment. While phosphorus abatement has been in place since the late 1970s and phosphorus loadings have been reduced from >200kg d^{-1} to < 10kg d^{-1} at present, the bay is still subject to frequent algal blooms (Munawar et al., 2018; Minns et al., 2011; Nicholls and Carney, 2011). It should be reiterated however, that it is the upper bay, represented by the station at Belleville, that is in fact eutrophic, whereas the lower bay (Conway) is more oligotrophic and mixes with Lake Ontario. In previous work exploring the microbial-planktonic foodweb, we showed that heterotrophic micro-organisms at both sites (including bacteria, heterotrophic nanoflagellates and ciliates) were significant contributors to the organic carbon pool; up to 60% at Belleville and 85% at Conway (Munawar et al. 2011). It was not clear from that analysis however what role ciliates might play in the transfer of organic carbon through the foodweb.

Our results showed that ciliate biomass was significantly higher at Belleville than at Conway and that only Belleville had significant inter-annual variability. Late summer and early fall increases in the biomass of Haptoria at Belleville during 2007 (Sept. 18) and 2008 (Sept. 23; Oct. 8) account for virtually all of the observed increases in ciliate biomass. Species of *Mesodinium* were most common (≈90% of the Haptoria biomass) in September 2007 and 2008, whereas the October 2008 sample contained a mixture of *Askenasia* and *Mesodinium* which together accounted for more than 70% of *Haptoria* biomass. Both genera, Askenasia and Mesodinium, would be expected to feed on small flagellates and microalgae. In the current study, the peaks of Haptoria observed (123.9 – 678.1 mg m^{-3}) coincided with exceptionally large standing stocks of HNF (1292.9 – 3930.2mg m^{-3}) which could provide an ample food resource. Likewise, the biomass of predatory zooplankton (41.2 – 82.2mg m^{-3}; Bowen and Johannsson, 2011) was quite low at these times which would suggest limited grazing pressure on the ciliates. Overall, these results indicate that the complex predator/prey interactions of organisms within the microbial planktonic foodweb can at times have a very strong impact on ciliate biomass and composition.

Ciliates as a community tend to have very broad feeding characteristics and serve as herbivores, bacterivores and protistan grazers (Lynn, 2008); as such, they can be an important link in microbial-planktonic foodwebs. However, it can be a complex situation and therefore difficult to assess their precise impact without direct observations of feeding. Canonical Correspondence Analysis (CCA) has been used in other studies to tease out interactions between predator, prey and environment (e.g. Hyatt et al 2010, Kalin et al 2006). Most of the taxonomic groups, Halteria, Strobilidium, Tintinnina, Urotricha and Vorticella, cluster with surface temperature as the environmental variable. This relationship – increasing biomass associated with increasing temperature – is expected, given that planktonic organisms have strong seasonal gradients. Of the potential food resources for ciliates, autotrophic picoplankton (APP) are found within the same broad cluster whereas edible phytoplankton, bacteria and heterotrophic nanoflagellates fall outside of it but are still located nearby and close to the origin of the plot (Figure 3); the implication is that APP also maintain their biomass when ciliate biomass peaks, whereas bacteria, HNF and edible phytoplankton biomass decline.

With respect to ciliate feeding behavior, the clustering of APP with most of the ciliate taxa would suggest either a lack of grazing by ciliates on APP or that the grazing pressure exerted by ciliates has minimal impact. Given that APP standing crops were typically 2-3X greater than ciliate standing stocks during the 2003–2008 study period and the rapid turnover rates associated with APP (see Munawar et al., 2013), we feel that the APP can withstand any potential grazing pressure. Having said that, the upper bay is a chronically eutrophic environment which also has large standing stocks/crops of bacteria, HNF and 'edible' phytoplankton (including small flagellates as well as small centric diatoms) providing ample food resources for the ciliate community (ibid). The CCA shows some distance between the cluster and bacteria, HNF and edible phytoplankton (Figure 3). As bacteria and HNF both tend to have lower turnover rates than APP, this higher degree of dissimilarity could be indicative of general grazing pressure. However, the low biomass of ciliates (0.03 – 0.1 g m^{-3}) compared to the proportionately higher biomass of bacteria (0.1 – 0.3 g m^{-3}), edible phytoplankton (0.3 – 0.6 gm^{-3}) and heterotrophic nanoflagellates (1.4 – 7.3 g m^{-3}) suggest that ciliate grazing pressure is likely minimal.

In a previous study (Munawar et al., 2011), we hypothesized that HNF were likely outcompeting ciliates for food resources, but it is also possible that a top-down effect, direct grazing on ciliates from rotifers, metazooplankton and larval fish (Wickham and Gilbert, 1991; Arndt, 1993; Li et al. 2013; Zingel et al., 2019) could be limiting the overall size of the ciliate population. Zooplankton biomass during the 2003-2008 study period ranged from 0.6 – 1.3g m^{-3} (Bowen and Johannsson, 2011; assuming 10% dry weight conversion), the majority of which were herbivorous Cladocerans and Cyclopoids. Potential ciliate predators, including rotifers, omnivorous zooplankton and predatory zooplankton, accounted for only 10 – 30% of the total zooplankton biomass (i.e. 0.1 – 0.2g m^{-3}). Even if all of those fed entirely on ciliates, the biomass of planktonic grazers is probably too low to produce a top down effect on ciliate community structure.

Our lab began investigating the microbial foodweb of the Great Lakes in the late 1980s, including a study of the Lake Ontario ciliate community in 1990 (Lynn et al. 2003; Munawar et al. 2003). Comparing our current Bay of Quinte dataset to these historic Lake Ontario values shows considerable overlap. However, the mean biomass in the Bay of Quinte (both Belleville and Conway) from 2003 – 2008 and mean biomass of 64.3mg m^{-3} over those same years are significantly higher than Lake Ontario (29.8mg m^{-3}) ($F_{1, 312} = 15.5$, $p = 0.0001$) and also significantly higher than our mean estimates of biomass for the ciliate community for Lakes Superior (12.7mg m^{-3}), Huron (16.2mg m^{-3}) and Erie (42.2mg m^{-3}) as shown in Table 1. By way of comparison, Carrick and Fahnenstiel

Table 1. Comparison of Ciliate biomass (Mean ± 1 S.E.) in the North American Great Lakes. Data for Lakes Superior, Huron, Erie and Ontario is from Munawar and Lynn (2002).

Lake	Ciliate Biomass (mg m^{-3})
Superior	12 .7±1.1, n = 100
Huron	16.2±1.6, n = 239
Erie	42.2±8.8, n = 226
Ontario	29.8±4.0, n = 160
Bay of Quinte	64.3±7.9, n = 153

(1990) reported ciliate biomass in the range of approx. 15 – 80mg m^{-3} in the surface waters of Lakes Huron and Michigan during the May to October period of 1987. Our observations from the Bay of Quinte were generally higher but much more extreme (approx. 1 – 800mg m^{-3}) however our data set was also amassed over several years of study. Since the Bay of Quinte is eutrophic compared to most of the offshore open waters of the Great Lakes, the elevated biomass of ciliates in the bay is expected given the higher degree of potential food resources available.

Summary and conclusions

The ciliate community of the Bay of Quinte, Lake Ontario was assessed over a six year period from 2003 – 2008. One eutrophic site in the upper bay (Belleville) and one more oligotrophic site in the lower bay (Conway), were included in the survey. Both sites contained a mixture of Haptoria, Strobilidium, Halteria, Strombidium, and Vorticella, although the relative proportions varied among sites resulting in different community structures. Ciliate biomass was found to be significantly higher at Belleville (mean: 100.2±14.1mg m^{-3}) compared to Conway (mean: 37.1±3.5mg m^{-3}), but could be explained by sudden pulses of Haptoria, primarily species of *Mesodinium*, which peaked as high as 678.1mg m^{-3}. Canonical Correspondence Analysis showed a clustering of *Halteria, Strobilidium, Tintinnina, Urotricha and Vorticella* with surface temperature along with autotrophic picoplankton as a potential prey item. The results suggest that autotrophic picoplankton may be an important food resource for the ciliate community and that there is a strong seasonal gradient to community structure. Future research should be directed at understanding how ciliates influence the transfer of both primary and secondary production to higher trophic levels.

Our study examined the composition of the ciliate community of the Bay of Quinte over a six year period from 2003 – 2008 at one eutrophic site (Belleville) and one oligotrophic site (Conway). Overall, we found that ciliate biomass was significantly higher at Belleville compared to Conway. The dominant taxa at Belleville included a mixture of *Haptoria, Halteria, and Strobilidium* which reflects the eutrophic and largely benthic habitat. By comparison, the dominant taxa at Conway were *Strobilidium, Strombidium*, and to a lesser extent *Haptoria* which reflects incursions of oligotrophic waters from Lake Ontario and a more pelagic oriented foodweb. In particular, much of the differences between sites could be explained by late summer/early fall pulses of *Haptoria* (primarily

Mesodinium and Askenasia) at Belleville. However, considerable variability in both biomass and composition was observed among years and among sites indicating that there is no typical ciliate community. Canonical Correspondence Analysis showed that both food availability and grazing pressure, from other components of the microbial planktonic foodweb, could dramatically affect ciliate community structure. We recommended that future research be directed at how ciliates influence the transfer of both primary and secondary production to higher trophic levels.

Acknowledgements

We are grateful to the late Denis Lynn for his long-term collaboration in the study of Great Lakes ciliates and for his early contributions to this paper. A sincere thank you is also due to Dr. Michaela Strüder-Kypke for her diligent work in identifying and enumerating the ciliate communities, and for continuing the work of Dr. Lynn. This research was conducted under the banner of Project Quinte and we thank E.S. Millard, M. Frank (nee Burley), J. Gerlofsma, C. Timmins, R. Bonnell and A. Bedford for technical and scientific support. Thanks also to J. Weir (DFO) for assisting with the literature review. Funding was provided by Environment Canada through the Great Lakes Action Plan.

References

Arndt, H., 1993. Rotifers as predators on components of the microbial web (bacteria, heterotrophic flagellates, ciliates) – a review. Hydrobiol. 255/256, 231–246.

Azam, F., Fenchel, T., Field. J.G., Gray, J.S., Meyer-Reil, L.A., Thingstad, F., 1983. The ecological role of water-column microbes in the sea. Mar. Ecol. Prog. Ser. 10, 257–263.

Beaver, J.R., Crisman, T.L., 1989. The role of ciliated protozoa in pelagic freshwater ecosystems. Microb. Ecol. 17, 111–136.

Bowen, K.L., Johannsson, O.E., 2011. Changes in zooplankton biomass in the Bay of Quinte with the arrival of the mussels, Dreissena polymorpha and D. rostiformis bugensis, and the predatory cladoceran, Cercopagis pengoi: 1975 to 2008. Aquat. Ecosyst. Health Mgmt. 14(1), 44–55.

Carrias, J.F., Thouvenot, A., Amblard, C., Sime-Ngando, T., 2001. Dynamics and growth estimates of planktonic protists during early spring in Lake Pavin, France. Aquat. Microb. Ecol. 24, 163–174.

Carrick, H.J., 2005. An under-appreciated component of biodiversity in plankton communities: the role of protozoa in Lake Michigan (a case study). Hydrobiol. 551, 17–32.

Carrick, H.J. and Fahnenstiel, G.L., 1990. Planktonic Protozoa in Lakes Huron and Michigan: Seasonal Abundance and Composition of Ciliates and Dinoflagellates. Journal of Great Lakes Research, 16(2), 319–329.

Foissner, W., Berger, H., Schaumburg, J., 1999. Identification and ecology of limnetic plankton ciliates. des Bayer. Landesamtes für Wasserwirtschaft. (Bayer's information section. State Office for Water Management. in German). Heft 3/99.

Gilbert, J. J., 1994. Jumping behavior in the oligotrichs Strobilidium velox and Halteria grandinella, and its significance as a defense against rotifer predators. Microb. Ecol. 27, 189–200.

Hyatt, C. V., Paterson, A. M., Cumming, B. F., and Smol, J. P., 2010. Factors related to regional and temporal variation in the distribution of scaled chrysophytes in northeastern North America: Evidence from lake sediments. Nova Hedwigia, 87–102.

Jerome, C.A., Montagnes, D.J.S., Taylor, F.J.R., 1993. The effect of the quantitative protargol stain and Lugol's and Bouin's fixatives on cell size: a more accurate estimate of ciliate species biomass. J. Eukaryot. Microbiol. 40, 254–259.

Jürgens, K., Skibbe, O., Jeppesen, E., 1999. Impact of metazooplankton on the composition and population dynamics of planktonic ciliates in a shallow, hypertrophic lake. Aquat. Microb. Ecol. 17, 61–75.

Kalin, M., Wheeler, W. N., and Olaveson, M. M., 2006: Response of phytoplankton to ecological engineering remediation of a Canadian Shield Lake affected by acid mine drainage. Ecological Engineering, 28, 296–310.

Lischke, B., Weithoff, G., Wickham, S., Attermeyer, K., Grossart, H., Scharnweber, K., Hllt, S., Gaedke, U. 2016. Large biomass of small feeders: ciliates may dominate herbivory in eutrophic lakes. Journal of Plankton Research. 38(1), 2–15.

Lynn D.H. 2008. The ciliated protozoa – characterization, classification, and guide to the literature, Springer, Heidelberg.

Lynn, D.H., Munawar, M., 1999. Abundance, biomass, and diversity of planktonic ciliates (Ciliophora) in Lake Erie. In: M. Munawar, T. Edsall, If. Munawar (Eds.), *State of Lake Erie (SOLE) - Past, Present and Future*, pp. 155–168. Ecovision World Monograph Series, Backhuys Publishers, Leiden.

Lynn, D.H., Munawar, M., Carou, S., Niblock, H., Humby, P.L., 2003. Abundance, biomass and diversity of planktonic ciliates (Ciliophora) in Lake Ontario. In: M. Munawar (Ed.), *State of Lake Ontario (SOLO), past, present and future*, pp.171–186. Ecovision World Monograph Series. Aquatic Ecosystem Health and Management Society, Burlington, Ontario, Canada.

Minns, C.K., Munawar, M., Koops, M.A., Millard E.S., 2011. Long term ecosystem studies in the Bay of Quinte, Lake Ontario, 1972-2008: A prospectus. Aquat. Ecosyst. Health Mgmt. 14(1), 3–8.

Montagnes, D.J.S., Lynn, D.H. 1993. A quantitative protargol stain (QPS) for ciliates and other protists. In: P.F. Kemp, B.F. Sherr, E.B. Sherr, J.J. Cole (Eds.), *Handbook of methods in aquatic microbial ecology*, pp. 229–240. Lewis Publishers, Boca Raton.

Montagnes, D. J. S., Lynn, D. H, Roff, J. C., Taylor, W.D., 1988. The annual cycle of heterotrophic planktonic ciliates in the waters surrounding the Isles of Shoals, Gulf of Maine: An assessment of their trophic role. Mar. Biol. 99, 21–30.

Munawar, M., Lynn, D. H., 2002. Planktonic ciliates of the North American Great Lakes: Lakes Superior, Huron, Erie, and Ontario. Aquat. Ecosyst. Health and Mgmt. 5(3), 345–354.

Munawar, M., Legner, M., Munawar, I.F., 2003. Assessing the microbial foodweb of Lake Ontario. In: M. Munawar (Ed.), *State of Lake Ontario (SOLO), past, present and future*, pp.171–186. Ecovision World Monograph Series. Aquatic Ecosystem Health and Management Society, Burlington, Ontario, Canada.

Munawar, M., Fitzpatrick, M., Munawar, I.F., Niblock, H., 2010. Checking the pulse of Lake Ontario's microbial-planktonic communities: a trophic transfer hypothesis. Aquat. Ecosyst. Health Mgmt. 13(4), 395–412.

Munawar, M, Fitzpatrick, M., Niblock, H. and Lorimer, J. 2011. The relative importance of autotrophic and heterotrophic microbial communities in the planktonic foodweb of the Bay of Quinte, Lake Ontario 2000-2007. Aquat. Ecosyst. Health Mgmt.14(1), 21–32.

Munawar, M., Munawar, I.F., Fitzpatrick, M., 2013. Microbial foodweb comparison of the Laurentian Great Lakes during the summers of 2001-2004. Aquat. Ecosyst. Health Mgmt.16(3), 267–278.

Munawar, M., Fitzpatrick, M., Niblock, H., Kling, H., Lorimer, J., Rozon, R., 2018. Phytoplankton ecology in the Bay of Quinte: spatial distribution, dynamics and heterogeneity. Aquat. Ecosyst. Health Mgmt. 21(2), 213–226.

Nicholls, K.H. and Carney, E., 2011. The phytoplankton of the Bay of Quinte, 1972–2008: point-source phosphorus loading control, Driessenid mussel establishment, and a proposed community reference. Aquat. Ecosyst. Health Mgmt. 14(1), 33–43.

Porter, K., Pace, M.L., Battey, J.F., 1979. Ciliated protozoans as links in freshwater food chains. Nature 277, 563–565.

Roff, J.C., Hopcroft, R.R., 1986. High precision microcomputer based measuring system for ecological research. Can. J. Fish. Aquat. Sci. 43, 2044–2048.

Sanders, R., Wickham, S.A., 1993. Planktonic protists and metazoa: Predation, food quality and population control. Mar. Microb. Foodwebs 7, 197–223.

Sherr, E. B., Sherr, B. F., 1994. Bacterivory and herbivory: Key roles of phagotrophic protists in pelagic foodwebs. Microb. Ecol. 28, 223–235.

Šimek, K., Jürgens, K., Nedoma, J., Comerma, M., Armengol, J., 2000. Ecological role and bacterial grazing of *Halteria* spp.: small freshwater oligotrichs as dominant pelagic ciliate bacterivores. Aquatic Microb. Ecol. 22, 43–56.

Taylor, W.D., Johansson, O.E. 1991. A comparison of estimates of productivity and consumption by zooplankton and planktonic ciliates in Lake Ontario. J. Plankton Res. 13(2), 363–372.

ter Braak, C., Verdonschot, P., 1995. Canonical correspondence analysis and related multivariate methods in aquatic ecology. Aquatic Sciences 57(3), 1015–1621.

Weisse, T., Müller, H., 1998. Planktonic protozoa and the microbial foodweb in Lake Constance. Lake Constance: Characterization of an ecosystem in transition, 223–254.

Wickham, S. A., Gilbert, J. J., 1991: Relative vulnerabilities of natural rotifer and ciliate communities in cladocerans: laboratory and field experiments. Freshwat. Biol. 26, 77–86.

WoRMS Editorial Board, 2020, World Register of Marine Species. Available from http://www.marinespecies.org at VLIZ. Accessed 2020-01-16.

Zingel, P., Agasild, H., Karus, K., Buholce, H., Noges, T., 2019. Importance of ciliates as food for fish larvae in a shallow sea bay and a large shallow lake. European Journal of Protistology 67, 59–70.

The above article is a reprinted version of the following previously published article:

Ciliate ecology in the eutrophic Bay of Quinte, Lake Ontario: Community structure and feeding characteristics†

Munawar; M., Niblock, H., M. Fitzpatrick; M., Lorimer, J., 2020.

Aquatic Ecosystem Health & Management (2020) 23 (1): 35–44.

https://doi.org/10.1080/14634988.2020.173068

Submerged aquatic vegetation in the Bay of Quinte: Response to decreased phosphorous loading and Zebra Mussel invasion

K.E. Leisti[1,#] S.E. Doka[1,*] and C.K. Minns[2,1]

[1]Fisheries and Oceans Canada, 867 Lakeshore Road, Burlington, ON L7S 1A1
[2]Department of Ecology and Evolutionary Biology, University of Toronto, 25 Willcocks St., Toronto, ON M5S 3B2
**Corresponding author: susan.doka@dfo-mpo.gc.ca*
#retired

Keywords: retired macrophytes, time-series, eutrophication, Dreissenid Mussels

Introduction

Submerged aquatic vegetation (SAV) is an important component of aquatic ecosystems affecting light, temperature, nutrients and biota (Carpenter and Lodge, 1986). As a buffer between the terrestrial and pelagic environment, SAV reduce the effects of wave action (Keddy,1982), and retard water flow resulting in increased sedimentation and greater water clarity (Madsenetal., 2001; Horppila and Nurminen, 2003).SAV act as nutrient sinks (Blindow, 1992) and upon senescence, serveas nutrient sources (Best, 1982). SAV provide substrate for epiphytes that are food for zooplankton, which in turn are consumed by planktivorous fish (Keast, 1984). SAV provides spawning and nursery habitat for numerous fish species and have been found to influence fish abundance (Chick and McIvor,1994), growth rates (Wernеretal.,1983) and community structure (Weaver et al., 1997). Randall et al. (1996) estimated that sites with SAV had higher densities of fish and higher species richness than non-vegetated sites in three bays in southern Ontario. In addition, dense SAV had a production index (accounts for both biomass and fish size) that was over 2 times higher than found in areas where SAV was sparse or absent.

Project Quinte, a multi-agency research collaboration, was established in 1972 to assess the ecosystem impacts of a planned reduction in phosphorus Point-source loadings in the Bay of Quinte. In 1985,

Ecology of the Bay of Quinte: Health, Management and Global Implications
Editors: C.K. Minns, M. Munawar and M.A. Koops

Aquatic Ecosystem Health & Management Society, Canada

this multi-basin bay was designated as one of 43 Areas of Concern (AOC) around the Great Lakes by the International Joint Commission (IJC). Both the Canadian and United States governments committed to restore these AOCs under Annex 2 of the 1987 Great Lakes Water Quality Agreement (GLWQA). A common Remedial Action Plan (RAP) process was outlined in the Annex (IJC, 1987) to define beneficial use impairments (BUI) (Bay of Quinte RAP Coordinating Committee, 1990) and develop remedial action plans (Bay of Quinte RAP Coordinating Committee, 1993). In the Bay of Quinte, some BUIs were caused by excessive algae and loss of fish habitat, which included the loss of extensive submergent aquatic vegetation (SAV) beds, particularly in the upper bay. Ongoing research has continued in the Bay of Quinte to address many beneficial use impairments under the Remedial Action Plan (see http://www.bqrap.ca for the time sequence of the RAP beginning in 1986). These efforts have created a valuable dataset for the Bay of Quinte in terms of both time series and breadth of ecosystem assessment.

Historically, the Bay of Quinte was slightly to moderately oligotrophic prior to the late 1800s with a shift to a eutrophic state by 1890(Sly, 1986). Water quality deteriorated noticeably after 1930 and by the1950s, the bay was considered hypereutrophic (Johnson, 1986). In the1960s, algal blooms occurred in the upper and middle bays and sec chidisc depths <0.3m were recorded (Johnson and Hurley, 1986). SAV beds were described aslush in the1950s, particularly in the Trenton area, with sub sequent declines during the 1960s to half of their estimated extent (Crowder and Bristow, 1986). In the 1970s, macrophytes were restricted to a euphotic zone of 2 m in the upper bay and 4 m in the lower bay. Since 1972, two major changes have occurred: first, a major reduction in Point-source phosphorous (P) loadings from sewage treatment plants implemented in the winter of 1977–1978 and, second, the 1993 invasion by *Dreissena* spp.

Several studies have demonstrated major effects of Point-source P and *Dreissena* invasions on macrophyte density and distribution. In an assessment of the bay's nutrient budgets, Minns et al. (1986) proposed that the collapse of macrophytes in the 1960s was represented by a cusp-catastrophe model wherein macrophytes persisted in abundance until Point-source P-loading reached very high levels and excess periphyton growth curtailed their growth. Minns et al. (1986) then inferred from the model that if Point-source P loadings declined to very low levels and accumulated internal loading dissipated. Higgins and Vander Zanden (2010) showed in their extensive meta-analysis of the effects of Dreissenid Mussels on fresh-water ecosystems that littoral primary producers responded positively to the establishment of *Dreissena*. On average, increases of 182.1% in SAV cover, 39.4% in the maximum depth of SAV colonization, and 170.5% in periphyton biomass occurred. These changes were concurrent with increased water clarity (Secchi increases of 50.5% in littoral areas and 38.5% lake-wide) which was primarily attributed to the filtering capacity of Dreissena spp. (Higgins and Vander Zanden, 2010).

Given the potential strong effect of phosphorous and *Dreissenia* on SAV, the objective of this article was to evaluate the response of SAV in each of the sub-basins of the Bay of Quinte to these two changes. For the upper bay, it was anticipated that SAV response to P reductions may be slow due to internal P loading while increases in water clarity from filtration by *Dreissenia* may result in a rapid increase in SAV density and extent. It is predicted that SAV response in the middle and lower bays will not be as strong as in the upper bay. This is because the initial SAV condition was not as degraded as that found in the upper bay and basin morphometry limits SAV extent to a relatively small area adjacent to the shoreline. Measures of SAV response that were assessed included percent bottom cover, the distance SAV beds extended offshore, maximum depth of colonization, plant biomass and species composition.

Methodology

Location

The Bay of Quinte is a 254 km2, Z-shaped bay that opens out to the northeastern shore of Lake Ontario (Figure 1). Researchers typically divide the bay into three sections due to differences in basin morphometry and trophic gradients (Johnson and Hurley, 1986). The upper bay is shallowest with a mean depth of 3.5 m, whereas the middle and lower bay mean depths are 5.2 and 24.4 m, respectively. Water generally flows west-to-east in the bay and flushing is dominated by flow from the Trent River near Trenton with annual flushing rates in the upper bay ranging from 9.2 to 14.6 and May to September flushing rates from 1 to 4.6 in the years 1965 to 1981 (Johnson, 1986). Significant water exchange occurs between Lake Ontario and the lower bay through the Upper Gap. During summer, there can be backflow events at depth by Glenora with upwelling into the middle bay (Minns et al., 1986).

Datacollection

Reference transects were established in five locations in 1972 as seen in Figure 1: three in the upper bay at Trenton, Belleville and Big Bay, one in the middle bay at Hay Bay outflow and one at Conway in the lower bay (Bristow et al., 1977). A total of 10 transects crossed the bay with sampling starting at 0.5 m water depth and ending 90 m beyond the last macrophyte. Seasonal survey timing varied from mid-July to mid-September with a sampling interval between one to six years. From 1972 to 1988, the survey protocol used a 30 m rope with knots placed at 1 m intervals. Divers then noted SAV absence or presence at each of the 29 knots and the rope was re-positioned further offshore until three successive lengths of rope yielded no macrophytes. During this time, three full surveys of all ten transects were conducted in 1972, 1979 and 1988, whereas partial surveys in 1973, 1974, 1979, 1982 and 1985 were conducted at a subset of these sites. Percent cover was calculated for every 29 m segment based on the % presence of SAV for that segment interval. The depth at the mid-point of each 29 m segment was also recorded along with species composition. The distance that SAV beds extended offshore was estimated by multiplying the number of times the rope was re-positioned until the last SAV was reached, by rope length.

Survey methods were revised when hydroacoustic equipment was obtained. The same reference transects were sampled with hydro- acoustics in surveys in 1988, 1994, 2000, 2004 and 2007. A Raytheon DE-719C recording fathometer was used in 1988 and a Lowrance X-16 echosounder was used in 1994 and 2000. Paper echograms produced from these surveys were visually interpreted for percent cover and SAV depth at 3 mm intervals which approximately equals 1 m in actual length. The last macrophyte on the transect was interpreted from the echogram and the SAV bed extent was scaled from the echogram using marks corresponding to GPS coordinates taken in the field.

The last two surveys used a BioSonics DT-X digital echosounder with a 430 kHz single-beam transducer with a pulse length of 0.1 ms and a ping rate of 5 pps. EcoSav V1 and 2 software (BioSonics, 2000) was used to estimate percent cover and SAV depth. Results were mapped using ArcView 3.3 (Environmental Systems Research Institute Inc., 1992) and SAV extent was estimated by measuring from a fixed coordinate that serves as the start point on the transect to the last macrophyte as estimated in EcoSAV. In 1988, 1994 and 2000, divers sampled for species composition and above-ground biomass using 0.25 m^2 quadrats at 3 to 6 points along each transect. These points were used to ground-truth the acoustic data while in 2004 and 2007, a combination of rake tosses and underwater video was used for ground-truthing. Leisti et al. (2006) provides detailed information on sampling methodologies for all surveys.

Numerous water quality parameters were measured as part of Project Quinte, including offshore Secchi depth, ε_{par} (the vertical light extinction rate (m^{-1}) for photosynthetically active radiation), total phosphorous (TP) and Chla. Transparency is a strong determinant of depth colonization by SAV (Chambers and Kalff, 1985; Hudon et al., 2000). Surveys were conducted on a weekly (1972 to 1982) and bi-weekly (1982 to 2007) basis from mid-May to mid-October at Belleville, Hay Bay and Conway (Figure 1). Measurements were recorded for the Napanee offshore station, but there is a gap from 1983 to 1989 when this site was not sampled. Seasonal means by sub-bay were estimated for this dataset with both the Belleville and Napanee stations included in the upper bay.

Statistical analysis

Datasets were divided into three time stanzas; (1) the pre-phosphorous period (pre-P: 1972 to 1977), (2) a period after phosphorous control was implemented in sewage treatment plants in the upper bay (post-P: 1978 to 1994), and (3) a period after *Dreissena* was established in all bays (post-D: 1995 to 2007). Additionally, data were separated into upper, middle and lower bays. Kruskal-Wallis tests were conducted to determine differences among time stanzas within each bay. Non-parametric tests were used because transformation of the datasets did not improve normality. Post-hoc comparisons were calculated using the non-parametric multiple comparison (npmc) method in R (Gentleman and Ihaka, 1997), as described in Munsel and Hothorn (2001).

Results

Time series for seasonal mean trophic (TP and Chla) and transparency (Secchi and ε_{par}) variables had substantial interannual variability, but improving trends were evident through the three time stanzas (Figure 2). Comparisons among bays showed that lower bay trophic values were consistently and considerably lower and water clarity much higher than the upper and middle bays. Greater similarity existed between the middle and upper bays although the upper bay generally exhibited the highest eutrophic conditions.

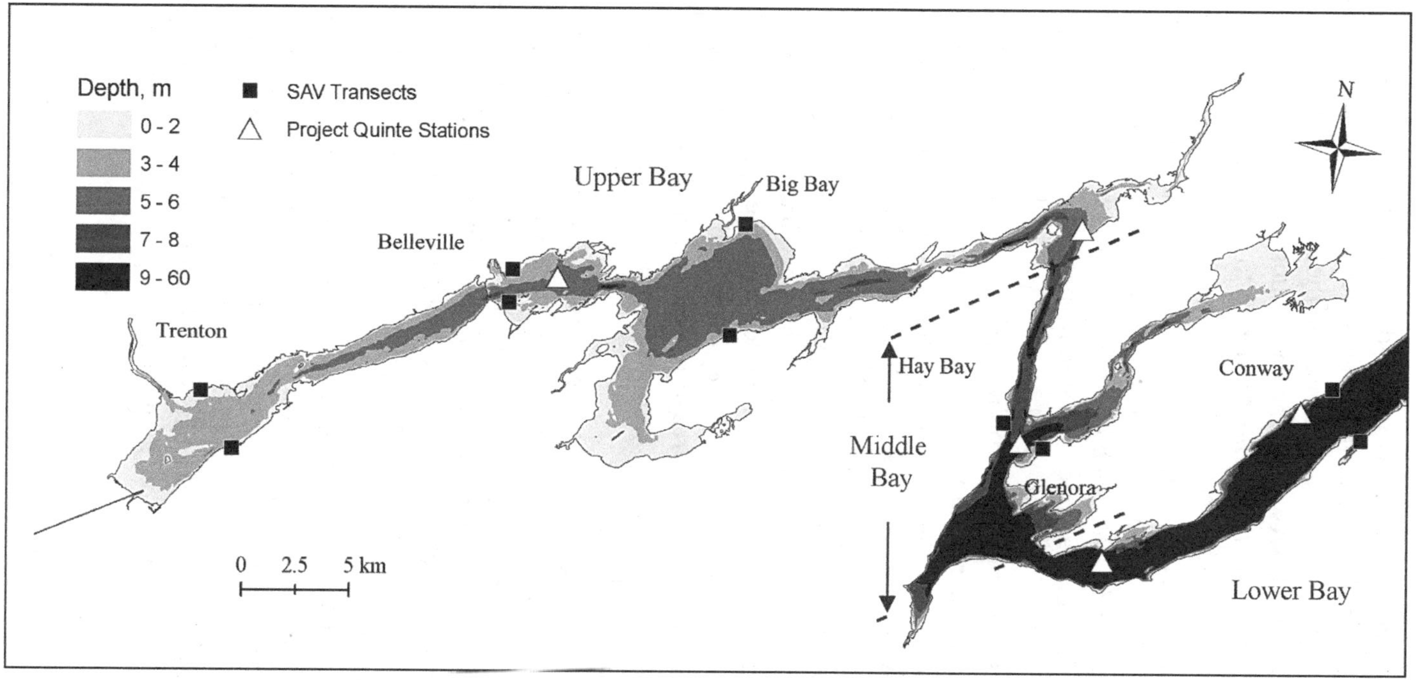

Figure 1. Location of submerged aquatic vegetation transects and Project Quinte water quality monitoring stations from 1972 to 2007. Depths are referenced to IGLD 1985.

A time series of % SAV cover-by-depth interval was calculated for each bay (Figure 3). SAV cover in the upper bay was generally sparser than the middle or lower bays throughout the time series, but especially in the pre-P period. SAV cover increased substantially in the upper bay after the invasion by *Dreissena.* Moderate-to-high SAV densities were recorded in the middle and lower bays from the beginning to the end of the time series. All bays experienced fluctuations in the maximum depth of SAV in the post-P period, with a substantial increase in Z_c occurring in the lower bay in 1988. Maximum SAV depths were consistent in the post-D period for the upper and middle bays, with depths of 4 and 3 m respectively.

Time stanza means of trophic variables (Table 1) indicated that Secchi depths increased whereas ε_{par} decreased through time for all three bays. Kruskal-Wallis tests in conjunction with post-hoc comparisons show both Secchi and ε_{par} differed significantly among all time stanzas in the upper bay. The largest increase in water clarity in the upper bay occurred between the post-P and post-D periods, where Secchi depths increased by 46% and ε_{par} decreased by 31%. Between the pre-P and post-P period, Secchi depths in the upper bay increased by 8% and ε_{par} decreased by 18%. The middle and lower bays also experienced the largest increase in clarity between the post-P and post-D time stanzas with a significant difference occurring only with the post-D period.

In the upper bay, SAV percent cover was significantly different over the three time stanzas, more than doubling between the pre- and post-P period, then a smaller 23% increase between the post-P and -D period when mean cover reached 51%. SAV cover in the middle and lower bays were already at 50 and 56% respectively in the pre-P period, did not change significantly in the post-P period and increased by 25% in the post-D period.

Mean bed extent more than tripled in the upper bay and doubled in the middle bay between the post-P and -D periods. Bed extent decreased in the upper bay between the pre- and post-P periods whereas the middle bay extent increased slightly, but both of these changes were not significant. In the lower bay, small changes in bed extent over the three time stanzas were not significant.

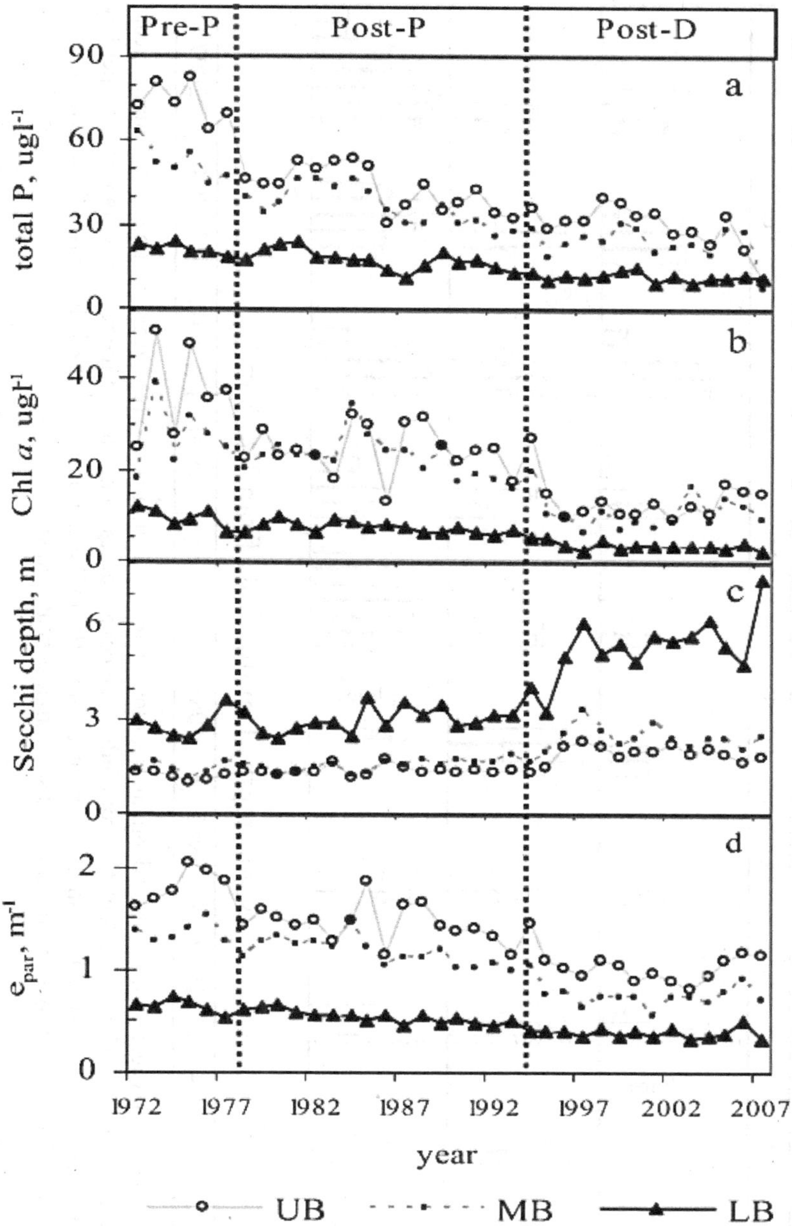

Figure 2. May to October means for (a) total phosphorous, (b) total chlorophyll *a* (uncorrected for phaeopigments), (c) Secchi depth and (d) ε_{par} from 1972 to 2007. Upper bay (UB) values have been calculated using data from the Belleville and Napanee stations, middle bay (MB) from the Hay Bay station and lower bay (LB) from the Conway station.

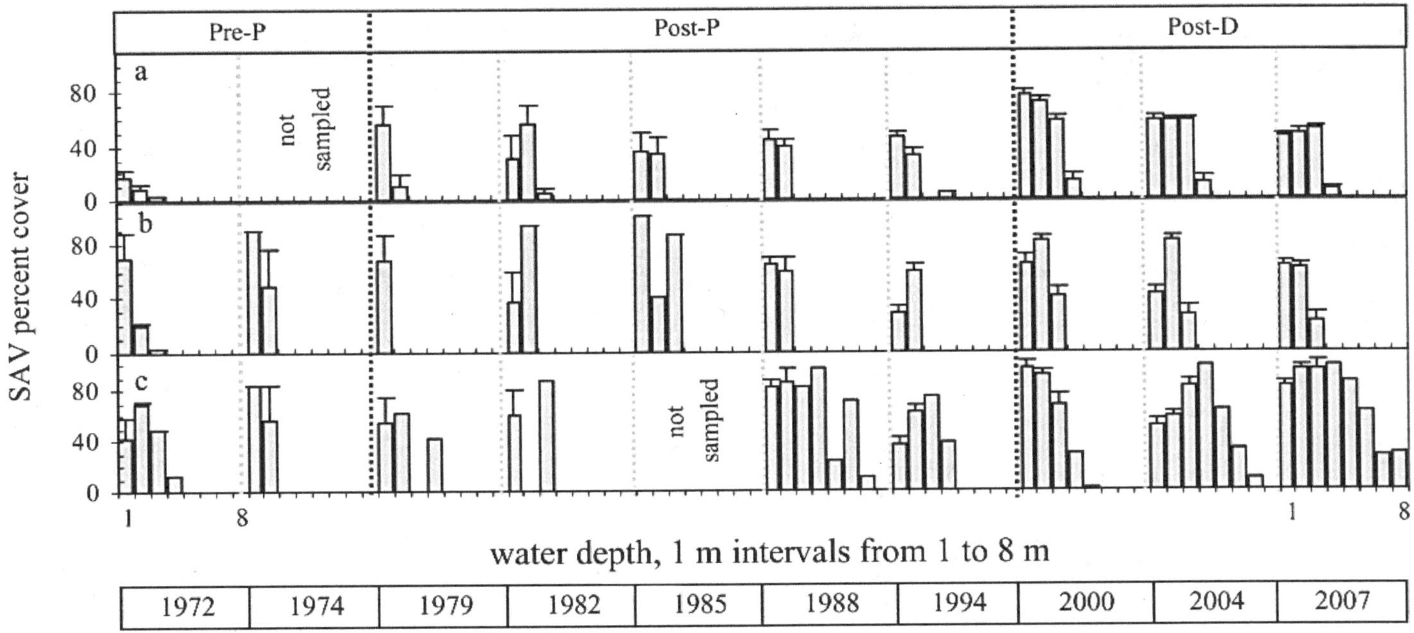

Figure 3. Submerged aquatic vegetation cover by depth interval for (a) upper, (b) middle and (c) lower bay areas in the survey years between 1972 and 2007. Mean cover was calculated from individual cover values along each of the reference transects. Upper bay cover was based on data from the north and south transects at Trenton, Belleville and Big Bay; middle bay cover from east and west transects at Hay Bay and lower bay cover from north and south transects at Conway.`

Table 1. Time stanza means by bay for transparency (Secchi, ε_{par}) and submerged aquatic vegetation variables. Significance was tested prior to averaging with Mann-Whitney tests for SAV biomass and Kruskal-Wallis tests for remaining variables. The differences in superscript letters indicate significant differences ($\alpha = 0.05$) were found in the post-hoc comparisons. ε_{par} is the vertical light extinction rate (m^{-1}) for photosynthetically active radiation, N is the number of samples, Z_c is the maximum depth of colonization.

					Submerged Aquatic Vegetation Variables							
	Secchi mean	Secchi N	ε_{par} mean	ε_{par} N	% Cover mean	% Cover N	Extent mean (m)	Extent N	Z_c mean (m)	Z_c N	Wet Biomass mean (g m^{-2})	Wet Biomass N
Upper Bay												
Pre-P	1.2^{A}	324	1.82^{A}	323	17^{A}	71	197^{AB}	10	2.3^{A}	10		
Post-P	1.3^{B}	397	1.50^{B}	374	42^{B}	301	114^{A}	28	1.6^{A}	27	50	33
Post-D	1.9^{C}	338	1.04^{C}	338	51^{C}	2227	417^{B}	18	3.5^{B}	18	962	18
Significance (P)	<0.0001		<0.0001		<0.0001		0.01		<0.0001		<0.0001	
Middle Bay												
Pre-P	1.4^{A}	162	1.40^{A}	160	50^{AB}	14	102^{AB}	4	2.6^{A}	4		
Post-P	1.5^{A}	237	1.24^{B}	224	49^{A}	119	115^{A}	10	2.2^{A}	10	226	14
Post-D	2.3^{B}	169	0.81^{C}	169	61^{B}	394	276^{B}	6	3.7^{B}	6	903	9
Significance (P)	<0.0001		<0.0001		0:01		0.03		0		0.07	
Lower Bay												
Pre-P	2.9^{A}	162	0.63^{A}	162	56^{AB}	17	123	4	3.7^{A}	4		
Post-P	3.0^{A}	236	0.55^{B}	230	57^{A}	53	91	8	4.3^{A}	8	429	9
Post-D	5.5^{B}	167	0.37^{C}	166	72^{B}	170	93	6	6.5^{B}	6	1363	5
Significance (P)	<0.0001		<0.0001		0		0.31		0.02		0.01	

The mean maximum depth of SAV colonization (Z_c) occurred during the post-D period, and was significantly different than the other two time stanzas in all three bays. The greatest increase in mean Z_c for all three bays occurred between the post-P and -D time stanzas, with the largest increase of 123% occurring in the upper bay. Substantial increases in the mean wet biomass were recorded for all three bays between the post-P and post-D period, although the middle bay difference was not significant due to high variability. The greatest increase occurred in the upper bay where biomass increased from 50 to 962 g m^{-2}.

The relative frequency of occurrence for the five most dominant species in the Bay of Quinte changed considerably over the time period (Figure 4). Not all species were detected in each of the surveys; *Ceratophyllum demersum* was absent in 1979 in the upper bay and was absent through most surveys in the middle and lower bays. *Myriophyllum spicatum* was not detected in the middle bay in 1982, and in the lower bay, *Zosterella dubia* was absent in 1972 and 1982 and *Elodea canadensis* absent in 1982.

Typically, the relative frequency for all species in all three bays for the entire time period was less than 30%. The exception in the upper bay was the dominance of *M. spicatum* in the pre-P period, *E. canadensis* in 1979 and *Z. dubia* in 1982. In the middle bay, both *Z. dubia* and *V. americana* had a relative frequency of 40 in 2004, while in 2007, *Z. dubia* dominated at 65% while *V. americana* decreased. In 1972 in the lower bay, *E. canadensis* dominated, while in 1979, 1982 and 1994, *V. americana* recorded relative frequencies of over forty percent.

Discussion

SAV responses to perturbations differed considerably among the Bay of Quinte sub-basins due principally to trophic gradients and morphometric differences. Point-source P loadings were highest in the shallow upper bay, likely stemming from the high human population density in this area. Major tributaries are also located in the upper bay region and serve to dilute P concentrations downstream when water levels and flushing rates are high. The upwelling that occurs in the middle bay provides further dilution, because Lake Ontario waters are more oligotrophic than the Bay of Quinte (Johannsson et al., 1998). Higher P loading, less dilution, shallower water and higher temperatures all combined to make the upper bay more eutrophic than the middle or lower bays throughout the study period. At the onset of the study, the lower bay

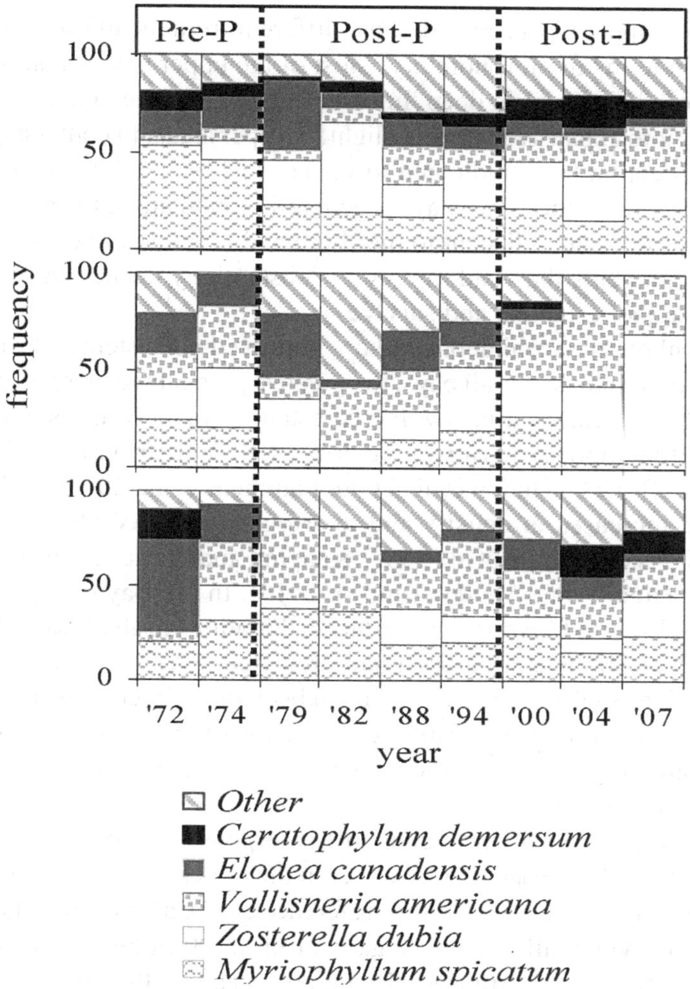

Figure 4. Time series of the frequency of occurrence for the five most abundant species in the (a) upper, (b) middle and (c) lower bay areas between 1972 and 2007. Time series were based on vegetated point data from north transects at Trenton, Belleville and Big Bay for the upper bay, Hay Bay east transect for the middle bay and the Conway north transect for the lower bay.

was considered mesotrophic and did not experience the hypereutrophic conditions of the upper bay before P-control was implemented. Nicholls (2012) provided a detailed analysis of the TP and Chla time series for the Bay of Quinte.

The upper bay is considerably different from the middle and lower bays in terms of morphometry. Most of the upper bay is shallow, with 87% of the area <5 m deep. In addition, much of the upper bay slopes gently into expansive but only slightly deeper areas. In contrast, the area <5 m deep in the middle and lower bays is 39 and 22% respectively. SAV habitat in these two bays is generally found along the narrow strip of shallow water adjacent to the shore and is limited to these areas by steep gradients that lead rapidly to substantially deeper waters where SAV cannot establish and survive.

Although Secchi depth and ε_{par} improved consistently through the three time stanzas for all bays, the greatest percentage increase for both variables occurred after *Dreissena* establishment. Many studies have noted the importance of water clarity on SAV distribution (Chambers and Kalff, 1985; Hudon et al., 2000; Lammens et al., 2004), and several found increases in density and distribution of SAV after *Dreissena* invasions (Skubinna et al., 1995; Knapton and Petrie, 1999; Zhu et al., 2006; Higgins and Vander Zanden, 2010). In the Bay of Quinte, SAV showed limited response to reduced P loading in the period 1978 to 1994. As Minns et al. (1986) had predicted, SAV recovery following reductions in P loadings was delayed because of P reflux from the sediment pool. SAV response after the invasion of Dreissenid Mussels was dramatic, particularly in the upper bay where extent and Z_c increased by more than 125%.

One exception to the higher post-D response of SAV parameters was a substantial increase in SAV cover in the upper bay between the pre-P and post-P period. Much of this difference can be attributed to the survey protocol which allowed SAV gaps of up to 90 m before determining the last SAV on the transect. In the upper bay, both mean SAV extent and Z_c were greater during the pre-P period than the post-P period, but cover was significantly lower. Percent cover was lower because the Trenton north transect in the pre-P period had many gaps before the end of the transect was reached. In the post-P period, the number of gaps and location of the last SAV had been reduced, thus substantially increasing cover and decreasing both extent and Z_c. The sparse SAV found furthest offshore in the pre-P period may have been from the seed bed of the earlier, extensive SAV beds. In the post-D period, the considerable increase in extent indicates that SAV are regaining previously occupied territory. By 2007, the Trenton transect had doubled in extent from the 2004 survey, but this also introduced areas with sparser SAV so the cover average did not increase substantially.

In 1972 in the middle and lower bays, SAV was in better condition in terms of density, Z_c and extent than compared with the upper bay. The greatest increase in density and Z_c for these two bays occurred after the invasion by *Dreissena*, but increases in extent were limited due to the steep slope in some areas, particularly in the lower bay. Steep slopes presented challenges in assessing both extent and Z_c in the lower bay because SAV in this area had already advanced to the edge of the slope in 1972, so assessment of potential expansion beyond the edge had to occur within a very small area along the steep slope. In the middle bay in 1972, SAV still had room for expansion before reaching the drop-off in bathymetry in that area. It achieved that extent on the east transect in the survey following Dreissenid Mussel establishment and the concurrent increase in water clarity. Steeper slopes can influence sediment stability (Hakanson, 1982) and the deposition of nutrient-rich material. Although slopes <3% had little effect on the distribution of fine-grained sediments, Rowan et al. (1986) found that these fine sediments were generally absent on slopes >7.5%. Duarte and Kalff (1986), noted that there was relatively low SAV biomass on slopes >2.2% which they attributed to slope effects on substrate type. In a 1990 study, Duarte and Kalff estimated that a slope of 15% formed the upper threshold for SAV presence.

Of the five most dominant species in the Bay of Quinte, all but *V. americana* were partially to highly tolerant to degraded water quality (Albert and Minc, 2004; Croft and Chow-Fraser, 2007). With improving water quality in the upper bay, *V. americana* increased in relative frequency from two in 1972 to a maximum of 24 in 2007. Additionally, less tolerant species such as *Potamogeton richardsonii*, *Potamogeton zosteriformis* and *Najas flexilis* all had higher occurrences in the post-D period. *M. spicatum* decreased substantially in the upper bay after 1974; declines in this invasive have been reported elsewhere (Carpenter, 1980; Trebitz et al., 1993) and are attributed to multiple factors. No clear trends in species composition were evident in either the middle or lower bays.

The absence or presence of SAV in shallower lakes can have a major impact on whether the lake experiences shifts to a turbid or clear-water regime (Genkai-Kato and Carpenter, 2005; Scheffer and van Nes, 2007). This is due to SAV ability to stabilize sediment resuspension and prevent P-recycling from the sediments, although many other factors, such as climate, lake size and bathymetry also play a role.

Additional factors, including depth, exposure, temperature and water levels can impact the distribution and abundance of SAV. Depth affects several factors, but the principal impact is light attenuation with increasing water depth through scattering and absorption (Wetzel, 1983). Wave exposure has direct effects through removal of plant biomass, displacement of seedlings and transport of propagules. Indirect effects of exposure include the erosion, transport and deposition of sediments and their associated nutrients and organic matter (Keddy, 1982; Cyr, 1998) and increased turbidity from resuspension of the sediments in shallow lakes (Madsen et al., 2001; Horppila and Nurminen, 2003). Water temperature affects SAV physiological responses such as seed germination, development, photosynthesis rates, oxygen consumption, dormancy and turion formation (Pip, 1989, Rooney and Kalff, 2000). High water levels during the early growing season reduce light available to plants; low water levels can damage SAV through ice scour or freezing in the winter, and by wave action or desiccation in summer. SAV may also be affected by herbivory (Lodge, 1991) by invertebrates, fish, mammals and waterfowl. The impact of one or a combination of these factors on SAV will vary depending on their magnitude, timing, duration and the physiology and life history of the SAV species.

A major rationale for listing loss of fish habitat as a beneficial use impairment (BUI) under the Remedial Action Plan in the Bay of Quinte was the substantial reduction in the offshore SAV in the upper bay in the 1960s. An interim target was developed for the delisting criteria for this IBU which requires both an areal and density increase in SAV so that 30% of the bay has a SAV cover of >50%; conditions that were believed to have existed in the early 1900s. This target was based on the anticipated increase in water clarity through reduction of algal densities due to decreased phosphorous Point-source loadings (Johnson, 1986). The final determination of whether the target has been met will be based on findings from an SAV model to be developed for Bay of Quinte using data from 1972 to 2004.

An interim assessment of the BUI target can be expressed as the percentage occupied by SAV of the total across bay width of the 3 transects in the upper bay (Figure 5). Extent in the upper bay had increased substantially in the post-D period, primarily due to changes at the Trenton site. Littoral slope influenced the magnitude of these changes because the low sloping site allowed rapid expansion in bed extent as SAV moved into slightly deeper waters. This interim assessment of the delisting target indicated that the extent criterion had been met by 2007.

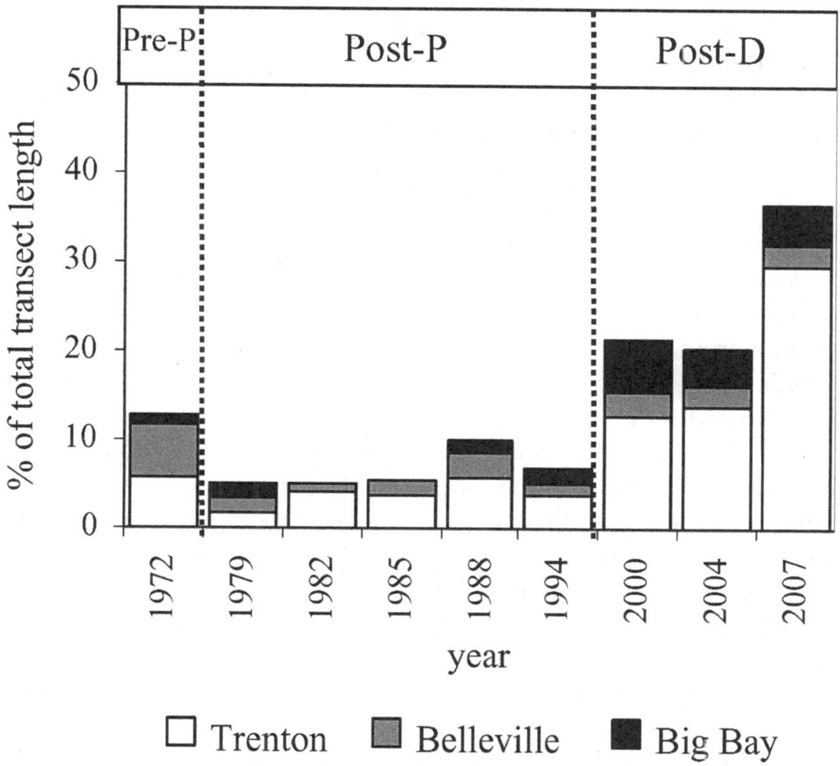

Figure 5. Time series of submerged aquatic vegetation (SAV) bed extent for the three upper bay locations between 1972 and 2007, expressed as a percentage of the total distance across the bay that was occupied by SAV at Trenton, Belleville and Big Bay.

Summary and conclusions

Originally mesotrophic, the Bay of Quinte ecosystem has experienced eutrophication since the 1940s, which resulted in the decline of once-lush submerged aquatic vegetation (SAV) beds in the upper bay by the mid-1960s. Since 1972, twelve SAV surveys have been conducted along ten index transects, recording: % cover, distance SAV beds extended from shore (extent), maximum depth of colonization (Z_c), species composition, and, in later years, wet plant biomass. Offshore Secchi depth and ε_{par}, (the vertical light extinction rate (m^{-1}) for photosynthetically active radiation), were also recorded either weekly or bi-weekly during the growing season since 1972. During this time, two major changes occurred within the bay:

the reduction in Point-source phosphorus (P-control) loadings in 1978 and the 1993 invasion by Dreissenid Mussel. SAV response to these changes varied temporally and spatially, with the shallow upper bay showing the greatest response, particularly after Zebra Mussels establishment. In the upper bay, mean Secchi depth increased by 8% from 1.2 m prior to P-control (pre-P), to 1.3 m after P-control (post-P) and further increased by 46% to 1.9 m after *Dreissena* establishment (post-D). Upper bay SAV responded to these invasive species with increases in the means of three variables: Z_c from 1.6 to 3.5 m, extent from 114 m to 417 m and wet biomass from 50 g m^{-2} to 962 g m^{-2}. SAV in the middle and lower bays were in better condition in 1972, with pre-P cover in excess of 50% and Z_c of 2.6 and 3.7 m, respectively. SAV cover did increase in the post-D (1994 to 2007) period by approximately 25% and Z_c increased to 3.7 and 6.5 m, but the narrow fringing strip of shallower water along the shore in these two deeper bays limited substantial increases in bed extent. Both water clarity and basin morphometry strongly influenced SAV distribution and abundance within the Bay of Quinte.

As reported elsewhere, the SAV in the Bay of Quinte experienced a greater response in bed extent and Z_c to the invasion by Zebra Mussel rather than the reduction in P source loadings. In addition to increases in water clarity, basin morphometry and the sheltered nature of the Bay of Quinte played a key role in the magnitude of the response. Although the lowest SAV density in the post-D period was in the upper bay, the gently sloping sites allowed for substantial expansion of SAV beds when waters became sufficiently clear. This 35-year dataset will be used to estimate if the SAV component of the delisting criteria has been met; an important step towards designating the Bay of Quinte as an Area in Recovery. This assessment has documented changes in the bay showing that excessive eutrophication and invasive species had profound effects on the SAV community. Maintaining current, low P loading levels and preventing invasions of plants or animals that could be detrimental to SAV will be essential if a healthy SAV community is to be sustained in the future.

Acknowledgements

Funding for this project was provided by the Great Lakes Action Plan. The authors gratefully acknowledge all the people who contributed to this 35 year dataset, particularly Adele Crowder, Jeff Warren and GLLFAS field crews. Thanks also to Erin Gertzen who ran the post-hoc comparisons in R. Both Erin and Nick Lapointe provided most helpful comments which substantially improved this manuscript.

References

Albert, D.A., Minc, L.D., 2004. Plants as regional indicators of Great Lakes coastal wetland health. Aquatic Ecosyst. Health Mgmt. 7(2), 233–247.

Bay of Quinte Remedial Action Plan. www.bqrap.ca Accessed January 2011.

Bay of Quinte RAP Coordinating Committee, 1990. Stage 1: Environmental Settings and Problem Definition. Belleville, Canada.

Bay of Quinte RAP Coordinating Committee, 1993. Stage 2: Time to Act. Belleville, Canada.

Best, E.P.H., 1982. The aquatic macrophytes of Lake Vetchen. Species composition, spatial distribution and production. Hydrobiologia 95, 65–77.

BioSonics, 2000. BioPlant Version 1.0. Seattle, WA, USA. Blindow, I., 1992. Long- and short-term dynamics of submerged macrophytes in two shallow eutrophic lakes. Freshwater Biology 28, 15–27.

Bristow, J.M., Crowder, A.A., King, M.R., 1977. The growth of aquatic macrophytes in the Bay of Quinte prior to phosphate removal by tertiary sewage treatment (1975–1976). Naturaliste Can. 104, 465–473.

Carpenter, S.R., 1980. The decline of *Myriophyllum spicatum* in a eutrophic Wisconsin lake. Can. J. Bot. 58, 527–535.

Carpenter, S.R., Lodge, D.M., 1986. Effects of submerged macrophytes on ecosystem processes. Aquat. Bot. 26, 341–370.

Chambers, P.A., Kalff, J., 1985. Depth distribution and biomass of submersed aquatic macrophytes communities in relation to Secchi depth. Can. J. Fish. Aquat. Sci. 42, 701–709.

Chick, J. H., McIvor, C.C., 1994. Patterns in the Abundance and Composition of Fishes Among Beds of Different Macrophytes - Viewing a Littoral Zone As a Landscape. Can. J. Fish. Aquat. Sci. 51(12), 2873–2882.

Croft, M.V., Chow-Fraser, P., 2007. Use and development of the wetland macrophyte index to detect water quality impairment in fish habitat of great lakes coastal marshes. J. Great Lakes Res. 33, 172–197.

Crowder, A., Bristow, M., 1986. Aquatic macrophytes in the Bay of Quinte, 1972–1982. Can. Spec. Publ. Fish. Aquat. Sci 86, 114–127.

Cyr, H., 1998. Effects of wave disturbance and substrate slope on sediment characteristics in the littoral zone of small lakes. Can. J. Fish. Aquat. Sci. 55(4), 967–976.

Duarte, C. M., Kalff, J., 1986. Littoral slope as a predictor of the maximum biomass of submerged macrophyte communities. Limnol. Oceanogr. 31, 1072–1080.

Duarte, C. M., Kalff, J., 1990. Patterns in the Submerged Macorphyte Biomass of Lakes and the Importance of the Scale of Analysis in the Interpretation. Can. J. Fish. Aquat. Sci. 47, 357–363.

Environmental Systems Research Institute, 1992. ArcView GIS 3.3.

Genkai-Kato, M., Carpenter, S.R., 2005. Eutrophication due to phosphorous recycling in relation to lake morphometry, temperature, and macrophytes. Ecology, 86(1), 210–219.

Gentleman, R., Ihaka, R., 1997. *The R Project for Statisticalm Computing*. University of Auckland, New Zealand.

Hakanson, L., 1982. Bottom dynamics in lakes. Hydrobiologia 91, 9–22.

Higgins, S.N., Vander Zanden, M.J., 2010. What a difference a species makes: a meta-analysis of Dreissenid Mussel impacts on freshwater ecosystems. Ecol. Monogr. 80(2), 179–196.

Horppila, J., Nurminen, L., 2003. Effects of submerged macrophytes on sediment resuspension and internal phosphorus loading in Lake Hiidenvesi (southern Finland). Water Research. 37, 4468–4474.

Hudon, C., Lalonde, S., Gagnon, P., 2000. Ranking the effects of site exposure, plant growth form, water depth, and transparency on aquatic plant biomass. Can. J. Fish. Aquat. Sci. 57, 31–42.

International Joint Commission (IJC), 1987. Revised Great Lakes Water Quality Agreement of 1978, as amended by protocol, signed November 18, 1987. United States and Canada.

Johannsson, O.E., Millard, E.S., Ralph, K.M., Myles, D.D., Graham, D.M., Taylor, W.D., Giles, B.G., and Allen, R.E., 1998. The changing pelagia of Lake Ontario (1981 to 1995): A report of the DFO Long-Term Biomonitoring (Bioindex) Program. Can. Tech. Rep. Fish. Aquat. Sci. 2243.

Johnson, M.G., 1986. Phosphorus loadings and environmental quality in the Bay of Quinte, Lake Ontario. Can. Spec. Publ. Fish. Aquat. Sci. 86, 247–258.

Johnson, M.G., Hurley, D.A., 1986. Overview of Project Quinte - 1972–82. Can. Spec. Publ. Fish. Aquat. Sci. 86, 1–6.

Keast, A., 1984. The introducted aquatic macrophyte, *Myriophyllum spicatum*, as a habitat for fish and their invertebrate prey. Canadian Journal of Zoology 62, 1289–1303.

Keddy, P.A., 1982. Quantifying within-lake gradients of wave energy: interrrelationships of wave energy, substrate particle size and shoreline plants in Axe Lake, Ontario. Aquatic Botany 14, 41–58.

Knapton, R.W., Petrie, S.A., 1999. Changes in distribution and abundance of submersed macrophytes in the Inner Bay at Long Point, Lake Erie: Implications for foraging waterfowl. J. Great Lakes Res. 25(4), 783–798.

Lammens, E.H.R.R., van Nes, E.H., Meijer, M.L., van den Berg, M.S., 2004. Effects of commercial fishery on the bream population and the expansion of *Chara aspera* in Lake Veluwe. Ecol. Model. 177, 233–244.

Leisti, K.E., Millard, E.S., Minns, C.K., 2006. Assessment of Submergent Macrophytes in the Bay of Quinte, Lake Ontario, August 2004, Including Historical Context. Can. MS Rpt. Fish. Aquat. Sci. 2762.

Lodge, D.M., 1991. Herbivory on freshwater macrophytes. Aquatic Botany 41, 195–224.

Madsen, J. D., Chambers, P.A., James, W.F., Koch, E.W., Westlake, D.F., 2001. The interaction between water movement, sediment dynamics and submersed macrophytes. Hydrobiologia 444 (1–3), 71–84.

Minns, C.K., Owen, G.E., Johnson, M.G., 1986. Nutrient Loads and Budgets in the Bay of Quinte, Lake Ontario, 1965–1981. Can. Spec. Publ. Fish. Aquat. Sci. 86, 59–76.

Munsel, U., Hothorn, L.A., 2001. A Unified Approach to Simultaneous Rank Test Procedures in the Unbalanced One-way Layout. Biometrical Journal 43(5), 553–569.

Nicholls, K. H., 2012. Phosphorous and chlorophyll in the Bay of Quinte: A time-series/ intervention analysis of 1972–2008 data. Aquat. Ecosyst. Health Mgmt. 15(4), 421–429.

Pip, E., 1989. Water Temperature and Freshwater Macrophyte Distribution. Aquatic Botany 34, 367–373.

Randall, R.G., Minns, C.K., Cairns, V.W., Moore, J.E., 1996. The relationship between an index of fish production and submerged macrophytes and other habitat features at three littoral areas in the Great Lakes. Can. J. Fish. Aquat. Sci. 53, 35–44.

Rooney, N., Kalff, J., 2000. Inter-annual variation in submerged macrophyte community biomass and distribution: the influence of temperature and lake morphometry. Aquatic Botany 68, 321–335.

Rowan, D.J., Kalff, J., Rasmussen, J.B., 1992. Estimating the Mud Deposition Boundary Depth in Lakes from Wave Theory. Can. J. Fish. Aquat. Sci. 49(12), 2490–2497.

Scheffer, M., van Nes, E.H., 2007. Shallow lakes theory revisited: various alterantive regimes driven by climate, nutrients, depth and lake size. Hydrobiologia 584, 455–466.

Skubinna, J.P., Coon, T.G., Batterson, T.R., 1995. Increased abundance and depth of submersed macrophytes in response to decreased turbidity in Saginaw Bay, Lake Huron. J. Great Lakes Res. 21(4), 476–488.

Sly, P.G., 1986. Review of postglacial environmental changes and cultural impacts in the Bay of Quinte. Can. Spec. Publ. Fish. Aquat. Sci. 86, 7–26.

Trebitz, A.S., Nichols, S.A., Carpenter, S.R., Lathrop, R.C., 1993. Patterns of vegetation change in Lake Wingra following a *Myriophyllum spicatum* decline. Aquatic Botany 46, 325–340.

Weaver, M.J., Magnuson, J.J., Clayton, M.K., 1997. Distribution of littoral fishes in structurally complex macrophytes. Can. J. Fish. Aquat. Sci. 54(10), 2277–2289.

Werner, E.E., Gilliam, J.F., Hall, D.J., 1983. An experimental test of the effects of predation risk on habitat use in fish. Ecology 64, 1540–1548.

Wetzel, R.G., 1983. *Limnology*. Harcourt Brace College Publishers, Orlando, FL.

Zhu, B., Fitzgerald, D.G., Mayer, C.M., Rudstam, L.G., Mills, E.L., 2006. Alteration of Ecosystem Function by Zebra Mussels in Oneida Lake: Impacts on Submerged Macrophytes. Ecosystems 9, 1017–1028.x

The above article is a reprinted version of the following previously published article:

Submerged aquatic vegetation in the Bay of Quinte: Response to decreased phosphorous loading and Zebra Mussel invasion

K. E. Leisti; S. E. Doka; C. K. Minns

Aquatic Ecosystem Health & Management (2012) 15 (4): 442–452.

https://doi.org/10.1080/14634988.2012.736825

Changes in zooplankton biomass in the Bay of Quinte with the arrival of the Mussels, *Dreissena polymorpha* and *D. rostiformis bugensis*, and the predatory cladoceran, *Cercopagis pengoi*: 1975 to 2008

Kelly L. Bowen* and Ora E. Johannsson

Fisheries and Oceans Canada,
Burlington, ON, L7S 1A1
**Corresponding author: kelly.bowen@dfo-mpo.gc.ca*

Keywords: cyclopoid, rotifer, Lake Ontario, Zebra and Quagga Mussels

Introduction

The Bay of Quinte is a narrow embayment 64 km long and 254 km^2, on the north-eastern shore of Lake Ontario (Johnson and Hurley, 1986; Figure 1). The Bay was officially listed as an Area of Concern in the 1987 amendment of the Great Lakes Water Quality Agreement (International Joint Commission, 1988). However, concerns over cultural eutrophication started in the 1960s, and led to the development of "Project Quinte" in 1972 (Minns et al., 1986). This multi-agency, multi-year research program was largely developed to evaluate ecological responses to phosphorus reductions implemented in 1978. This monitoring has continued to the present (2008), providing one of the longest running and most complete limnological data sets focusing on water quality, zooplankton, phytoplankton and fish.

The Remedial Action Plan developed for Quinte listed zooplankton as one of its impaired beneficial uses (Bay of Quinte RAP Coordinating Committee, 1993). The zooplankton community changes in structure from the shallow, eutrophic habitat of the upper bay, through less eutrophic conditions in the middle section, to a meso-oligotrophic environment in the mouth of the bay. Zooplankton community structure and biomass are controlled not only by the physical environment but also by the quality and quantity of available food and level and type of predation. This is determined by nutrient conditions, the structure of the fish community

Ecology of the Bay of Quinte: Health, Management and Global Implications
Editors: C.K. Minns, M. Munawar and M.A. Koops

Aquatic Ecosystem Health & Management Society, Canada

which is the primary source of mortality in this system, and the introduction of new species which either compete for food or alter predation on or within the zooplankton community. A few strongly invasive species colonized the Bay in the 1990s. Zebra Mussels (*Dreissena polymorpha*) and Quagga Mussels (*D. rostriformis bugensis*) were established in the lower bay by 1993, and *D. polymorpha* was found in the upper bay by 1994 (Dermott et al., 2003). *Cercopagis pengoi,* a predatory cladoceran, arrived in Lake Ontario in 1998 (MacIsaac et al., 1999) and were found in the lower bay in 1999.

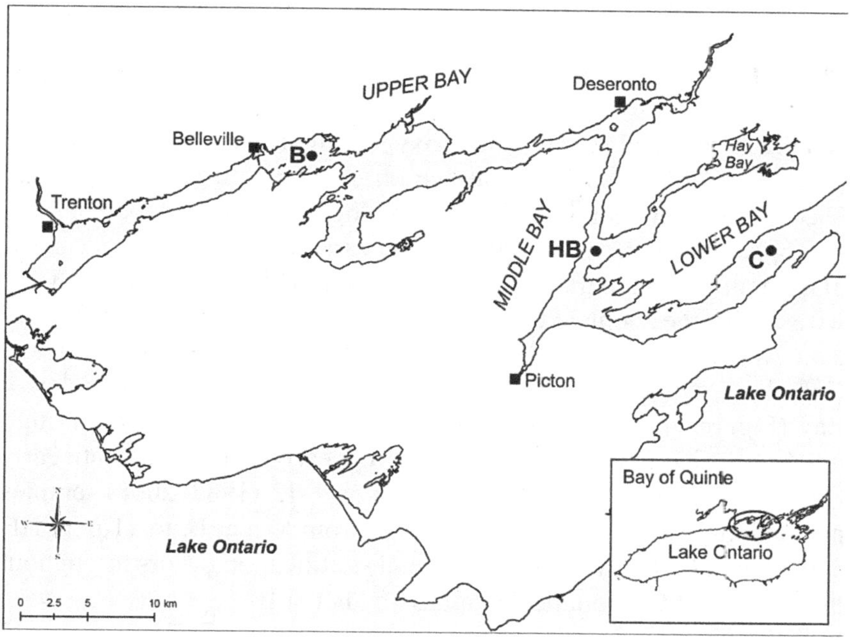

Figure 1. Location of zooplankton sampling stations in the Bay of Quinte, Lake Ontario. B = Belleville, HB = Hay Bay, and C = Conway.

At the beginning of Project Quinte, it was postulated that phosphorus reductions, and resulting lowered algal abundance would lead to a drop in zooplankton standing crop and production, and changes in species composition (Cooley et al., 1986). However, monitoring results comparing two years (1975–1976) prior to phosphorus control with the five years following phosphorus control (1979–1983) indicated that these changes did not occur. In the 1960s and early 1970s, small-bodied cladocerans such as *Bosmina*, *Eubosmina* and *Chydorus* dominated the upper Bay, and this dominance continued throughout the Bay in the early post-phosphorus control period (Cooley et al., 1986). Cyclopoid copepods, particularly juvenile nauplii and copepodids, increased in relative abundance down the length of the bay toward Lake Ontario. As a follow-up to Cooley et al. (1986), the purpose of this paper is to provide an overview of trends in zooplankton community composition, abundance, and biomass from 1975 to 2008, with emphasis on the changes that occurred following invasion by Dreissenid Mussels and *C. pengoi*. Our goal was not to examine interactions with planktivorous fish or with changes in the algal community in the Bay of Quinte; those analyses will be discussed in future publications.

Methods

A suite of physical (temperature, oxygen, light attenuation), chemical (total phosphorus, soluble reactive phosphorus, nitrate+nitrite, total Kjedhal nitrogen, conductivity) and biological (benthos, zooplankton, phytoplankton and, as of 2000, rotifers) variables were collected as part of the monitoring program in the Bay of Quinte (Minns et al., 1986). Here we examine the zooplankton and rotifer data, touching on other relevant data. At least three primary stations, Belleville (B), Hay Bay (HB), and Conway (C) which fall along a trophic gradient from the upper to lower Bay (Figure 1), have been sampled every one or two weeks through much of the ice-free season, generally from early to mid-May until early October: a total of 17–23 (1975–1982) or 9–12 (1982–2008) samples per station-year. Station depth increases from 5 m at B, to 11 m at HB and 32 m at C. This means that thermal stratification occurs throughout the summer at C, sporadically at HB and not at B.

Zooplankton sample collection

Discrete samples for zooplankton were collected through the water column with a 41-L Schindler-Patalas trap fitted with 64-µm mesh. Three depths were sampled at B (1 m, 2 m and 3 m), five at HB (1 m, 2 m, 3 m, 6 m and 9 m), and eight at C (1 m, 3 m, 5 m, 8 m, 10 m, 15 m, 20 m and 25 m). Samples were preserved with 4% sugar, buffered formalin. A single composite sample was constructed for each station-date by combining 50% of the sample from each depth.

Rotifer sampling began in 2000 by collecting 1 L of water from each depth indicated above using a Van Dorn sampler. For each station date, water was pooled and filtered through 20-µm mesh. Rotifers were narcotized using carbonated water and preserved as above. Each year, a seasonal composite sample for each station was made by combining 50% of the sample from each date.

Sample enumeration

Zooplankton were enumerated using a weighted counting procedure where enumeration of dominant species ceased before that of rarer species (Cooley et al., 1986). The count was considered complete when a minimum of 400 individuals had been identified. Loose eggs were counted in all aliquots examined. A minimum of 200 organisms were counted in a sample dominated by only one to two species with extremely few other species present, as can occur in the early spring. For numerically dominant species, a minimum 50 individuals were measured. The exotic predatory cladoceran *C. pengoi* was found at B and C in 1999, and at all three stations by 2000. After its arrival, each of the discrete depth samples was filtered through 400-µm mesh, and all *C. pengoi* enumerated. This species cannot be sub-sampled because individuals form clumps as their caudal spines become entangled.

Seasonally-weighted mean (SWM) abundance and biomass of zooplankton were calculated over the May 1 to October 6 sampling season, weighting for time between each sample, including the start (May 1) and end (October 6) dates, in order to standardize the time period and adequately account for missing samples etc. Starting in 1995, zooplankton lengths were measured on each date using a Summa Sketch III digitizing system and dissecting microscope equipped with a camera lucida. A program written by Russ Hopcroft (University of Alaska) recorded the lengths and calculated the biomass using length-weight relationships from Johannsson et al. (2000), except for *Bosmina* and *Eubosmina* where a relationship developed for Lake Erie zooplankton was employed. This is $W = 10.715*L^{2.12}$, where weight (W) is in µg and length (L) is in mm. *C. pengoi* weights were estimated using equations in Grigorovich (2000). Cladoceran taxa not listed in Johannsson et al. (2000) were rare and most closely resembled *Daphnia*, thus the *Daphnia* equation was applied. Starting in 1995, biomass was calculated for each species on each station-date by multiplying the abundance and estimated mean individual weight for that date.

Prior to 1995, a single individual mean weight was initially assigned to each species based on average measured dry weights of individuals caught in Lakes Superior, Huron and Ontario in the early 1970s. These weights were originally determined by J.B. Wilson and N.H.F. Watson (Fisheries and Oceans Canada, Burlington, ON, Canada, unpublished data) and Cooley et al. (1986). Biomass was initially estimated as abundance times these weights. However, it was necessary to standardize our methodology across all years to examine temporal trends and to assess the impact of this change in methodology on biomass estimates in the Bay. We collated and analysed archived samples across the season for each station-year from 1975–1994 in order to get species lengths and estimates of species annual mean weights. SWM biomass of each species was then calculated as the product of the station-year specific weight and SWM abundance. For rare taxa, for which there were few or no measurements during these years, mean weights were assigned depending on available data in this order: (1) the average for the 1975–1994 period, (2) the average for the 1995–2008 period, and (3) a mean weight determined for a similar Great Lakes' species (mean GL) determined by Wilson and Watson; that is, so rare that it did not occur in any of the collated samples, nor was it found after 1994.

Enumeration and identification of rotifers were similarly performed using a weighted counting method and limits similar to that of the zooplankton. In cases where samples contained few rotifers, a maximum of 25% of the sample by volume was analysed. The biovolume (mm^3) for each individual was calculated by using formulae of Ruttner-Kolisko (in McCaulley, 1984). Biovolume equals wet weight, assuming a density of 1. Formula used for *Polyarthra* species are given in Johannsson et al. (2000). Dry weights were calculated by multiplying wet weight by 0.1, or in the case of *Asplanchna* sp. 0.036 (Dumont et al., 1975). Mean weights of rotifer taxa from Quinte were compared to those from Lake Erie (1993, 1994) (Graham et al., 1996; O. Johannsson and J. LeBlanc, Fisheries and Oceans Canada, Burlington, ON, Canada, unpublished data).

Statistical analyses

Unless otherwise indicated, comparisons were considered significantly different at $p < 0.05$. The pre-phosphorus control (pre-P) period from 1975 and 1976 was excluded from statistical analyses as the effect of phosphorus control was described by Cooley et al. (1986). The years 1992 and 1993 were also omitted from the analyses because of their unusually low zooplankton biomass associated with the cold summer of the Pinatubo eruption (Newhall and Punongbayan, 1996) and slow return of copepod abundance. 1994 was also omitted because it was a transition year for dreissenid colonization. For each zooplankton group (total zooplankton, cladocerans, cyclopoids and calanoids), the remaining 27 year data set was log-transformed (no zero values present) and split into the following time stanzas: pre-dreissenid (pre-DM): 1979 to 1991 and post-dreissenid (post-DM): 1995 to 2008.

For each group, differences in SWM biomass were examined using general linear models (GLM) in SystatV11, with station, stanza and station*stanza as factors. Within each stanza, differences among stations were examined using ANOVA and Bonferroni post-hoc tests ($p < 0.025$). At each station, differences among the time stanzas were examined using t-tests. Cladoceran mean length, an index of fish predation, was also compared between stanzas at each site using t-tests.

In order to test for impacts associated with the *C. pengoi* invasion, the post-DM data were divided into pre-CP (1995–1998) and post-CP (1999–2008) periods. Differences in biomass of each zooplankton group and dominant taxa (veligers, *Bosmina* spp., *Eubosmina coregoni*, *Daphnia galeata mendotae*, *Daphnia retrocurva*, *Ceriodaphnia* sp., *Chydorus* sp., cyclopoid nauplii and copepodids and calanoid nauplii) were assessed using t-tests on the transformed data.

Results

Total zooplankton densities and biomass

Zooplankton biomass peaked at all three stations between 1981–83 and 1991 and appeared to be lower and more variable thereafter, although means at HB have been relatively stable since 2004 (Figure 2A– C). High values were again observed in 1994 and 2001 at B, and in 1996 and 2008 at C. Biomass reached unusually low levels at all three stations in 1992 (and again in 1993 at C and HB). The eruption of Mount Pinatubo in the Philippines in June 1991 (Newhall and Punongbayan, 1996) cooled the earth and was reflected in unusually low summer temperatures in the Bay of Quinte in 1992. The June to August mean water temperature at B was the lowest observed during the study: 20.3°C compared with the 1975–2008 mean of 22.3°C (excluding 1992).

GLM analyses revealed that zooplankton densities and biomass decreased significantly between the pre-DM and post-DM time stanzas (Figure 3A). The patterns were similar across stations as indicated by a lack of significant interaction terms (stanza*station) in the GLM models. Mean total densities in the pre-DM stanza averaged 207 ± 16, 275 ± 28 and 92 ± 7 individuals l^{-1} at B, HB and C, respectively - values which

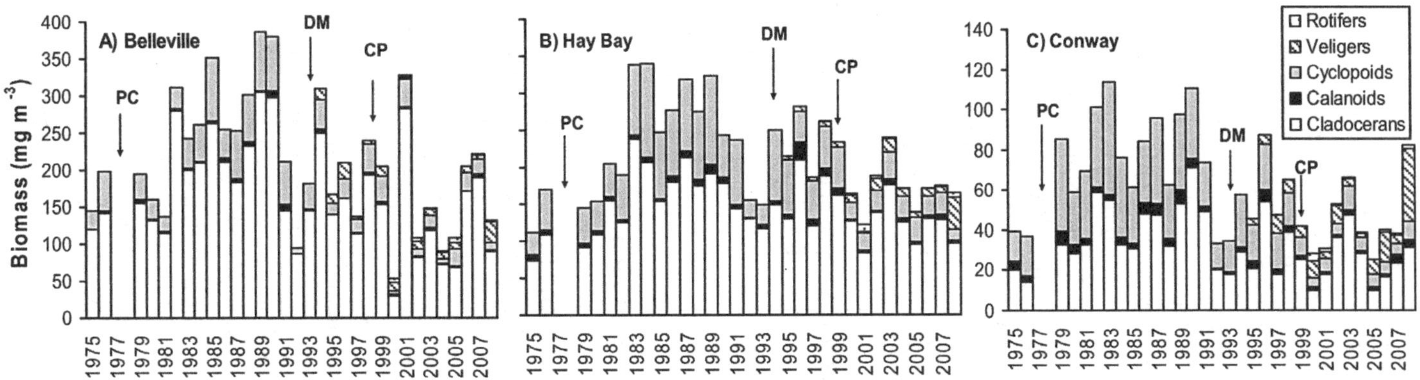

Figure 2. Trends in seasonally-weighted mean dry biomass (A-C) of zooplankton groups at Belleville, Hay Bay and Conway. These are volumetric estimates based on measured lengths from May 1 to October 6. Time stanzas are indicated by the arrows: implementation of phosphorus control (PC), invasion by Dreissenid Mussels (DM) and invasion by *Cercopagis pengoi* (CP). Note that C uses a different scale. No samples were collected in 1977 or 1978. Rotifer sampling began in 2000. One *Bosmina* value from May 1991 at HB was unrealistically high and replaced with a more realistic value (0.1 * original value).

were all higher than the corresponding values in the post-DM stanza (146 ± 9, 167 ± 9 and 55 ± 6 individuals l^{-1}). In the pre-DM period, mean total biomass (±1 S.E.) was higher at B (265 ± 21 mg m^{-3}) and HB (253 ± 19 mg m^{-3}) compared to C (84 ± 5 mg m^{-3}). In the post-DM stanza, B (166 ± 20 mg m^{-3}) and HB (192 ± 13 mg m^{-3}) were again higher than C (48 ± 5 mg m^{-3}). Declines in total biomass were 37% at B, 24% at HB, and 42% at C (Figure 3A).

Compared to the pre-DM period, cladoceran mean length increased in the post-DM period at C (0.356 ± 0.011 mm to 0.425 ± 0.009 mm) and HB (0.401 ± 0.014 mm to 0.450 ± 0.013 mm), but remained unchanged at B (0.439 ± 0.018 mm to 0.434 ± 0.017 mm). Within the post-DM period, cladoceran mean length decreased post-CP at HB (0.503 ± 0.025 to 0.423 ± 0.033 mm) but remained unchanged at B and C.

GLM analyses revealed that zooplankton densities and biomass decreased significantly between the pre-DM and post-DM time stanzas (Figure 3A). The patterns were similar across stations as indicated by a lack of significant interaction terms (stanza*station) in the GLM models. Mean total densities in the pre-DM stanza averaged 207 ± 16, 275 ± 28 and 92 ± 7 individuals l^{-1} at B, HB and C, respectively - values which were all higher than the corresponding values in the post-DM stanza (146 ± 9, 167 ± 9 and 55 ± 6 individuals l^{-1}). In the pre-DM period, mean total biomass (±1 S.E.) was higher at B (265 ± 21 mg m^{-3}) and HB (253 ± 19 mg m^{-3}) compared to C (84 ± 5 mg m^{-3}). In the post-DM stanza, B (166 ± 20 mg m^{-3}) and HB (192 ± 13 mg m^{-3}) were again higher than C (48 ± 5 mg m^{-3}). Declines in total biomass were 37% at B, 24% at HB, and 42% at C (Figure 3A).

Compared to the pre-DM period, cladoceran mean length increased in the post-DM period at C (0.356 ± 0.011 mm to 0.425 ± 0.009 mm) and HB (0.401 ± 0.014 mm to 0.450 ± 0.013 mm), but remained unchanged at B (0.439 ± 0.018 mm to 0.434 ± 0.017 mm). Within the post-DM period, cladoceran mean length decreased post-CP at HB (0.503 ± 0.025 to 0.423 ± 0.033 mm) but remained unchanged at B and C.

Zooplankton group biomass

With the exception of cladocerans and calanoids at HB, biomass of each zooplankton group declined after the dreissenid invasion (Figures 3B– D). The patterns were similar across stations as indicated by a lack of significant interaction terms (stanza*station) in the GLM models. Cladocerans were dominant at all stations (Figure 2), and values were higher at B and HB relative to C in both time stanzas (Figure 3B). In the pre-DM stanza, percent cladocerans (by biomass) was highest at B (79%), intermediate at HB (66%) and lowest at C (52%). In the post-DM stanza, the proportion of cladocerans remained relatively stable, with both B (80%) and HB (71%) higher than C (56%). Based on t-tests, dominant cladoceran taxa also showed significant differences following dreissenid invasion, but these were often station-specific (Table 1). For example, *Bosmina* spp. showed a decline only at C, but *E. coregoni* dropped at all three stations following invasion.

Table 1. Changes in biomass of dominant cladoceran taxa at Belleville (B), Hay Bay (HB) and Conway (C) during the pre- and post-DM time stanzas, based on t-tests ($p < 0.05$). NS = no significant change, + represents an increase following dreissenid invasion (post-DM) relative to pre-DM, and – represents a decrease.

Taxa Name	B	HB	C
Total Cladocerans	–	NS	–
Bosmina longirostris	NS	NS	–
Eubosmina coregoni	–	–	–
Daphnia galeata mendotae	NS	+	–
Daphnia retrocurva	–	–	+
Chydorus sp.	–	+	–
Ceriodaphnia sp.	NS	NS	NS

In both stanzas, cyclopoid biomass was higher at HB relative to B and C. The proportion of cyclopoids in the pre-DM stanza was highest at C (42.7%), intermediate at HB (31.8%) and lowest at B (19.3%). These values dropped at all stations in the post-DM period, with corresponding values of 21.4%, 19.6% and 12.5%. Calanoid biomass tended to be highest at HB in both stanzas. Percentages of calanoids were highest at C, intermediate at HB and lowest at B in both time stanzas, and ranged from 1.2% to 5.2%.

Following dreissenid invasion, the proportion of veliger biomass in the zooplankton community largely replaced the proportion lost by cyclopoids. Veligers were more dominant at C (17.5%) than at HB (5.9%), and B (6.2%). Veliger biomass showed no significant differences among stations.

Only a few *Bythotrephes* have been found at Quinte, and these have been restricted to the lower bay in recent years. *C. pengoi* has only

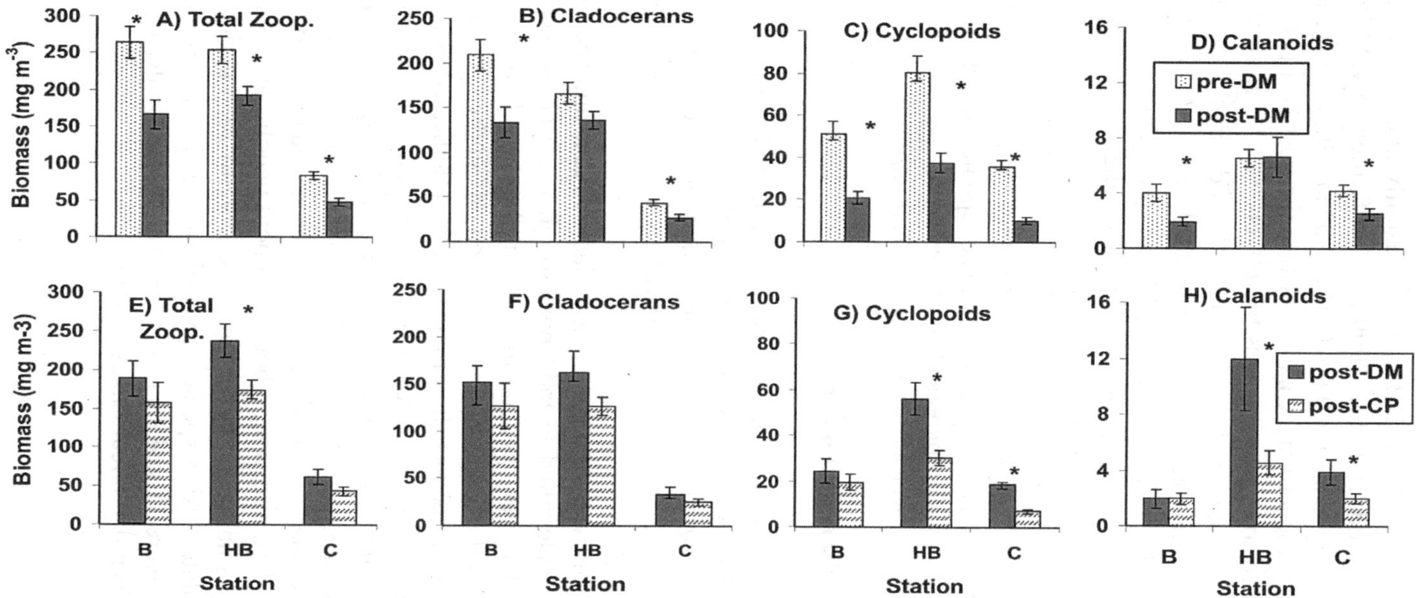

Figure 3. Effect of time stanza on mean (±1 SE) SWM dry biomass (A-D)for dominant zooplankton groups in the pre-dreissenid (pre-DM) and post-Dreissenid (post-DM) periods at Belleville (B), Hay Bay (HB) and Conway (C) from May 1 to October 6. Figures E-H compare groups in the post-DM/pre-*C. pengoi* (pre-CP; 1995–1998) to the post-DM/post-*C. pengoi* (post-CP; 1999–2008) periods. Stations denoted with an * showed a significant decline. Please note the different scales.

rarely been found at B and HB, and was usually abundant at C for a short summer period. Between 1999 and 2008, mean SWM biomass of *C. pengoi* averaged 0.01 ± 0.00, 0.04 ± 0.02 and 0.58 ± 0.12 mg m^{-3} at B, HB and C, respectively. Changes within the post-DM period were further examined to determine possible effects of *C. pengoi* invasion. At C, biomass of total cyclopoids, total calanoids, cyclopoid nauplii, cyclopoid copepodids and *D. retrocurva* declined significantly in the post-CP period (1999–2008) compared to the pre-CP, post-DM period (1995–1998) (Figure 3E–H). Total cladocerans, veligers, calanoid nauplii and the cladocerans *Bosmina* spp., *E. coregoni*, *D. galeata mendotae*, *Chydorus* and *Ceriodaphnia* remained unchanged. At HB, where *C. pengoi* were much less abundant, total biomass, total cyclopoids, total calanoids, cyclopoid nauplii and cyclopoid copepodids also declined over this period, and veligers increased. At B, where *C. pengoi* were rarely found, the only change was an increase in *Ceriodaphnia*.

Species composition and mean sizes

The following zooplankton taxa were dominant at all stations (>5% of biomass): the cladocerans *Bosmina* spp., *D. galeata mendotae*, *D. retrocurva* and *E. coregoni*, cyclopoid copepodids and dreissenid veligers (Table 2, in bold). The cyclopoid *Diacyclops thomasi* was a dominant species at C.

Individual dry weights of many zooplankton taxa and species, estimated from mean length data and length-weight regressions, were lower than the mean GL weights determined from the early 1970s (Table 2). Mean weights of the dominant taxa are based on hundreds or thousands of measurements, whereas values for the rare taxa (indicated by an *) may be based on only a few measurements. *Daphnia* and *Ceriodaphnia* taxa, *Diaphanosoma birgei*, *Holopedium gibberum*, and *Sida crystallina* were notably smaller in Quinte than in the open waters of the Great Lakes in the early 1970s. Species individual dry weights tended to be more similar for the copepod taxa, although both cyclopoid nauplii and copepodids were generally smaller in Quinte. Over the 1995 to 2008 period, SWM biomass calculated using species dry weights determined from length measurements were 38 to 45% lower than biomass estimates calculated using the 1970s mean Great Lakes weights.

Rotifers

Seasonal mean dry rotifer biomass (±S.E.) over the 2000 to 2008 period was 2.87 ± 0.66 mg m^{-3} at B, 2.83 ± 0.92 mg m^{-3} at HB, and 1.69 ± 0.54 mg m^{-3} at C (Figure 4). Mean rotifer biomass was only 1.7, 1.5 and 3.4% of zooplankton + rotifer SWM biomass at the three stations, respectively. The dominant rotifer taxa (>5% of biomass) were *Asplanchna priodonta, Asplanchna* sp., *Keratella cochlearis, Polyarthra dolichoptera, Polyar thra major, Polyarthra vulgaris, Trichocera cylindrica* and *Trichocera multicrinis* (Table 3). Estimated measured weights for a number of rotifer taxa were smaller than the Lake Erie means based on the same length weight equations, including *Asplanchna* sp., *P. dolichoptera, P. major* and *T. multicrinis*. A few others, including A. *priodonta* and T. *cylindrica* were larger.

Discussion

Biomass values in this study were calculated from length-weight relationships. Such estimates of zooplankton biomass are preferable to estimates based on abundance and a set species weight because they capture more variation in the populations associated with growth patterns and levels and types of predation. These estimates based on length measurements in the Bay of Quinte were 38 to 45% lower than the initial estimates based on weights of offshore Lake Superior, Huron and Ontario animals from the early 1970s. The length-weight equations used at Quinte are widely recognized relationships from the literature and were selected because they best represented actual length-weight data determined for *Daphnia retrocurva, Diacyclops thomasi* (narrow cyclopoid) and *Mesocyclops edax* (wide cyclopoid) from Lakes Ontario and Erie from the late 1980s and early 1990s (Johannsson, unpubl. data). The 1970s measured mean species weights may be higher that the means observed at Quinte due to differences in predation amongst the different regions. Nearshore, nursery areas, such as the Bay of Quinte, tend to have higher levels of planktivory than more offshore regions, which would select for smaller body sizes (Hall et al., 2003). Importantly, the method has now been standardized across all years, allowing trend analyses.

Zooplankton biomass has significantly declined in the Bay of Quinte as of the mid-1990s, with the largest change seen in cyclopoid copepods. Declines in zooplankton biomass may be driven by a number of causes: unusually cold weather (to witness, 1992), increased predation, declines in food supplies and alteration of habitat. Reduced phosphorus loading to the Bay in the late 1970s did not result in lower zooplankton biomass or production (Cooley et al, 1986). Rather the zooplankton increased in biomass and reached peak levels during the 1980s (Figure 2). Any effect of reduced nutrients was overshadowed, likely by a decline in planktivory associated with changes to the fish community (Johannsson and Nicholls, 2002).

A number of observations suggest that dreissenids were implicated in the decline of zooplankton in the mid-1990s. Dreissenids are ecological engineers, causing change that ripples through foodwebs. Dreissenids were observed through out the Bay starting in the early 1990s (Dermott et al., 2003). Veliger larvae of dreissenids contributed 7.0, 2.6 and 6.8% of total zooplankton biomass at B, HB and C by 1995, indicating a strong presence in these regions by this time. Veligers have continued to be a significant part of the zooplankton communities up until the present time (2008). Declines in chlorophyll-*a*, (43 to 55% across the three stations) and increases in water clarity occurred at all sites as of 1995 (Bedford and Gerlofsma, 2010) and rapid expansion of macrophytes were observed in the upper and middle Bay (Leisti et al., 2006). Similar changes have been observed at other sites of invasion (Dahl et al., 1995; Fahnenstiel et al., 1995). Such changes could also be caused by decreases in phosphorus loading with consequent declines in total phosphorus levels (TP); however, TP levels were lower in 1995 but did not decline across the 1990s except perhaps at C (Nicholls, 2010; E.S. Millard and M. Burley, Fisheries and Oceans, Burlington, unpubl. data). Chlorophyll-*a*/TP ratios, an indication of dreissenid grazing often observed after the dreissenid invasion (Nicholls et al., 1999; Hall et al., 2003), plummeted between 1995 and 1996 at B and C, and between 1994 and 1996 at HB (Figure 5). Levels of the cyanobacterium *Microcystis* increased as of 1995 in the upper Bay (Nicholls, 2002). Phytoplankton biomass declined in 1995 at all stations, increased for a few years in the late 1990s and remained low again as of 2000 (Nicholls, 2010). Nicholls et al. (in press) consider the ecosystem changes that occurred in the upper Bay in 1995 to be a regime shift. Six of the top ten drivers which defined this regime shift were phytoplankton groups. All these changes suggest that dreissenids were altering the pelagic ecosystem throughout the Bay as of 1995/1996.

Table 2. Mean individual dry weights (μg) of zooplankton species from 1995 to 2008 at B, HB and C. Mean Great Lakes (GL) values were based on unpublished data from Watson and Wilson[a] or from Cooley et al. (1986)[b]; otherwise they were estimated from the literature or weight of a similar species. Dominant taxa are shown in bold (>1% by biomass at one or more stations). Rare species (5–20 individuals counted) are denoted by *, very rare species (1–5 individuals counted) by **.

Taxa Name/life stage	measured weight (μg)				GL (μg)
	B	HB	C	mean	
Cladocera					
*Alona affinis***	2.6	2.3	—	2.4	5.5[a]
*Alona rectangula***	1.1	—	—	1.1	5.5
Alona sp.	3.4	2.5	1.3	3.2	4[a]
***Bostnina* sp.**	0.9	1	0.8	0.9	1.2*
*Bythotrephes longimanus**	—	96.8	90.2	69.1	140
Cercopagis pengoi	3.7	4.6	4.6	4.6	—
Ceriodaphnia lacustris	0.6	0.6	0.6	0.6	1.9*
*Ceriodaphnia quadrangula***	3.6	—	—	3.6	1.9*
***Ceriodaphnia* sp.**	0.5	0.5	0.5	0.5	4
Chydorus sphaericus	0.4	0.4	0.5	0.4	0.6[b]
*Daphnia ambigua***	—	3.9	—	3.9	4
*Daphnia catawba***	0.9	—	1.4	1.1	4
Daphnia galeata mendotae	5.3	5.8	4.4	5.5	10[b]
*Daphnia longiremis***	—	1	1.1	1.1	5[a]
*Daphnia parvula**	1.4	3.6	1.8	2.3	—
Daphnia pulex	.	9.3	.	9.3	.
*Daphnia pulicaria**	3.7	7	4.5	6.5	10
Daphnia retrocurva	2.7	2.7	2.4	2.7	4.2[s]
Daphnia sp.	1.5	5.6	0.6	1.9	4[a]
Diaphanosoma birgei	1.5	1.5	1.5	1.5	4[a]
Diaphanosoma brachyurum	1.1	1.1	1.2	1.1	4
*Disparalona hamata**	1.7		1.2	1.7	—
Eubostnina coregoni	1.5	1.5	1.6	1.5	1.3*
*Eurycercus lamellatus**	4.7	14.3	4.1	8.4	3[a]
Holopedium gibbenjm	3.7	4	2.2	3	14.6[a]
Leptodora kindtii	6.9	7	6.4	6.9	14.8[a]
*Ophryoxus gracilis***	—	13	—	13	4
*Paralona pigra***	—	5.2	—	5.2	—

Table 2. Continued...

Taxa Name/life stage	measured weight (µg) B	HB	C	mean	GL (µg)
Polyphemus pediculus	1.9	1.5	1.9	1.8	28[a]
*Scapholeberis kingi***	0.5	—	—	0.5	—
*Sida ctystallina**	8.4	6.4	8	7.7	20[a]
Cvclopoida					
cyclopoid nauplii	0.1	0.1	0.1	0.1	0.2[a]
cyclopoid copepodid	1	1.3	1.1	1.1	1.9*
*Cyclops scutifer***	—	4.2	6	4.9	4
Cyclops vemalis	3.2	3.6	2.6	3.3	3.5[a]
Diacyclops thornasi	3.7	3.7	3.4	3.5	3.2[a]
Eucyclops agilis	2.4	2	1.7	2.2	4
*Eucyclops speratus***	4	6.3	4.9	6.1	4
Mesocyclops edax	5.5	6.3	5.1	5.9	6.7[a]
Tropocyclops extensus	0.8	0.8	0.9	0.8	1.7[a]
Calanoida					
calanoid nauplii	0.1	0.1	0.1	0.1	0.3[b]
calanoid copepodid	1.6	2.2	1.9	2	1.9*
*Epischura copepodid**	—	12.5	3.5	6.3	—
*Epischura lacustris**	—	19.7	16.1	16.5	9[s]
Eurytemora affinis	5.8	7	7.4	6.9	6.6[a]
*Leptodiaptomus ashlandi***	2.9	—	—	2.9	4.1[a]
*Leptodiaptomus minutus**	3.5	4.4	4.2	4.1	4.5[a]
*Leptodiaptomus sicilis**	—	16	11.1	12.1	12.3[a]
Leptodiaptomus siciloides	6.4	6.4	6.4	6.4	7.5[a]
Limnocalanus copepodid	5.6	4.2	6.8	5.8	8
*Limnocalanus macrurus**	27.9	39.7	38.4	36.5	55[b]
*Limnocalanus nauplii**	—	—	0.3	0.3	0.4[a]
Skistodiaptomus oregonensi	7.4	8.4	7.7	8.1	9.3[a]
Harpacticoida					
Harpacticoid nauplii	0.1	0	0.4	0.1	0.3
Harpacticoid adults	0.7	1.2	0.7	0.8	3
Mollusca					
Dreissena veligers	0.4	0.4	0.5	0.4	0.8

Dreissenid-mediated declines in zooplankton could occur through changes in phytoplankton/seston quantity and quality and/or changes in predation associated with the increases in water clarity which improves detection by predators. Lack of decrease in cladoceran mean length between the pre- and post-DM periods indicates that the decline in zooplankton was not due to increases in fish predation. The observed reductions in chlorophyll/phytoplankton suggest that food limitation may be one of the causes of zooplankton decline, as in other systems (MacIssac et al., 1995; Jack and Thorp, 2000; Nicholls et al., 2002; Thorpe and Casper, 2003). Mussels consume particles in a wide size range including microzooplankton such as protozoa, ciliates, rotifers, veligers and nauplii (MacIssac et al., 1995; Miller and Watzin, 2007) as well as phytoplankton. This could explain the low rotifer biomass in the Bay (1.7–3.4% of total zooplankton biomass): normally rotifers constitute slightly <30% of total zooplankton biomass in shallow eutrophic lakes in Europe where dreissenids are not a major driver (data from 50-µm mesh samples; Gyllström et al., 2005). Omnivorous copepod populations were likely affected directly through predation on nauplii and indirectly through competition for prey (rotifers, ciliates, small cladocerans, copepod juveniles and phytoplankton).

That dreissenids could impact the zooplankton in the Bay of Quinte is further suggested by the documented impacts of dreissenid invasions on zooplankton in other areas. Zooplankton, particularly copepod nauplii, and rotifers significantly declined in the shallow, unstratified western basin of Lake Erie following mussel establishment in the late 1980s (MacIsaac et al., 1995). In the eastern basin of Lake Erie (Outer Long Point Bay), biomass declined by 60% between 1984–1987 (prior to dreissenids) and 1993–1994 (post-dreissenids) (calculated from Table 3; Johannsson et al., 1999). Densities of cladocerans and copepods in Lake St. Clair, another shallow system somewhat comparable to the upper Bay of Quinte, fell by about 50% following the arrival of dreissenids, and rotifers dropped by 86% (David et al., 2009). In the Hudson River, total zooplankton biomass declined by more than 70% following invasion (Pace et al., 1998), and rotifers and nauplii were particularly impacted. Declines in zooplankton upon dreissenid invasion do not always occur (Idrisi et al., 2001), but the general experience has been a marked decline. Unfortunately, any decrease in rotifers following invasion cannot be assessed due to the lack of pre-invasion data.

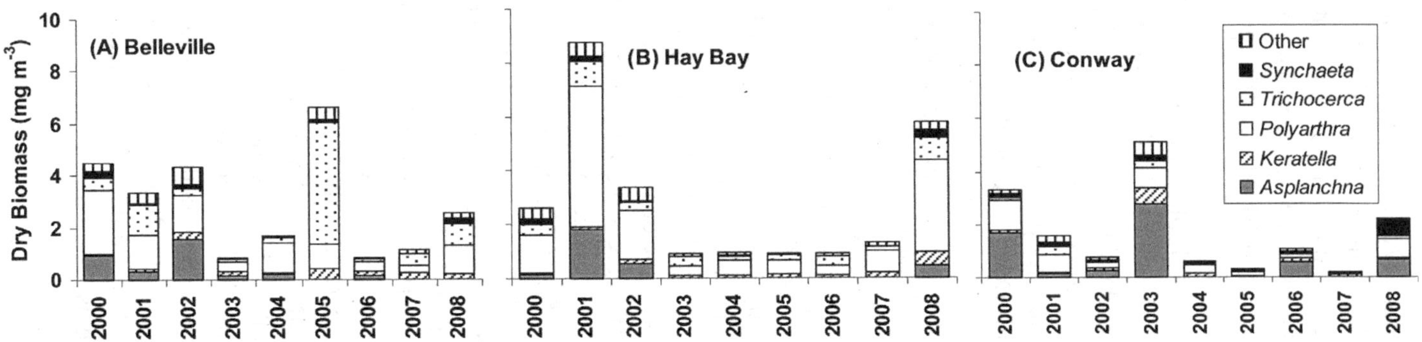

Figure 4. Seasonal (May–October) mean rotifer dry biomass at Belleville, Hay Bay and Conway, 2000 to 2008, showing the dominant genera (A–C).

Table 3. Mean individual dry weights (μg) of rotifer species in seasonal composite samples from 2000 to 2008 at B, HB and C. Mean Lake Erie weights (see text) are given for comparison. Rare species (6–20 individuals counted) are denoted by *; very rare species (1–5 individuals counted) by **; dominant taxa (>5.0% by biomass at one or more stations) are in bold.

Taxa Name/life stage	B	HB	C	mean	L. Erie
Rotifera					
Anuraeopsis f/ssa**	0.002		0.003	**0.002**	
Ascomorpha ovalis	0.011	0.009	0.010	**0.010**	0.024
Ascomorpha sp.*	0.017	0.004	0.016	**0.008**	
Asplanchna priodonta	0.818	1.360	1.576	**0.976**	0.698
***Asplanchna* sp.**		0.502	0.347	**0.420**	0.788
*Bnachionus angularis***	0.177		0.007	**0.145**	
Conochilus unicomis	0.010	0.015	0.015	**0.012**	0.008
Euchlanis sp.*	0.172	0.130		**0.166**	
*Filinia brachiata***		0.007		**0.007**	
*Filinia longiseta***	0.013			**0.013**	
*Filinia terminalis***		0.014		**0.014**	
*Gastropus stylifer**	0.016	0.023	0.022	**0.020**	
Kellicottia longispina	0.003	0.005	0.005	**0.005**	0.012
Keratella cochlearis	0.002	0.004	0.003	**0.003**	0.004
Keratella cochlearis tecta	0.001	0.001	0.001	**0.001**	0.004
Keratella quadrata	0.049	0.060	0.060	**0.053**	0.068
*Keratella sermlata**	0.001			**0.001**	
*Notholca squamula***	0.005			**0.005**	0.016
*Ploesoma hudsoni**		0.267	0.188	**0.241**	0.058
*Ploesoma lenticulare**		0.022		**0.022**	0.043
*Ploesoma truncatum**		0.015	0.100	**0.024**	0.047
Polyarthra dolichoptera	0.012	0.010	0.010	**0.011**	0.024
Polyarthra major	0.040	0.052	0.037	**0.046**	0.075
Polyarthra remata	0.004	0.005	0.005	**0.005**	0.010
Polyarthra vulgaris	0.013	0.014	0.013	**0.013**	0.017
Pompholyx sulcata	0.010	0.009	0.011	**0.010**	
*Synchaeta grandis***		0.015	0.176	**0.060**	0.015
Synchaeta kitina	0.007	0.009	0.008	**0.008**	0.004

Table 3. Continued...

Taxa Name/life stage	B	HB	C	mean	L. Erie
Synchaeta pectinata	0.022	0.028	0.025	**0.026**	0.015
Synchaeta sp.	0.011	0.041	0.012	**0.013**	0.042
*Synchaeta stylata**		0.018	0.021	**0.020**	0.004
Trichocerca cylindrica	0.119	0.110	0.125	**0.119**	0.014
*Trichocerca elongata**	0.118	0.142		**0.131**	
Trichocerca longiseta	0.042	0.021	0.010	**0.035**	
Trichocerca multicrinis	0.051	0.048	0.051	**0.050**	0.089
Trichocerca porcellus	0.014	0.017	0.009	**0.015**	0.055
Trichocerca pusilla	0.010	0.012	0.009	**0.010**	0.022
Trichocerca rousseleti	0.007	0.006	0.006	**0.007**	0.007
Trichocerca similis	0.016	0.016	0.016	**0.016**	0.019
*Tylotrocha monopus***	0.007			**0.007**	

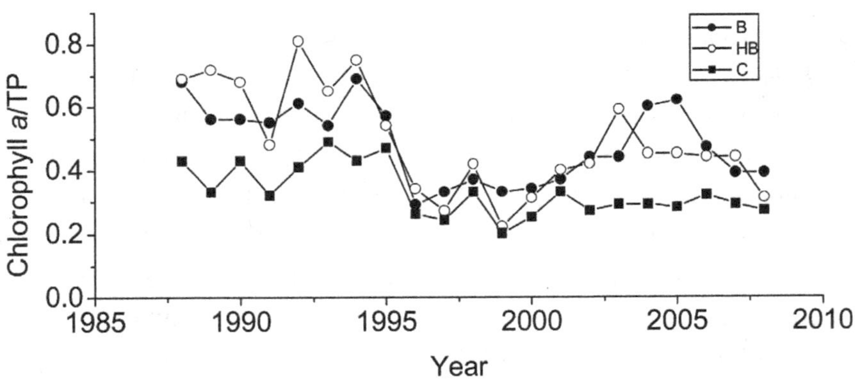

Figure 5. Trends in the ratio of chlorophyll-*a* to total phosphorus at Belleville (B), Hay Bay (HB), and Conway (C). Phosphorus data were provided by E.S. Millard (Fisheries and Oceans, Burlington) and chlorophyll-*a* data were from Bedford and Gerlofsma (2010).

The chlorophyll-*a*/TP ratio increased again at B and HB between 2002 and 2006/7 suggesting some relaxation of the impact of dreissenids (Figure 5). Round Goby (*Neogobius melanostomus*) invaded the Bay in the early 2000s and reached average catches >1 per gillnet set averaged across the three stations between 2002 and 2005 (Hoyle, 2010). Young gobies are produced throughout the season and consume zooplankton, while adult gobies consume benthos including dreissenids (Jude et al., 1995; Ray and Corkum, 1997). Thus a series of interactions will occur amongst gobies, dreissenids and zooplankton. Zooplankton biomass at B was low between 2002 and 2005 suggesting that gobies may play a noticeable role along with dreissenids in regulating zooplankton biomass in the shallow upper Bay.

C. pengoi was present for a short period each summer at C. During that summer period, the abundance of copepod nauplii and copepodids declined (Benoit et al., 2002). Similar seasonal patterns were observed in nearshore Lake Ontario (Warner et al., 2006). *C. pengoi* was very rare at B and no zooplankton declines were observed between the pre- (1995–1998) and post- (1999–2008) *C. pengoi* periods. However, in the middle (HB) and lower (C) Bay total cyclopoid and total calanoid biomass declined in synchrony (Figure 2) with the *C. pengoi* invasion. It is unknown how much of this decline was due to this invasive cladoceran. Although *C. pengoi* biomass was an order of magnitude lower at HB compared to C, it is possible that their effects were greater at HB due to the lack of a hypolimnetic refuge for zooplankton prey. *C. pengoi* at C were most abundant in the epilimnion and metalimnion. Contrary to expectations based on observations in the literature (Benoit et al., 2002; Laxon et al., 2003; Warner et al., 2006), small species, like *Bosmina*, did not decline. Either they are less susceptible to *C. pengoi* predation and/ or other factors are important in the copepod declines. There were no increases in cladoceran mean length between the pre-DM and post-DM period which might have supported an increase in the importance of invertebrate predation. To balance this observation, it must be remembered that cyclopoids themselves are ominvores/invertebrate predators and the loss of one predator may partly offset the impact of the other.

Summary and conclusions

Long term monitoring of Lake Ontario's Bay of Quinte provides the opportunity to examine the impact of dreissenid invasion on the zooplankton community. Weekly or biweekly zooplankton samples have been collected from 1975 to 2008 at 3 stations: Belleville (B), Hay Bay (HB), and Conway (C) along a trophic and depth gradient down the bay. Rotifers have been collected since 2000. Biomass estimates based on measured zooplankton lengths started in 1995. Archived seasonal composite samples prior to 1995 were reanalysed and biomass recalculated from length-weight equations to allow for comparable data in trend analysis. Mean May 1– October 6 zooplankton biomass was low from 1975 into the early 1980s and peaked between 1982–1983 and 1991. Biomass fell during the cold summer of 1992 associated with the Pinatubo eruption and was low after the invasion of Dreissenid Mussels. From 1979–1991 (after phosphorus control and prior to dreissenid invasion), biomass averaged 265, 253 and 84 mg m^{-3} at B, HB and C, respectively. Seasonal biomass of most zooplankton groups, as well as total biomass, was significantly lower at all stations after the dreissenid invasion. Cladocerans still dominated zooplankton biomass after the invasion, averaging 56% to 80% of the total. Cyclopoid numbers and biomass fell dramatically.

After the invasion of *Cercopagis pengoi* in 1999, calanoid and cyclopoid biomass at HB and C decreased by approximately 50%. *Bosmina* biomass did not change. Seasonal mean rotifer biomass over the 2000 to 2008 period was 1.9% to 4.4% of total zooplankton biomass.

Reductions in zooplankton following dreissenid and *C. pengoi* invasions are thought to be caused by both direct predation of micro-zooplankton (e.g. nauplii) by mussels and copepods by *C. pengoi* (at C and HB), and competition between zooplankton and dreissenids for food resources.

Although zooplankton populations/communities are governed by a number of drivers, a dominant structuring force in the Bay of Quinte since 1995 has been *Dreissena* spp. The continued decline of copepods in the middle and lower Bay may be due to the invasion of *C. pengoi.* The trends in chlorophyll-*a*/TP ratios suggest that the effect of Dreissenid Mussel grazing has declined, possibly because of predation by Round Goby. This needs further examination.

Acknowledgements

Foremost, we acknowledge our indebtedness to the Fisheries and Oceans Canada field and laboratory crews who have made this work possible: J. Gerlofsma, R. Bonnell, A. Bedford, M. Frank (M. Burley), H. Niblock, M. Fizpatrick, B. Timmins, T. Hollister, K. Bonnell and others too countless to name. Thanks to C. Tudorancea, W. Geiling, D. Geiling and J. Moore, among others, for zooplankton and rotifer sample enumeration, taxonomy and data analysis. We thank J. Cooley, J. Moore and K. Minns who started and ran the zooplankton data collection and analysis prior to us, and to the Directors of GLLFAS (DFO) whose continuous support of the program since its inception helped to ensure its viability. Total phosphorus data were provided by Scott Millard and Michele Burley (M.

Frank), Fisheries and Oceans, Burlington. We acknowledge the support of the Bay of Quinte RAP Restoration council and other leads on Project Quinte over the years, including M. Munawar, R. Randall, R. Dermott, K. Minns, K. Nicholls, M, Johnson, J, Christie, D. Hurley and others. Thanks also to Environment Canada's Technical Operations Services for assistance with moorings, and Ontario Ministry of Natural Resources for use of their Glenora lab. Financial support was provided by the Great Lakes Action Plan.

References

Bay of Quinte RAP Coordinating Committee, 1993. The Bay of Quinte Remedial Action Plan Stage 2 Report—Time to Act. 271 pp. http://www.bqrap.ca/publications/annualreports/ Accessed 14 April 2010

Bedford, A., Gerlofsma, J., 2010. Photosynthesis, chlorophyll-*a* and light extinction. In: *Project Quinte Annual Report 2008*. pp. 41–62. Bay of Quinte RAP Restoration Council/Project Quinte. Kingston, ON.

Cooley, J. M., Moore, J. E., Geiling, W. T., 1986. Population dynamics, biomass and production of the macrozooplankton in the Bay of Quinte during changes in phosphorus loadings. In: C.K. Minns, D.A. Hurley, K.H. Nichols (Eds.), *Project Quinte: point-source phosphorus control and ecosystem response in the Bay of Quinte, Lake Ontario,* p. 166–176. Can. Spec. Publ. Fish. Aquat. Sci. 86.

Dahl, J. A., Graham, D. M. Dermott, R., Johannsson, O. E., Millard, E. S., Myles, D. D., 1995. Lake Erie 1993, western west central and eastern basins: Changes in trophic status, and assessment of the abundance, biomass and production of the lower trophic levels. Can. Tech. Rep. Fish. Aquat. Sci. 2070.

David K. A., Davis, B. M., Hunter, R. D., 2009. Lake St. Clair zooplankton: Evidence for post-Dreissenid changes. J. Freshwat. Ecol. 24, 199–209.

Dermott, R., Bonnell, R., Carou, S., Dow J., Jarvis, P., 2003. Spatial distribution and population structure of the mussels *Dreissena polymorpha* and *Dreissena bugensis* in the Bay of Quinte, Lake Ontario, 1998 and 2000. Can. Tech. Rep. Fish. Aquat. Sci. 2479.

Dumont, H. J., Van de Velde I., Dumont, S., 1975. The dry weight estimate of biomass in a selection of Cladocera, Copepoda, and Rotifers from the plankton, periphyton, and benthos of continental waters. Oecologia 19, 75–97.

Fahnenstiel, G. L., Lang, G. A., Nalepa, R. F., Johengen, T. H., 1995. Effects of zebra mussel (*Dreissena polymorpha*) colonization on water quality parameters in Saginaw Bay, Lake Huron. J. Great Lakes Res. 21, 435–448.

Graham, D. M, Dahl, J. A., Millard, E. S., Johannsson, O. E., White, L. L, 1996. Assessment of abundance, biomass and production of the lower trophic levels in the eastern basin of Lake Erie, 1994. Can. Tech. Rept. Fish. Aquat. Sci. No. 2110.

Grigorovich, I. A., MacIsaac, H. J., Rivier, I. K., Aladin, N. V., Panov, V. E., 2000. Comparative biology of the predatory cladoceran *Cercopagis pengoi* from Lake Ontario, Baltic Sea and the Caspian Lake. Arch. Hydrobiol. 149, 23–50.

Gyllström, M., Hansson, L.–A., Jeppesen, E., Garcí-Criado, F., Gross, E., Irvine, K., Kairesalo, T., Kornijow, R., Miracle, M. R., Nykänen, M., Nõges, T., Romo, S., Stephen, D., Van Donk, E., Moss, B., 2005. Zooplankton community structure in shallow lakes: interaction between climate and productivity. Limnol. and Oceanogr. 50, 2008–2021.

Hoyle, J. A., 2010. Bay of Quinte Fish and Fisheries, 2008. In: *Project Quinte Annual Report 2008*. pp. 109–118. Bay of Quinte RAP Restoration Council / Project Quinte. Kingston, ON.

Idrisi, N., Mills, E. L., Rudstam, L. G., Stewart, D. J., 2001. Impact of zebra mussels (*Dreissena polymorpha*) on the pelagic lower trophic levels of Oneida Lake, New York. Can. J. Fish. Aquat. Sci. 58(7), 1430–1441.

International Joint Commission, 1988. Great Lakes Water Quality Agreement of 1978, as amended by protocol signed November 1987.

Jack, J. D., Thorp, J. H., 2000. Field experiments on the regulation of potamoplankton in a large river by a benthic suspension-feeding mussel, *Dreissena polymorpha*. Freshwat. Biol. 44, 569–579.

Johannsson, O. E., Nicholls K. H., 2002. Bay of Quinte Remedial Action Plan Assessment of the State of Impairment of Beneficial Uses: II. Zooplankton. Prepared for the Department of Environment, Ontario Region, Bay of Quinte RAP.

Johannsson, O. E., Graham, D. M., Einhouse D. W. E., Mills, E. L., 1999. Historical and recent changes in the Lake Erie zooplankton community and their relationship to ecosystem function. In: M. Munawar and T. Edsall (Eds.), *The State of Lake Erie Ecosystem (SOLE) – past, present and future*. pp. 169–196. Backhuys Publishers, the Netherlands.

Johannsson, O. E., Dermott, R., Graham, D. M., Dahl, J. A., Millard, E. S., Myles, D. D., LeBlanc, J., 2000. Benthic and pelagic secondary production in Lake Erie after the invasion of *Dreissena* spp. with implications for fish production. J. Great Lakes Res. 26, 31–54.

Johnson, M. G, Hurley, D. A., 1986. Overview of Project Quinte – 1972–82. In: C.K. Minns, D.A. Hurley, K.H. Nicholls (Eds.), *Project Quinte: point-source phosphorus control and ecosystem response in the Bay of Quinte, Lake Ontario,* pp. 1–6. Can. Spec. Publ. Fish. Aquat. Sci. 86.

Jude, D. J., Janssen, J., Crawford. G., 1995. Ecology, distribution, and impact of the newly introduced round and tubenose gobies on the biota of the St. Clair and Detroit Rivers In: M. Munawar, T. Edsall and J. Leach (Eds.), *The Lake Huron Ecosystem: Ecology, Fisheries and Management,* pp. 447–46. Ecovision World Monograph Series, Academic Publishing, Amsterdam, the Netherlands.

Hall, S. R., Pauliokonis, N. K., Mills, E. L., Rudstam, L. G., Schneider, C. P., Lary, S. J., Arrhenius, F., 2003. A comparison of total phosphorus, chlorophyll-*a*, and zooplankton in embayment, nearshore and offshore habitats of Lake Ontario. J. Great Lakes Res. 29, 54–69.

Laxson, C. L., McPhedran, K. N., Makarewicz, J. C., Telesh, I. V., MacIsaac, H. J., 2003. Effects of *Cercopagis pengoi* on the lower foodweb of Lake Ontario. Freshwater Biology 48, 2094–2106.

Leisti, K. E., Millard, E. S., Minns, C. K., 2006. Assessment of submergent macrophytes in the Bay of Quinte, Lake Ontario, August 2004, including historical context. Canadian Manuscript Report of Fisheries and Aquatic Sciences 2762. MacIsaac, H. J., Lonnee C. J., Leach, J. H., 1995. Suppression of microzooplankton by zebra mussels: importance of mussel size. Freshwat. Biol. 34, 379–387.

MacIsaac, H. J., Grigorovich I. A., Hoyle J. A., Yan N. D., Panov V. E., 1999. Invasion of Lake Ontario by the Ponto-Caspian predatory cladoceran *Cercopagis pengoi*. Can. J. Fish. Aquat. Sci. 56, 1–5.

McCaulley, E., 1984. The estimation of the abundance and biomass of zooplankton in samples. In: J.A. Downing and F. H. Rigler (Eds.), *A Manual on Methods for the Assessment of Secondary Productivity in Fresh Waters, Second Edition*, pp. 228–265. Blackwell Scientific Publications, Oxford.

Miller, E. B., Watzin, M. C., 2007. The effects of zebra mussels on the lower planktonic foodweb in Lake Champlain. J. Great Lakes Res. 33, 407–420.

Minns, C. K., Hurley, D. A., Nicholls, K. H. (Eds.), 1986. *Project Quinte: point-source phosphorus control and ecosystem response in the Bay of Quinte, Lake Ontario*, Can. Spec. Publ. Fish. Aquat. Sci. 86.

Newhall, C. G. and Punongbayan, R. S., 1996. *Fire and Mud: Eruptions and Lahars of Mount Pinatubo*, Philippines. University of Washington Press, Seattle. Nicholls, K. H., 2010. Total phosphorus, nitrogen chlorophyll and phytoplankton in the Bay of Quinte. In: *Project Quinte Annual Report 2008*, pp. 29–40. Bay of Quinte RAP Restoration Council/Project Quinte. Kingston, ON.

Nicholls, K. H., Hopkins, G. J., Standke, S. J., 1999. Reduced chlorophyll to phosphorus ratios in nearshore Great Lakes waters coincide with the establishment of dreissenid mussels. Can. J. Fish. Aquat. Sci. 56, 153–161.

Nicholls, K. H., Heintsch, L., Carney, E., 2002. Univariate step-trend and multi-variate assessments of the apparent effects of P loading reductions and zebra mussels on the phytoplankton in the Bay of Quinte, Lake Ontario. J. Great Lakes Res. 28, 15–31.

Pace M. L., Findlay, S. E. G., Fischer, D., 1998. Effects of an invasive bivalve on the zooplankton community of the Hudson River. Freshwat. Biol. 39, 103–116. Ray, W. J, Corkum, L. D., 1997. Predation of zebra mussels by round goby, *Neogobius melanostomus*. Environ. Biol. Fishes 50, 267–273.

Thorpe, J. H., Casper, A. F., 2003. Importance of biotic interactions in large rivers: an experiment with planktivorous fish, dreissenid mussels and zooplankton in the St. Lawrence River. River Res. Applic. 19, 265–279.

The above article is a reprinted version of the following previously published article:

Bowen; K.L., Johannsson, O.E., 2011
Changes in zooplankton biomass in the Bay of Quinte with the arrival of the mussels, Dreissena polymorpha and D. rostiformis bugensis, and the predatory cladoceran, Cercopagis pengoi: 1975 to 2008
Aquatic Ecosystem Health & Management (2011) 14 (1): 44–55.
https://doi.org/10.1080/14634988.2011.550528

Changes in the benthic community of the Bay of Quinte, Lake Ontario, over a 40-year period

R. Dermott,* D. Martchenko, and M. Johnson

Fisheries and Oceans Canada,
867 Lakeshore Road, Burlington, ON L7S 1A1,
**Corresponding author: dermott1@lara.on.ca*

Keywords: benthic species composition, clustering analysis, ordination analysis

Introduction

Composition of the benthic community is structured by water quality, temperature, depth, substrate and biotic interactions, while population size is a function of the productivity of the environment (Alley and Powers, 1970). Changes to the benthic community occur following degraded water quality (eutrophication), shifts in fish community or abundance of aquatic macrophytes, presence of new competitive species, and in response to climatic changes (Scheffer and van Nes, 2004; Beaugrand, 2004). Extensive efforts were made to reduce the inputs of nutrients to the Great Lakes, especially phosphorus, following the Great Lakes Water Quality Agreement in 1972 (IJC, 1987). To address pollution problems in the Bay of Quinte, in eastern Lake Ontario, a Remedial Action Plan (RAP) was created, listing the degraded ecosystem components, and establishing goals for restoration (Bay of Quinte RAP, 1993). One goal of the Bay of Quinte RAP was to restore the benthic community, degraded by eutrophication, to one which would demonstrate a positive change in the benthic macroinvertebrate community structure, with increased species and diversity in a community typical of mesotrophic conditions.

The Bay of Quinte has 3 sections (Figure 1): a shallow upper bay between Trenton and Deseronto with mid channel depths <7 m, a middle bay of <12 m deep, and a lower bay with mid channel depths of 19 to 60 m. The lower section extends from Picton to Lake Ontario, and is influenced by water flow from the oligotrophic lake. A study site typical of the shallow upper bay, and 3 sites in the deeper lower bay

Ecology of the Bay of Quinte: Health, Management and Global Implications
Editors: C.K. Minns, M. Munawar and M.A. Koops

Aquatic Ecosystem Health & Management Society, Canada

and open Lake Ontario were established by Johnson and Brinkhurst (1971). After 1972, abundance and species composition of benthic macroinvertebrates were monitored, initially to study the response of the degraded community to improvements in water quality following reduced phosphorus loading begun in 1978 (Johnson and Hurley, 1986; Koops et al., 2009). By 1984, a decline in productivity of the bay and a decrease of pollution tolerant species were measured (Johnson and McNeil, 1986). The general model of gradual transition between trophic states was found to be accelerated by periodic catastrophic events (Johnson and Hurley, 1986).

After arriving in 1992, both Zebra Mussels (*Dreissena polymorpha*) and Quagga Mussels (*D. rostriformis bugensis*) became common by 1994. As water clarity drastically improved due to their filter feeding, the bay was forced toward a macrophyte-dominated environment. This caused a regime shift in the fish community as the dominant predator Walleye (*Sander vitreus*) was replaced by centrarchids, which are more adapted to clear and weedy habitats (Chu et al., 2004; Hoyle et al., 2008). By 1997, the community in the lower bay became subordinate to Quagga Mussel (Dermott et al., 2003). The invasive Round Goby (*Apollonia melanostomus*) arrived in the bay between 1999 and 2001, and quickly became the dominant fish in the nearshore (Taraborelli et al., 2009), feeding on the benthic

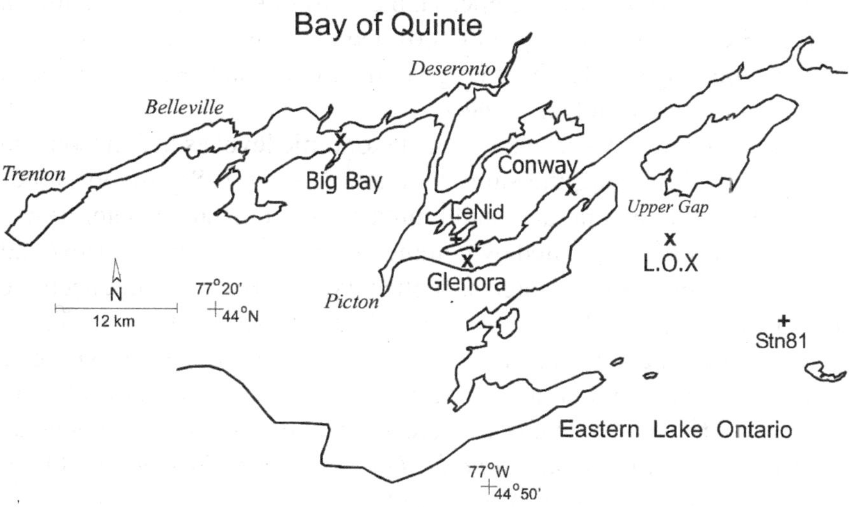

Figure 1. Location of benthic index sites at Big Bay, Glenora, Conway and Lake Ontario (LOX), as well as the extra sites at LeNid Point and CCIW station 81 in eastern Lake Ontario. The upper bay is between Trenton and Deseronto, the lower bay extends from Picton to the Upper Gap.

fauna including *Dreissena*. Other pelagic and benthic invasive species also arrived (Mills et al., 2003), resulting in an aquatic community now dramatically altered from that at the start of the restoration project.

Objectives of this study were to examine the changes to the composition of the benthic community and trends in total abundance and biomass of the macroinvertebrates present over three decades. This time period spanned a series of environmental changes to the bay. Data were divided into 4 time stanzas that represented the major drivers of change: pre-Phosphorus reduction (1967–1978); post-P reduction (1982–1991); post-*Dreissena* (1992–2001); and the post-Goby period (2002–2006).

Methods

Benthic collections used the same methods and locations over the entire study period and followed protocols established in 1967 (Johnson and Brinkhurst, 1971; Johnson and McNeil, 1986). Ekman dredge samples (0.05 m^{-2}) were screened through a 0.58 mm (# 30 mesh) sieve. Five to six replicates were collected at each site during late July to early August of each year (1967, 1968, 1972, 1977, 1982–2006). Benthic index sites were located at: Big Bay (6.5 m) in the upper bay; Glenora (21 m) and Conway (32 m) in the lower bay; and at the LOX site (30 m depth) located 8 km south of the Upper Gap exit to Lake Ontario (Figure 1). These sites occur along a gradient of depth and water quality with the fauna reflecting the depth, average temperature, and substrate at each site (Johnson and McNeil, 1986).

Identifications were to the same taxonomic level as in Johnson and McNeil (1986), using the same taxonomist after 1985. Invertebrates were identified, enumerated and biomass including shells determined. Wet biomass was estimated by volume displacement between 1967 and 1983, but after 1983 measured as blotted wet mass on a balance accurate to 0.1 mg. The data-base was recorded as number per Ekman sample to be consistent with that in Johnson and McNeil (1986). Numeric abundance and wet biomass of the major groups were used to compare population trends. These groups were: total *Oligochaeta, Chironomidae, Sphaeriidae, Amphipoda, Isopoda, Dreissena* and *Gastropoda.* Other minor groups were included in the totals.

Average diversity at each index site was compared over the years. Data were normalized using log x+1 prior to statistical treatment. Abundance and diversity metrics included abundance of the *Dreissena*. CUSUM (cumulative sum control chart) of total abundance (cumulative sum of differences between grand total and yearly averages) was examined to indicate break-points and amplify trends of abundance (Nicholls, 2001). Total non-dreissenid biomass was used to measure trends in macroinvertebrate biomass over time as biomass of the *Dreissena.* oftened exceeded the remaining biomass by a factor of 10. The Margalef (1958) diversity index was calculated for each sample as: D = S-1 / log n (N); where S is number of taxa present and N is number of individuals in the sample.

Percent Similarity of Community (PSC - Whittaker and Fairbanks, 1958; Johnson and McNeil, 1986) was calculated from average abundance data each year for individual species at the sites. Years when oligochaetes were not identified to species were excluded (1988, 1989, 1997, 1998, 2001, 2002). PSC = 100–0.5 Sum [a′- b′], where a′ and b′ are the respective percentage of each species in the total from site A and B. Absolute value of the difference was summed over all species. PSC values calculated for each site and year combination were used for cluster analysis in place of the more commonly used Bray-Curtis Index of Similarity. The R software package (R Development Core Team, 2009) was used to create a dissimilarity matrix (1- percent similarity), and to do the cluster analysis plotted by Ward's minimum variance algorithm, often used for quantitative variables such as density. Clustering was done for all sites, as well as separately for only the upper bay (Big Bay) and for Conway representing the lower bay.

Non-metric multidimensional scaling (NMDS) was also used to assess similarities at a single site over time. Distances between pairs reflect the community similarities, thus years closer together in ordination had more similar community composition. The same dissimilarity matrix calculated for cluster analysis was used. To ease interpretation, ordination was done separately for Big Bay and the Conway data set. As reference sites, data from LeNid Point near the Glenora Ferry, and an eastern Lake Ontario station were also included. LeNid was used as a reference to Big Bay, with same depth (6.5 m) and sediment characteristics, but had the better water quality of the lower bay. Data from LeNid existed for 1966, 2005 and 2006. As a reference for Conway, an eastern Lake Ontario site (Canada Centre Inland Waters station 81) was used which had similar depth and substrate as Conway, but at 15 km offshore was further removed from anthropogenic and terrestrial influences than even the LOX index site. Substrate at the LOX site was a thin layer of silt over a glacial till of brown clay and coarse sand, very different from the black silt at the Glenora and Conway sites. Data from Stn 81 were selected for 1981, 1985, 1990, 1995, 2000, 2002 and 2003.

Results

The benthic community at Big Bay was dominated by *Oligochaeta* and *Chironomidae* between 1967 and 2006. Initially Oligochaeta and *Sphaeriidae* dominated the community at the 21 m deep Glenora site. Beyond 25 m depth, the community of the lower bay and Lake Ontario had been dominated by the deepwater amphipod *Diporeia* spp. After 1993, *Dreissena polymorpha* (Zebra Mussel) and *D. rostriformis bugensis* (Quagga Mussel) became common in the upper and lower bay respectively, and major changes to the bay's ecosystem followed. Between 1993 and 1995, *Diporeia* was extirpated from the lower bay and eastern Lake Ontario, replaced by a much greater mass of Quagga Mussel on the soft

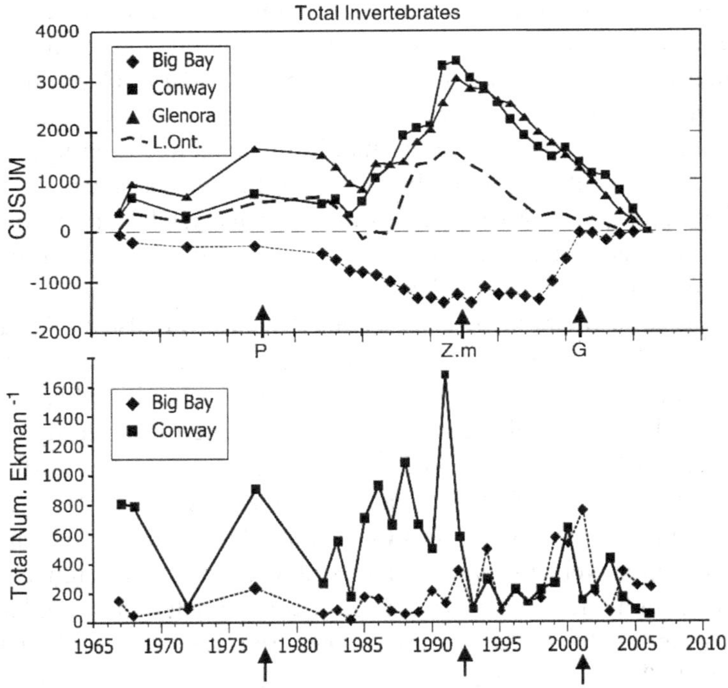

Figure 2. CUSUM for total invertebrate density at the Bay of Quinte benthic index sites, 1967 to 2006, and corresponding time trend of density at Big Bay and Conway. Arrows mark years of phosphate reduction (P), arrival of *Dreissena* (Z.m) and Round Gobies (G).

sediments. Data on abundances of the macroinvertebrates are provided in Johnson and McNeil (1986), Dermott (2001), Dermott et al. (2003), and in the appendices.

The CUSUM function for total invertebrate abundance showed different trends for Big Bay and the lower bay sites (Figure 2). At Big Bay, the negative CUSUM indicated a gradual trend, corresponding to increased abundance between 1977–1990, reflecting a gradual improvement in water quality. There was little change between 1988 and 1998, but a step-change occurred between 1999 and 2001 (Figure 2). The opposite trend in CUSUM occurred at the lower bay sites (Glenora, Conway and Lake Ontario). At these sites, there was a dramatic breakpoint after 1992 reflecting the loss of *Diporeia* from the lower bay. After initial P reductions and crash of the White Perch (*Morone americana*) population in 1978, *Diporeia* density increased in the lower bay to peak in 1991. The change point in CUSUM at Glenora and Conway between 1982–1985 likely reflected a delayed response to the P reduction.

While *Dreissena* increased throughout the Bay of Quinte, total non-dreissenid biomass at Big Bay changed little over the years. Although

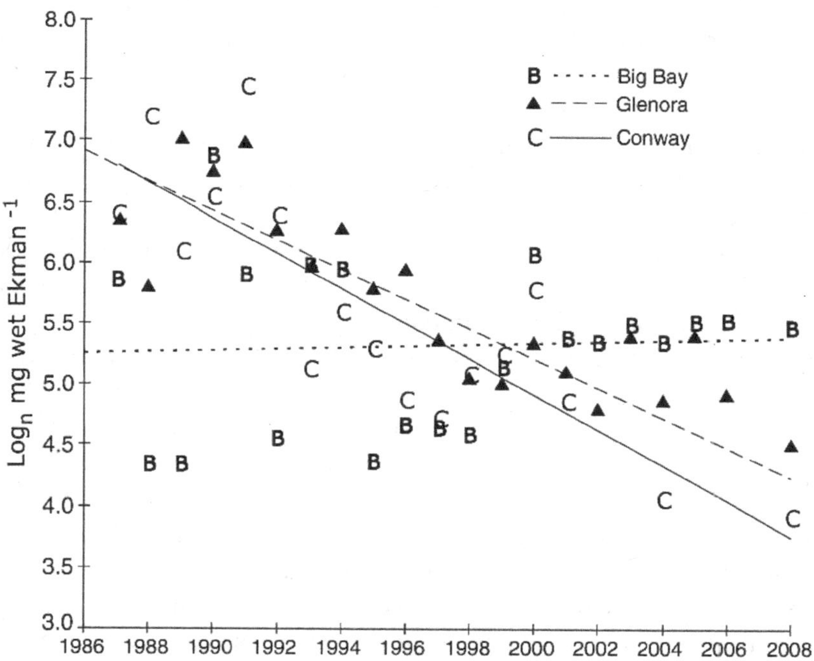

Figure 3. Trend in non-dreissena biomass (natural log scale) at Big Bay, Glenora and Conway.

variable between 1986 and 2000, biomass at Big Bay was very stable between 2001 and 2006 with a non-significant increasing trend (p =

0.38, $r^2 = 0.04$). In the lower bay, total non-dreissenid biomass greatly decreased at all 3 sites ($p = 0.0001$, all $r^2 > 0.6$), between 1986 and 2006 (Figure 3). Average wet non-dreissena biomass decreased from 18.77 g m^{-2} at Conway during the pre-*Dreissena* time stanza to 1.74 in the post-Goby period. During the later time stanza, *Dreissena* biomass at Conway averaged 324.6 g m^{-2} (wet + shells).

Community composition

Although total biomass remained similar at Big Bay, biomass composition showed a gradual shift between 1972 and 1987, followed by an increased proportion of biomass from less common macroinvertebrates (Figure 4). Larva of the large *Chironomus plumosus* var. *semireductus* dominated the biomass at Big Bay (>280 wet mg per sample or 5.6 g m^{-2}), followed by tubificid oligochaetes during the period before P reduction. Oligochaete composition remained similar at Big Bay and Glenora after 1977, but eutrophic tolerant species became less abundant as water quality improved (Johnson and McNeil 1986). After 1987, during the later years of the post-P reduction period, *Pisidium* spp. and *Amphipoda* (*Gammarus fasciatus*) became more common in the upper bay. Biomass of the epibenthic *Gastropoda* increased at Big Bay during the post-*Dreissena* period. The filter feeding polychaete *Manayunkia* became abundant in Big Bay (>1000 m^{-2}) between 1997 and 2002, but was very rare after 2003 (<10 m^{-2}). In the post-*Dreissena* period, several small chironomid genera such as the *Tanytarsini* appeared at Big Bay. After 2000 when Round Goby appeared, Amphipoda, *Pisidium* and Gastropoda again were minor contributors to total non-dreissenid biomass on the mud at the Big Bay site (Figure 4). Shallow-water amphipods in the upper bay were mostly *Gammarus fasciatus*, and rarely *Hyalella azteca*, with a few exotic *Echinogammarus ischnus* present after 1998.

Zebra Mussels were rare on the soft organic mud at the Big Bay site representing <5% of total biomass between 2000 and 2006. Yet they remained the dominant dreissenid in the upper bay where they were very common on hard substrates above 3 m depth (Dermott et al., 2003).

In the lower bay, the amphipod *Diporeia* increased following P reduction and remained dominant until 1993 when the community composition changed dramatically (Figure 5). Between 1988 and 1991 *Diporeia* density peaked at 12,000 m^{-2}, when the species attained 65% of the total benthic biomass (11.4 g m^{-2} wet) at Conway. In 1993, *Diporeia* suddenly disappeared at Conway and no specimens were collected from the lower bay and eastern Lake Ontario after 1997. Since 1997, all amphipods found were shallow-water species. The isopod *Caecidotea racovitzai* has also been absent since 1995 at all sites, including Glenora and Conway where it was very common between 1968 and 1992. At Conway, isopods peaked at 13,000 ± 1500 m^{-2} in 1991, before disappearing in 1994.

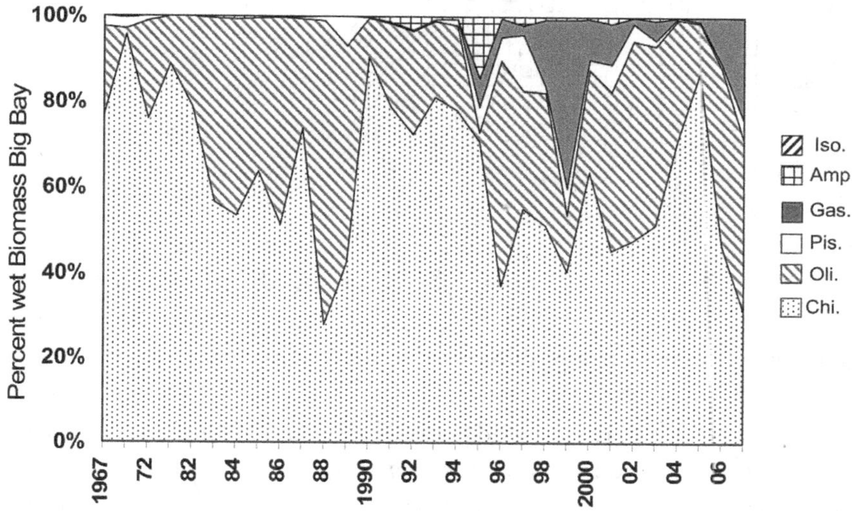

Figure 4. Biomass composition of non-dreissenid macroinvertebrates at Big Bay, 1967–2006.

After 1994, Quagga Mussel dominated the community in the lower bay. Extensive Quagga Mussel colonies formed on the soft muds at Conway and Lake Ontario beyond 30 m depth, with average densities of 1500 m^{-2} in 2002–2006 and wet biomass (+shells) of 240 g m^{-2} (wet tissue = 140.1 g m^{-2} shell-free). Maximum biomass of dreissenids occurred in 2003 at Conway (412 g m^{-2} + shells) and in 2005 at LOX (858 g m^{-2}). After 1993, oligochaetes dominated the non-dreissenid biomass of the lower Bay (Figure 5). Between 1990 and 2006, the large mesotrophic *Chironomus atritibia* continued to be the dominant chironomid at Glenora. Chironomids remained at low density (<650 m^{-2}) at Conway and in Lake Ontario but their biomass represented an increasing percent of the remaining non-dreissenid biomass. After *Dreissena* appeared, the *Sphaerium* and *Musculium* species disappeared at all the benthic sites in Quinte, the last specimen collected in 1999. Biomass of the smaller *Pisidium* species slowly decreased after 1996 in the lower bay and at LOX, where *Pisidium* became rare by 2006 (Figure 5).

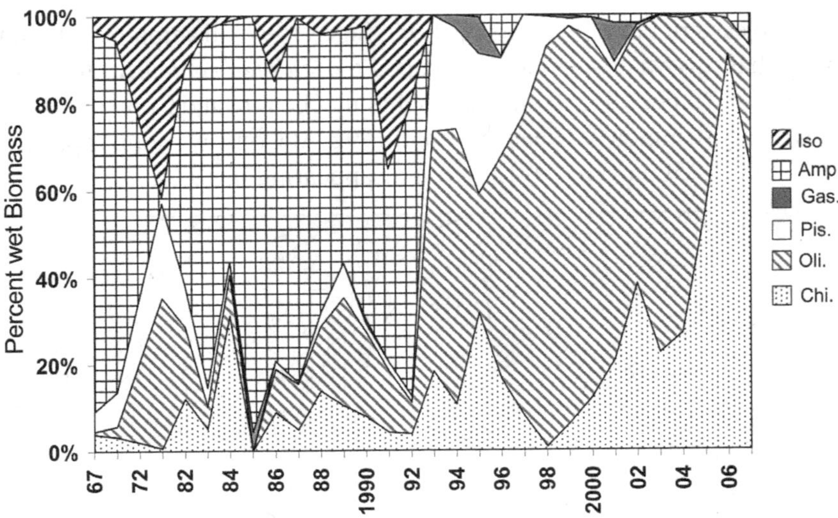

Figure 5. Biomass composition of non-dreissenid macroinvertebrates at Conway, 1967–2006.

Margalef diversity increased between 1982 and 1999 at all sites except LOX. As more species returned to Big Bay and new species appeared, the diversity index increased from less than 0.6 in 1977 to 3.9 in 1999. At all sites except LOX, diversity values exceeded 3 as common species became less dominant.

Clustering and ordination

Cluster analysis of the benthic community at all sites is displayed in Figure 6. Seven clusters were identified: (I) Glenora data for 1982–1996; (II) Conway and Lake Ontario data between 1982–1993; (III) Glenora, Conway and Lake Ontario data before 1977; (IV) Conway and Lake Ontario data 1999–2006; (V) Conway and Ontario 1994–1996; (VI) Glenora, LeNid and Big Bay after 1990; and (VII) early Big Bay and LeNid data with outliers of Glenora 1972 and Lake Ontario 1985. The Big Bay data did not group with any of the Conway data, because the community in the warm shallow (-6 m) upper bay was very different from that in the lower bay at depths beyond 21 m, especially in years when *Diporeia* dominated the lower bay. Most years during pre-P reduction (<1978) separated from the post-Zebra Mussel years (>1992). Later years of Big Bay (B03, B94) formed a cluster with the last years of Glenora (G03, G04), and LeNid data from 2005–2006 (cluster VI). The Glenora community became more similar to Big Bay following the presence of

Dreissena, as chironomids became the dominant group at Glenora.

Reference site LeNid in 1966 (N66) clustered with the earliest pre-P reduction years from Big Bay (B68) in the Big Bay dendrogram (Figure 7). The latest years (N05, N06) also clustered with the corresponding Big Bay data (B05, B06). Clustering of the Conway data (Figure 8), separated the earliest pre-P reduction years from the later years. The earliest data at Station 81 (S81) clustered with the Conway 1982 community, while the later community from the reference site (S00 - S03) likewise grouped with the Conway community from 1999 and 2000.

NMDS ordination of the Big Bay community showed a clear separation of pre- and post-P reduction years, with the earliest plotted near the bottom (Figure 7). LeNid in 1966 was widely separated from the early Big Bay community, yet in 2005 LeNid plotted close to Big Bay for the same year. The community of 1990–1992 plotted far from the pre-P reduction community of 1967 (Figure 7). After the arrival of Zebra Mussels, 1995 separated widely from 1992. Years since 2003 plotted close to the data of LeNid, as the communities became very similar at the two sites. In 1966 the LeNid site supported gastropods and *Gammarus* that were rare at Big Bay until 1990. After 1999, the community shifted back to be more similar to that present at Big Bay before 1991, prior to arrival of *Dreissena* in the bay.

Conway showed less separation between years of pre- and post-P reduction than at Big Bay (Figure 8), as water quality at Conway was very good before 1978 when P reduction began. Greater separation occurred between 1992 and post-1993, representing the pre- and post-*Dreissena* communities when the dominant *Diporeia* disappeared from the community. Ordination of Conway data put the earliest years at the lower left, and years after 1999 to the right on axis 1. The community at Station 81 during 1981–1985 plotted nearest the Conway 1991–1992 data (Figure 8). After 2000, changes to the community at Station 81 paralleled that which occurred at Conway, their communities became very similar to each other but were widely separated from that of earlier years, as Quagga Mussels became the dominant biomass at Conway, LOX and the Station 81 reference site.

Discussion

Long term shifts in the benthic community occurred between the 3 defined time periods: #1 post-P reduction (1982–1991); #2 post-*Dre issena* (1992–2001); and #3 post-Goby (2002–2006). Both CUSUM

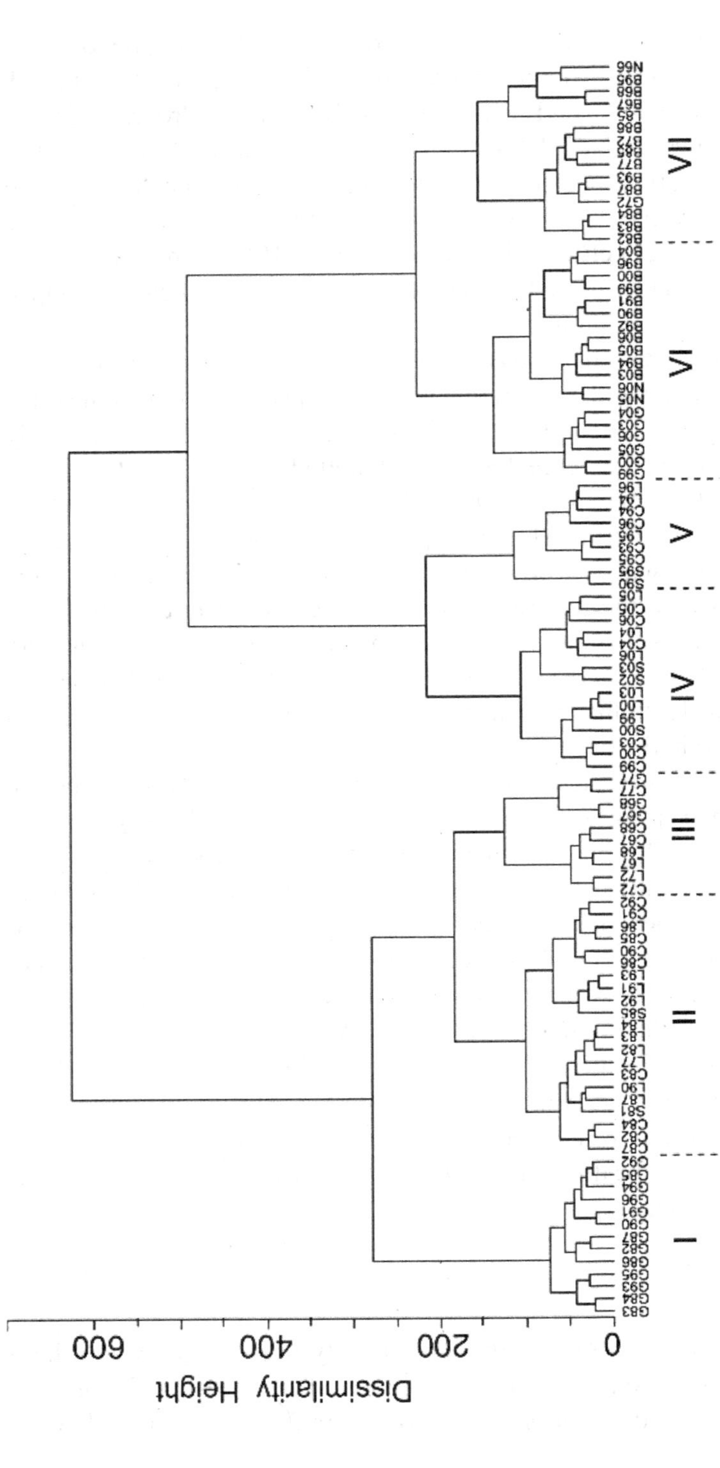

Figure 6. Cluster of benthic community (Percent Similarity of Community from log transformed annual average density). Data of 1967–2006 for Bay of Quinte sites: Big Bay (B–), Glenora (G–), Conway (C–) and Lake Ontario (L–), and reference sites at LeNid (N–1966, 2005–2006) and Station 81 (S–1981, 1985, 1990, 1995, 2000, 2002–2003).

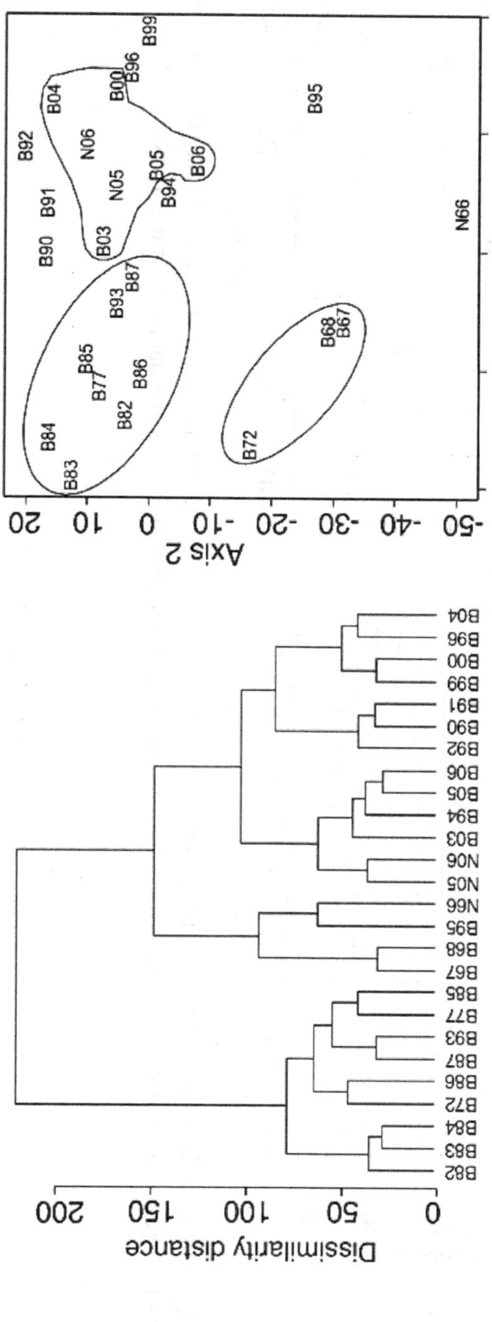

Figure 7. Cluster of benthic community at Big Bay from 1966 to 2006, including LeNid (N–) reference site (using PSC and Ward's algorithm). NMDS ordination of the same data (dimensional stress = 0.16).

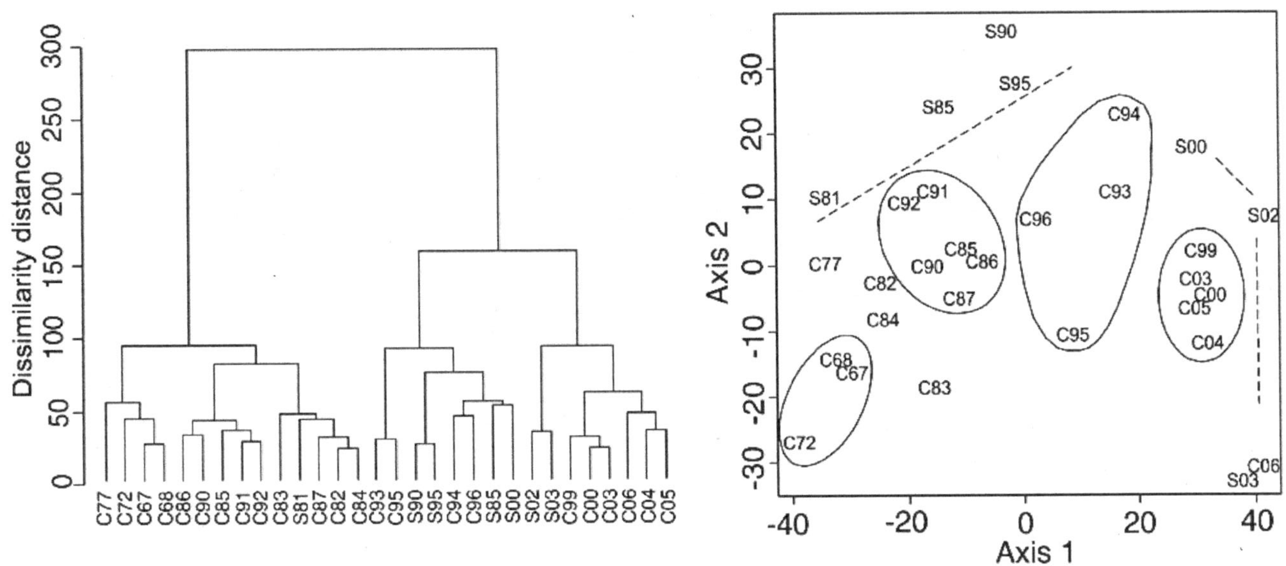

Figure 8. Cluster dendrogram of benthic community at Conway (1967–2006), including the eastern Lake Ontario station 81 (S–) reference site (using PSC and Ward's algorithm). NMDS ordination of the same data (stress was 0.18).

and ordination revealed a greater change to the community post-*Dreissena,* then during the post-P reduction stanza, and that change was most pronounced in the lower bay. The upper bay community at Big Bay responded in an opposite direction than the community of the lower bay, without loss of major taxa or reduced overall biomass. However, the Big Bay community did show a slow shift in composition with an increase in less common taxa, and hence an increase in overall diversity. In addition to changes resulting from the 3 major events, there were 5 additional changes to the Bay of Quinte ecosystem, some of which were cascade effects initiated by the arrival of *Dreissena*. Some of these events were gradual and predictable such as the effect of P reduction, while others were more abrupt.

P reduction

During the post-P reduction stanza, the shift in composition and biomass was initially slow due to phosphorus recycling from the sediment pool (Minns et al., 1986). Ordination of Big Bay data showed the benthic community in the post-P reduction years were clearly dissimilar from the pre-reduction years (1967–1978). Phosphorus reduction led to a slow shift in species composition, when *Pisidium*, *Amphipoda*, *Gastropoda* and other epibenthic species reappeared at Big Bay between 1978–1991. In response to improved water quality over the 40 years, composition of the oligochaetes and chironomids changed, especially at Big Bay (Johnson and Brinkhurst 1971; Johnson and McNeil, 1986). Smaller species such as the *Tanytarsini*, and various *Chironomini* which lack haemoglobin became more common, including genera often associated with macrophyte beds. These smaller species require higher oxygen concentrations than the large *Chironomus* spp. that use haemoglobin to store oxygen for later use.

Change accelerated during the post-*Dreissena* stanza, (1992–2001) when *Dreissena* impacted the bay by their filtering and nutrient cycling (Hecky et al., 2004). The increased water clarity and resurgence of macro phytes in the upper bay changed the dominant fish community (Hoyle et al., 2008). Yet in Big Bay, the benthic community did not undergo major changes in composition or total non-dreissenid biomass as occurred in the lower bay. In the shallow upper bay, the loss of phytoplankton production due to dreissenid filtering activities was replaced by increased benthic algae and submerged macrophyte growth as the water cleared. Competition with *Dreissena* reduced the filter feeding polychaetes and

Sphaerium clams which thrived during the initially improved conditions. The sudden and complete disappearance of both the isopod *Caecidotea* and the amphipod *Diporeia* (Dermott, 2001) occurred within 2 years after the arrival of *Dreissena*. *Caecidotea* prefers eutrophic conditions, so its decrease may have been be partly due to improved water quality. However, *Diporeia* should have increased with improved water quality as it is a cold oligotrophic indicator (Johnson and Brinkhurst, 1971). The oligotrophic indicator worm *Stylodrilus heringianus*, (Milbrink, 1983), also became very rare in the lower bay and at LOX after 1997 as the lake became more oligotrophic. Declines in these two species were likely mediated by the presence of Quagga Mussels.

Post-Goby

Species composition and ordination analysis indicated that the community in the post-Goby stanza (2002–2006) was clearly dissimilar to that which existed at Big Bay during post P-reduction (1982–1991) or during the post-*Dreissena* period, excluding 1994. Extensive predation by high abundances of Centrarchidae (mostly *Lepomis gibbosus)* and later by the Round Goby (OMNR, 2005), was likely responsible for the reduction of epibenthic species that had reappeared at Big Bay (1992–2001). Barton et al. (2005), and Lederer et al. (2008) found that predation by Round Goby greatly reduced littoral invertebrate populations in Lake Erie and Lake Michigan, respectively. Benthic densities are generally positively related to nutrients and food availability, providing oxygen levels remain adequate, and negatively related to predation pressure (Miracle et al., 2006). In the Bay of Quinte, abundances of both cold-water (Diporeia) and warm-water (Gammarus) amphipods appeared to be more influenced by predation than by nutrient reduction. Johnson and McNeil (1986) found that the *Diporeia* population in the lower bay was controlled by White Perch predation, until the fish's winter-kill in the cold winter of 1978. The reduced predation pressure by White Perch in 1978–1982 resulted in a large increase in *Diporeia* density between 1977 and 1983 (Johnson and McNeil, 1986). This cold climatic event in 1978 also shifted the fish community of the bay to one dominated by Walleye, a large piscivore (Johnson and Hurley, 1986). Other periods of climate variability have also affected the fish community of the Bay of Quinte (Casselmann et al., 1999).

By far, the greatest changes to the benthic community of the Bay of Quinte was the major regime shifts created by the arrival of *Dreis sena* in the bay which increased water clarity and resulted in a macro phyte-adapted community in the upper bay. This in turn, led to a shift in the fish community from Walleye to centrarchids. *Dreissena* may have been responsible for the extirpation of the dominant *Diporeia* from eastern Lake Ontario. These changes that were related directly or indirectly to *Dreissena* had cascading effects on the plankton (Nicholls, 2001), fish community (Hoyle at al., 2008), and sediment bacteria (Dermott and Legner, 2002). After 2001, the benthic community biomass adjusted to the progressive oligotrophication of the lower bay and another shift in fish community as Round Goby increased.

A number of species became more common or appeared in response to the changing environment over the decades of this study. At all sites except LOX (Table S3), the Margalef Diversity index increased and exceeded 3 by 1999 before decreasing again in the last time stanza. Although a simple index, it provides a measure of taxa richness from less than 1 in stressed communities dominated by a few species, to values over 4 in diverse communities such as in Georgian Bay (Munawar et al., 2001). Nalepa et al. (2003) found a similar increase in diversity in Saginaw Bay, Lake Huron where oligochaetes shifted to more mesotrophic species after the arrival of *Dreissena*. However, they found both a decrease in oligochaete biomass at shallow sites (6–12 m depth) and the loss of *Diporeia* biomass from the deeper sites in outer Saginaw Bay.

Summary and conclusions

The benthic community structure in the Bay of Quinte, eastern Lake Ontario, was examined over a 40 year period (1967–2006). Changes in species composition were examined using clustering and ordination analysis at a shallow upper bay site (6 m) and in the deeper lower bay site (32 m). *Tubificidae* and *Chironomidae*, which dominated the upper bay, responded to phosphorus reductions between 1978 and 1991 with increased diversity, and a return of pollution sensitive species. This response accelerated in 1993, after the arrival of *Dreissena*. Epibenthic *Amphipoda*, *Gastropoda* and *Sphaeriidae* increased in the upper bay between 1986–1999, but decreased following the arrival of the Round Goby (*Apollonia melanostomus*). The lower bay underwent a major change after 1994, when Quagga Mussel replaced the amphipod *Diporeia* as the dominant organism. Since 1990, non-dreissenid biomass decreased in the lower bay but has not significantly changed in the upper bay. Ordination showed that the community in the upper bay changed over time, and remained different from the lower bay community. In the lower bay and Lake Ontario, the benthic community greatly changed to one very different from the historical community of the Great Lakes.

Over time, the additive impacts of nutrient reduction, competition with invasive species and changing fish predation have shifted the benthic community in the Bay of Quinte away from the initial community present in the late 1960s. Reduced eutrophication shifted the community in the upper bay back toward the "target" community similar to that at the mesotrophic LeNid site, with an increase in species diversity, reduced dominance by a few eutrophic tolerant species and a stable biomass. In the lower bay, composition of the benthic community of the post-*Dreissena* and post-Goby periods shifted far from the composition that existed initially in eastern Lake Ontario, but diversity changed less. Important species disappeared and total non-dreissenid biomass continued to decline, not due to water quality but due to invading species, as *Dreissena* consumed much of the bay's algal production. The benthic community of the lower bay and Lake Ontario remains in a state of flux very dissimilar to what existed 40 years ago, and the resulting effects have cascaded into the fish communities (Hoyle et al., 2008). It is unlikely the benthic community in the lower bay, or Lake Ontario will ever return to one similar to that present during the post-P control/pre-*Dreissena* period.

The value of long term studies such as Project Quinte was displayed by showing the range of annual variation in the biota relative to their responses to both management actions and several uncontrolled events. The value of sampling each year was revealed as several sudden changes occurred within 1–2 years and would have been misinterpreted or even missed had sampling been spaced at 3 or 5 year intervals thus blurring the perceived causes and effects.

Acknowledgements

Special thanks are extended to Bud Timmins, Robert Bonnell, Jocelyn Gerlofsma, and M. Frank (Michele Burley). Bill Morton did most of the taxonomical identifications and Tanya Trivedi created the database. Our appreciation is given to the dozens of students and crew members on the research vessels *Surf, Shark, Bayfield* and the *Leslie J.* over the years of sampling. We thank the staff of the Ontario Ministry of Natural Resources at Glenora for assistance and lab space.

References

Alley, W.P., Powers, C.F., 1970. Dry weight of the macrobenthos as an indicator of eutrophication of the Great Lakes. Proc. Internat. Assoc. Great Lakes Res. 13, 595–600.

Barton, D.R., Johnson, R.A., Campbell, L., Petruniak, J., Patterson, M., 2005. Effects of Round Gobies (*Neogobius melanostomus*) on Dreissenid Mussels and other invertebrates in eastern Lake Erie, 2002–2004. J. Great Lakes Res. 31, 252–261.

Bay of Quinte RAP, 1993. "Time to Act, The Bay of Quinte Remedial Action Plan Stage 2 Report." Bay of Quinte RAP Coordinating Committee, September, 1993. Ontario Ministry of the Environment, Kingston, Ontario.

Beaugrand, G., 2004. The North Sea regime shift: Evidence, causes, mechanisms and consequences. Progress in Oceanography 60, 245–262.

Casselman, J.M., Scott, K.A., Brown, D.M., Robinson, C.J., 1999. Changes in relative abundance, variability, and stability of fish assemblages of eastern Lake Ontario and the Bay of Quinte – the value of long-term community sampling. Aquat. Ecosyst. Health Mgmt. 2(3), 255–269.

Chu, C., Minns, C.K., Moore, J.E., Millard, E.S., 2004. Impact of oligotrophication, temperature, and water levels on Walleye habitat in the Bay of Quinte, Lake Ontario. Trans. Amer. Fish. Soc. 133, 868–879.

Dermott, R., 2001. Sudden disappearance of the amphipod *Diporeia* from eastern Lake Ontario: 1993 - 1995. J. Great Lakes Res. 27, 423–433.

Dermott, R., Legner, M., 2002. Dense mat-forming bacterium *Thioploca ingrica* (Beggiatoaceae) in eastern Lake Ontario: implications to the benthic foodweb. J. Great Lakes Res. 28, 688–697.

Dermott, R., Bonnell, R., Carou, S., Dow, J., Jarvis, P., 2003. Spatial distribution and population structure of the mussels *Dreissena polymorpha* and *Dreissena bugensis* in the Bay of Quinte, Lake Ontario, 1998 and 2000. Can. Tech. Rep. Fish. and Aquat. Sci., No. 2479.

Hecky, R.E., Smith, R.E.H., Barton, D.R., Guildford, S.J., Taylor, W.D., Charlton, M.N., Howell, T., 2004. The nearshore phosphorus shunt: a consequence of ecosystem engineering by dreissenids in the Laurentian Great Lakes. Can. J. Fish. Aquat. Sci. 61, 1285–1293.

Hoyle, J.A., Bowlby, J.N., Morrison, B.J., 2008. Lake Whitefish and Walleye population responses to Dreissenid Mussel invasion in Eastern Lake Ontario. Aquat. Ecosyst. Health Mgmt.11(4), 403–411.

International Joint Commission, 1987. Great Lakes Water Quality Agreement of 1978, Agreement, with Annexes and Terms of Reference, between the United States and Canada signed at Ottawa, November 22, 1978 and Phosphorus Load Reduction Supplement signed October 7, 1983 as amended by Protocol signed November 18, 1987. Consolidated by the International Joint Comm., Windsor, Ontario.

Johnson, M.G., Brinkhurst, R.O., 1971. Associations and species diversity in benthic macroinvertebrates of Bay of Quinte and Lake Ontario. J. Fish. Res. Bd. Canada 28, 1683–1697.

Johnson, M.G., Hurley, D.A., 1986. Overview of Project Quinte - 1972–82. Can. Spec. Publ. Fish. Aquat. Sci. 86, 1–6.

Johnson, M.G., McNeil, O.C., 1986. Changes in abundance and species composition in benthic invertebrate communities of the Bay of Quinte, 1966–84. Can. Spec. Publ. Fish. Aquat. Sci. 86, 177–189.

Koops, M.A., Dermott, R.M., Leisti, K.E., Johannsson, O.E., Millard, E.S., Minns, C.K., Munawar, M., Nicholls, K.H., Hoyle, J.A. 2009. The Bay of Quinte: a model for large lake ecosystem management. Verh. Internat. Verein. Limnol. 30, 1024–1029.

Lederer, A.M., Janssen, J., Reed, T., Wolf, A. 2008. Impacts of the introduced Round Goby (*Apollonia melanostoma*) on dreissenids (*Dreissena polymorpha* and *Dreissena bugensis*) and on macroinvertebrate community between 2003 and 2006 in the littoral zone of Green Bay, Lake Michigan.

Margalef, R., 195. Information theory in ecology. Gen. Syst. 3, 36–71.

Milbrink, G. 1983. An improved environmental index based on the relative abundance of oligochaete species. Hydrobiol. 102, 89–97.

Mills, E.L., Casselman, J.M., Dermott, R., Fitzsimons, J.D., Gal, G., Holeck, K.T., Hoyle, J.A., Johannsson, O.E., Lantry, B.F., Makarewicz, J.C., Millard, E.S., Munawar, I.F., Munawar, M., O'Gorman, R., Owens, R.W., Rudstam, L.G., Schaner, T., and Stewart, T.J., 2003. Lake Ontario: foodweb dynamics in a changing ecosystem (1970–2000). Can. J. Fish. Aquat. Sci. 60(4), 471–490.

Minns, C.K., Owen, G.E., Johnson, M.G., 1986. Nutrient loads and budgets in the Bay of Quinte, Lake Ontario, 1965–81. Can. Spec. Publ. Fish. Aquat. Sci. 86, 59–76. Miracle, M.R., Moss, B., Vicente, E., Romo, S., Rueda, J., Bécares, E., Fernàndez-Alàez, C., Fernàndez-Alàez, M., Hietala, J., Kairesalo, T., Vakkilainen, K., Stephen, D., Hansson, L.A., Gyllström, M., 2006. Response of macroinvertebrates to experimental nutrient and fish additions in European localities at different latitudes. Limnetica 25, 585–612.

Munawar, M., Munawar, I.F., Dermott, R., Munawar, S.F., Norwood, W., Wenghofer, C., Lynn, D., Johannsson, O.E., Niblock, H., Carou, S., Fitzpatrick, M. Gaasenbeek, K., Weisse, T., 2001. Aquatic ecosystem health of Fathom Five National Marine Park: a structural and functional assessment. In: S. Parker and M. Munawar (Eds.), *Ecology, culture, and conservation of a protected area: Fathom Five National Marine Park, Canada*. pp. 99–149. Backhuys Publishers, Leiden, The Netherlands.

Nalepa, T.F., Fanslow, D.L., Lansing, M.B., Lang, G.A., 2003. Trends in the benthic macroinvertebrate community of Saginaw Bay, Lake Huron, 1987 to 1996: responses to phosphorus abatement and the Zebra Mussel, *Dreissena polymorpha*. J. Great Lakes Res. 29, 14–33.

Nicholls, K.H., 2001. CUSUM phytoplankton and chlorophyll functions illustrate the apparent onset of dreissenid mussel impacts in Lake Ontario. J. Great Lakes Res. 27, 393–401.

Ontario Ministry of Natural Resources (OMNR), 2005. Lake Ontario fish communities and fisheries: 2004 annual report of the Lake Ontario management unit. Ontario Ministry of Natural Resources, Picton, Ontario, Canada.

R Development Core Team, 2009. A language and environment for statistical computing. R Foundation for Statistical Computing, Vienna, Austria. ISBN 3-900051-07-0. URL http://www.R-project.org.

Scheffer, M., van Nes, E. H., 2004. Mechanisms for marine regime shifts: Can we use lakes as microcosms for oceans? Progress in Oceanography 60, 303–319.

Taraborelli, A.C., Fox, M.G., Schaner, T., Johnson, T.B. 2009. Density and habitat use by the Round Goby (*Apollonia melanostoma*) in the Bay of Quinte, Lake Ontario. J. Great Lakes Res. 35, 266–271.
Whittaker, R.H., Fairbanks, C.W., 1958. A study of plankton copepod communities in the Columbia Basin, southeastern Washington. Ecology 39, 46–56.

The above article is a reprinted version of the following previously published article:

Changes in the benthic community of the Bay of Quinte, Lake Ontario, over a 40 year period
R. Dermott; R., Martchenko; D., Johnson, M., 2012.
Aquatic Ecosystem Health & Management (2012) 15 (4): 410–420.
https://doi.org/10.1080/14634988.2012.728480

Appendix

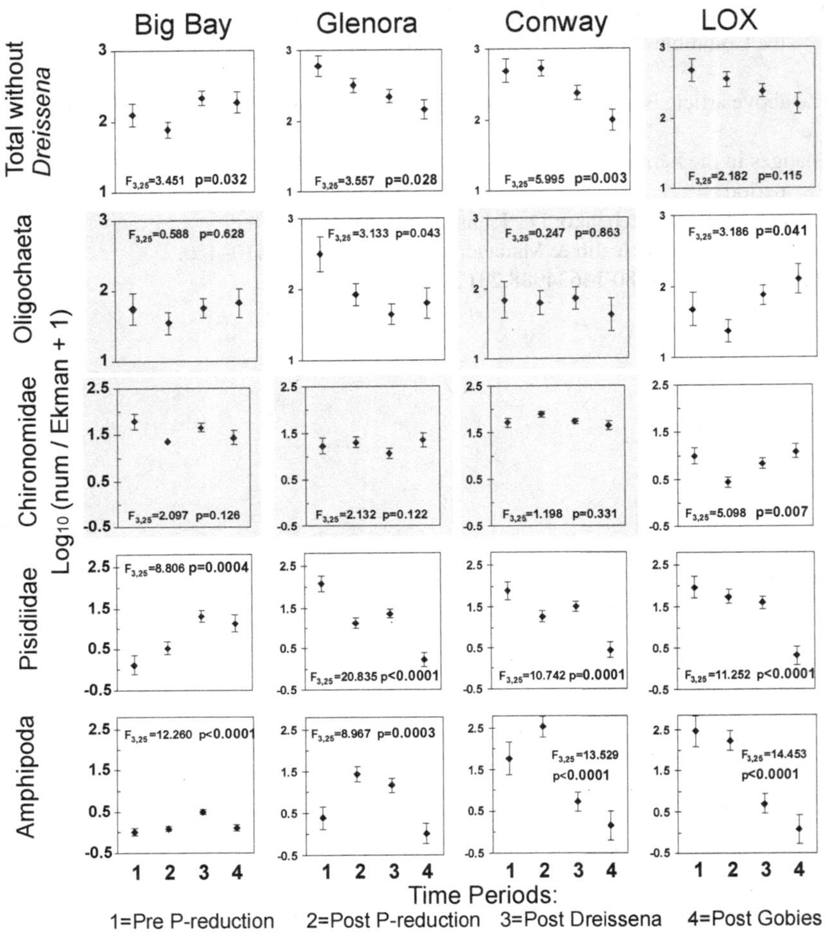

Figure S1. ANOVA for average abundances of five taxonomic groups in 4 time periods (Pre-Phosphorus reduction (1967-1978); Post-P reduction (1982-1991), Post-Zebra Mussels (1992-2001) and Post-Goby (2002-2006)) at the Bay of Quinte index sites (Big Bay, Glenora, Conway and Lake Ontario Exit site). Grey panels are not significant at $p > 0.05$

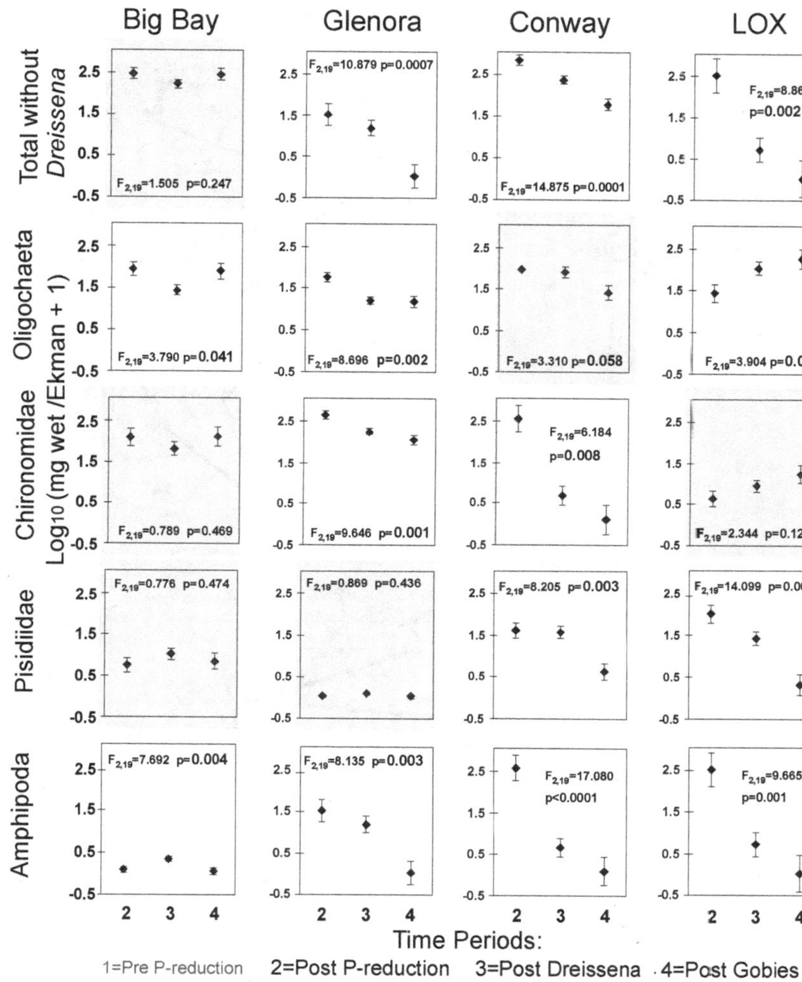

Figure S2. ANOVA for average wet biomass of five taxonomic groups in the Bay of Quinte as in figure A1. Indirect biomass measurements in years 1967-1978 restricted the analysis to 3 time periods: (Post-P reduction (1982-1991), Post-Zebra Mussels (1992-2001) and Post-Goby (2002-2006).

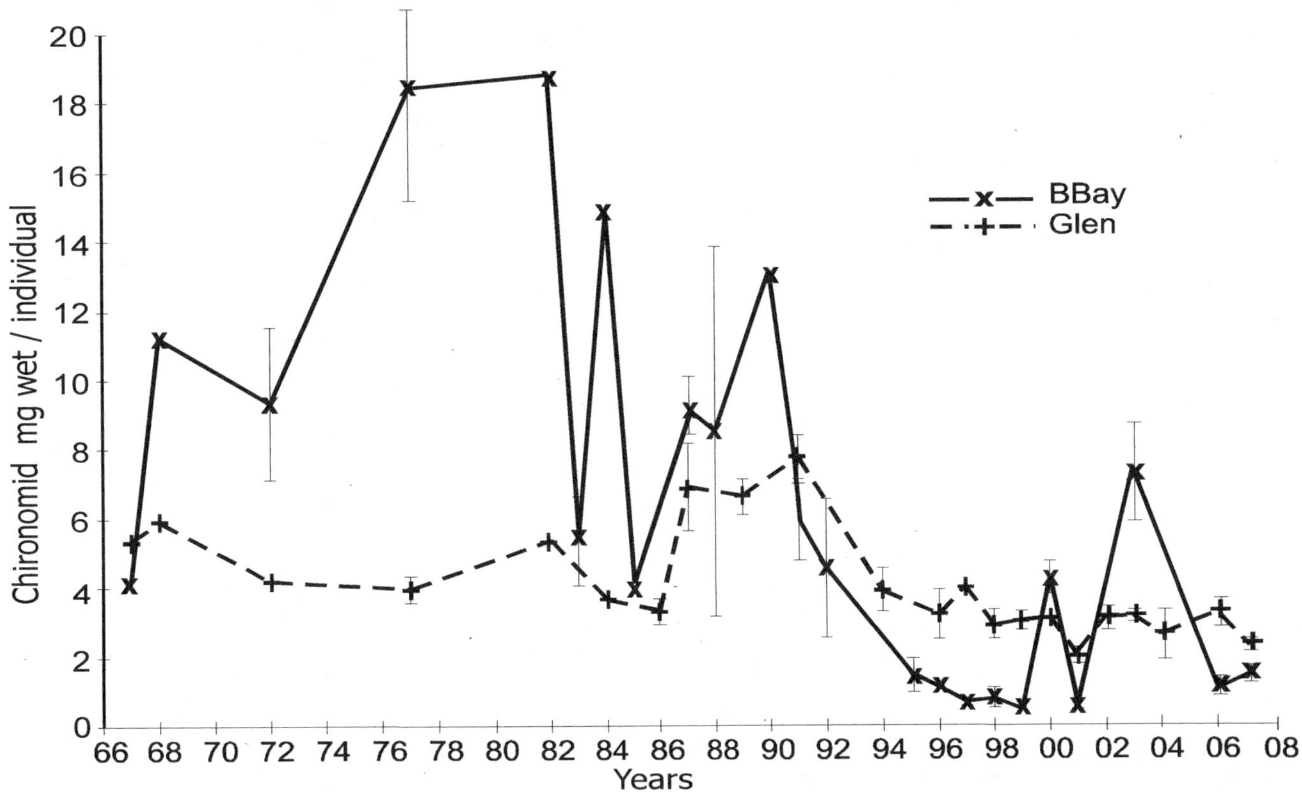

Figure S3. Trend in average individual weight of chironomids at Big Bay and Glenora.

Table S1. Average abundance of benthic macroinvertebrates (per Ekman $0.05m^2$) during the timeline of changes in the Bay of Quinte. Totals include other minor invertebrate groups.

Time Period	Years	Group	Big Bay		Glenora		Conway		Lake Ontario	
			Avg.	SE.	Avg.	SE.	Avg.	SE.	Avg.	SE.
Pre-sewage	4	Total	147.77	38.51	827.03	255.62	665.13	180.85	559.58	124.42
treatment		*Oligochaeta*	73.50	19.71	502.95	191.52	209.46	156.66	57.08	15.14
(to 1978)		*Chironomidae*	70.33	19.20	60.96	21.82	21.68	8.38	10.74	1.45
		Sphaeriidae	0.58	0.20	194.13	60.11	92.75	26.43	112.58	33.01
		Amphipoda	0.04	0.04	2.04	0.54	314.09	179.99	371.33	90.56
		Isopoda	0.00	0.00	65.21	37.34	26.50	13.39	5.25	2.86
		Dreissena	0	0	0	0	0	0	0	0
		Gastropoda	0.13	0.13	0.25	0.05	0.12	0.08	1.39	0.48
Post–Phosphorus	10	Total	119.17	26.42	590.79	168.82	891.33	276.51	580.19	137.22
Control		*Oligochaeta*	50.45	10.17	191.89	67.39	143.26	41.26	53.98	12.77
& Pre- *Dreissena*		*Chironomidae*	40.79	10.78	95.25	12.46	24.68	4.28	3.93	1.39
(1982-1991)		*Sphaeriidae*	9.15	3.57	20.91	5.15	37.03	10.39	133.61	45.82
		Amphipoda	0.66	0.26	67.53	25.46	457.17	80.03	381.42	90.43
		Isopoda	0.08	0.08	60.67	26.41	151.63	120.75	4.23	1.91
		Dreissena	0	0	0	0	0	0	0	0
		Gastropoda	0.03	0.02	0.05	0.04	0.06	0.03	3.27	3.21

Table S1 Continued…

Time Period	Years	Group	Big Bay		Glenora		Conway		Lake Ontario	
			Avg.	SE.	Avg.	SE.	Avg.	SE.	Avg.	SE.
Post- *Dreissena* (1992-2001)	10	Total	357.96	79.99	268.54	73.89	279.42	59.77	283.79	32.72
		Oligochaeta	115.68	32.14	84.88	21.93	93.58	18.66	131.20	29.72
		Chironomidae	57.83	9.60	55.37	6.38	18.89	6.87	8.90	2.91
		Sphaeriidae	36.96	8.10	34.03	8.84	44.30	11.01	67.77	22.19
		Amphipoda	3.74	1.04	31.88	12.22	66.26	63.14	28.68	19.46
		Isopoda	0.03	0.03	0.95	0.52	2.96	2.96	0.53	0.48
		Dreissena	47.58	47.40	1.37	0.68	77.05	56.14	55.88	46.04
		Gastropoda	1.31	0.46	0.26	0.12	0.41	0.14	0.51	0.32
Post- Round Goby (2002-2006)	5	Total	222.55	42.36	154.30	22.69	200.48	66.24	381.52	109.29
		Oligochaeta	89.29	21.20	78.51	18.90	69.69	21.39	181.35	42.75
		Chironomidae	43.86	18.14	48.73	10.32	23.08	4.71	18.79	6.29
		Sphaeriidae	20.78	7.65	1.72	1.62	3.62	1.82	1.96	1.05
		Amphipoda	0.43	0.15	0.10	0.06	0.83	0.60	0.35	0.22
		Isopoda	0	0	0	0	0	0	0	0
		Dreissena	1.15	0.86	0.16	0.10	75.08	39.51	156.52	112.66
		Gastropoda	2.53	1.28	0.08	0.08	0.28	0.12	0.68	0.40

Table S2. Average biomass (mg wet + shells per Ekman) of macroinvertebrates during the timeline of changes in the Bay of Quinte.

Time Period	Yrs	Group	Big Bay		Glenora		Conway		Lake Ontario	
			Avg.	S.E.	Avg.	S.E.	Avg.	S.E.	Avg.	S.E.
Pre-Phosphorus	4	Totals	518.57	99.66	2873.99	937.96	1854.07	686.08	2518.17	228.86
Controls		*Oligochaeta*	73.5	24.28	573.05	1.64	155.65	123.60	83.74	43.82
(to 1978)		*Chironomidae*	435.63	92.35	579.67	117.54	55.38	29.50	30.07	18.07
		Sphaeriidae	9.13	3.61	522.01	45.35	178.38	64.21	172.96	26.87
		Amphipoda	0.26	0.26	12.89	2.48	1242.56	713.48	2182.69	203.19
		Isopoda	0.00	0.00	1186.38	969.85	222.12	137.75	48.74	20.07
		Dreissena	0.00	0.00	0.00	0.00	0.00	0.00	0.00	0.00
Post-Phosphorus	10	Totals	430.03	120.43	740.12	114.47	938.82	175.35	1402.43	263.18
control &		*Oligochaeta*	119.64	35.21	57.40	20.56	127.57	25.40	65.78	23.97
Pre- *Dreissena*		*Chironomidae*	299.98	110.58	508.76	83.81	82.03	18.98	11.69	4.46
(1982-1991)		*Sphaeriidae*	5.27	1.83	41.75	9.56	47.18	11.74	178.65	55.77
		Amphipoda	1.07	0.63	86.42	28.34	544.32	81.80	1071.43	268.41
		Isopoda	0.16	0.16	40.03	9.39	128.93	77.18	15.52	12.61
		Dreissena	0.00	0.00	0.00	0.00	0.00	0.00	0.00	0.00

Table S1 Continued...

Time Period	Yrs	Group	Big Bay		Glenora		Conway		Lake Ontario	
			Avg.	S.E.	Avg.	S.E.	Avg.	S.E.	Avg.	S.E.
Post- *Dreissena* (1992-2001)	10	Totals	221.54	46.85	302.45	45.02	3713.34	1823.37	563.49	147.03
		Oligochaeta	39.09	8.77	20.14	3.66	90.66	21.11	218.62	50.06
		Chironomidae	112.7	36.15	176.31	23.86	17.75	3.41	10.12	2.46
		Sphaeriidae	16.10	3.49	54.44	16.74	56.55	12.97	51.16	17.52
		Amphipoda	2.15	0.74	38.91	14.27	40.81	39.13	34.62	22.66
		Isopoda	0.01	0.01	0.18	0.10	11.83	11.83	1.66	1.13
		Dreissena	7.20	6.59	2.16	1.80	3481.55	1821.88	238.07	142.03
Post- Round Goby (2002-2006)	5	Totals	344.36	91.65	173.48	26.24	16315.94	2186.05	12441.90	7733.88
		Oligochaeta	78.70	5.48	18.85	3.51	47.68	16.89	301.32	82.08
		Chironomidae	201.16	91.24	134.72	28.31	27.55	5.91	33.07	14.70
		Sphaeriidae	10.54	3.90	2.43	2.40	2.23	1.20	1.84	0.96
		Amphipoda	0.41	0.32	0.12	0.10	0.57	0.48	0.09	0.06
		Isopoda	0.00	0.00	0.00	0.00	0.00	0.00	0.00	0.00
		Dreissena	16.57	15.67	0.47	0.46	16229.02	2168.08	12095.03	7780.99

Table S3. Average Benthic Diversity (Margalef) in the Bay of Quinte.

Year	Big Bay (7m)	Glenora (21m)	Conway (32m)	L.Ontario (31m)
1967	1.17	1.99	1.39	1.33
1972	1.05	1.32	1.39	0.99
1977	0.55	1.77	1.79	1.82
1982	0.78	1.66	1.27	1.42
1988	1.48	1.86	1.46	0.76
1985	1.14	2.04	1.13	1.51
1992	1.79	1.54	1.1	1.08
1995	2.33	2.39	1.85	1.44
1996	2.04	1.79	1.98	1.08
1999	3.87	2.74	2.62	1.23
2000	3.3	2.49	2.82	2.16
2001	3.06	2.99	2.45	1.88
2002	2	3.05	2.78	1.85

Effects of water levels and water level regulation on the supply of suitable spawning habitat for eight fish guilds in the Bay of Quinte, Lake Ontario

E. L. Gertzen,[1,*] S. E. Doka,[1] C. K. Minns,[1,2] J. E. Moore,[3] and C. Bakelaar[1]

[1]*Fisheries and Oceans Canada, 867 Lakeshore Road, Burlington, ON L7S 1A1*
[2]*Department of Ecology and Evolutionary Biology, University of Toronto, 25 Willcocks St., Toronto, ON, M5S 3B2*
[3]*JEMSys Software Systems Inc., 22 Marion Crescent, Dundas, ON L9H 1J1*
**Corresponding author: erin.gertzen@dfo-mpo.gc.ca*

Keywords: habitat suitability, habitat supply, water level regime, wetlands, Great Lakes

Introduction

The Canadian Policy for the Management of Fish Habitat (DFO, 1986) is guided by the principle of no net loss of the productive capacity of fish habitat. Site-specific development proposals affecting physical habitat features are assessed against that principle. However, fish habitats are also increasingly being affected by pervasive multiple stressors including climate change, cumulative land use change, aquatic invasive species, nutrient loading and cycling, non-point source pollution and water level regulation (Schindler, 2001). Given the potential magnitude of the impact of these stressors on fishes and fish habitat, rigorous assessment methods are needed.

Water level regulation is one important stressor affecting the biological communities of the Laurentian Great Lakes (Hudon et al., 2006; Wilcox and Xie, 2007; Farrell et al., 2010). Beginning in the 1960s with the construction of the Moses-Saunders Dam and the establishment of the St. Lawrence Seaway, water levels in Lake Ontario have been regulated in accordance with the International Joint Commission's (IJC) Order of Approval, using regulation plan 1958D with Deviations (1958DD). Plan 1958DD aims to regulate water levels between 74.2 and 75.4 m; in practice, it has dampened water level fluctuations by approximately half of the 1.5 m range water levels would fluctuate under a natural flow regime (Farrell et al., 2010). Such natural fluctuations are a phenomenon to which Great

Ecology of the Bay of Quinte: Health, Management and Global Implications
Editors: C.K. Minns, M. Munawar and M.A. Koops

Aquatic Ecosystem Health & Management Society, Canada

Lakes communities, particularly coastal wetlands, have adapted during the post-glacial period (Keddy and Reznicek, 1986; Wilcox and Xie, 2007). Wetlands provide important habitat for rich biota, including breeding fishes and birds. The frequency and duration of flooding events strongly influence survival and development of plant assemblages and the ability of fishes to access wetlands (Mortsch, 1998; Keddy, 2000). Thus, understanding how water level regime and management affect coastal wetlands is of utmost importance to the management of fish habitat.

The majority of wetlands along the north shore of Lake Ontario are concentrated in the Bay of Quinte (Randall et al., 2011). In 1985, the Bay was designated as an Area of Concern (AOC) by the IJC, under the Great Lakes Water Quality Agreement, and has since been the subject of substantial restoration efforts. In recent decades, the quality and quantity of fish habitat in the Bay has been impacted by phosphorus loadings, the introduction of non-indigenous species such as Dreissenid Mussels, shoreline changes due to urbanization, and the loss and degradation of coastal wetlands (see Johanson and McNevin, 2007 for further details). Water level regulation has been implicated as one factor contributing to habitat changes, and specifically to the degradation of coastal wetlands (Johanson and McNevin, 2007). Importantly, more than half of the fish species found in the Bay use the shallow nearshore vegetation for spawning (Jude and Pappas, 1992). While the Bay provides important non-vegetated habitat for key species such as Walleye (*Sander vitreum*) and Smallmouth Bass (*Micropterus dolomieu*) (Ontario Ministry of Natural Resources, 2010), the immense areas of vegetation that exist in the Bay are unique within Lake Ontario (Randall et al., 2011). Given the importance of the Bay of Quinte in terms of providing vegetated and non-vegetated fish habitat, an assessment of the effects of water level management on the availability of this habitat is essential.

The IJC-commissioned Lake Ontario-St. Lawrence River Study (2000–2005; LOSL Study Board, 2006) sought to re-examine current water level and flow regulations in terms of economic and environmental benefits to various stakeholders. During the study, a Lake Ontario Habitat Model (LOHM) was developed to assess the links between water levels and suitable fish habitat supply and fisheries production (Limno-Tech, 2005; LOSL Study Board, 2006). At the core of the LOHM, habitat suitability models provided a quantitative and mechanistic approach for assessing habitat availability under different regulation regimes. In the current study, we apply the LOHM to the Bay of Quinte AOC and expand upon the model to assess fish habitat supply at a finer spatial and temporal resolution. The objectives of our study are to (1) evaluate how the Bay of Quinte nearshore and coastal areas contribute to spawning habitat supply (as an example for all life stages) by comparing select regions using simulation models, and (2) assess how the Bay of Quinte's nearshore contribution changes over significant periods of high-low water cycles within regulated and unregulated water level scenarios.

Materials and Methods

Lake Ontario habitat model structure and inputs

The LOHM has a structural base of habitat units with four major fish habitat components assigned to each–substrate, depth, vegetation cover and temperature. We divided the entire Bay of Quinte AOC into 1 km shoreline segments and subdivided these segments into 1 m elevation bands. The result was 14,876 unique habitat units, which varied in size, depending on bathymetry. Each habitat unit belonged to one of the Bay of Quinte's three naturally divided sub-sections–upper, middle and lower bays (Figure 1). Each habitat unit was assigned a shoreline type, substrate composition, and area by elevation profile. All units within each 1 km shoreline segment were assigned a substrate composition based on the segment's nearshore geology and shoreline characteristics (Table 1; Stewart, 2002). The 1 m elevation bands covered elevations between 52 and 80 m above sea level based on the International Great Lakes Datum of 1985 (IGLD85). This range included the 0 to 20 m water depth zone for all water level scenarios; 0–20 m was chosen to be our definition of nearshore because this is the area most affected by water fluctuations. Daily depth for each 0.1 m elevation contour within the 1 m bands was calculated by subtracting the contour's elevation from the daily surface water level. Daily surface water level was computed using linear interpolation of quarter monthly water level data for two water level scenarios in Lake Ontario (LOSL Study Board, 2006). Each habitat unit was also assigned a daily temperature profile. Modelled historical daily surface water temperatures (Moin and Thompson, 2006) were adjusted for location, as the upper bay tends to be warmer than the lower bay, and for season and depth. Temperature-depth relationships for each season were calibrated using three years of continuous temperature monitoring data in the Bay of Quinte (Doka, Fisheries and Oceans Canada, Burlington, Ontario, Canada, unpublished data) and a daily time series of water temperatures from the Belleville water treatment plant (J. Casselman, pers. comm.).

Vegetation cover was estimated on an annual basis using rule based models. Each habitat unit was assigned a percent cover of emergent, submergent and no vegetation. Emergent vegetation was predicted using the WS-USGS Emergent Vegetation Model (EVM) (Wilcox and Xie, 2007) and submerged aquatic vegetation was estimated using the DFO-GLLFAS Submerged Aquatic Vegetation Model (SAVM) (Seifried, 2002; Doka et al., 2006; Bakelaar et al., Fisheries and Oceans Canada, Burlington, Ontario, Canada, in progress). The EVM used depth and time since last exposure to predict percent cover for various wetland types. The SAVM was calibrated for Lake Ontario water clarity and was applied where (i) water depth < = 5 m and (ii) effective fetch <5 km (averaged across 16 compass directions; Scheffer et al., 1992). We applied both the EVM and SAVM in sequence for wetland habitat units, and only the SAVM for non-wetland habitat units.

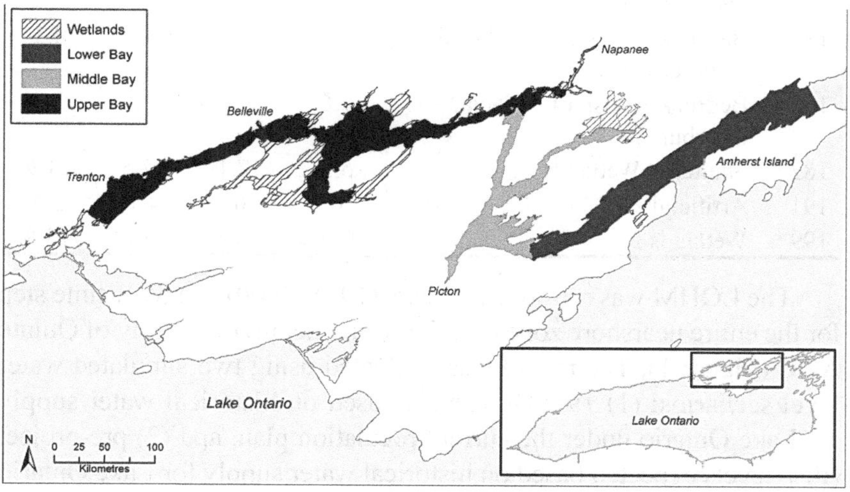

Figure 1. Map of the Bay of Quinte Area of Concern showing the upper, middle and lower bays, and their associated wetlands. Inset map shows the location of the Bay of Quinte in Lake Ontario.

Table 1. Total area associated with each shoreline type in the Bay of Quinte. Proportion (%) by shoreline type, with total area (km^2) for the Bay of Quinte and each sub-bay individually. Total area encompasses elevations 52–80 m above sea level (IGLD85), including land areas above low water datum; therefore, it is not comparable to the commonly used 254 km^2 which represents the average wetted area of the Bay of Quinte.

Shore type	Description	Lower Bay 50.9 km^2	Middle Bay 98.2 km^2	Upper Bay 237 km^2	Whole Bay 386 km^2
121	Glacial Till Low Bank/Low Plain	43.1	56.6	49.1	50.2
122	Composite Low Bank	4.8	0	0	0.6
125	Creek Bank	0	0	0.3	0.2
143	Sand Beach (Natural Deposition)	1.5	0	0	0.2
161	Bedrock (Resistant) - No Overburden	3.8	0	0	0.5
162	Bedrock (Resistant) - Glacial Overburden	31.0	0	0	4.1
172	Bedrock (Erosive) - Glacial Overburden	9.8	7.8	1.0	3.9
182	Sheltered Wetlands	5.3	17.1	27.5	21.9
191	Artificial	0	0	4.1	2.5
199	Wetlands	0.6	18.5	18.0	15.8

The LOHM was run for fifty years (1951–2000) at a daily time step for the entire nearshore zone (0–20 m water depth) of the Bay of Quinte AOC (Figure 1). The model was evaluated using two simulated water level scenarios: (1) 1958DD–levels based on historical water supply for Lake Ontario under the current regulation plan, and (2) pre-project (PP)–levels expected based on historical water supply for Lake Ontario in the absence of a regulation plan (LOSL Study Board, 2006).

Fish guilds and habitat suitability

The suitability of habitat in the Bay of Quinte for spawning was assessed using habitat associations compiled by Lane et al. (1996), spawning associations of Lake Ontario fishes reviewed by Cudmore-Vokey and Minns (2002), temperature preferences compiled by Eakins (2011), and the Habitat Alteration Assessment Tool (HAAT) habitat suitability assessment model of Minns et al. (2001). Bay of Quinte fish species that spawn in shallow water (0–5 m) during the spring and summer were included in the analysis; these are the species that will be most sensitive to water level regime (51 species; Hurley, 1986; Minns et al., 2006). Species that spawn in deeper water or exclusively in rivers were excluded; the Bay of Quinte AOC does not extend into the streams and tributaries and habitat data for these locations are lacking. Modelled species were split among eight guilds based on their vegetation and temperature preferences during spawning: vegetation vs. no vegetation crossed with four temperature categories (8–12°C, 12–16°C, 16–20°C and 20–28°C; Eakins, 2011). Guilds are described throughout this article by vegetation-temperature codes – for example, guild 'high-cold' is the guild that prefers high vegetation and temperatures of 8–12°C for spawning (see Table 2 for full guild code descriptions).

Habitat suitability was calculated using HAAT (Minns et al., 2001). In HAAT, species-specific habitat suitability, ranging from 0 to 1, is based on literature-derived species-habitat associations with depth, vegetation and substrate; for each habitat unit, habitat suitability is simply the product of suitability values for these three habitat components. The habitat categories used in the calculations were described by: depth (% 0–1 m, 1–2 m, 2–5 m, 5–10 m, 10–20 m), vegetation cover (% emergent, submergent, no cover) and substrate composition (% bedrock, boulder, rubble, cobble, gravel, sand, silt, clay, hardpan) for each habitat unit. Guild-based suitabilities take the average habitat suitability of a given habitat unit across all species in the guild. This results in a matrix of spawning suitability by habitat unit and guild for use in the LOHM.

In the LOHM, we further specified the daily spawning suitability of a given habitat unit by incorporating a temperature suitability component. Daily temperatures were compared to cited temperature ranges for spawning by guild; each habitat unit was assigned a temperature suitability value of either 0 (outside range) or 1 (inside range). Temperature suitability was multiplied by habitat suitability from HAAT to give an overall suitability value for each unit. Daily weighted suitable area (WSA) in km^2 was calculated by multiplying the unit's suitability value by the wetted area of the unit on the current day. We used both spawning suitabilities and WSAs to assess how annual water levels (high and low years) and water level scenarios (regulated and pre-project flows) affected the habitat available to the Bay of Quinte fish guilds.

Table 2. The species membership of the eight spawning guilds. In the guild codes, high and low refer to high and low vegetation preference, respectively, and cold, cold/cool, cool/warm and warm refer to spring spawning temperature (°C) preference ranges in ascending order of preference.

Guild No.	Guild Code	No. Spp.	Species in Spawning Guild	Vegetation Preference	Temperature Low	Temperature High
1	high-cold	5	Grass Pickerel, Muskellunge, Northern Pike, White Perch, Yellow Perch	High	8	12
2	low-cold	9	Longnose Dace, Mottled Sculpin, Northern Hog Sucker, Quillback, Round Goby, Shorthead Redhorse, Silver Redhorse, Walleye, White Sucker	Low	8	12
3	high-cold/cool	7	Bigmouth Buffalo, Black Crappie, Brook Stickleback, Central Mudminnow, Common Carp, Fathead Minnow, Spottail Shiner	High	12	16
4	low-cold/cool	7	Alewife, Johnny Darter, Lake Chub, Lake Sturgeon, Logperch, Smallmouth Bass, Threespine Stickleback	Low	12	16
5	high-cool/warm	10	Blackchin Shiner, Blacknose Shiner, Bluegill, Bluntnose Minnow, Bowfin, Bridle Shiner, Brook Silverside, Gizzard Shad, Largemouth Bass, Longnose Gar	High	16	20
6	low-cool/warm	3	Common Shiner, Greater Redhorse, Rock Bass	Low	16	20
7	high-warm	6	Banded Killifish, Black Bullhead, Brown Bullhead, Golden Shiner, Pumpkinseed, Tadpole Madtom	High	20	28
8	low-warm	4	Channel Catfish, Sand Shiner, Spotfin Shiner, Stonecat	Low	20	28

Analyses

Because of considerable spatial variation in basin morphometry, we divided the Bay of Quinte into its three sub-bays (Figure 1). We computed composite hypsometric curves for each bay and coastal wetlands. Hypsometric curves show the accumulation of area across elevations and provide an indication of how slopes, habitat area and depths are altered by changing water levels. For example, increasing water levels up a bluff will not significantly increase wetted area; however it will change the depth regime of already wetted areas. Contrarily, increasing water levels across a gradually sloping area will result in major gains in wetted area with relatively little depth change. Based on the hypsometric curves, we calculated the average wetted wetland area by bay for each water level scenario.

As a first step to examining overall differences between scenarios, we used a cumulative sum of suitable habitat (represented by WSA in km^2) across all days and years, where:

$$CUSUM = \frac{\sum_{Y=1951}^{2000} (WSA_{PP,Y} - WSA_{1958DD,Y})}{\sum_{Y=1951}^{2000} WSA_{1958DD,Y}} \tag{1}$$

This ratio gives the mean difference in suitable habitat area between scenarios. When CUSUM > 0, pre-project (PP) provides more suitable habitat; when CUSUM < 0, 1958DD provides more suitable habitat. In the absence of knowledge on the degree of difference that is biologically significant, we defined significance as a 5% statistic difference between the two scenarios (Zar, 2010). We computed CUSUM by guild for the entire Bay of Quinte and for each sub-bay separately. CUSUM provides a general indication of how water level regime affects suitable habitat area in the long-term; however, it does not provide an indication of how the mean and variance in WSA change across time and space at finer intervals. A more detailed analysis is provided below.

We compared spawning suitability spatially. For each water level plan, we used a MANOVA to examine whether spawning suitability for each guild differed among bays and shore types. The response variables were average spawning suitability for each guild over the fifty-year period by habitat unit and the factors were shore type and bay. If the MANOVA was significant, we conducted post-hoc analyses on the univariate guild models using Tukey's HSD to determine which groups differed.

We used two approaches to assess how spawning habitat supply changed temporally with water levels. First, we examined the relationship between average spring spawning suitability across all usable habitat units in each year and average water level over the spawning period using a linear regression. We repeated this analysis for each guild and both scenarios, applying the Bonferroni correction ($\alpha = 0.05/16$) in order to account for the increased probability of obtaining a Type I error when conducting multiple comparisons on a single dataset. Water level was calculated separately for each guild, as spawning is temperature-dependent and guilds spawn at different times, thus having differing average water levels. Secondly, we explored summary statistics of spawning WSA over time. We calculated average annual WSA (daily WSA aver aged across all days in the season where WSA > 0) for each guild on average (1951–2000) and during the lowest (1964) and highest (1976) water years which existed during the time period of our model. We observed how changes in habitat supply in low- and high-water years differed between the two water level scenarios. For demonstrations to management and public interest groups, the results could also be presented as daily time-lapse movies of the spatial changes in suitabilities over seasons and years.

Results

For the period between 1951 and 2000, annual water level was decreased by 0.25 m by regulation (simulated PP mean of 74.98 m to the 1958DD mean of 74.73 m). The average range in water levels was suppressed by 0.89 m, from 1.71 m in the PP scenario to 0.82 m under 1958DD. The Bay of Quinte levels were lowest in 1964, with water levels reaching 73.81 m under PP and 73.78 m under 1958DD, and highest in 1976, with a high of 76.17 m under PP and 75.69 m under 1958DD. Water levels occurring over the modelled period were notably higher than the long-term historical average for Lake Ontario; model results should be considered within this context.

The hypsometric curves of the bays and wetlands show varying bathymetry (Figure 2). The upper bay has a minimum elevation of 65.2 m above mean sea level; the middle and lower bays have minimum elevations below 52 m, where the hypsometric model stops. Bottom slope is gradual in the upper bay, moderate in the middle bay and steep in the lower bay. Maximum wetland area in the 52–80 m elevation range was minimal in the lower bay (0.31 km^2), as compared to the middle (18.2 km^2) and upper (42.7 km^2) bays (Figure 1; Table 1). If water level was raised 1 m above low water datum (74.2 m for Lake Ontario), the total area gained in the wetted zone (down to 52 m above mean sea level) would be small for non-wetland areas (2.3%) and large for wetlands (24.7%). If water level was reduced to 1 m below low water datum, the total wetted area would be diminished by 12.3% for non-wetland areas and 22.5% for wetlands. Sixty percent of the wetland area in the entire Bay of Quinte exists above low water datum (Figure 2). In general, PP provided greater or equal relative wetland area than 1958DD under average, low and high water levels in all bays (Table 3).

The CUSUM (Equation 1) suggests that, for the Bay of Quinte as a whole, PP provided significantly greater cumulative weighted suitable habitat (CUWSA) than 1958DD for guilds high-cold (see Table 2 for guild code descriptions), low- cold/cool and low-warm, and marginally greater habitat for three other guilds; 1958DD provided marginally greater CUWSA than PP for the other two guilds (Figure 3). At the scale of individual bays, PP provided significantly or marginally more CUWSA than 1958DD in the upper bay across all guilds, whereas it provided significantly less CUWSA than 1958DD across most guilds in the lower bay (Figure 3). Nevertheless, PP provided significantly greater CUWSA to three guilds despite losses in the lower bay. Because CUSUM for the whole of Bay of Quinte was not significantly different between the plans for the other five guilds, the gain of CUWSA in the larger near shore area of the upper bay during PP conditions was generally greater or proportional to the significant reduction in CUWSA in the smaller nearshore area of the lower bay under 1958DD (Table 1).

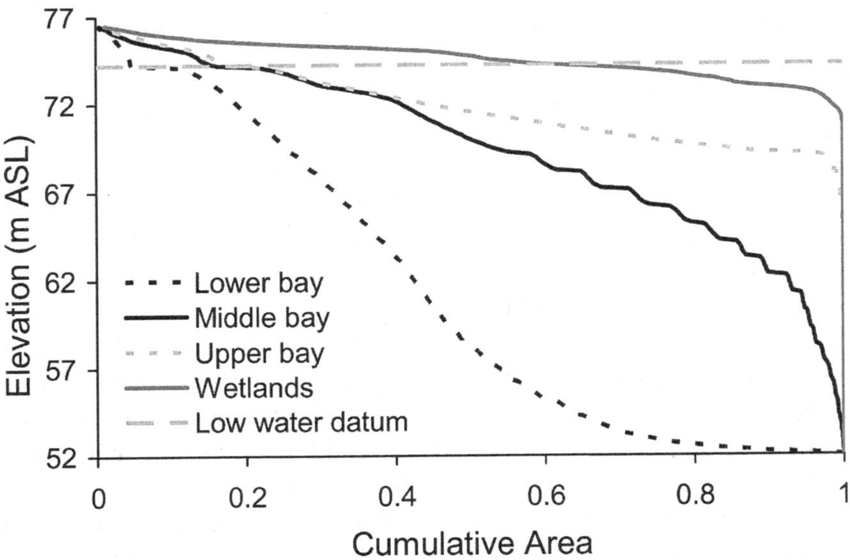

Figure 2. Hypsometric curves for area by bay and for wetlands in the Bay of Quinte. Hypsometric curves show proportion of area (cumulative) at a given elevation. Low water datum (74.2 m) is shown for reference.

Table 3. Proportion (%) of total wetted area that is wetlands during an average water level year (1951–2000), low year and high year, 1964 and 1976, for the 1958DD (DD) and pre-project (PP) scenarios.

	Average 1951-2000		**Low water 1964**		**High water 1976**	
Bay	DD	PP	DD	PP	DD	PP
Lower	0.5	0.6	0.2	0.2	0.8	1.0
Middle	13.1	14.7	8.1	8.1	19.5	20.3
Upper	8.6	10.0	5.0	5.0	13.8	14.7

The MANOVA examining guild suitabilities was significant under both water level scenarios for the factors: bay (1958DD: $F = 57.5$, $df = 2$, $p < 0.001$; PP: $F = 86.7$, $df = 2$, $p < 0.001$), shore type (1958DD: $F = 63.0$, $df = 6$, $p < 0.001$; PP: $F = 97.3$, $df = 6$, $p < 0.001$) and the interaction between bay and shore type (1958DD: $F = 2.2$, $df = 6$, $p < 0.001$; PP: $F = 3.4$, $df = 6$, $p < 0.001$). Post-hoc pair-wise comparisons on the univariate models showed significant differences in average spawning suitability between the lower and upper bays for all guilds under the PP scenario and for six guilds under 1958DD; differences between the lower and middle bays were significant for six and five guilds under

PP and 1958DD, respectively (Tables S2 and S3). Differences between the upper and middle bays were non-significant for all guilds, with the exception of guild low-warm under PP. Pair-wise comparisons between shoreline types showed that wetlands were significantly more suitable than all other shore types 100% of the time for high vegetation guilds across both scenarios and 64% of the time for low vegetation guilds; wetlands were significantly less suitable than other shore types 6% of the time for low vegetation guilds (Figure 4; Figure S1 and Tables S4 and S5). Under both water level scenarios, the average spawning suitability for all high vegetation guilds was highest among wetland habitat units; low vegetation guilds generally had lower wetland suitabilities than high vegetation guilds but higher suitabilities across other shore types (Figure 4; Figure S1). The importance of other shore types for low vegetation guilds was particularly evident in the lower bay, where wetland area is much reduced as compared to the other sub-sections of the Bay of Quinte (Tables 1 and 3).

For the whole Bay of Quinte AOC, the relationship between average annual spawning suitability and average water level during the spawning window was only significant and positive under 1958DD for guild high cold ($r2 = 0.17$, $df = 48$; $p = 0.003$); all other relationships were non-significant ($p = 0.004–0.87$; Bonferroni-corrected $\alpha = 0.003125$; Table S1). While a linear regression was used to analyse relationships, actual patterns were not completely linear, due in part to the shape of the elevation profiles in the different sub-bays (Figure 2).

Generally, average spawning WSA varied between the current regulation plan and PP scenario during average (1951–2000), low (1964) and high (1976) water years (Table 4). The low water year provided an average of 11.4% more WSA than the 1951–2000 mean across all guilds and scenarios. Contrarily, the high water year provided more variable results; it provided greater average WSA than the mean for five guilds under 1958DD (range: 3.3–15.8%), but only one guild under PP (8.2%; Table 4). Although the low water year provided an average of 13.9% greater WSA than the high water year across all guilds, the number of days during which WSA was available for spawning during low water years was significantly reduced for several guilds (Table 4). Not all habitat units provide WSA across the same days; as water levels fluctuate, habitats may become dry or too deep for spawning, and temperature mismatches may occur. The high- and low-cold/cool guilds experienced only 2.0 and 2.5 days, respectively, where WSA was available in the low water year under both water level scenarios (average temperature 12.9°C). In contrast, these guilds experienced a period of 19.8–25.2 days with WSA (average temperature 13.8°C) during the high water year across both scenarios and approximately 16 days with WSA (average temperature 13.9°C) on average across all years under 1958DD and PP, respectively (Table 4). The number of days when WSA was available for spawning for cool/warm guilds revealed an opposite pattern: they experienced more days with WSA during the low water year (approximately 36 days) than during the high water year (10.8–13.5 days; Table 4). Generally, the number of days available for spawning was similar between water level scenarios, varied between guilds and years, and increased as temperature preference increased.

Across all years, mean differences between the two water level scenarios were not large: PP provided more habitat area for five guilds (range: 2.9–9.8% increase in average WSA) whereas 1958DD provided more habitat area for two guilds (range: 0.97–1.1%). During the low water year, PP provided more habitat area for seven guilds (range: 1.9–6.3% increase in WSA); during the high water year, PP provided marginally increased habitat area for one guilds (1.6%) as compared to 1958DD, which provided increased habitat area for the other guilds (range: 1.6–12.7%). An animation of differences in spawning suitability across years between the two regulation plans demonstrates these variable results. Across the Bay of Quinte and across years, each plan provides more suitable habitat in under certain water level conditions as other interacting factors such as depth and vegetation change with water levels.

Discussion

The effects of water levels and fluctuations on the availability of suitable habitat for fishes are complex and not necessarily intuitive. The degree to which human regulation of water levels will be detrimental or beneficial to fish habitat and the communities that use that habitat depends on various interacting factors across space, scale and time. We found that habitat for some guilds responded positively to natural fluctuations whereas it responded positively to regulation for other guilds during a high water supply time period (1951–2000). Further, we found that water levels in a given season affected not only the amount of suitable area available for spawning but also the length of time over which this habitat was available.

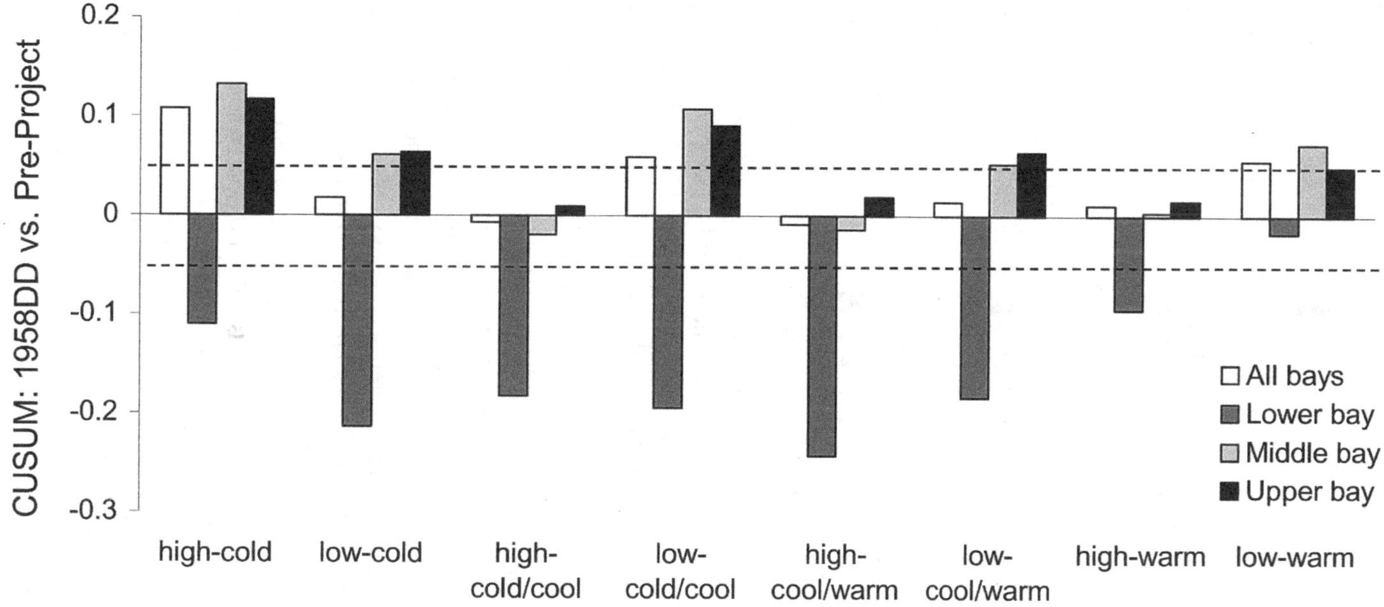

Figure 3. Cumulative sum (CUSUM) difference (Equation 1) in spawning WSA for 1951–2000 between 1958DD and pre-project (PP) plans for the eight fish guilds in the Bay of Quinte. When CUSUM > 0, pre-project provides greater WSA; when CUSUM < 0, 1958DD provides greater WSA. Dashed lines denote a significant difference (5%) in cumulative sums between the two scenarios.

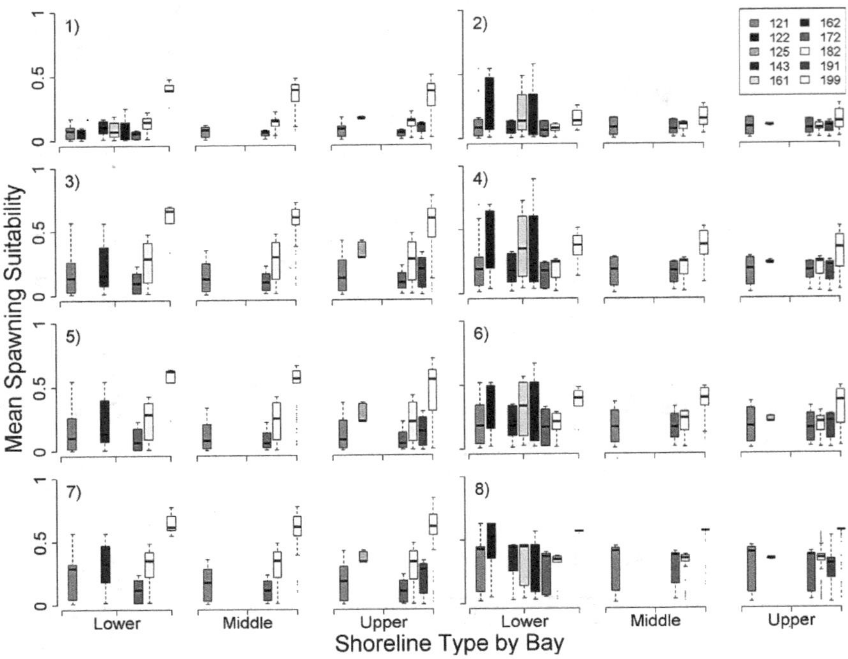

Figure 4. A representation of the results from the MANOVA of mean spawning suitability for each guild over the entire 1951–2000 period by bay and shoreline type for 1958DD. Panels 1–8 represent guilds: (1) high-cold, (2) low-cold, (3) high-cold/cool, (4) low- cold/cool, (5) high-cool/warm, (6) low-cool/warm, (7) high-warm and (8) low-warm. See Table 1 for shoreline type IDs (wetlands are white boxes) and Table 2 for guild code descriptions.

Water level regulation alters both the mean and variance of water levels, which in turn affect where wetland vegetation will grow and the types of plant communities that will be present. Studies have found that diverse communities germinate when seed banks are moist and that *Typha*-dominated communities germinate when seed banks remain flooded for long periods of time (Farrell et al., 2010). Diverse communities, including meadow marshes, provide highly suitable spawning habitat to species such as Northern Pike (*Esox lucius*) (LOSL Study Board, 2006); *Typha* communities tend to grow in dense stands that provide fringe habitat but are difficult for adult fishes to penetrate (Leira and Cantonati, 2008; Farrell et al., 2010). Given this, one might expect habitat availability for high vegetation guilds to respond negatively to regulation because diverse wetland area is reduced. In fact, in terms of cumulative habitat supply, the response to regulation was negative for only one high vegetation guild and weak across others.

Table 4. Average weighted suitable area (WSA in km2) by habitat unit, summed across habitat units; by water level scenario. Standard deviation in WSA is reported in italics under the 1951–2000 average conditions. The average number of days where spawning habitat is thermally available in each year is in parentheses.

	Average 1951–2000		Low water 1964		High water 1976	
Guild	1958DD	Pre-project	1958DD	Pre-project	1958DD	Pre-project
High-cold	10.1 (14.9) *0.55*	11.2 (15.4) *1.10*	10.4 (20.9)	10.4 (20.9)	12.0 (16.8)	12.2 (14.0)
Low-cold	6.6 (14.9) *0.57*	6.8 (15.4) *0.58*	7.3 (20.9)	7.5 (20.9)	6.9 (16.8)	6.7 (14.0)
High-cold/cool	17.1 (16.3) *1.49*	16.9 (16.4) *1.53*	20.4 (2.0)	20.9 (2.0)	17.1 (25.2)	16.8 (20.4)
Low-cold/cool	14.3 (16.2) *0.83*	15.1 (16.3) *1.04*	15.4 (2.5)	15.7 (2.5)	14.9 (24.2)	14.5 (19.8)
High-cool/warm	14.6 (18.6) *1.57*	14.6 (19.2) *1.63*	17.9 (36.2)	19.1 (36.2)	15.3 (11.3)	14.1 (10.8)
Low-cool/warm	14.6 (19.1) *1.10*	15.0 (19.7) *1.19*	16.8 (35.8)	18.1 (35.8)	15.1 (13.5)	13.4 (12.7)
High-warm	20.8 (38.2) *0.94*	20.6 (42.4) *1.32*	23.5 (36.0)	24.6 (36.0)	20.4 (61.2)	19.9 (48.5)
Low-warm	25.1 (36.9) *1.31*	25.9 (41.0) *1.60*	27.9 (34.4)	29.6 (34.4)	24.6 (61.2)	23.4 (48.5)

Conversely, one might expect the response of habitat supply to regulation to be minimal or positive for low vegetation guilds. Low vegetation habitat is important for key species such as Walleye and Smallmouth Bass (Table 2). Where diverse coastal wetlands are replaced by *Typha*-dominated communities, little impact on low vegetation guild habitat would be expected; however, in areas where vegetation is lost, non-vegetated habitat could increase. Further, the difference in water level mean and variance between regulated and pre-project flows would translate into changes in habitat availability due to flooding and habitat suitability due to changes in depth. The model results suggest that cumulative habitat area response to regulation was negative for two low vegetation guilds and weak for the other two. Nevertheless, in certain years and under certain conditions, each scenario was able to provide more WSA than the other scenario. These variable responses emphasize the complexity of the system and suggest that other inter acting factors are important. Additionally, the responses may differ for the modelled scenarios and other flow regimes across other time periods, given that the current model runs occur during a particularly high water supply period.

The model in this study focused on habitat for species that depend largely on the 0–2 m zone for spawning (Lane et al., 1996); thus, assessment of this small band is critical to evaluating spawning habitat avail ability. Our results suggest that a low-water year provides significantly more spawning WSA than an average or high-water year for most guilds under high supply periods. This non-intuitive finding is largely due to the bathymetry of the Bay of Quinte (Figure 2). A steep bluff exists above low water datum. This is particularly evident in the lower bay. Increases in water level up this bluff will have little effect on overall wetted area; however, as wetted areas become deeper, reductions in area within the critical 0–2 m zone occur.

In contrast, wetlands exhibit a shallow slope across the same elevations. Small increases in water level across these zones greatly increase the habitat area available to species that spawn in very shallow water, such as Northern Pike (Farrell et al., 2006). On the other hand, small decreases in water level in such shallow-sloping areas may lead to great losses in habitat and strand the eggs and fry of shallow spawning fishes (Leira and Cantonati, 2008). The contrast in bathymetry between bays likely explains why, under pre-project conditions, where the mean water level was 0.25 m higher than 1958DD, significant reductions in WSA were observed in the lower bay while increases were observed in the upper bay, where shallowly sloping habitats are concentrated. This observation provides strong evidence that, at a given water level, important trade-offs in habitat supply exist spatially between the upper and lower bay.

Another important water level management consideration is that fishes depend on acquiring appropriate temperatures for spawning. Temperature regimes change with water levels. The very low number of days where WSA was available for the cold/cool guilds during the low water year may be attributed to a mismatch in temperatures during the spawning season. In this year, the average temperature of habitat units when spawning WSA was available was 12.9°C, which is at the lower end of the 12–16°C range of suitable temperatures for cold/cool guilds. Other years and guilds generally experienced average temperatures closer to the mid-point of their ranges. Temperature mismatches may result in the loss of optimal spawning habitat during years when the appropriate thermal habitat lasts as few as 2–3 days. Spawning outside of the thermal window would likely be sub-optimal and mismatches in thermal habitat occurring during consecutive years could affect population dynamics. The fact that the number of days when WSA was available increased as temperature preference increased suggests that low-temperature spawning guilds may be most susceptible to annual losses in optimal habitat due to temperature mismatches. The observed trends in habitat supply across temperature guilds are consistent with field data from the Bay of Quinte showing that the Bay provides better habitat for warmwater fishes (Hurley, 1986).

In this study, spawning habitat was modelled as an example of the habitat-based approach for all life stages. In general, habitat supply should provide an indication of relative population sizes within the fish community; however, we recognize that habitat limitations during another life stage could limit population sizes and responses. Future studies could link habitat supply across life stages to population models in order to evaluate population responses to changes in habitat supply.

We have shown that water level regime and its interaction with depth, vegetation and temperature can have complex effects on fish habitat. Habitat supply in the Bay of Quinte is likely also affected by other perturbations that are not explicitly modelled here. The reduction of nutrient loadings through point source phosphorus control and the intro duction of Dreissenid Mussels have affected water clarity and vegetation cover (Leisti et al., 2012). Dreissenid Mussels themselves have altered the composition of the physical substrate and continued urbanization pressure is changing the structure of coastal habitats.

While these cumulative impacts and their relative importance should be modelled in the future, assessing the impact of water level regime is an important step in understanding how habitat in the Bay of Quinte responds to such perturbations at an ecosystem-wide scale. Furthermore, it provides a framework by which future studies can assess a host of cumulative impact scenarios.

While scenario-based comparisons are an effective tool for assessing fish habitat supply, gaps and uncertainties remain. To use the tool in a predictive capacity (e.g. to estimate actual fish community density and composition based on habitat supply), calibration and validation of the model results would be required. Nevertheless, the approach remains valid for the goal of comparing WSA between different stressor regimes on a relative basis.

Spatial resolution is one important modelling concern. The coarse scale we used could lead to a potential loss of detail. When we aggregate across large areas, it becomes impossible to differentiate small highly suitable areas within a generally unsuitable landscape from landscapes with broad areas of low suitability; both cases give the same overall WSA estimates, making it difficult to prioritize where to manage critical habitat (Minns and Bakelaar, 1999). Although appropriate spatial resolution is a major consideration, data availability is often a limiting factor (Graniero, 2001). Additionally, over large areas such as Lake Ontario, there is a trade-off between spatial resolution and computation power.

Our approach also requires adequate knowledge of the habitat requirements of fishes, as well as adequate knowledge of spatial habitat information – bathymetry, substrate, vegetation and temperature (Minns and Bakelaar, 1999). Information on species' requirements is often lacking for less studied species or difficult to obtain and parameterize because of plasticity; we used the best available information (Lane et al., 1996; Cudmore-Vokey and Minns, 2002). Information on habitat is often not very detailed. The bathymetry data used here had a 1 m contour resolution. While finer-scale bathymetry data would be ideal, such data are rarely available. The substrate data consisted of generalized percent age-composition information, based on dominant nearshore geology and shoreline characteristics along 1 km shoreline segments (Stewart, 2002). Although more detailed spatial information would be beneficial, such information is not available for all areas (especially the lower bay). The EVM and SAVM were calibrated using field data. Nevertheless, the SAVM did not use substrate as an input variable because the substrate deposition in the Bay results in areas that are good for vegetation growth accumulating between rocks at a very fine spatial scale. Incorporating the generalized substrate classes that are known to exist in the Bay of Quinte in the SAVM would have led to severe underestimates of vegetation cover. Temperature relations were also calibrated using field data; however, improving upon temperature models could provide greater insight into the linkages between temperature, water level inputs and spatial structure in regulating habitat availability. Although the variables discussed above are known to represent important habitat features, other potentially important habitat factors such as water quality and contaminants were not considered.

Summary and conclusions

Lake Ontario water levels are regulated under the International Joint Commission's Order of Approval. We assessed the availability of spawning habitat for eight fish guilds (grouped by vegetation and spawning temperature preferences) that use nearshore habitat in the Bay of Quinte, Lake Ontario, under regulated and pre-project water level scenarios. Our model used habitat characteristics of depth, substrate, vegetation cover and temperature across 14,876 habitat units and guild based habitat suitabilities to assess the weighted suitable area (WSA) available for spawning. During the modelled 1951–2000 period, regulated water levels were on average 0.25 m lower than pre-project levels and varied about half as much. Overall, the large and shallow upper bay within the Bay of Quinte benefited from pre-project flows whereas the smaller and deeper lower bay benefited from regulated flows. Higher variance in water levels increased highly suitable wetland area and higher mean water levels increased the availability of shallow flooded area in the upper and middle bays, where the bathymetry is gradually sloping. The number of days over which WSA was available was generally greater for high temperature guilds, and appeared to decrease when temperature mismatches occurred in a season. Results emphasize the complexity of the relationship between fish habitat supply and water level regime.

Given that factors regulating habitat availability are complex and depend on a host of intermediate and interacting factors, it is important to recognize that variability drives succession and diversity in wetland communities (Keddy, 2000; Wilcox and Xie, 2007), and that these diverse communities provide critical habitat for many spawning fishes. In Lake Ontario, much of the diverse vegetated habitat is located in the Bay of Quinte and is intermixed with important non-vegetated habitat. We observed that in all cases, pre-project flows provided more critical wetland area than did regulated flows, and that bathymetry interacted with water levels in important ways. Until a full assessment of tradeoffs can occur, natural flows should be the default water level scenario to promote habitat needs. This recommendation has been echoed by other studies (Wilcox and Xie, 2007; Farrell et al., 2010). Further assessments should involve scenario-based comparisons to determine the relative impact and cumulative effects of multiple stressors, including but not limited to: water level regulation, water quality, habitat quality, wetland and shoreline changes, and climate change.

Acknowledgements

We thank N. Lapointe and two anonymous reviewers for helpful comments on the manuscript. We also thank C. Chu, numerous field and laboratory staff, and everyone involved in the International Joint Commission Lake Ontario St. Lawrence River Study.

References

Cudmore-Vokey, B., Minns, C.K., 2002. Reproductive ecology and vegetation association databases for Lake Ontario Fishes. Can. MS Rpt. Fish. Aquat. Sci. 2607.

DFO, 1986. The Department of Fisheries and Oceans. Policy for the Management of Fish Habitat. DFO, Ottawa, ON.

Doka, S., Bakelaar, C., Bouvier, L., 2006. Coastal wetland fish community assessment of climate change in the Lower Great Lakes. In: L. Mortsch, J. Ingram, A. Hebb and S. Doka (Eds.), 2006. *Great Lakes Coastal Wetland Communities: Vulnerabilities to Climate Change and Response to Adaptation Strategies*. pp. 101–127. Final report submitted to the Climate change Impacts and Adaptation Program, Natural Resources Canada. Environment Canada and the Department of Fisheries and Oceans, Toronto, Ontario, http://www.environment.uwaterloo.ca/research/aird/aird_pub/Great_Lakes_ Coastal_Wetlands_Report_2006.pdf

Eakins, R.J., 2011. Ontario Freshwater Fishes Life History Database. Version 4.01. On-line database. (http://www.fishdb.ca), accessed 19 October 2011.

Farrell, J.M., Mead, J.V., Murry, B.A., 2006. Protracted spawning of St. Lawrence River Northern Pike (*Esox lucius*): simulated effects on survival, growth, and production. Ecol. Freshw. Fish. 15, 169–179.

Farrell, J.M., Murry, B.A., Leopold, D., Halpern, A., Rippke, M.B., Godwin, K.S., Hafner, S.D., 2010. Water-level regulation and coastal wetland vegetation in the upper St. Lawrence River: inferences from historical aerial imagery, seed banks, and *Typha* dynamics. Hydrobiologia 647, 127–144.

Graniero, P.A., 2001. The effect of spatiotemporal sampling strategies and data acquisition accuracy on the characterization of dynamic ecological systems and their behaviours. PhD thesis, University of Toronto.

Hudon, C., Wilcox, D., Ingram, J., 2006. Modelling wetland plant community response to assess water-level regulation scenarios in the Lake Ontario – St. Lawrence River Basin. Environ. Monit. Assess. 113, 303–328.

Hurley, D.A., 1986. Fish populations of the Bay of Quinte, Lake Ontario, before and after phosphorus control. 201–214. In: C. K. Minns, D. A. Hurley, K. H. Nicholls (Eds.), *Project Quinte: point-source phosphorus control and ecosystem response in the Bay of Quinte, Lake Ontario*. Can. Spec. Publ. Fish. Aquat. Sci. 86.

Johanson, P., McNevin, B., 2007. A fish habitat management plan for the Bay of Quinte. Bay of Quinte Remedial Action Plan. http://www.bqrap.ca/cms_lib/Fish%20 Habitat%20April_13_2007.pdf

Jude, D.J., Pappas, J., 1992. Fish utilization of Great Lakes coastal wetlands. J. Great Lakes Res. 18, 651–672.

Keddy, P.A., 2000. *Wetland ecology: principles and conservation*. Cambridge University Press, Cambridge.

Keddy, P.A., Reznicek, A.A., 1986. Great Lakes vegetation dynamics: the role of fluctuating water levels and buried seeds. J. Great Lakes Res. 12, 25–36. Lake Ontario – St. Lawrence River Study Board, 2006. Options for managing Lake Ontario and St. Lawrence River water levels and flows. Final Report to the International Joint Commission. www.losl.org/PDF/report-main-e.pdf

Lane, J.A., Portt, C.B., Minns, C.K., 1996. Spawning habitat characteristics of Great Lakes fishes. Can MS Rep Fish Aquat Sci. 2368.

Leira, M., Cantonati, M., 2008. Effects of water-level fluctuations on lake: an annotated bibliography. Hydrobiologia 613, 171–184.

Leisti, K.E., Doka, S.E., Minns, C.K., 2012. Submerged aquatic vegetation in the Bay of Quinte: Response to decreased phosphorus loading and Zebra Mussel invasion. Aquat. Ecosyst. Health Mgmt. 15(4), 442–452.

Limno-Tech, Inc., 2005. Development of an integrated ecological response model (IERM) for the Lake Ontario – St. Lawrence River Study. Prepared for the International Joint Commission. http://www.limno.com/ierm/IERM report draft-042905.pdf

Minns, C.K., Bakelaar, C.N., 1999. A method for quantifying the supply of suitable habitat for fish stocks in Lake Erie. In: M. Munawar, T. Edsall, I.F. Munawar (Eds.), *State of Lake Erie: Past, Present and Future*, pp. 481–496. Backhuys Publishers, the Netherlands.

Minns, C.K., Moore, J.E., Stoneman, M., Cudmore-Vokey, B., 2001. Defensible Methods of Assessing Fish Habitat: Lacustrine Habitats in the Great Lakes Basin – Conceptual Basis and Approach Using a Habitat Suitability Matrix (HSM) Method. Can. MS Rpt. Fish. Aquat. Sci. 2559.

Minns, C.K., Bernard, A., Bakelaar, C.N., Ewaschuk, M., 2006. A fish habitat classification model for the Upper and Middle Sections of the Bay of Quinte, Lake Ontario. Can. MS Rep. Fish. Aquat. Sci. 22748.

Moin, S.M.A., Thompson, A., 2006. Long term water temperature modelling in the Bay of Quinte, Ontario. Proceedings of the 7th International Conference on HydroScience and Engineering. 2006 September 10–13. Philadelphia, PA.

Mortsch, L.D., 1998. Assessing the impact of climate change on the Great Lakes shoreline wetlands. Climate Change. 40, 391–416.

Randall, R.G., Koops, M.A., Minns, C.K., 2011. A comparison of approaches for integrated management of coastal marine areas of Canada with the historical approach used in the Great Lakes (Bay of Quinte). Aquat. Ecosyst. Health Mgmt. 14(1), 104–113.
Ontario Ministry of Natural Resources, 2010. Bay of Quinte Fisheries Management Plan. Final Version. http://www.mnr.gov. on.ca/stdprodconsume/groups/lr/@mnr/@lets fish/ documents/document/251958.pdf.
Scheffer, M., de Redelijkhied, M.R., Noppert, F., 1992. Distribution and dynamics of submerged vegetation in a chain of shallow eutrophic lakes. Aquat. Bot. 42, 199–216.
Schindler, D.W., 2001. The cumulative effects of climate warming and other human stresses on Canadian freshwaters in the new millennium. Can. J. Fish. Aquat. Sci. 58, 18–29.
Seifried, K.E., 2002. Submerged macrophytes in the Bay of Quinte: 1972 to 2000. M.Sc. Thesis, Department of Zoology, University of Toronto.
Stewart, C.J., 2002. A revised geomorphic, shore protection and nearshore classification of the Canadian and United States shorelines of Lake Ontario and the St. Lawrence River. Coastal task working group summary report to the International Joint Commission. http://www.cjscons.com/ downloads/IJCONT/2002%20Shore%20
Classification%20Report.pdf.
Wilcox, D.A., Xie, Y., 2007. Predicted wetland plant community responses to proposed water-level-regulation plans for Lake Ontario: GIS-based modelling. J. Great Lakes Res. 33, 751–773.
Zar, J.H., 2010. *Biostatistical Analysis. 5th Ed.* Prentice Hall Inc., Saddle River, N.J.

The above article is a reprinted version of the following previously published article:

Gertzen; E.L., Doka; S.E., Minns; C.K., Moore; J.E., Bakelaar, C., 2012.
Effects of water levels and water level regulation on the supply of suitable spawning habitat for eight fish guilds in the Bay of Quinte, Lake Ontario
Aquatic Ecosystem Health & Management (2012) 15 (4): 397–409.
https://doi.org/10.1080/14634988.2012.728418

Supplemental Materials

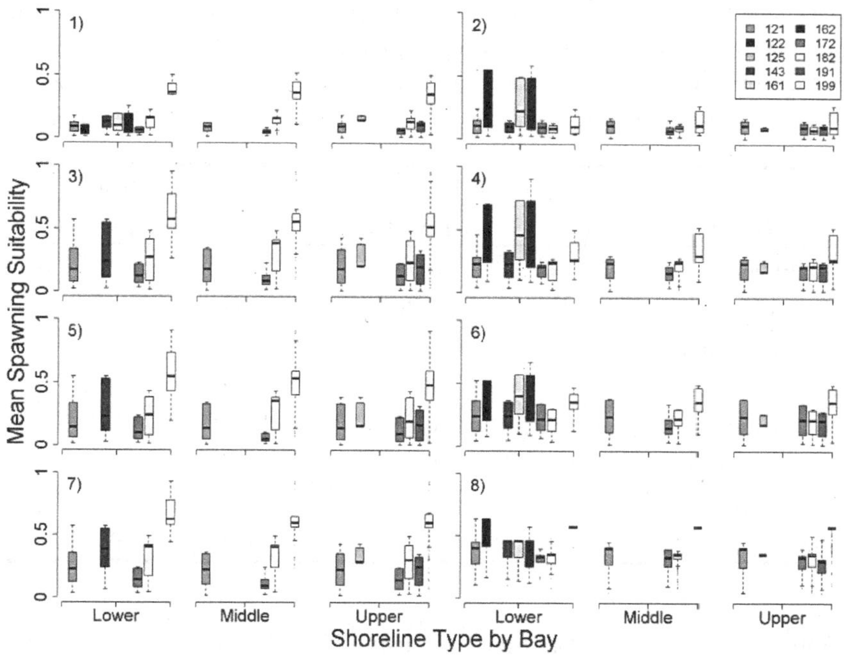

Figure S1. A representation of the results from the MANOVA of mean spawning suitability for each guild over the 1951-2000 period by bay and shoreline type for the Pre-Project (PP) plan. Panels 1-8 represent the eight guilds: (1) high-cold, (2) low-cold, (3) high-cold/cool, (4) low-cold/cool, (5) high-cool/warm, (6) low-cool/warm, (7) high-warm and (8) low-warm. See Table 1 in the main article for shoreline type descriptions (wetlands are white boxes) and Table 2 in the main article for guild code descriptions.

Table S1. Results of linear regression between mean spring water level and mean spawning suitability for eight fish guilds and for water level scenarios 1958DD and Pre-Project. r^2 and p-values are reported. Degrees of freedom = 48 for all analyses. The direction of the relationship (- or +) is shown in brackets for significant relationships (p-value<0.003125 after Bonferroni correction for multiple comparisons). See Table 2 in the main article for guild code descriptions.

	1958DD		Pre-Project	
Guild	r^2	**p-value**	r^2	**p-value**
high-cold	0.173	0.003 (+)	0.038	0.173
low-cold	0.024	0.281	0.163	0.004
high-cold/cool	0.001	0.871	0.007	0.560
low-cold/cool	0.009	0.519	0.030	0.225
high-cool/warm	0.073	0.058	0.028	0.244
low-cool/warm	0.031	0.220	0.096	0.028
high-warm	0.001	0.802	0.074	0.056
low-warm	0.020	0.332	0.010	0.497

Table S2. Results of post-hoc analyses using Tukey's HSD for the univariate models comparing fish guild suitability across bays and shoreline types for the 1958DD plan. Pair-wise comparisons between bays are shown for the eight spawning guilds. Significant differences (p-value<0.05) are denoted by an asterisk (*). See Table 2 in the main article for guild code descriptions.

Guild	Upper-Middle	Upper-Lower	Middle-Lower
high-cold	0.337	0*	0*
low-cold	0.083	0*	0*
high-cold/cool	0.468	0.001*	0.014*
low-cold/cool	0.428	0*	0*
high-cool/warm	0.599	0.132	0.438
low-cool/warm	0.294	0*	0*
high-warm	0.116	0.005*	0.180
low-warm	0.234	0.430	0.999

Table S3. Results of *post-hoc* analyses using Tukey's HSD for the univariate models comparing fish guild suitability across bays and shoreline types for the Pre-Project plan. Pair-wise comparisons between bays are shown for the eight spawning guilds. Significant differences (p-value<0.05) are denoted by an asterisk (*). See Table 2 in the main article for guild code descriptions.

Guild	Upper-Middle	Upper-Lower	Middle-Lower
high-cold	0.341	0*	0*
low-cold	0.242	0*	0*
high-cold/cool	0.458	0.002*	0.037*
low-cold/cool	0.485	0*	0*
high-cool/warm	0.564	0.036*	0.202
low-cool/warm	0.258	0*	0*
high-warm	0.087	0*	0.037*
low-warm	0.004*	0.009*	0.917

Table S4. Results of post-hoc analyses using Tukey's HSD for the univariate models of each dependent fish guild suitability in the MANOVA across bays and shoreline types for the 1958DD plan. Pair-wise comparisons between shore types are shown for the eight spawning guilds. Significant differences (p-value<0.05) are denoted by an asterisk (*). n/a appears in cases where shoreline types are always unsuitable. See Table 1 in the main article for shore type descriptions and Table 2 in the main article for guild code descriptions.

Guild high-cold	**121**	**122**	**125**	**143**	**161**	**162**	**172**	**182**	**191**
122	0.999								
125	0.001*	0.195							
143	0.214	0.849	0.998						
161	0.359	0.972	0.921	1					
162	0*	0.576	0.896	1	1				
172	0.445	0.824	$1.0 \bullet 10^{-4}$*	0.044*	0.072	0*			
182	0*	0.774	0.626	1	1	0.993	0*		
191	0.994	1	0.016*	0.499	0.742	0.002*	0.395	0.001*	
199	0*	0*	0*	0*	0*	0*	0*	0*	0*

Table S4 Continued...

Guild low-cold	**121**	**122**	**125**	**143**	**161**	**162**	**172**	**182**	**191**
122	0*								
125	1	0.006*							
143	0.541	$2.0 \cdot 10^{-6}$*	0.697						
161	0.181	0.414	0.877	0.030*					
162	0*	0.498	0.131	$5.2 \cdot 10^{-5}$*	0.997				
172	0.222	0*	0.939	0.981	0.017*	0*			
182	0.012*	0*	0.989	0.881	0.031*	0*	0.994		
191	0.973	0*	0.997	0.867	0.089	0*	0.995	1	
199	0*	0.014*	0.713	0.002*	1	0.178	0*	0*	0*
Guild high-cold/cool	**121**	**122**	**125**	**143**	**161**	**162**	**172**	**182**	**191**
122	n/a								
125	0.028*	n/a							
143	0.557	n/a	0.990						
161	n/a	n/a	n/a	n/a					
162	n/a	n/a	n/a	n/a	n/a				
172	0.034*	n/a	0.001*	0.096	n/a	n/a			
182	0*	n/a	0.546	0.997	n/a	n/a	0*		
191	0.999	n/a	0.031*	0.514	n/a	n/a	0.521	0.006*	
199	0*	n/a	$1.9 \cdot 10^{-4}$*	$2.0 \cdot 10^{-5}$*	n/a	n/a	0*	0*	0*

Table S4 Continued...

Guild low-cold/cool	**121**	**122**	**125**	**143**	**161**	**162**	**172**	**182**	**191**
122	$2.1 \cdot 10^{-5}$*								
125	0.881	0.478							
143	0.953	0.002*	0.601						
161	0.017*	0.998	0.951	0.037*					
162	0*	1	0.397	$2.2 \cdot 10^{-4}$*	1				
172	0.280	$7.0 \cdot 10^{-7}$*	0.362	1	0.001*	0*			
182	0.997	$5.8 \cdot 10^{-5}$*	0.936	0.918	0.030*	0*	0.145		
191	0.999	$4.1 \cdot 10^{-5}$*	0.799	0.993	0.015*	0*	0.967	0.990	
199	0*	1	0.263	$5.9 \cdot 10^{-5}$*	0.999	1	0*	0*	0*
Guild high-cool/warm	**121**	**122**	**125**	**143**	**161**	**162**	**172**	**182**	**191**
122	n/a								
125	0.126								
143	0.483	n/a	1						
161	n/a	n/a	n/a	n/a					
162	n/a	n/a	n/a	n/a	n/a				
172	0.046*	n/a	0.008*	0.080	n/a	n/a			
182	0*	n/a	0.751	0.970	n/a	n/a	$4.0 \cdot 10^{-7}$*		
191	1	n/a	0.136	0.466	n/a	n/a	0.522	0.047*	
199	0*	n/a	$2.3 \cdot 10^{-4}$*	$3.7 \cdot 10^{-4}$*	n/a	n/a	0*	0*	0*

Table S4 Continued...

Guild low-cool/warm	**121**	**122**	**125**	**143**	**161**	**162**	**172**	**182**	**191**
122	0.457								
125	1	0.996							
143	0.998	0.578	0.984						
161	0.591	1	0.997	0.628					
162	$1.0 \cdot 10^{-7}$*	1	0.887	0.151	1				
172	0.545	0.109	0.923	1	0.185	$2.0 \cdot 10^{-7}$*			
182	0.267	0.183	0.983	1	0.296	0*	0.995		
191	0.904	0.197	0.970	1	0.296	$4.6 \cdot 10^{-6}$*	1	1	
199	0*	0.828	0.207	0.006*	0.891	0.262	0*	0*	0*

Guild high-warm	**121**	**122**	**125**	**143**	**161**	**162**	**172**	**182**	**191**
122	n/a								
125	0.018*	n/a							
143	0.109	n/a	1						
161	n/a	n/a	n/a	n/a					
162	n/a	n/a	n/a	n/a	n/a				
172	0.002*	n/a	$1.2 \cdot 10^{-4}$*	0.003*	n/a	n/a			
182	0*	n/a	0.678	0.901	n/a	n/a	0*		
191	1	n/a	0.039*	0.162	n/a	n/a	0.065	0.002*	
199	0*	n/a	$5.5 \cdot 10^{-5}$*	$2.2 \cdot 10^{-4}$*	n/a	n/a	0*	0*	0*

Table S4 Continued...

Guild low-warm	121	122	125	143	161	162	172	182	191
122	0.014*								
125	1	0.658							
143	1	0.542	1						
161	1	0.338	1	1					
162	0.994	0.105	1	1	1				
172	0.639	0.002*	0.964	0.999	1	0.503			
182	0.630	0.004*	0.996	1	1	0.747	0.989		
191	0.376	0.001*	0.921	0.996	0.998	0.307	1	0.906	
199	0*	0.980	0.018*	0.020*	0.002*	0*	0*	0*	0*

Table S5. Results of post-hoc analyses using Tukey's HSD for the univariate models of each dependent fish guild suitability in the MANOVA across bays and shoreline types for the Pre-Project plan. Pair-wise comparisons between shore types are shown for the eight spawning guilds. Significant differences (p-value<0.05) are denoted by an asterisk (*). n/a appears in cases where shoreline types are always unsuitable. See Table 1 in the main article for shore type descriptions and Table 2 in the main article for guild code descriptions.

Guild high-cold	**121**	**122**	**125**	**143**	**161**	**162**	**172**	**182**	**191**
122	0.997								
125	0.007*	0.366							
143	0.045*	0.607	1						
161	0.006*	0.509	1	1					
162	0*	0.289	0.999	1	1				
172	0.093	0.538	$2.1 \cdot 10^{-4}$*	0.002*	0*	0*			
182	0*	0.528	0.960	0.997	0.997	0.971	0*		
191	0.989	1	0.058	0.194	0.072	$7.2 \cdot 10^{-5}$*	0.108	$1.8 \cdot 10^{-5}$*	
199	0*	0*	0*	0*	0*	0*	0*	0*	0*

Table S5 Continued...

Guild low-cold	**121**	**122**	**125**	**143**	**161**	**162**	**172**	**182**	**191**
122	0*								
125	1	$1.7 \cdot 10^{-5}$*							
143	0.136	0*	0.605						
161	$1.7 \cdot 10^{-6}$*	0.879	0.010*	$7.0 \cdot 10^{-7}$*					
162	0*	0.107	0.007*	0*	0.998				
172	0.009*	0*	0.962	0.932	0*	0*			
182	0.012*	0*	1	0.428	0*	0*	0.526		
191	0.992	0*	1	0.397	$2.0 \cdot 10^{-6}$*	0*	0.668	1	
199	0*	$1.0 \cdot 10^{-7}$*	0.870	0.001*	0.009*	$1.0 \cdot 10^{-7}$*	0*	0*	$3.3 \cdot 10^{-4}$*
Guild high-cold/cool	**121**	**122**	**125**	**143**	**161**	**162**	**172**	**182**	**191**
122	n/a								
125	0.635	n/a							
143	0.241	n/a	0.998						
161	n/a	n/a	n/a	n/a					
162	n/a	n/a	n/a	n/a	n/a				
172	0.007*	n/a	0.064	0.012*	n/a	n/a			
182	0*	n/a	1	0.945	n/a	n/a	0*		
191	0.997	n/a	0.559	0.206	n/a	n/a	0.330	0.002*	
199	0*	n/a	$2.0 \cdot 10^{-7}$*	$4.0 \cdot 10^{-5}$*	n/a	n/a	0*	0*	0*

Table S5 Continued...

Guild low-cold/cool	**121**	**122**	**125**	**143**	**161**	**162**	**172**	**182**	**191**
122	0*								
125	0.999	0.022*							
143	0.783	$7.6 \cdot 10^{-6}$*	0.758						
161	0*	1	0.014*	$4.7 \cdot 10^{-6}$*					
162	0*	1	0.012*	$5.0 \cdot 10^{-7}$*	0.997				
172	0.019*	0*	0.638	1	0*	0*			
182	0.967	$1.0 \cdot 10^{-7}$*	1	0.666	$1.0 \cdot 10^{-7}$*	0*	0.004*		
191	0.999	$1.0 \cdot 10^{-7}$*	0.995	0.935	$1.0 \cdot 10^{-7}$*	0*	0.620	0.974	
199	0*	0.702	0.126	$1.7 \cdot 10^{-5}$*	0.571	0.348	0*	0*	0*
Guild high-cool/warm	**121**	**122**	**125**	**143**	**161**	**162**	**172**	**182**	**191**
122	n/a								
125	0.907	n/a							
143	0.264	n/a	0.980						
161	n/a	n/a	n/a	n/a					
162	n/a	n/a	n/a	n/a	n/a				
172	0.008*	n/a	0.212	n/a	n/a	0.014*			
182	0*	n/a	1	n/a	n/a	0.884	0*		
191	0.996	n/a	0.841	n/a	n/a	0.219	0.367	0.017*	
199	0*	n/a	$3.0 \cdot 10^{-7}$*	n/a	n/a	$3.8 \cdot 10^{-4}$*	0*	0*	0*

Table S5 Continued…

Guild low-cool/warm	**121**	**122**	**125**	**143**	**161**	**162**	**172**	**182**	**191**
122	0.201								
125	1	0.759							
143	0.982	0.256	0.999						
161	0.012*	0.999	0.310	0.051					
162	0*	1	0.271	0.027*	1				
172	0.037*	0.005*	0.987	1	$1.3 \cdot 10^{-4}$*	0*			
182	0.014*	0.031*	1	1	0.001*	0*	0.783		
191	0.635	0.038*	1	1	0.001*	0*	0.996	1	
199	0*	0.989	0.061	0.003*	1	0.964	0*	0*	0*
Guild high-warm	**121**	**122**	**125**	**143**	**161**	**162**	**172**	**182**	**191**
122	n/a								
125	0.168	n/a							
143	0.008*	n/a	0.979						
161	n/a	n/a	n/a	n/a					
162	n/a	n/a	n/a	n/a	n/a				
172	$3.2 \cdot 10^{-5}$*	n/a	0.001*	n/a	n/a	$1.3 \cdot 10^{-5}$*			
182	0*	n/a	0.997	n/a	n/a	0.608	0*		
191	1	n/a	0.230	n/a	n/a	0.016*	0.010*	$5.2 \cdot 10^{-5}$*	
199	0*	n/a	0*	n/a	n/a	$4.2 \cdot 10^{-6}$*	0*	0*	0*

Table S5 Continued…

Guild low-warm	121	122	125	143	161	162	172	182	191
122	$9.0 \cdot 10^{-7}$*								
125	0.999	0.077							
143	1	0.043*	1						
161	0.988	0.049*	1	1					
162	0.970	0*	1	1	1				
172	0.069	0*	0.740	0.938	0.423	0.054			
182	0.122	0*	0.974	0.999	0.851	0.336	0.731		
191	0.005*	0*	0.522	0.815	0.213	0.007*	1	0.213	
199	0*	0.829	$1.1 \cdot 10^{-7}$*	$6.0 \cdot 10^{-6}$*	$2.0 \cdot 10^{-7}$*	0*	0*	0*	0*

Fish community structure in the Bay of Quinte, Lake Ontario: The influence of nutrient levels and invasive species

J. A. Hoyle,[1,*] J. N. Bowlby,[1] C. M. Brousseau,[2]

T. B. Johnson,[1] B. J. Morrison,[1] and R. G. Randall[2]

[1]*Ontario Ministry of Natural Resources and Forestry 41 Hatchery Lane, Picton, ON K0K 2T0, Canada*

[2]*Fisheries and Oceans Canada, 867 Lakeshore Road, Burlington, ON L7R 4A6, Canada*

**Corresponding author: jim.hoyle5@gmail.com*

Keywords: Yellow Perch, White Perch, Walleye, piscivores, centrarchids, fish populations, ecosystem change, Dreissenid Mussels

Introduction

The Bay of Quinte is a very productive system for fish, including species that support important recreational and commercial fisheries both in the bay and eastern Lake Ontario. The Bay of Quinte fish community has experienced significant ecological stress associated with excessive anthropogenic nutrient input and invasion of non-native species. Hurley and Christie (1977) documented the depreciation of the fish community from the 1930s through the mid-1970s, recognizing cultural eutrophication as a major driver of observed fish community change. Excessive nutrient input from municipal sewage led to increased algae production, and decreased oxygen and water clarity; growth of aquatic vegetation was impaired over broad expanses of the bay. These factors suppressed top predators such as Walleye (*Sander vitreus*), Longnose Gar (*Lepisosteus osseus*), Bowfin (*Amia calva*), Northern Pike (*Esox lucius*), and Largemouth Bass (*Micropterus salmoides*), as well as Centrarchid species such as Pumpkinseed (*Lepomis gibbosus*), Bluegill (*L. macrochirus*) and Black Crappie (*Pomoxis nigromaculatus*). With piscivore levels low, the abundance of small-bodied species tolerant of hyper-eutrophic water quality conditions, Alewife (*Alosa pseudoharengus*), White Perch (*Morone americana*), and Gizzard Shad (*Dorosoma cepedianum*) exploded. Hurley

Ecology of the Bay of Quinte: Health, Management and Global Implications
Editors: C.K. Minns, M. Munawar and M.A. Koops

Aquatic Ecosystem Health & Management Society, Canada

(1986a) updated trends in fish abundance in the context of reduced nutrient loading (phosphorus control) that occurred in 1977–1978. Hurley (1986a) found that partial recovery of the fish community was related to over-winter mortality of the abundant Alewife and White Perch, and improved water quality following reduced nutrient loading. Predator-prey interactions were identified as critical to restore healthy ecosystem function. In the early 1990s, Dreissenid Mussels invaded the Quinte area (Dermott et al., 2012). Immediately following their establishment, water clarity increased and submerged aquatic vegetation (SAV) re-appeared over vast areas of the Bay (Leisti et al., 2006, 2012). Round Goby (*Neogobius melanostomus*) invaded the Bay of Quinte in 1999 (Dietrich et al., 2006) and was first captured in our surveys in 2001; their abundance increased rapidly over the next two years.

Fish community sampling programs have existed in the Bay of Quinte since the 1950s, initially using gill nets and later adding trap nets, bottom trawls, and boat electrofishing. While each gear has unique bias with respect to habitat and species selectivity, collectively, these programs provide a very comprehensive data set within which long-term trends in the Bay of Quinte fish community can be examined. The first objective of this article was to report on the long-term dynamics of fishes in the Bay of Quinte fish community through 2009 and relate major changes in species dominance and abundance to large-scale ecological events associated with nutrient levels and invasive species. Specifically, results were interpreted relative to four major ecological time-stanzas: (1) phosphorus: prior to implementation of phosphorus control in municipal sewage treatment plants during the winter of 1977–1978; (2) post-phosphorus: after Point-source phosphorus control (1978–1994); (3) dreissenid: after Dreissenid Mussel invasion and proliferation (1995–2002); and (4) goby: after Round Goby invasion and proliferation. Our second objective, complementary to and focusing our first objective, was to repeat an analysis presented by Hurley and Christie (1977) and thereby compare and contrast the proportion of piscivores (Minns et al., 1994) in the fish community in both offshore and nearshore habitats and among time-stanzas.

Methods

Study area

The Bay of Quinte is a long (64 km), narrow and productive embayment extending from Trenton in the west and to Kingston and eastern

Lake Ontario in the east (see map in Bowlby and Hoyle, 2011). The Trent, Moira, Salmon and Napanee Rivers and numerous warm-water tributaries enter the bay. Physical and biological gradients span shallow and eutrophic conditions in the upper bay and deep and mesotrophic conditions in the lower bay near Lake Ontario. Fish movements are unrestricted in and out of the bay and include annual spawning migrations of Walleye, Alewife, and Lake Whitefish (*Coregonus clupeaformis*).

Sampling programs

Long-term and intensive fish community sampling programs were first initiated on the Bay of Quinte in the late 1950s (Hurley 1986a). Our analysis draws upon trend-through-time catch data from four fishing gears (gill nets, trap nets, bottom trawls, electrofishing) that collectively span five decades of sampling and provide intensive geographic coverage from the mouth of the Trent River in the upper bay to Lake Ontario in the lower bay. Gill nets and bottom trawls sampled offshore habitats while trap nets and boat electrofishing were used nearshore. The nearshore is distinguished from the offshore at approximately 5 m depth. Bottom trawls and electrofishing sampled all sizes of fish, particularly smaller ones, whereas gill nets and trap nets more effectively sampled large-bodied fish. Our analysis only included summer (June-September) data, and therefore makes no attempt to address seasonal variation in species abundance. Detailed sampling methods have been reported elsewhere (Hurley, 1986a, 1992; Bowlby et al., 1991; Casselman and Scott, 2003; Hoyle et al., 2008) but brief summaries are provided here.

Gill net sampling began in 1958, but only one site (at Hay Bay in the middle bay of Quinte) was sampled until 1972, when more sites (along with a second gear type, bottom trawl, see below) were added to extend coverage to the entire Bay of Quinte. We report annual gill net catches from 1972–2009; earlier catches are summarized by Hurley (1986a). All gill net samples come from bottom sets at fixed offshore locations. Seasonal gill net sets were conducted in some years but here only summer catches were included. Generally, gill nets consisted of a graded series of mesh sizes from 38–127 mm stretched mesh in 13 mm increments for a total of eight panels in a single gill net gang. The range of mesh sizes facilitated capture of a wide variety of fish species and sizes but was not effective at capturing very small species (e.g. cyprinids) or the young-of-the-year (YOY) of large species. Gill net mesh was made from multifilament material prior to 1992 and with monofilament thereafter. Paired gill net sets in 1991 using both materials, allowed conversion factors to be calculated (simple linear regression) for major species. Also, prior to 1992 individual mesh panels were 22.9 m (small mesh sizes) or 45.8 m (large mesh sizes) in length but all panels were shortened to 15.2 m thereafter. Catches were adjusted to standardize catch to 15.2 m of each mesh size for the entire time-series. Since 1992, 140 and 152 mm mesh sizes were added to the gill net gang but catches in these mesh sizes were not reported here. Catches for each gill net gang were summed across mesh sizes.

Bottom trawling was conducted every year since 1972 (except 1989) using a ¾ Western bottom trawl at several fixed offshore site locations, generally corresponding to the gill net sites. The trawl has a 13 mm mesh cod-end and most effectively samples small fish species and the young of larger species. As for gill nets, only summer bottom trawl catches were reported here. Catches were standardized for the entire time series as total number of individuals per 6 min trawl duration, covering approximately 400 m linear distance and 0.2453 ha in area (Hurley, 1992).

Beginning in 1969, standard 6-ft (1.83-m) trap nets were used to sample fish in the nearshore at two fixed site locations: Trenton (44° 06.0' N, 77° 31.8' W) in the upper bay and Hay Bay (44° 05.4' N, 77° 03.6' W) in the middle bay. Sampling occurred, primarily in July, in 13 of 20 years between 1969 and 1988; thereafter the program was discontinued. From 2001–2009, excluding 2006, a trap net program that employed a provincially standardized sampling protocol (Nearshore Fish Community Index Netting; Stirling, 1999) was conducted. This program used gear similar to that of the earlier years but sampling sites were chosen at random throughout the Bay of Quinte, in the month of September.

Boat electrofishing was conducted following a standardized protocol (Brousseau et al., 2005). Catch data were collected from 100 m transects at 1.5 m depth at several fixed locations representing a variety of coastal habitat types. Sampling was conducted periodically (1989–1990, 1992, 1999, 2001, 2007 and 2009), depending on Remedial Action Plan (RAP) needs (Brousseau et al., 2011).

Data analysis

For each catch, fish were identified to species and counted. Often fish were weighed in aggregate by species; if not, the weight of the catch was estimated based on either a sub-sample of individual fish weights or a sub-sample of individual fish lengths in combination with a length-weight regression. This resulted in a species-specific description of catch by numbers and biomass. In a few cases, in the recent bottom trawl data, YOY Pumpkinseed and Bluegill were recorded collectively as "Sunfish." Random or length-stratified sub-samples of fish were processed for biological attributes such as weight, sex, state of maturity, and diet. All species-specific catch results were initially summarized (arithmetic means) by year, gear type and geographic areas of the bay: upper, middle and lower as described in detail by Hurley (1986a).

The index of relative importance (IRI; Pinkas et al., 1971; Cortes, 1998) was used to quantify the relative dominance status of each fish species in the Bay of Quinte fish community (IRI = (% by number + % by weight) x (% frequency of occurrence)). This index was developed for diet studies to compare the contribution of different components but has also been adopted more recently in fish community studies (Karpov et al., 1995; Jutagate et al., 2001; Zhu et al., 2008). IRI was calculated for each species, time-stanza and sampling gear combination to examine differences and patterns among time-stanzas and gear types, and averaged across all time-stanzas and gear types to yield an overall fish species importance index.

For selected dominant species, mean catches by number was examined between adjacent time-stanzas with a two-way analysis of variance (ANOVA; STATISTICA 8.0), by gear type, with geographic region and time-stanzas as factors. All test statistics were considered significant at $p < 0.05$.

The community biomass consisting of piscivores (PPB) was determined for offshore (gill net) and nearshore (trap nets) habitats. PPB values reported by Hurley and Christie (1977) were re-calculated to include Smallmouth Bass and exclude Channel Catfish. Only data from upper and middle Bay of Quinte were used. A PPB greater than 0.20 was considered indicative of a balanced fish community trophic structure following Hurley et al. (1986), Brousseau et al. (2004) and Brousseau and Randall (2008).

Results and Discussion

A detailed account of fish community trends spanning four decades, utilising four gear types, and covering a diverse geographic area (at both nearshore and offshore habitats) is not possible within the confines of the present article. We instead focus on major highlights, re-examining those first described by Hurley and Christie (1977) and Hurley (1986b) and, particularly, describing responses of the community associated with the return of SAV to vast expanses of the Bay of Quinte fish after Dreissenid Mussel invasion. Preliminary responses of this system to Round Goby invasion are also considered.

Over a 41-year period, 3,227 samples from four gear types (Table 1) caught 65 species (53 native and 12 non-native) of fish representing 27 families (Table 2). Yellow Perch (native invertivore), White perch (non-native invertivore) and Alewife (non-native planktivore) were the dominant species. The most important piscivore was Walleye followed distantly by Largemouth Bass. Species dominance and composition varied across the four major time-stanzas and gear types (Table 3).

Table 1. Number of samples taken (total = 3,227) by ecological time-stanza and gear type in the Bay of Quinte, Lake Ontario.

Time-stanza	Years	Gill net	Bottom trawl	Trap net	Electrofishing
Phosphorus	1969–1977	102	95	59	–
Post-phosphorus	1978–1994	376	314	107	136
Dreissenid	1995–2002	300	417	108	138
Goby	2003–2009	236	364	360	115
Totals	41 years	1,014	1,190	634	389

Chronology of fish community changes among time-stanzas

The phosphorus time-stanza was dominated by White Perch, Alewife and Gizzard Shad (herbivore) (Table 3). The Bay of Quinte ecosystem was affected during this stanza by excessive Point-source nutrient input that degraded water quality to the point where SAV was severely limited due to light-limitation associated with excessive algal production (Hurley and Christie, 1977; Leisti et al., 2006). In the absence of controlling interactions by top predators and competitors, these small-bodied, tolerant species thrived in the hyper-eutrophic conditions. Reproduction of top predators, especially Walleye, may have been impaired by the poor water quality conditions. Also, the hyper-abundant intermediate predators may have contributed to poor top predator survival through predation on their larval stages (Hurley and Christie, 1977). Nearshore centrarchid species, such as Pumpkinseed, Bluegill, Black Crappie and Largemouth Bass, which depend on abundant SAV for reproduction, feeding and protection, had relatively low IRI values during the phosphorus time-stanza (Table 3).

Table 2. Index of relative importance (IRI) for fish species (n = 65) caught in the Bay of Quinte, Lake Ontario and pooled for four sampling gears (gill net 44 species; bottom trawl 44 species; trap net 34 species and electrofishing 44 species) from 1969–2009. Origin, thermal regime, trophic class and environment inhabited are indicated. Species are ranked by IRI and those with an IRI less than 0.01 are denoted with "+".

Common name	Scientific name	Origin	Thermal regime	Trophic class	Environment	IRI
Yellow Perch	*Perca flavescens*	Native	Cool	Invertivore	Benthopelagic	3428.10
White Perch	*Morone americana*	Non-native	Warm	Invertivore	Benthopelagic	2639.65
Alewife	*Alosa pseudoharengus*	Non-native	Cool	Planktivore	Pelagic	2082.51
Brown Bullhead	*Ameiurus nebulosus*	Native	Warm	Invertivore	Benthic	1159.24
Pumpkinseed	*Lepomis gibbosus*	Native	Warm	Invertivore	Benthopelagic	863.39
Walleye	*Sander vitreus*	Native	Cool	Piscivore	Benthopelagic	760.18
Bluegill	*Lepomis macrochirus*	Native	Warm	Invertivore	Benthopelagic	617.30
White Sucker	*Catostomus commersoni*	Native	Cool	Invertivore/Detritivore	Benthic	559.84
Gizzard Shad	*Dorosoma cepedianum*	Native	Warm	Herbivore	Pelagic	471.72
Largemouth Bass	*Micropterus salmoides*	Native	Warm	Piscivore	Benthopelagic	318.79
Freshwater Drum	*Aplodinotus grunniens*	Native	Warm	Invertivore	Benthic	308.67
Rainbow Smelt	*Osmerus mordax*	Non-native	Cold	Invertivore	Pelagic	276.80
Trout-perch	*Percopsis omiscomaycus*	Native	Cool	Invertivore	Benthopelagic	135.00
Rock Bass	*Ambloplites rupestris*	Native	Warm	Invertivore	Benthopelagic	111.90
Black Crappie	*Pomoxis nigromaculatus*	Native	Cool	Invertivore	Benthopelagic	107.62
Spottail Shiner	*Notropis hudsonius*	Native	Cool	Invertivore/Planktivore	Benthopelagic	104.29
Channel Catfish	*Ictalurus punctatus*	Native	Warm	Invertivore/Piscivore	Benthic	101.45

Table 2. ***(Cont)***

Common name	Scientific name	Origin	Thermal regime	Trophic class	Environment	IRI
Smallmouth Bass	*Micropterus dolomieu*	Native	Warm	Invertivore/Piscivore	Benthopelagic	96.78
Northern Pike	*Esox lucius*	Native	Cool	Piscivore	Benthopelagic	90.65
Common Carp	*Cyprinus carpio*	Non-native	Warm	Invertivore/Detritivore	Benthopelagic	76.09
American Eel	*Anguilla rostrata*	Native	Cool	Invertivore/Piscivore	Benthic	57.49
Round Goby	*Neogobius melanostomus*	Non-native	Cool	Invertivore	Benthic	51.53
Bowfin	*Amia calva*	Native	Warm	Piscivore	Benthopelagic	43.44
Golden Shiner	*Notemigonus crysoleucas*	Native	Cool	Invertivore/Herbivore	Benthopelagic	36.86
Lake Trout	*Salvelinus namaycush*	Native	Cold	Piscivore	Benthopelagic	36.81
Logperch	*Percina caprodes*	Native	Warm	Invertivore	Benthic	21.05
Brown Trout	*Salmo trutta*	Non-native	Cold	Piscivore	Benthopelagic	15.39
Longnose Gar	*Lepisosteus osseus*	Native	Warm	Piscivore	Benthopelagic	13.81
Brook Silverside	*Labidesthes sicculus*	Native	Warm	Planktivore/Invertivore	Pelagic	11.30
Lake Whitefish	*Coregonus clupeaformis*	Native	Cold	Invertivore	Benthic	3.84
Silver Redhorse	*Moxostoma anisurum*	Native	Cool	Invertivore	Benthic	3.33
Emerald Shiner	*Notropis atherinoides*	Native	Cool	Planktivore	Benthopclagic	3.22
Lake Herring	*Coregonus artedi*	Native	Cold	Planktivore/Invertivore	Pelagic	2.99
Bluntnose Minnow	*Pimephales notatus*	Native	Warm	Detritivore	Benthopelagic	1.46

Table 2. *(Cont)*

Common name	Scientific name	Origin	Thermal regime	Trophic class	Environment	IRI
White Bass	*Morone chrysops*	Native	Warm	Invertivore/Piscivore	Benthopelagic	1.36
Johnny Darter	*Etheostoma nigrum*	Native	Cool	Invertivore	Benthic	1.30
Shorthead Redhorse	*Moxostoma macrolepidotum*	Native	Warm	Invertivore	Benthic	0.58
Fathead Minnow	*Pimephales promelas*	Native	Warm	Detritivore/Invertivore	Benthopelagic	0.42
Banded Killifish	*Fundulus diaphanus*	Native	Warm	Invertivore/Planktivore	Benthopelagic	0.40
River Redhorse	*Moxostoma carinatum*	Native	Cool	Invertivore	Benthic	0.36
Blackchin Shiner	*Notropis heterodon*	Native	Cool	Invertivore	Benthopelagic	0.15
Golden Redhorse	*Moxostoma erythrurum*	Native	Warm	Invertivore	Benthic	0.06
Greater Redhorse	*Moxostoma valenciennesi*	Native	Warm	Invertivore	Benthic	0.04
Burbot	*Lota lota*	Native	Cold	Piscivore	Benthic	0.02
Chinook Salmon	*Oncorhynchus tshawytscha*	Non-native	Cold	Piscivore	Pelagic	0.02
Common Shiner	*Luxilus cornutus*	Native	Cool	Invertivore	Benthopelagic	0.02
Mottled Sculpin	*Cottus bairdi*	Native	Cold	Invertivore	Benthic	0.02
Slimy Sculpin	*Cottus cognatus*	Native	Cold	Invertivore	Benthic	0.01
Mooneye	*Hiodon tergisus*	Native	Warm	Invertivore	Pelagic	0.01
Grass Pickerel	*Esox americanus vermiculatus*	Native	Warm	Piscivore	Benthopelagic	0.01
Sea Lamprey	*Petromyzon marinus*	Non-native	Cold	Parasite	Benthopelagic	+

Table 2. ***(Cont.)***

Common name	Scientific name	Origin	Thermal regime	Trophic class	Environment	IRI
Splake	*Salvelinus fontinalis x S. namaycush*	Non-native	Cold	Piscivore	Benthopelagic	+
Black Bullhead	*Ameiurus melas*	Native	Warm	Invertivore	Benthic	+
Blacknose Shiner	*Notropis heterolepis*	Native	Cool	Invertivore/Herbivore	Benthopelagic	+
Central Mudminnow	*Umbra limi*	Native	Cool	Invertivore	Benthic	+
Goldfish	*Carassius auratus*	Non-native	Warm	Invertivore/Herbivore	Benthopelagic	+
Rainbow Trout	*Oncorhynchus mykiss*	Non-native	Cold	Piscivore	Benthopelagic	+
Lake Sturgeon	*Acipenser fulvescens*	Native	Cool	Invertivore	Benthic	+
Brook Stickleback	*Culaea inconstans*	Native	Cool	Planktivore/Invertivore	Benthopelagic	+
Rudd	*Scardinius erythrophthalmus*	Non-native	Cool	Invertivore/Herbivore	Benthopelagic	+
Atlantic Salmon	*Salmo salar*	Native	Cold	Piscivore	Benthopelagic	+
Quillback	*Carpiodes cyprinus*	Native	Cool	Invertivore	Benthic	+
Stonecat	*Noturus flavus*	Native	Warm	Invertivore	Benthic	+
Threespine Stickleback	*Gasterosteus aculeatus*	Native	Cool	Invertivore	Benthopelagic	+
Silver Lamprey	*Ichthyomyzon unicuspis*	Native	Cool	Parasite	Benthopelagic	+

Note: Lake Trout and Atlantic Salmon are native but extirpated and re-introduced; Golden Redhorse and Black Bullhead are not verified (would represent range expansion); Splake is an introduced artificial hybrid.

The post-phosphorus time-stanza was dominated by Yellow Perch, White Perch, and Alewife. Yellow Perch increased in importance but White Perch, Alewife and Gizzard Shad declined compared to the phosphorus time-stanza. Walleye, the key top predator in the system, increased in importance by 2–3 orders of magnitude in three gear types (Table 3). Immediate improvement in water quality, following a 50% reduction in phosphorus loading at local sewage treatment plants in the winter of 1977–1978 (Robinson, 1986), along with massive over-winter die-offs of Alewife in 1977 and White Perch in 1978 set the stage for exceptionally strong percid (Yellow Perch and Walleye) year-classes in 1978 (Hurley, 1986a). The Walleye population subsequently recovered to very high levels (Bowlby et al., 1991) and effectively controlled the formerly hyper-abundant prey fish species (e.g. Alewife) through predation (Hurley, 1986b; Ridgeway et al., 1990). Abundant Walleye may also have prevented Yellow Perch from continuing to increase in the upper bay following production of its own large 1978 year-class. Thus, by the late 1980s to early 1990s a predator-prey balance had been re-established in the offshore portion of the fish community. However, water transparency had not improved to the extent required to promote expansion of SAV (Leisti et al., 2006). As a result, the nearshore fish community (e.g. as shown by trap nets) resembled that of the offshore (gill nets and bottom trawls), where Alewife, White Perch, Yellow Perch and Walleye dominated (Table 3).

Large changes in the fish community occurred in the dreissenid time-stanza (Table 3). Yellow Perch IRI more than doubled in the offshore gill nets and bottom trawls. Walleye, White Perch and Alewife declined in importance in most gears. Brown Bullhead, Pumpkinseed, Bluegill, Black Crappie and Largemouth Bass increased in importance in both nearshore and offshore sampling gear. A key consequence of the Dreissenid Mussel invasion to the Bay of Quinte in the early to mid-1990s was an increase in water transparency (Leisti et al., 2012). As water clarity improved, both SAV cover and extent significantly increased in the upper bay, vegetation bed extent expanded in the middle bay, and there was little response in the lower bay likely due to basin morphometry (Leisti et al., 2006). The re-emergence of aquatic macrophytes in the nearshore allowed for a dramatic increase in the abundance of fish species associated with vegetation such as Pumpkinseed, Bluegill, Black Crappie and Largemouth Bass. Concurrently, White Perch and Alewife declined to low levels in nearshore habitats. Bowlby et al. (2010) suggested that Walleye abundance likely declined due to the habitat and associated fish assemblage changes in the nearshore areas.

Table 3. IRI for fish species caught in the Bay of Quinte separated by sampling gear and ecological time-stanza. Only those species with an IRI greater than 1, in Table 2, are listed. The total number of species caught in each gear and time-stanza (including those species not listed) is indicated. "np" indicates a species was not sampled by that gear in that time-stanza. "+" denotes species with an IRI < 0.01.

	Phosphorus			Post-phosphorus			
Species	Gill net	Trawl	Trap net	Gill net	Trawl	Trap net	Electro-fishing
Yellow Perch	2001.08	453.14	336.86	4138.49	1356.10	706.12	5166.24
White Perch	5039.05	6368.02	9792.26	1831.43	1340.43	6069.93	102.75
Alewife	6221.33	5975.89	5590.88	3465.30	4974.13	906.84	355.14
Brown Bullhead	115.50	353.40	328.80	48.41	461.29	1938.96	284.59
Pumpkinseed	7.78	3.52	665.36	1.48	23.14	1353.01	432.26
Walleye	36.82	0.25	11.87	1642.78	775.41	2115.14	1715.17
Bluegill	+	np	4.28	np	+	5.90	3.36
White Sucker	239.10	8.89	46.18	1858.33	861.21	217.36	163.36
Gizzard Shad	4828.00	55.93	2.79	702.22	546.37	20.60	58.84
Lgmouth Bass	0.02	0.08	0.01	0.00	np	0.49	59.21
Freshwater Drum	5.27	0.02	0.02	168.20	85.87	26.19	229.76
Rainbow Smelt	5.87	2890.69	np	0.55	1141.13	np	0.04
Trout-Perch	0.03	327.04	np	+	879.67	np	np
Rock Bass	10.39	3.17	49.91	34.68	0.86	748.73	260.77
Black Crappie	0.02	0.73	231.22	0.04	5.93	65.18	1.23
Spottail Shiner	+	140.88	np	np	418.76	np	116.07

Table 3. ***(Cont.)***

Species	Phosphorus			Post-phosphorus			
	Gill net	Trawl	Trap net	Gill net	Trawl	Trap net	Electro-fishing
Channel Catfish	5.78	0.02	95.15	4.21	0.42	658.67	0.13
Smallmouth Bass	65.94	0.21	107.76	92.29	0.45	980.94	46.46
Northern Pike	30.17	0.01	3.10	23.62	2.04	40.66	379.69
Common Carp	6.03	65.86	1.38	0.31	242.69	1.81	537.61
American Eel	+	414.04	75.83	0.00	144.23	164.22	53.26
Round Goby	np	np	np	np	np	np	np
Bowfin	57.06	0.01	92.64	0.37	np	109.68	35.34
Golden Shiner	0.09	+	0.98	0.05	np	0.57	37.82
Lake Trout	0.03	np	np	346.42	0.14	np	np
Logperch	np	np	np	np	0.30	np	188.56
Brown Trout	0.04	np	np	223.34	0.02	np	np
Longnose Gar	2.20	0.02	1.42	6.24	1.63	5.18	0.66
Brook Silverside	np	np	np	np	np	np	0.23
Lake Whitefish	0.01	np	np	7.63	33.80	np	0.00
Silver Redhorse	np	np	np	+	np	0.05	0.03
Emerald Shiner	np	np	np	np	np	np	6.24
Lake Herring	0.03	np	np	0.47	0.96	np	np
Bluntnose Minnow	np	np	np	np	np	np	np
White Bass	1.81	np	0.61	2.56	3.21	1.52	1.11
Johnny Darter	+	5.16	np	np	6.56	np	0.10
No. of species	32	24	23	34	33	27	32

Table 3. (*Cont.*)

Species	Dreissenid				Goby			
	Gill net	Trawl	Trap net	Electro-fishing	Gill net	Trawl	Trap net	Electro-fishing
Yellow Perch	8935.15	5478.01	86.95	6431.35	5683.76	4904.10	134.44	5609.76
White Perch	630.28	1725.05	26.99	1.33	2812.53	3539.41	268.11	47.23
Alewife	452.15	1114.13	np	2.71	627.82	1495.18	np	56.15
Brown Bullhead	62.67	1263.40	6897.92	960.72	31.04	883.54	3109.52	648.90
Pumpkinseed	45.58	436.52	4006.33	3014.38	14.63	268.34	1582.04	1096.52
Walleye	1392.51	288.81	658.69	45.49	986.49	414.38	924.74	394.21
Bluegill	4.02	15.37	3491.51	399.13	11.05	13.35	3880.17	1431.28
White Sucker	1165.14	1899.30	316.27	47.51	551.02	535.74	459.68	28.50
Gizzard Shad	24.01	442.85	44.86	42.19	26.41	187.98	83.89	8.91
Lgmouth Bass	+	1.07	243.06	1148.09	0.01	1.66	362.85	2965.24
Freshwater Drum	404.51	395.93	524.51	30.61	506.67	1174.92	841.26	236.30
Rainbow Smelt	0.19	104.24	np	np	0.28	9.01	np	np
Trout-Perch	np	565.36	np	np	np	252.89	np	np
Rock Bass	8.44	0.20	19.66	153.85	4.42	0.11	146.51	236.87
Black Crappie	0.07	2.27	442.66	2.79	1.13	3.03	856.67	1.38
Spottail Shiner	0.00	318.86	np	92.85	+	440.38	+	36.59
Channel Catfish	0.38	0.09	409.51	np	1.36	0.16	345.92	np
Smallmouth Bass	12.57	1.28	37.32	6.05	0.65	1.15	34.34	64.26
Northern Pike	50.53	0.87	136.93	398.80	40.21	0.38	160.02	92.66

Table 3. (*Cont.*)

Species	Dreissenid				Goby			
	Gill net	Trawl	Trap net	Electro-fishing	Gill net	Trawl	Trap net	Electro-fishing
Common Carp	0.09	12.40	8.79	49.28	1.28	66.47	54.79	92.50
American Eel	np	7.89	2.39	np	np	np	0.44	np
Round Goby	0.08	0.21	np	np	38.55	728.75	0.00	5.36
Bowfin	np	np	82.87	121.55	np	np	144.62	7.43
Golden Shiner	np	np	0.08	168.63	np	+	0.38	344.30
Lake Trout	113.86	+	np	np	91.61	0.02	+	np
Logperch	np	1.16	np	34.69	np	10.69	np	80.38
Brown Trout	0.81	+	np	np	6.54	0.03	np	np
Longnose Gar	25.19	+	8.12	0.17	84.59	0.01	53.03	18.73
Brook Silverside	np	0.01	np	124.72	np	0.01	np	44.46
Lake Whitefish	5.10	3.90	np	np	3.32	3.88	+	np
Silver Redhorse	np	np	np	0.93	np	np	26.89	22.05
Emerald Shiner	np	+	np	1.21	np	np	np	40.81
Lake Herring	41.43	0.25	np	np	0.21	1.49	np	np
Bluntnose Minnow	np	0.00	np	20.16	np	np	np	1.75
White Bass	0.02	1.22	0.32	0.10	0.02	5.40	2.42	0.14
Johnny Darter	np	7.10	np	0.46	np	0.10	np	np
No. of species	33	37	25	34	30	36	29	39

Yellow Perch abundance increased immediately and dramatically in the upper bay following Dreissenid Mussel invasion but their abundance then declined somewhat as the nearshore centrarchid species became more abundant.

Round Goby were first reported in the Bay of Quinte in 1999 (Dietrich et al., 2006) and first captured in our sampling gear in 2001. Their abundance was high in offshore areas of the bay by 2003, declined rapidly after 2004 in gill net catches but stabilized at relatively high levels in the bottom trawls. This observation suggests that while large Round Goby, vulnerable to gill net gear, were included in the goby size structure soon after goby invasion, after 2004 the goby size structure included primarily smaller individuals. Goby abundance may have been high in nearshore areas prior to 2003 but the nearshore trap nets did not catch them and nearshore electrofishing was not conducted between 2001 and 2007. Round Goby were absent in the electrofishing gear in 2001 and present in 2007. Goby also appeared in the diet of many piscivores in the Bay of Quinte in 2003 (Taraborelli et al., 2010), and were a significant component of the Walleye diet in the Bay of Quinte during 2004–2006 (Bowlby et al., 2010). Walleye growth increased during the goby time-stanza (Bowlby et al., 2010) but whether or not this is due to the diet change is not known. Yellow Perch remained the most important species in the bay during the goby time-stanza (Table 3). White Perch increased in offshore sampling gear compared to the dreissenid time-stanza. Bluegill, Largemouth Bass and Black Crappie continued to increase while Brown Bullhead and Pumpkinseed declined.

Fish community changes during the goby time-stanza did not appear to be as extensive as for the earlier time-stanzas. On the other hand, more time may be required for Round Goby impacts to fully materialize. Fish recruitment responses (e.g. increased fecundity or fitness due to increased growth) may take several generations to become evident. Certainly, Round Goby invasion has potential to dramatically alter the Bay of Quinte foodweb. The species, by being one of the few benthivores that exploit Dreissenid Mussels as prey, provides a new energy vector from dreissenids to higher levels (i.e. piscivores) in the foodweb (Johnson et al., 2005; Dietrich et al., 2006; Campbell et al., 2009).

Trends in dominant species

Yellow Perch was the dominant species in the Bay of Quinte fish community, except during the pre-phosphorus control time-stanza (Table 3). Yellow Perch were prominent in all gears except trap nets which did not effectively capture this species. Yellow Perch abundance trends through time varied with gear type and geographic area of the bay (Figure 1; Appendix). In gill nets, abundance generally increased to a peak in the middle and lower bays during the post-phosphorus time-stanza, and then gradually declined to the present. In contrast, Yellow Perch abundance in the upper bay remained suppressed through the post-phosphorus period, despite production of a very large year-class in 1978, increased dramatically to very high numbers following Dreissenid Mussel invasion, then declined somewhat. Presumably, predation by the resurging Walleye and/or interactions with Alewife (Hoyle, 1993) inhibited Yellow Perch production in the upper bay after 1978. But after the Dreissenid Mussel invasion in the early 1990s, Yellow Perch abundance increased rapidly in the upper bay. Yellow Perch were scarce in bottom trawl catches, except for the large YOY catch in 1978, in all geographic areas until after Dreissenid Mussel invasion when catches increased in the upper and middle bays. We suggest that the apparent increased survival of young Yellow Perch after dreissenid establishment was associated with expanded SAV (Leisti et al., 2006) at that time. Yellow Perch catches in trap nets were lower in recent time-stanzas compared to past while electrofishing catches remained consistently high through time (Figure 1).

White Perch were important in all gear types; especially offshore gill nets and trawls but also, prior to Dreissenid Mussel invasion, in the nearshore trap nets and electrofishing surveys (Table 3; Appendix). White Perch invaded the Bay of Quinte in the 1950s (Scott and Christie, 1963), quickly rose to prominence in the fish community due to much reduced populations of potential predators and competitors, and were the dominant species in the post-phosphorus time-stanza. The hyper-abundant White Perch were implicated as one of the accumulating factors suppressing Walleye by preying upon Walleye fry (Hurley and Christie, 1977). A catastrophic die-off of the White Perch occurred in 1978 associated with a severely cold winter that year (Minns and Hurley, 1986). In the years following, White Perch recovered but not to increased levels of abundance (Figure 1). Predation by the recovering Walleye population was presumably sufficient to limit the White Perch recovery. Overall, White Perch importance in the fish community reached a low point during the post-Dreissenid Mussel years, but YOY catches in bottom trawls increased in the post-dreissenid and goby time-stanzas. Increased abundance in gill nets lagged that of the bottom trawls and it now appears that the species has regained a high level of prominence in the Bay of Quinte fish community. By way of contrast, White Perch have remained low in nearshore areas (Figure 1; Appendix) suggesting the return of SAV and the associated changes in the nearshore fish community are not conducive to high White Perch abundance.

Walleye was the most important large-bodied, piscivore in the Bay of Quinte fish community. Walleye was important in all gear types, habitats, and time-stanzas except during pre-phosphorus control (Table 3; Figure 1). At that time (late 1960s and 1970s) the population was reduced to remnant levels in the face of degraded water quality conditions and fry predation from the hyper-abundant White Perch population (Hurley and Christie, 1977). The 1978 Walleye year-class was the largest in recent times and was coincident with improved water quality and the White Perch "winter-kill." Although water clarity had improved, even by the early 1990s it had not improved to the extent required to promote significant expansion of SAV (Leisti et al., 2006). Therefore, the nearshore fish community remained depressed. Under these conditions, the Walleye population returned to historically high levels of abundance. Dreissenid Mussel invasion changed these conditions quickly and dramatically, and Walleye populations declined precipitously (Figure 1). A 75% decline in abundance of YOY Walleye occurred at this time (Bowlby et al., 2010) contributing to population levels about 25% of their peak abundance. Potential hypotheses explaining the decline in YOY abundance include reduced food supply/availability linked to dreissenid-induced increases in macrophytes and increased levels of predation and/or competition linked to dreissenid-induced clearing of the water column (Hoyle et al., 2008; Bowlby et al., 2010). Most recently, bottom trawl catches indicate that Walleye year-class strength was similar or somewhat higher in the goby era compared to the dreissenid time-stanza (Table 3; Figure 1). Walleye growth increased and remained higher in the goby time-stanza (Bowlby et al., 2010).

Centrachid species underwent dynamic changes in their contribution to the catch for all gears and time-stanzas (Table 3; Figure 1; Appendix). Centrarchid species dependent on SAV, such as Pumpkinseed, Bluegill, Black Crappie and Largemouth Bass, were depressed greatly during phosphorus and post-phosphorus time-stanzas. Smallmouth Bass and Rock Bass prefer hard substrate devoid of vegetation, and therefore increased in abundance during the post-phosphorus time-stanza and declined after Dreissenid Mussel invasion. Vegetation-dependent species abundance increased significantly after the return of SAV during the Dreissenid Mussel time-stanza (Table 3; Figure 1; Appendix).

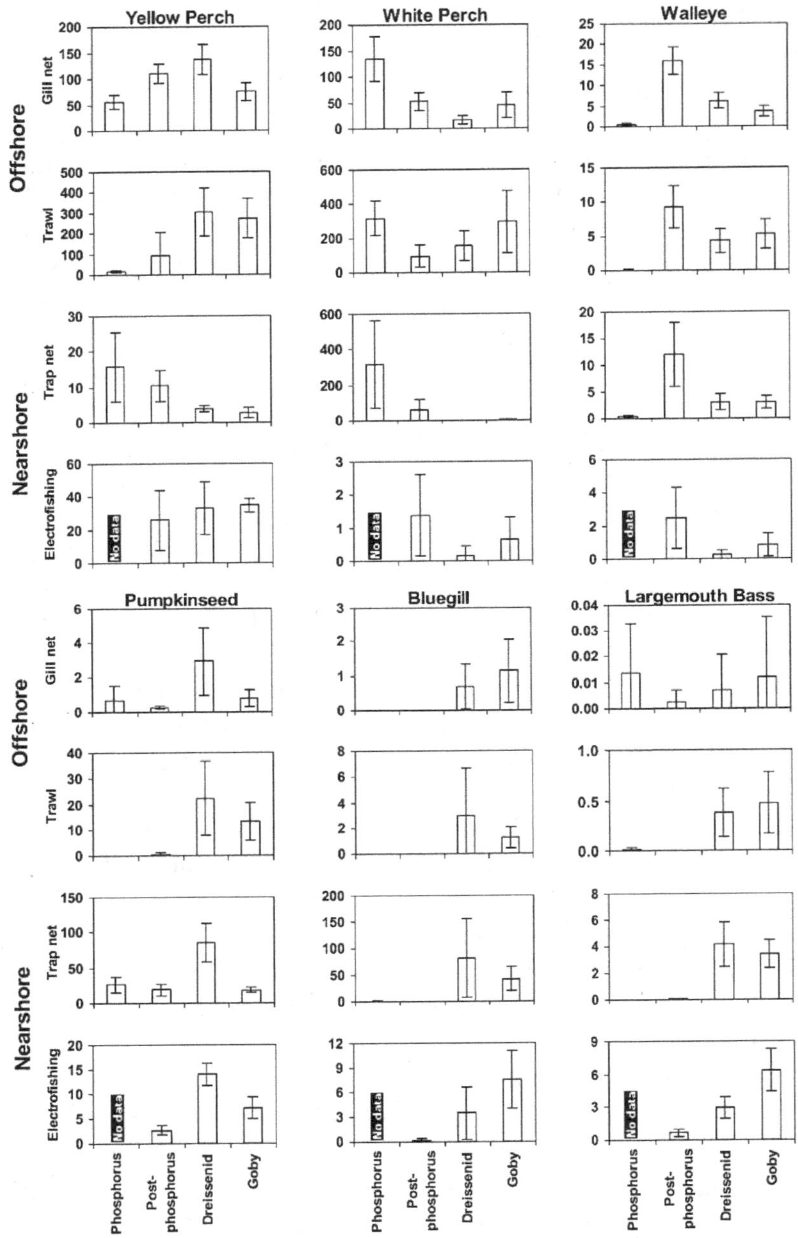

Figure 1. Mean abundance (by number) of selected key species during four ecological time stanzas (phosphorus, post-phosphorus, dreissenid and goby) in two offshore (gill net and trawls) and two nearshore (trap net and electrofishing) gear types. Error bars depict ±2 standard errors.

Piscivore biomass (PPB)

PPB increased from 0.04 to 0.30 in the offshore and from 0.08 to 0.35 in the nearshore between the phosphorus and post-phosphorus control time-stanzas concomitant with the resurgence of Walleye (Figure 2). For the dreissenid time-stanza, PPB declined to 0.21 in the offshore, due to a decline in Walleye, and to 0.15 in the nearshore, due primarily to declines in Walleye and Smallmouth Bass. PPB remained steady in the offshore but increased to 0.26 in the nearshore for the goby time-stanza. PPB was generally higher in the nearshore. Piscivore species richness was much greater in the nearshore habitat than in the offshore zone. In the nearshore, six piscivores are now relatively common (Figure 2), while in the offshore, Walleye dominated the piscivore assemblage in all time-stanzas. Overall, in the Bay of Quinte, Walleye remain the most important piscivore. Unlike other piscivores, their life history includes annual migrations of adult fish to feed on the abundant Alewife prey in eastern Lake Ontario (Bowlby and Hoyle, 2011); therefore a significant component of the population's production may come from feeding and growth in the lake.

PPB in the Bay of Quinte nearshore (0.26) is now considerably greater than that reported for the 1969–1976 time-period by Hurley and Christie (1977). These authors argued that PPB in the Bay of Quinte should be more like that of West Lake (0.25), a Lake Ontario embayment with similar morphometry and limnology, but not the same level of eutrophication stress. By way of comparison, the PPB in Hamilton Harbour, a degraded embayment habitat, averaged 0.08 during 1988–2008 (Brousseau and Randall, 2008). Similar to the Bay of Quinte, the PPB in Penetang Bay (Severn Sound) increased from 0.12 in 1990 to 0.34 in 2002, as degraded habitat was restored, and this PPB closely matched nearby Hog Bay which averaged 0.35 during this period (Brousseau et al., 2004).

Management implications

Management of aquatic resources in the Bay of Quinte has been challenging because of dynamic ecosystem change caused by the chronological sequence of large-scale ecological stressors: eutrophication, phosphorus control, dreissenid colonization and most recently, the establishment of Round Goby. The Bay of Quinte was designated a Great Lakes Area of Concern (AOC) in 1986 largely due to issues associated with excessive nutrient loadings. One of the impaired beneficial uses was degradation of fish populations. The Bay of Quinte fish community

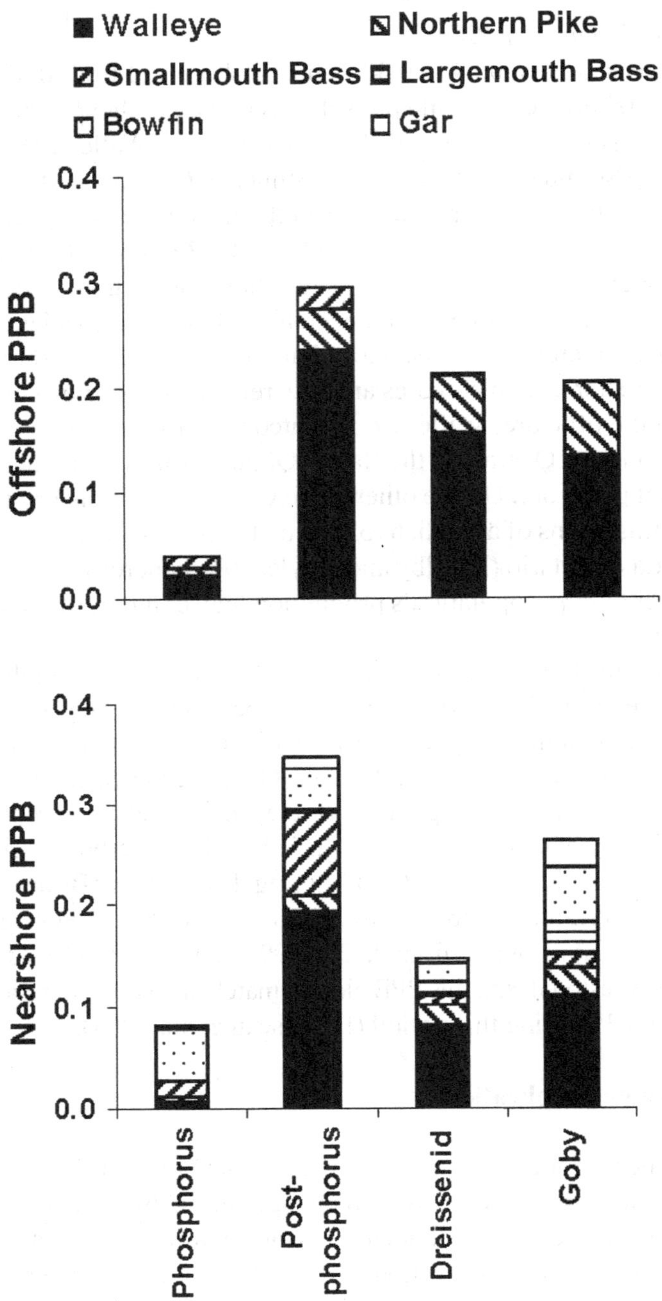

Figure 2. Proportion of total fish community biomass (PPB) represented by different piscivore species in the offshore (gill net) and nearshore (trap net) of the Bay of Quinte in four ecological time-stanzas.

was highly impaired during the pre-phosphorus control period (Hurley and Christie, 1977). By 1985 and post-phosphorus control, the fish community had partially recovered. Walleye, the key top predator in the system, increased in abundance, effectively controlling the formerly hyper-abundant small fish species and restoring a predator-prey balance in the offshore fish community (Hurley, 1986a). But, because SAV remained low in nearshore areas, fish species dependent on this habitat remained suppressed. Continued efforts to control nutrient input along with the Dreissenid Mussel-induced increases in water clarity in the mid 1990s led to recovery of SAV and the fish community associated with that habitat. The Bay of Quinte's fish community trophic structure, measured here as PPB, is now similar to other, non-impaired, Lake Ontario embayments. PPB is also greater than 0.20, in nearshore and offshore habitats, a benchmark value identified by Hurley et al. (1986) below which severe depreciation of the Bay of Quinte fish community occurred. Consistent with Brousseau et al. (2011), the Bay of Quinte fish community currently appears to be indicative of a healthy ecosystem. Looking forward, nutrient input issues and new invasions, in addition to and in the context of broad-scale environmental stressors such as climate change, will mean continued management challenges for the Bay of Quinte ecosystem.

Summary and conclusions

Intensive, long-term sampling in the Bay of Quinte with multiple gears (i.e. gill nets, bottom trawls, trap nets and boat electrofishing) allowed examination of the fish community and major fish populations in the context of key stressors up to 2009. Excessive nutrient input and hyper-abundant non-native fish species, White Perch and Alewife, shaped the depreciated fish community of the 1970s. After implementation of phosphorus input control measures and simultaneous winter-kills of the hyper-abundant non-native fish in the late 1970s, Walleye recovered and served to restore a predator-prey balance to the fish community by the late 1980s. However, in the absence of a significant recovery of submerged aquatic vegetation (SAV) in littoral areas, off-shore species (e.g. Alewife and White Perch) still tended to dominate; even in littoral areas. Following establishment of Dreissenid Mussels in the mid-1990s, water transparency increased and SAV increased significantly in littoral areas. This pivotal event led to a shift in the fish community that included an overall decline in Walleye, an increase followed by a decrease in Yellow Perch, and dominance by centrarchids (i.e. Bluegill, Pumpkinseed, Black Crappie and Largemouth Bass) in the nearshore. Round Goby invaded in 1999, proliferated and became important in the diet of piscivores by 2003. The current species assemblage, including the piscivores, is diverse and indicative of a healthy fish community.

Long-term fish monitoring using a variety of gears was invaluable for tracking the changes in the fish community, in both nearshore and offshore habitats, in different areas of the Bay of Quinte. Fish populations responded to the different environmental and ecological conditions in each of these time-stanzas in a unique manner, as documented in this study. The proportion and diversity of piscivore species was a good indicator of ecosystem health, both spatially and temporally throughout the monitoring period.

Acknowledgements

We gratefully acknowledge past and on-going contributions by field and laboratory crews at the Ontario Ministry of Natural Resources, Glenora Research Station in Picton, and at the Fisheries and Oceans Canada office in Burlington. Funding was provided by the Ontario Ministry of Natural Resources and Fisheries and Oceans Canada Science. We also thank two anonymous reviewers for their thoughtful suggestions on an earlier version of this manuscript.

References

Bowlby, J.N., Hoyle, J.A., 2011. Distribution and movement of Bay of Quinte Walleye in relation to temperature, prey availability and dreissenid coloniztion. Aquat. Ecosyst. Health Mgmt. 14(1), 56–65.

Bowlby, J.N., Mathers, A., Hurley, D.A., Eckert, T.H., 1991. The resurgence of Walleye in Lake Ontario. In: P.J. Colby, C.A. Lewis, R.L. Eshenroder, (Eds.), Status of Walleye in the Great Lakes: Case Studies Prepared for the 1989 Workshop. Great Lakes Fish. Comm. Spec. Publ. 91–1.

Bowlby, J.N., Hoyle, J.A., Lantry, J.R., Morrison, B.J., 2010. The status of Walleye in Lake Ontario 1988–2006. In: Status of Walleye in the Great Lakes. Proceedings of the 2006 Symposium. Great Lakes Fish. Comm. Tech. Rep. 69.

Brousseau, C.M., Randall, R.G., 2008. Assessment of long-term trends in the littoral fish community of Hamilton Harbour using an Index of Biotic Integrity. Can. Tech. Rep. Fish. Aquat. Sci. 2811.

Brousseau, C.M., Randall, R.G., Clark, M.G., 2004. Comparison of Severn Sound fish assemblages in Hog Bay and Penetang Bay, 1990 and 2002, using two indices of productive capacity. Can. Manuscr. Rep. Fish. Aquat. Sci. 2686.

Brousseau, C.M., Randall, R.G., Clark, M.G., 2005. Protocol for boat electrofishing in near shore areas of the lower Great Lakes: transect and point survey methods for collecting fish and habitat data, 1988 to 2002. Can. Manuscr. Rep. Fish. Aquat. Sci. 2702.

Brousseau, C.M., Randall, R.G., Hoyle, J.A, Minns, C.K., 2011. Fish community indices of ecosystem health: How does the Bay of Quinte compare to other coastal sites in Lake Ontario? Aquat. Ecosyst. Health Mgmt. 14(1), 75–84.

Campbell, L.M., Thacker, R., Barton, D., Muir, D.C.G., Greenwood, D., Hecky, R.E., 2009. Re-engineering the eastern Lake Erie littoral foodweb: the trophic function of non-indigenous Ponto-Caspian species. J. Great Lakes Res. 35, 224–231.

Casselman, J.M, Scott, K.A., 2003. Fish community dynamics of Lake Ontario—long-term trends in the fish populations of eastern Lake Ontario and the Bay of Quinte. In: M. Munawar (Ed.) *State of Lake Ontario: past, present and future*, pp. 349–383. Ecosystem World Monograph Series, Aquatic Ecosystem Health and Management Society, Burlington, ON.

Cortes, E., 1998. A critical review of methods of studying fish feeding based on analysis of stomach contents: application to elasmobranch fishes. Can. J. Fish. Aquat. Sci. 54, 726–738.

Dermott, R., Martchenko, D., Johnson, M., 2012. Changes in the benthic community of the Bay of Quinte, Lake Ontario, over a 40 year period. Aquat. Ecosyst. Health Mgmt. 15(4), 410–420.

Dietrich, J.P., Morrison, B.J., Hoyle, J.A., 2006. Alternative ecological pathways in the eastern Lake Ontario foodweb—Round Goby in the diet of Lake Trout. J. Great Lakes Res. 32, 395–400.

Hoyle, J.A., 1993. Yellow perch population dynamics near Brighton, Lake Ontario. In: Lake Ontario Fisheries Unit 1992 Annual Report. Ontario Ministry of Natural Resources. http://www.glfc.org/lakecom/loc/mgmt unit/LOA%2093.01.pdf

Hoyle, J.A., Bowlby, J.N., Morrison, B.J., 2008. Lake Whitefish and Walleye population responses to Dreissenid Mussel invasion in eastern Lake Ontario. Aquat. Ecosyst. Health Mgmt. 11(4), 403–411.

Hurley, D.A., 1986a. Fish populations in the Bay of Quinte, Lake Ontario, before and after phosphorus control. In: C.K. Minns, D.A. Hurley, K.H. Nichols (Eds.), Project Quinte: point source phosphorus control and ecosystem response in the Bay of Quinte, Can. Spec. Publi. Fish, Aquat. Sci., Lake Ontario 86.

Hurley, D.A., 1986b. Growth, diet, and food consumption of Walleye (*Stizostedion vitreum vitreum*): an application of bioenergetics modelling to the Bay of Quinte, Lake Ontario, population. In: C.K. Minns, D.A. Hurley, K.H. Nichols (Eds.), Project Quinte: Point Source Phosphorus Control and Ecosystem Response in the Bay of Quinte, Can. Spec. Pub. Fish. Aquat. Sci., Lake Ontario 86.

Hurley, D.A., 1992. Feeding and trophic interactions of White Perch (*Morone americana*) in the Bay of Quinte, Lake Ontario. Can. J. Fish. Aquat. Sci. 49, 2249–2259.

Hurley, D.A., Christie, W.J., 1977. Depreciation of the warm water fish community in the Bay of Quinte, Lake Ontario. J. Fish. Res. Bd. Can. 34, 1849–1860.

Hurley, D.A., Christie, W.J., Minns, C.K., Millard, E.S., Cooley J.M., Johnson, M.G., Nicholls, K.H., Robinson, G.W., Owen, G.E., Sly, P.G., Geiling, W.T., Crowder, A.A., 1986. Trophic structure and interactions in the Bay of Quinte, Lake Ontario, before and after point-source phosphorus control. In: C.K. Minns, D.A. Hurley, K.H. Nichols (Eds.), Project Quinte: Point Source Phosphorus Control and Ecosystem Response in the Bay of Quinte, Can. Spec. Pub. Fish. Aquat. Sci., Lake Ontario 86.

Johnson, T.B. Bunnell, D.B., Knight, C.T., 2005. A potential new energy pathway in Central Lake Erie: the Round Goby connection. J. Great Lakes Res. 31 (Supplement 2), 238–251.

Jutagate, T., Lamkom, T., Satapornwanit, K., Naiwinit, W., Petchuay, C., 2001. Fish species diversity and ichthyomass in Pak Mun Reservoir, five years after impoundment. Asian Fisheries Science 14, 417–424.

Karpov, K.A., Albin, D.P., Van Buskirk, W.H., 1995. The marine recreational fishery in northern and central California a historical comparison (1958–86), status of stocks (1980–86), and effects of changes in the California current. California Department of Fish and Game. Fish Bulletin 176.

Leisti, K.E., Millard, E.S., Minns, C.K., 2006. Assessment of Submergent Macrophytes in the Bay of Quinte, Lake Ontario, August 2004, Including Historical Context. Can. MS Rpt. Fish. Aquat. Sci., 2762.

Leisti, K.E., Doka, S.E., Minns, C.K., 2012. Submerged aquatic vegetation in the Bay of Quinte: Response to decreased phosphorus loading and Zebra Mussel invasion. Aquat. Ecosyst. Health Mgmt 15(4), 442–452.

Minns, C.K., Hurley, D.A., 1986. Population dynamics and production of white perch, *Morone americana*, in the Bay of Quinte, Lake Ontario. In: C.K. Minns, D.A. Hurley, K.H. Nichols (Eds.), Project Quinte: Point Source Phosphorus Control and Ecosystem Response in the Bay of Quinte, Lake Ontario, Can. Spec. Pub. Fish. Aquat. Sci. 86.

Minns, C.K., Cairns, V.W., Randall, R.G., Moore, J.E., 1994. An Index of Biotic Integrity (IBI) for fish assemblages in the littoral zone of Great Lakes' Area of Concern. Can. J. Fish. Aquat. Sci. 51, 1804–1822.

Ridgeway, M.S., Hurley, D.A., Scott, K.A., 1990. Effects of winter temperature and predation on the abundance of Alewife (*Alosa pseudoharengus*) in the Bay of Quinte, Lake Ontario. J. Great Lakes Res. 16, 11–20.

Robinson, G.W., 1986. Water quality of the Bay of Quinte, Lake Ontario, before and after reductions in phosphorus loading. In: C.K. Minns, D.A. Hurley, K.H. Nichols (Eds.), Project Quinte: point source phosphorus control and ecosystem response in the Bay of Quinte, Lake Ontario, Can. Spec. Publ. Fish, Aquat. Sci. 86.

Scott, W.B., Christie, W.J., 1963. Invasion of the lower Great Lakes by the White Perch, *Roccus americanus* (Gemlin). J. Fish. Res. Board Can. 20, 1189–1195.

Stirling, M., 1999. Manual of Instructions: Nearshore Community Index Netting (NSCIN). Ontario Ministry of Natural Resources, Lake Simcoe Fisheries Assessment Unit. Sutton West, Ontario.

Taraborelli, A.C., Fox, M.G., Johnson, T.B., Schaner, T., 2010. Round Goby (*Neogobius melanostomus*) population structure, biomass, prey consumption and mortality from predation in the Bay of Quinte, Lake Ontario. J. Great Lakes Res. 36, 625–632.

Zhu, X., Johnson, T.B., Tyson, J.T., 2008. Synergistic changes in the fish community of western Lake Erie as modified by non-indigenous species and environmental fluctuations. In: M. Munawar and R. Heath (Eds.), *Checking the Pulse of Lake Erie*, pp. 439–474. Ecovision World Monograph Series, Aquatic Ecosystem Health and Management Society, Burlington, ON.

The above article is a reprinted version of the following previously published article:

Hoyle; J.A., Bowlby; J.N., Brousseau; C.M., Johnson; T.B., Morrison; B.J., Randall, R.G., 2012.

Fish community structure in the Bay of Quinte, Lake Ontario: The influence of nutrient levels and invasive species

Aquatic Ecosystem Health & Management (2012) 15 (4): 370–384.

https://doi.org/10.1080/14634988.2012.727732

Larval Lake Whitefish abundance, diet and growth and their zooplankton prey abundance during a period of ecosystem change on the Bay of Quinte, Lake Ontario

James A. Hoyle,[1]* Ora E. Johannsson,[2] and Kelly L. Bowen[2]

[1]*Ontario Ministry of Natural Resources and Forestry 41 Hatchery Lane, Picton, ON K0K 2T0*

[2]*Fisheries and Oceans Canada, 867 Lakeshore Road, Burlington, ON L7S 1A1*

**Corresponding author: jim.hoyle5@gmail.com*

Keywords: prey availability, cyclopoid copepods, Lake Whitefish recruitment, cladocerans, Dreissenid Mussels

Introduction

The Bay of Quinte Lake Whitefish (*Coregonus clupeaformis*) spawning stock formed the basis for a commercial food fishery on Lake Ontario for many years (Christie, 1963), and has undergone severe changes in stock abundance as a result of a variety of anthropogenic stressors (Christie, 1972; Casselman et al., 1996; Hoyle et al., 2003). Although scientific and commercial harvest monitoring programs of juvenile and adult life stages are on-going, studies of earlier life history stages have not been conducted since the 1920s (Hart, 1930). McKenna and Johnson (2009) and Johnson et al. (2009) recently studied larval Lake White fish from early April to early May (2004–2006) in Chaumont Bay, an open-embayment in New York waters of Lake Ontario. In this paper, we investigate larval Lake Whitefish abundance, diet composition and growth, and the abundance of their zooplankton prey spanning a period of dramatic change in the Bay of Quinte ecosystem from the early 1990s to the mid 2000s.

The study described here was initially undertaken because invasion of Dreissenid Mussels (Zebra Mussel, *Dreissena polymorpha* and later Quagga Mussel, *D. bugensis*) into the Bay of Quinte area was imminent by the early 1990s.

The mussels were expected to blanket the Lake Whitefish spawning substrate and, perhaps, impact egg or larval fish survival. In 1991 and

Ecology of the Bay of Quinte: Health, Management and Global Implications
Editors: C.K. Minns, M. Munawar and M.A. Koops

Aquatic Ecosystem Health & Management Society, Canada

1992 spawning sites were identified and egg deposition and fry emergence were described (Hoyle, 1992; Hoyle 1993; Hoyle and Melkic, 1991). Three sites were established to examine larval whitefish abundance and diet, and prey availability in 1991–1993 and 1995–1996—years that spanned mussel invasion and proliferation. Cyclopoid copepods and small-bodied cladocerans dominated the larval fish diet. The early life history studies were suspended after 1996.

Following Dreissenid Mussel invasion in the early 1990s, the Lake Whitefish population collapsed. Hoyle (2005) attributed the collapse to the loss of the burrowing amphipod *Diporeia* spp.—the primary diet of juvenile and adult whitefish (Hart, 1931; Ihssen et al., 1981)—which coincided with Dreissenid Mussel invasion. Correlates with the white fish population collapse included diet change, declines in body condition and growth, delayed age at maturity, changes in distribution and feeding patterns, and very poor reproductive success (Hoyle, 2005). Hoyle (2005) interpreted low juvenile abundance in August bottom trawls as an indication of poor reproductive success and hypothesized that the very low catches from 1998–2002 may have been linked to poor condition of adults and a consequence of low lipid reserves; i.e. reduced egg viability.

However, independent, long-term and annual monitoring of Bay of Quinte zooplankton showed that a 60% decline in abundance of cyclopoid copepodids was coincident with the Dreissenid Mussel invasion (Johannsson and Nicholls, 2002). Therefore, during 2003–2005, we repeated our larval whitefish studies, using the same sites and methodology as used in the earlier years, to investigate the possibility that declines in larval whitefish food resources may contribute to low recruitment to the juvenile stage in August. In a Lake Huron Lake Whitefish population, Freeberg et al. (1990) identified a bottleneck at the larval stage and found that prey availability, expressed as the number of zooplankton per fish ($z\ f^{-1}$), was positively linked to larval fish growth and survival and to recruitment to a later life stage. Here, we investigate the possibility of a similar bottleneck in the Bay of Quinte whitefish population.

Our objectives were to: (1) examine trends in the abundance of larval whitefish and their zooplankton prey and, (2) examine the relationships among prey availability ($z\ f^{-1}$), larval fish growth rate and the abundance of juvenile Lake Whitefish in August.

Methods

Study Area

Lake Whitefish early life history studies were focused at three sites: Trident Point, Sherman's Point and Indian Point in the Upper, Middle and Lower Bay of Quinte, respectively (Figure 1). All sites were adjacent to shore at or near points of land and characterized at Trident Point by bedrock with crevices and pockets of gravel and sand, at Sherman's Point by bedrock overlaid with medium-sized broken rock cobble, and at Indian Point by gravel and small cobble substrate close to shore and sand offshore. In each year, we visited each site 3–12 times, on a rotational basis, from "ice-out" in late March or early April until mid-May when larval Lake Whitefish disappeared from the shallow nearshore waters. Larval fish studies were carried out from 1991–1993, 1995–1996 (hereafter also referred to as early years), and 2003–2005 (hereafter also referred to as later years).

Juvenile Lake Whitefish (age approximately 4 months) were sampled annually, in August, near Conway in the Lower Bay of Quinte (Figure 1) at a depth of about 20 m.

Larval fish sampling

Larval fish were sampled with a 0.5 m diameter tow net like that described in Loftus (1982) and Cucin and Faber (1985) and towed from a 6.1 m open aluminum boat, as close to shore as possible (water depth ranging from 0.2–2.0 m), at 1 knot (0.5 m s^{-1}) for 5–15 min. Larval fish abundance was expressed as number m^{-3}. One to three replicate tows were made during each visit to each site. Basic environmental data were collected during each visit including water temperature. After each tow, larval fish were identified, enumerated and a random sample of up to 100 individual Lake Whitefish were kept in ice-water for processing on the same day in the laboratory. All fish were measured to the nearest 0.1 mm and a smaller number were weighed to the nearest 0.0001 g wet weight. Three stomach samples were taken from each tow. Each sample consisted of the combined stomach contents of 10 fish and stored in 5% buffered formalin. Stomach samples were later analysed in the same manner as zooplankton samples (see below) and diet composition was expressed as proportions by number.

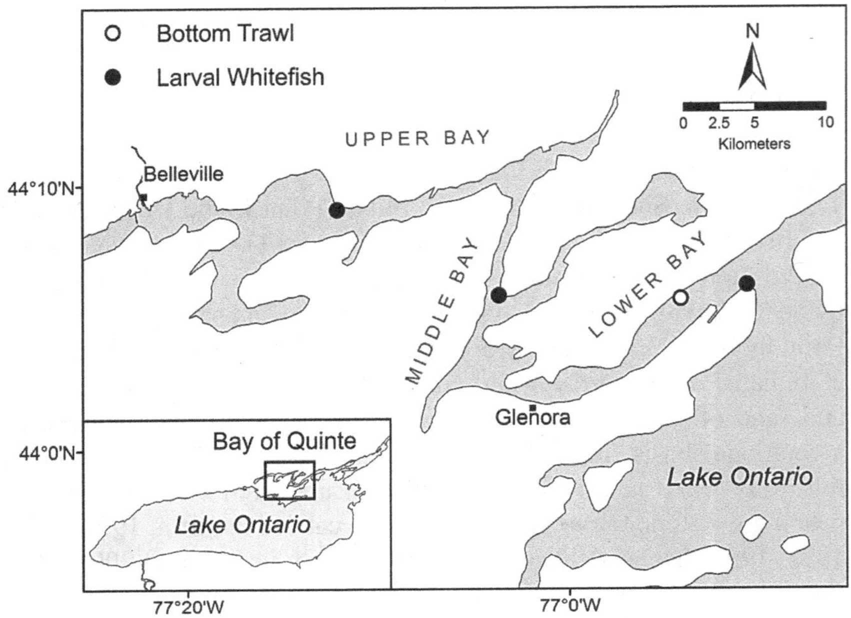

Figure 1. Map of the study area, Bay of Quinte, Lake Ontario. Solid circles locate the three larval whitefish and zooplankton sampling locations at Trident Point (Upper Bay: 44°09.35′N, 77°12.70′ W), Sherman's Point (Middle Bay: 44°06.28′N, 77°04.06′ W), and Indian Point (Lower Bay: 44°06.88′N, 76°51.04′W). Open circle locates the August bottom trawling location near Conway in the Lower Bay (44°06.39′N, 76°54.39′W).

Juvenile fish sampling

Juvenile Lake Whitefish were sampled annually in August (age approximately 4-mth) with a ¾ Western bottom trawl. Lake Whitefish do not inhabit the Upper or Middle reaches during summer because water temperatures are too warm; at this time the juvenile fish are found in the Lower Bay (Hoyle et al., 2008). Four replicate trawls of 6 min duration and a covering a distance of 463 m (¼ nautical mile) were made during each of two visits annually (see Hoyle et al., 2008 for more detailed methodology). Juvenile fish abundance was expressed as number trawl^{-1} and used to assess Lake Whitefish year-class strength since 1972 because this long term index of juvenile abundance has been shown to be significantly and positively correlated with age-3 population size three years later (Ward, 2007).

Zooplankton sampling

Zooplankton sampling was conducted concurrently with larval fish sampling using a 73 µm mesh Wisconsin-style plankton tow with a mouth opening of 0.115 m dia. The tow net was not metered and a 100% flow efficiency was assumed. Horizontal tows were conducted at 1 knot (0.5 m s^{-1}) for up to 360 seconds duration in earlier years of the study (1991 and 1992) and standardized to 80 seconds in all subsequent years. Samples were preserved in 5% buffered formalin. Individual zooplankton was identified and enumerated from a 2–100% fraction of each sample. Zooplankton abundance was expressed as number m^{-3}.

We also used an independent zooplankton dataset from Fisheries and Oceans, Canada (DFO) for offshore sites in the Upper, Middle, and Lower reaches of the Bay of Quinte (Figure 1). Every year since 1975, a vertical series of Shindler-Patalas trap samples (64-µm sleeve) are collected at each site bi-weekly starting in early May (Bowen and Johannsson, 2011, in this issue for more details regarding sampling and enumeration). Preliminary analysis indicated that these offshore zooplankton data during May (number m^{-3}) were not significantly correlated with the nearshore zooplankton abundance in April/early May. However, our expectation was that these offshore zooplankton data would reflect larval Lake Whitefish prey availability after the larval fish disappeared from nearshore waters and might be correlated with larval fish survival in August.

Zooplankton were grouped as cladocerans (primarily *Bosmina longirostris* and *Chydorus sphaericus*), calanoid copepods (excluding nauplii), calanoid nauplii, cyclopoid copepods (excluding nauplii), and cyclopoid nauplii. Zooplankton abundance was expressed as number m^{-3}. Diet composition of larval Lake Whitefish was expressed as proportion by number of each prey taxa.

Data analysis

Statistical analyses were used to test for differences between earlier (1991–1993, 1995–1996) and later (2003–2005) sampling time-periods for the dependent variables larval fish and edible zooplankton abundance (based on diet composition, see below) abundance for a window of sampling dates common to both time-periods (April 1–May 10 with water temperatures ranging approximately from 3–13°C). For comparing mean abundances between time-periods, analysis of variance (ANOVA; STATISTICA 8.0) was used on $\log_{10}$ transformed data. All test statistics were considered significant at $p < 0.05$. Mean annual juvenile Lake Whitefish abundance data were summarized for three time periods: 1991–1997, 1998–2002 (time-period of very poor recruitment reported by Hoyle, 2005), and 2003–2005. For comparing mean juvenile abundances among these three time-periods, analysis of variance was used on $\log_{10}$ transformed data. The Bonferoni post-hoc multiple comparison test was used to identify those pairs of means statistically different from each other ($p < 0.05$). Edible zooplankton taxa were grouped and an annual prey availability estimate determined as the ratio of the number edible zooplankton per individual larval fish ($z\ f^{-1}$) following Freeberg et al. (1990).

Initial inspection of larval fish length at age (days) data revealed a general pattern of little or no change in length for the first 2–3 wks of sampling each year followed by a rapid increase in length over the remainder of sampling (mid-May). For each year, we pooled all larval fish length data and calculated instantaneous daily growth during a 3-wk period of accelerated growth (approximately wks 3–5) using the approach of Freeberg et al. (1990):

$u = (\log e l_t - \log e l_{t-1})/T$;

u = instantaneous daily growth;

l_t = mean length (mm) at beginning of rapid growth phase; l_{t-1} = mean length (mm) 3 wks later; and

T = time interval in days (21 days).

Mean larval fish length was calculated annually for the period of time prior to the 3-wk accelerated growth phase.

Simple correlation coefficients were calculated between August juvenile (age approximately 4-mth) Lake Whitefish abundance in bottom trawls and spring larval fish and edible prey abundance, prey availability ($z\ f^{-1}$), mean larval fish length, instantaneous daily growth, and May cyclopoid copepod (excluding nauplii) abundance. Correlation coefficients were also calculated between larval fish length and abundance and between larval fish growth and prey availability ($z\ f^{-1}$) and mean larval fish length.

Results

A total of 269 larval fish tows and 193 zooplankton samples were taken during eight years of sampling (1991–1993, 1995–1996, 2003–2005) at the three Bay of Quinte sites. Over 17,000 larval Lake Whitefish were caught. Larval fish were first captured immediately after ice-out and as early as March 27 but more commonly in early April. Fish were caught as late as May 19. Highest catches tended to occur in late April and the first week of May corresponding to water temperatures of about 7–11°C.

Generally, larval Lake Whitefish abundance tended to be lower (83%) in later sampling years (2003–2005; mean = 0.055 m^{-3}, sd = 0.049) than in earlier years (1991–1993 and 1995–1996; mean = 0.325 m^{-3}, sd = 0.722; Table 3) but a two-way ANOVA with sampling time period, i.e. early years and later years, and sampling site as factors indicated that abundance was not significantly different ($F_{1,18}$ = 2.601, p = 0.124). Neither site ($F_{2,18}$ = 0.525, p = 0.601) nor period x site interaction ($F_{2,18}$ = 0262, p = 0.772) terms were significant.

The early spring zooplankton community was dominated by cyclopoid copepods (67% by number) and small-bodied cladocerans (30%; primarily *Bosmina longirostris* and *Chydorus sphaericus*). Larval fish stomachs contained food only after about 2–3 wks and these same zooplankton taxa groups, excluding cyclopoid nauplii and referred to hereafter as edible zooplankton, also dominated the diet composition (Table 1). Cyclopoid copepods were most common in the diet at 61% by number followed by cladocerans at 32%. Although cyclopoid nauplii were the most common group in the zooplankton community (36%) they were infrequently eaten by larval whitefish (4%). Zooplankton abundance declined dramatically (e.g. total zooplankton declined by 78% by number) from earlier sampling years to later sampling years (Tables 2 and 3). Edible zooplankton abundance was examined with a two-way ANOVA with sampling time period, and sampling site as factors. The results indicated that edible zooplankton abundance declined significantly (89%) from earlier (mean = 3871 m^{-3}, sd = 4537) to later sampling years (mean = 425 m^{-3}, sd = 208; $F_{1,18}$ = 34.408, p = 0.00002). Neither site ($F_{2,18}$ = 1.475, p = 0.255) nor period x site interaction ($F_{2,18}$ = 1.111, p = 0.351) terms were significant.

Table 1. Mean zooplankton abundance (number m^{-3} and proportion) available and proportion (by number) in larval Lake Whitefish stomachs by major taxa in the Bay of Quinte, Lake Ontario. Values are means for eight years sampled (1991–1993, 1995–1996, and 2003–2005) from April to mid-May.

	Available		
	No. m^{-3}	Proportion	Diet Proportion
Cladocerans	1253	0.30	0.32
Calanoid copepods	38	0.01	0.03
Calanoid naupli	94	0.02	0.00
Cyclopoid copepods	1287	0.31	0.61
Cyclopoid nauplii	1506	0.36	0.04

Mean annual larval fish total length prior to the period of accelerated growth ranged from 13.6 mm in 2003 to 14.9 mm in 1993 (Table 3). Mean length was not significantly correlated with larval fish abundance ($r = -0.150$, $p > 0.05$, Table 4). Growth in length accelerated as water temperatures increased above 7–8°C in late-April. During the period of accelerated growth, instantaneous daily growth ranged from 0.008 in 2004 to 0.028 in 1991 (Table 3). Growth was highly correlated with prey availability ($z\ f^{-1}$, $r = 0.813$, $p < 0.05$) but not with mean larval fish length prior to the period of accelerated growth ($r = 0.469$, $p > 0.05$, Table 4).

Table 2. Mean zooplankton abundance (numbers m^{-3}) by major taxa in the Bay of Quinte, Lake Ontario. Values are annual means for three sites sampled from April to mid-May. Edible zooplankton include cladocerans and calanoid and cyclopoid copepods (excluding nauplii).

	1991	1992	1993	1995	1996	2003	2004	2005
Cladocerans	2346	5239	393	774	1000	59	101	115
Calanoid copepods	42	22	51	37	80	59	9	5
Calanoid naupli	21	5	53	77	221	304	46	27
Cyclopoid copepods	1383	2761	985	1943	2299	339	327	261
Cyclopoid nauplii	705	318	501	3830	4474	862	513	849
Edible zooplankton	3771	8022	1430	2753	3380	457	437	381
Total zooplankton	4497	8345	1983	6660	8074	1622	997	1258

Table 3. Annual larval fish and zooplankton abundance, prey availability ($z f^{-1}$), larval fish mean length and instantaneous daily growth in the Bay of Quinte, Lake Ontario. Edible zooplankton include cladocerans and calanoid and cyclopoid copepods (excluding nauplii). Also shown is the abundance of cyclopoid copepod sampled in May by DFO and the abundance of juvenile Lake Whitefish caught in August bottom trawls at Conway in the Lower Bay of Quinte. See text for details of calculations.

		April/early May zooplankton (no. m^{-3})			May zooplankton (no. m^{-3})			
	Larval fish (number m^{-3})	Edible	Total	$z f^{1}$	Length (mm)	Growth	Cyclopoid copepods[1]	Juvenile fish
1991	0.0169	3771	4497	223560	14.3	0.028	47417	80.0
1992	0.0633	8022	8345	126825	14.4	0.023	12104	46.8
1993	0.3418	1430	1983	4182	14.9	0.020	2389	6.3
1995	0.1325	2753	6660	20782	14.3	0.016	26690	54.3
1996	1.0688	3380	8074	3162	13.8	0.011	6313	5.3
2003	0.0903	457	1622	5061	13.6	0.009	6028	16.2
2004	0.0446	437	997	9811	14.3	0.008	4780	0.0
2005	0.0287	381	1258	13287	13.7	0.017	4458	5.7

[1]excluding nauplii

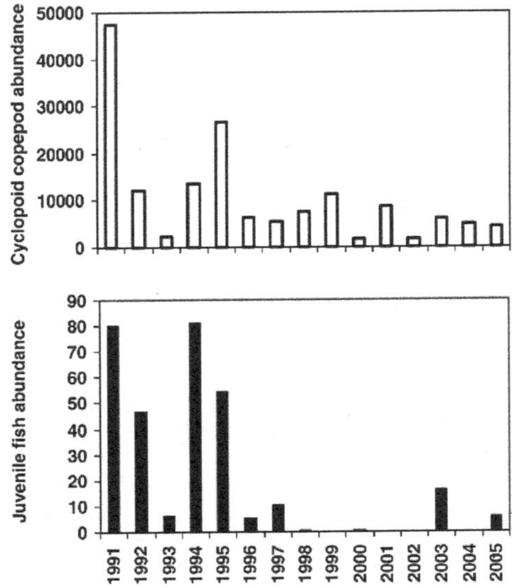

Figure 2. Upper panel, mean annual (1991–2005) cyclopoid copepod abundance (number m^{-3}, excluding nauplii) in DFO surveys in the Bay of Quinte (month of May data only). Lower panel, mean annual (1991–2005) juvenile Lake Whitefish abundance (number $trawl^{-1}$) in August bottom trawl surveys near Conway in the Lower Bay of Quinte, Lake Ontario.

Juvenile Lake Whitefish catches in August Bay of Quinte bottom trawls were variable, but generally high in the early 1990s nearly zero from 1998–2002 and relatively low thereafter (Figure 2). Mean abundance from 1991–1997 (18.95 fish $trawl^{-1}$, sd = 15.40) was significantly higher than that for 1998–2002 (0.15 fish $trawl^{-1}$, sd = 0.16, p = 0.001) but only weakly higher than for 2003–2005 (3.15 fish $trawl^{-1}$, sd = 3.28, p = 0.084). Mean abundance for 1998–2002 was not different than for 2003–2005 (p = 0.421).

May cyclopoid copepod abundance (excluding nauplii) was much higher in early years. Mean abundance from 1991–1995 was 20,447 m^{-3} while that for 1996–2005 was 5802, a decline of 72%. The four highest values occurred during the five early years of our larval Lake Whitefish study (Figure 2).

For years corresponding to the larval fish studies, juvenile Lake Whitefish abundance was significantly and positively correlated with larval Lake Whitefish prey availability ($z\ f^{-1}$, $r = 0.846$, $p < 0.05$) and instantaneous daily growth ($r = 0.740$, $p < 0.05$) and with May cyclopoid copepod abundance ($r = 0.932$, $p < 0.05$) but not with larval fish abundance ($r = -0.365$, $p > 0.05$), edible prey abundance ($r = 0.567$, $p > 0.05$)

or mean length ($r = 0.176$, $p > 0.05$) prior to the period of accelerated growth (Table 4).

Table 4. Simple correlation coefficients (r) between August juvenile Lake Whitefish abundance and spring larval fish and edible zooplankton abundance, prey availability ($z\ f^{-1}$), mean larval fish length, instantaneous daily growth, and May cyclopoid copepod abundance (excluding nauplii). Also shown are the correlation coefficients between larval fish length and abundance and between larval fish growth and prey availability ($z\ f^{-1}$) and mean larval fish length. The critical correlation coefficient $r_{0.05(2),6} = 0.707$.

	Juvenile fish	Length	Growth
Larval fish abundance	-0.365	-0.150	n/a
Edible zooplankton abundance	0.576	n/a	n/a
zf^{-1}	0.846*	n/a	0.813*
Length	0.176	n/a	0.469
Growth	0.740*	n/a	n/a
Cyclopoid copepod abundance	0.932*	n/a	n/a

*Significant at $p = 0.05$

Discussion

When early life history studies were resumed in spring 2003, we were very uncertain about larval whitefish status. Since the last larval fish study in 1996, the adult population had been subjected to severe stress—the loss of their preferred prey, *Diporeia*—possibly to the extent that reproduction was impaired because of poor nutritional status (Hoyle, 2005). Very few young-of-year whitefish were captured in August bottom trawl surveys between 1998 and 2002 (Figure 2). In spring 2003 we confirmed that larval fish were present at the same sites previously studied. However, in August 2003 juvenile abundance in bottom trawls had increased relative to those very low catches of 1998–2002. If reproductive success had been impaired during 1998–2002 by some factor such as poor adult nutritional status, this factor appeared to be alleviated by 2003. Larval catches tended to be lower in the later years (2003–2005) although the decline in abundance was not statistically significant due to high variation in catches.

Cyclopoid copepods and small-bodied cladocerans were the major prey of larval Lake Whitefish in the present study. This result was similar to that for Hart's (1930) feeding study in the 1920s on the same Bay of Quinte Lake Whitefish population. Johnson et al. (2009) found that cyclopoids were the primary prey of whitefish larvae over a 3-year period (2004–2006) in Chaumont Bay, Lake Ontario. We observed less calanoid copepods and more cladocerans in the diet than reported by Johnson et al. (2009). Johnson et al. (2009) found that, although larval whitefish feeding was more intense during the day, chironomid larvae were an important part of their diet at night. We did not sample larval fish at night and did not observe chironomids in the larval Lake Whitefish stomachs in our Bay of Quinte study.

Perhaps the most significant finding of our study was that the major prey types available to larval Lake Whitefish in nearshore waters had declined dramatically from the years studied in the 1990s to the later sampling years (2003–2005). A major gap in our data series between 1996 and 2003 impaired our ability to pinpoint the period of major zooplankton decline. Nevertheless, the reason for the decline is unclear. The May cyclopoid copepod abundance data collected annually in offshore waters by DFO suggest that Bay of Quinte zooplankton populations were impacted by the mid-1990s (Figure 2). At least two ecological changes that could have impacted zooplankton dynamics occurred over the course of our larval fish studies in the Bay of Quinte. First, Dreissenid Mussel establishment impacted water quality (e.g. lower phosphorus and chlorophyll-*a* and increased clarity) and phytoplankton communities by 1995 (Hoyle et al., 2003). Secondly, *Cercopagis pengoi* (a predatory cladoceran) invaded in 1999 and appeared to negatively impact small-bodied zooplankton types that year and in 2000 (Benoit et al., 2002). Bowen and Johannsson (this issue) discuss, in detail, potential causes for the decline in Bay of Quinte zooplankton. The reductions are thought to be caused by both direct predation of microzooplankton (e.g. nauplii) by mussels and *C. pengoi*, and competition between zooplankton and dreissenids for food resources.

Was larval Lake Whitefish prey abundance limiting whitefish recruitment to later life stages in the recent time period of our study? Larval fish prey abundance decreased significantly between earlier and later sampling years. We observed that prey availability ($z\ f^{-1}$) was correlated with larval fish growth and to juvenile abundance later in the summer, and recent year-class strength, measured as juvenile abundance in August, remains low relative to that in the early 1990s. Freeberg et al.

(1990) showed that a very abundant Lake Whitefish year-class (1984) at hatch was subsequently depleted by a prey availability ($z\ f^{-1}$) bottleneck only a few weeks later, despite prey (zooplankton) abundance that was similar to the previous year when many fewer fry hatched. The key was the ratio of zooplankton prey to individual larval fish. In our study, the positive relationships between prey availability, larval fish growth and juvenile abundance were largely driven by very high prey abundance in earlier sampling years. We contend that if the depressed zooplankton abundance levels observed in this study continue, recruitment of a strong Lake Whitefish year-class is unlikely because $z\ f^{-1}$ cannot be sufficiently high with low zooplankton abundance. High survival rate from larval to juvenile stages is possible at low zooplankton abundance but only at very low levels of larval fish abundance.

Without accounting for prey abundance, larval Lake Whitefish abundance was not correlated with juvenile abundance later in the summer. Mean length of larval fish, prior to accelerated growth, was also not correlated with juvenile abundance. Larval fish abundance and initial size, prior to exogenous feeding and accelerated growth, are likely related to number, quality and hatching success of eggs. We suspect that these and related factors (e.g. spawning stock size, nutritional status of adults, over-winter and early spring environmental conditions), possibly in addition to low zooplankton levels, were also likely limiting Lake Whitefish year-class strength during the 1998–2002 time period. Further consideration of these factors is beyond the scope of this study but could be the focus of future work. A laboratory study by Brown and Taylor (1992) concluded that both egg composition and prey availability significantly influence the growth and survival of larval Lake Whitefish. Claramunt et al. (2010) found that variation in larval density was best explained by larval length, spring wind intensity, and adult stock density.

Cyclopoid copepod abundance in May (DFO data) from offshore waters of the Bay of Quinte was positively correlated with juvenile abundance later in the summer. Larval Lake Whitefish moved offshore about mid-May in our study and, perhaps, zooplankton abundance there was also important for continued larval fish survival. At the very least, the same conditions that were important for high cyclopoid copepod abundance also favoured juvenile Lake Whitefish survival. This fortuitous result is useful because the zooplankton abundance data is collected annually and could be used by fisheries managers as a very early indicator of Lake Whitefish recruitment levels.

Summary and conclusions

Studies to examine larval Lake Whitefish (*Coregonus clupeaformis*) abundance, diet composition and growth, and the abundance of their zooplankton prey were conducted during eight years (1991–1993, 1995–1996 and 2003–2005) over the course of two decades that spanned a period of major ecosystem change—primarily Dreissenid Mussel related impacts—on the Bay of Quinte, northeastern Lake Ontario. Larval fish were captured in shallow, nearshore waters (0.2–2.0 m) from early April to mid-May each year. Larval Lake Whitefish fed primarily on cyclopoid copepods and small-bodied cladocerans. The key finding of our studies was that prey abundance declined by 89% from the earlier (1991–1993, 1995–1996) to the later (2003–2005) sampling years. Larval fish growth during spring was significantly correlated with prey availability. Recruitment to the juvenile stage in August was correlated with spring prey availability and larval fish growth. The observed decline in larval Lake Whitefish prey in the Bay of Quinte may be contributing to poor stock performance during and following a period of significant ecosystem change.

Hoyle (2005) reported a decline in adult Lake Whitefish lengthat-age in the mid-1990s and attributed it to the loss of *Diporeia*—the major food source for juvenile and adult whitefish. The results of the present study suggest that larval whitefish prey availability may also be a critical factor influencing Lake Whitefish recruitment and population dynamics in the Bay of Quinte. Here, we report an 89% decline in larval whitefish zooplankton prey that appears to be causing decreased larval fish growth and survival to the juvenile life stage.

Acknowledgements

We gratefully acknowledge past and on-going contributions by field and laboratory crews at the Ontario Ministry of Natural Resources, Glenora Research Station in Picton, and at the Fisheries and Oceans Canada office in Burlington. Thanks to W. Geiling, D. Geiling and C. Tudorancea for zooplankton sample enumeration, taxonomy and data analysis. Thanks to C. Lake for preparing the map of the Bay of Quinte. We also thank two anonymous reviewers for their thoughtful suggestions on an earlier version of this manuscript. Funding was provided by the Ontario Ministry of Natural Resources and Fisheries and Oceans Canada.

References

Benoit, H.P., Johannsson, O.E., Warner, D.M., Sprules, W.G., Rudstam, L.G., 2002. Assessing the impact of a recent predatory invader: The population dynamics, vertical distribution, and potential prey of *Cercopagis pengoi*, in Lake Ontario. Limnol. Oceanogr. 47, 626–635.

Bowen, K.L., Johannsson, O.E., 2011. Changes in zooplankton biomass in the Bay of Quinte with the arrival of the mussels, *Dreissena polymorpha* and *D. rostiformis bugensis*, and predatory cladoceran, *Cercopagis pengoi*: 1975 to 2008. Ecosyst. Health Mgmt. 14(1), 44–55.

Brown, R.W., Taylor, W.W., 1992. Effects of egg composition and prey density on the larval growth and survival of Lake Whitefish (*Coregonus clupeaformis*). J. Fish Biol. 40, 381–394.

Casselman, J.M., Hoyle, J.A., Brown, D.M., 1996. Resurgence of Lake Whitefish, *Core gonus clupeaformis*, in Lake Ontario in the 1980s. Great Lakes Res. Rev. 2, 20–28.

Christie, W.J., 1963. Effects of artificial propagation and the weather on recruitment in the Lake Ontario Whitefish fishery. J. Fish. Res. Board Can. 20, 597–646.

Christie, W.J., 1972. Lake Ontario: effects of exploitation, introductions, and eutrophication on the salmonid community. J. Fish. Res. Board Can. 29, 913–929.

Claramunt, R.M., Muir, A.M., Sutton, T.M., Peeters, P.J., Ebener, M.P., Fitzsimmons, J.D., Koops, M.A., 2010. Measures of larval Lake Whitefish length and abundance as early predictors of year-class strength in Lake Michigan. J. Great Lakes Res. 36, 84–91.

Cucin, D., Faber, D.J., 1985. Early life studies of Lake Whitefish (*Coregonus clupea formis*), cisco (*Coregonus artidii*) and yellow perch (*Perca flavescens*) in Lake Opeongo, Ontario. Ont. Fish. Tech. Rep. Ser. No. 16.

Freeberg, M.H., Taylor, W.W., Brown, R.W., 1990. Effect of larval survival on year-class strength of Lake Whitefish in Grand Traverse Bay, Lake Michigan. Trans. Amer. Fish. Soc. 119, 92–100.

Hart, J.L., 1930. The spawning and early life history of the whitefish (*Coregonus clupea formis* (Mitchill)), in the Bay of Quinte, Ontario. Contributions to Canadian Biology and Fisheries 6, 165–214.

Hart, J.L., 1931. The food of the whitefish, *Coregonus clupeaformis* (Mitchill), in the Bay of Quinte, Ontario. Contrib. Can. Biol. Fish. 6, 445–454.

Hoyle, J.A., 1992. Lake Whitefish early life history studies in the Bay of Quinte and eastern Lake Ontario, 1991. In: Lake Ontario Fisheries Unit 1991 Annual Report, Chapter 5, Ontario Ministry of Natural Resources, Picton ON.

Hoyle, J.A., 1993. Lake Whitefish early life history studies in the Bay of Quinte and eastern Lake Ontario, 1992. In: Lake Ontario Fisheries Unit 1992 Annual Report, Chapter 4, Ontario Ministry of Natural Resources, Picton ON.

Hoyle, J.A., 2005. Status of Lake Whitefish (*Coregonus clupeaformis*) in Lake Ontario and the response to the disappearance of *Diporeia* spp. In: Proceedings of a workshop on the dynamics of Lake Whitefish (*Coregonus clupeaformis*) and the amphipod (*Diporeia* spp.) in the Great Lakes. Edited by: Mohr, L.C., Nalepa, T.F. Great Lakes Fish. Comm. Tech. Rep. 66, 47–66.

Hoyle, J.A., Melkic, A., 1991. Lake Whitefish spawning shoal observations at four Bay of Quinte and eastern Lake Ontario sites. In: Lake Ontario Fisheries Unit 1990 Annual Report, Chapter 15. Ontario Ministry of Natural Resources, Picton, ON.

Hoyle, J.A., Casselman, J.M., Dermott, R., Schaner, T., 2003. Resurgence and decline of Lake Whitefish (*Coregonus clupeaformis*) stocks in eastern Lake Ontario, 1972 to 1999. In: M. Munawar (Ed.), *State of Lake Ontario: Past, present, and future*, pp. 475–492. Ecovision World Monograph Series, Aquatic Ecosystem Health and Management Society, Burlington, Ontario. Canada.

Hoyle, J.A., Bowlby, J.N., Morrison, B.J., 2008. Lake Whitefish and Walleye population responses to Dreissenid mussel invasion in eastern Lake Ontario. Aquat. Ecosyst. Health Mgmt. 11(4), 403–411.

Ihssen, P.E., Evans, D.O., Christie, W.J., Reckahn, J.A., and DesJardine, R.L., 1981. Life history, morphology, and electrophoretic characteristics of five allopatric stocks of Lake Whitefish (*Coregonus clupeaformis*) in the Great Lakes region. Can. J. Fish. Aquat. Sci. 38, 1790–1807.

Johannsson, O.E., Nicholls, K.H., 2002. Bay of Quinte Remedial Action Plan Assessment of the State of Impairment of Beneficial Uses: II. Zooplankton. Prepared for the Department of Environment, Ontario Region, Bay of Quinte RAP.

Johnson, J.H., McKenna, J.E. Jr., Chalupnicki, M.A., Wallbridge, T., Chiavelli, R., 2009. Feeding ecology of Lake Whitefish larvae in eastern Lake Ontario. J. Great Lakes Res. 35, 603–607.

Loftus, D.H., 1982. Larval fish sampling in Lake Huron, 1978. Ontario Ministry of Natural Resources. Ont. Fish. Tech. Rep. Ser. No. 2.

McKenna, J.E. Jr., Johnson, J.H., 2009. Spatial and temporal variation in distribution of larval Lake Whitefish in eastern Lake Ontario: Signs of recovery? Journal of Great Lakes Research 35, 94–100.

STATISTICA 8.0. StatSoft, Inc., 2007. STATISTICA (data analysis software system), version 8.0. www.statsoft.com.

Ward, C.L., 2007. Long-term correlates of Lake Whitefish recruitment in eastern Lake Ontario spanning an ecosystem regime shift. M. Sc. Thesis. University of Guelph, Guelph, ON.

The above article is a reprinted version of the following previously published article:

James A. Hoyle; J. A., Johannsson; O.E., Bowen, K.L., 2011.
Larval Lake Whitefish abundance, diet and growth and their zooplankton prey abundance during a period of ecosystem change on the Bay of Quinte, Lake Ontario
Aquatic Ecosystem Health & Management (2011) 14 (1): 66–74.
https://doi.org/10.1080/14634988.2011.548730

Bay of Quinte commercial and recreational fisheries: Harvest trends and current status

J.A. Hoyle

Ontario Ministry of Natural Resources, 41 Hatchery Lane, Picton, Ontario K0K 2T0, Canada
jim.hoyle5@gmail.com

Keywords: Walleye, Lake Ontario, commercial harvest

Introduction

The Bay of Quinte is a highly productive embayment of Lake Ontario. Fish harvest, both commercial and recreational, from the bay represents a substantial portion for all Lake Ontario. The Bay of Quinte fish community has experienced significant ecological stress associated with excessive anthropogenic nutrient input and invasion of non-native species (Hurley and Christie, 1977; Hoyle et al., 2012) but important fisheries persist. This chapter reports on Bay of Quinte fisheries.

The Bay of Quinte is located in northeastern Lake Ontario. It is a long (64 km), narrow and productive embayment extending from the mouth of the Trent River at Trenton in the west to eastern Lake Ontario and the St. Lawrence River in the east (Figure 1). Physical and biological gradients span shallow, warm and eutrophic conditions in the upper bay and deep, colder and mesotrophic conditions in the lower bay near Lake Ontario. The Trent, Moira, Salmon and Napanee Rivers enter the upper bay. The Bay of Quinte supports a diverse fish community. At least 65 species of fish representing 27 families reside in bay (Hoyle et al., 2012; see also Hurley, 1986). A variety of warm and cool water species are prevalent in the fish community including Yellow Perch (*Perca flavescens*), White Perch (*Morone americana*), Alewife (*Alosa pseudoharengus*), Brown Bullhead (*Ameiurus nebulosus*), Pumpkinseed (*Lepomis gibbosus*), Walleye (*Sander vitreus*), Bluegill (*Lepomis macrochirus*), White Sucker (*Catostomus commersoni*), Gizzard Shad (*Dorosoma cepedianum*), Largemouth Bass (*Micropterus salmoides*), and Freshwater Drum (*Aplodinotus grunniens*). Round Goby (*Neogobius melanostomus*) was first reported in the Bay of Quinte in 1999, and soon became a prominent member of

Ecology of the Bay of Quinte: Health, Management and Global Implications
Editors: C.K. Minns, M. Munawar and M.A. Koops

Aquatic Ecosystem Health & Management Society, Canada

the fish community (Hoyle et al., 2012). Fish movements are unrestricted in and out of the bay through "gaps" at the west and east ends of Amherst Island to the open-waters of eastern Lake Ontario and the St. Lawrence River. These movements include annual spawning migrations of Walleye, Alewife, and two cold water species that reside in the lower reaches of the Bay of Quinte and eastern Lake Ontario: Lake Whitefish (*Coregonus clupeaformis*), and Cisco (or Lake Herring *Coregonus artedi*). The Bay of Quinte, along with eastern Lake Ontario and the upper St. Lawrence River, also provides an important freshwater feeding and growth area in the complex life-cycle of American Eel (*Anguilla rostrata*), a catadromous species that spawns in the Sargasso Sea of the North Atlantic Ocean (COSEWIC, 2012).

The commercial fishery of northeastern Lake Ontario, including the Bay of Quinte, has had a long history that has helped to shape local culture. It is a multi-species based fishery. The fishery includes migratory species (e.g. Lake Whitefish) from Lake Ontario proper and resident species especially Yellow Perch, sunfishes (Pumpkinseed and Bluegill), Brown Bullhead, and the invasive White Perch. Intensive management of the commercial fishery began in the early 1980s with the establishment of individual, transferable harvest quotas. Individual license conditions were established that included harvest (species-specific quotas), season and gear restrictions (Smith and Edwards, 2003). Upper (Trenton to Deseronto) and middle (Deseronto to Glenora including Hay Bay) reaches of the Bay of Quinte were managed with somewhat different objectives compared to the lower bay (Glenora to east end Amherst Island; Figure 1). This was done to minimize conflict among multiple resource users in the more populated upper/middle bay region. For example, primarily only live-capture fishing gear (hoop and trap nets) and no Walleye harvest was permitted in the upper/middle bay.

The recreational fishery has had a shorter history having developed in the 1950s and based largely on Walleye. The Walleye population and fishery collapsed in 1970s, resurged in 1980s, declined in 1990s and have been relatively stable in recent years (Hoyle et al., 2008; Bowlby et al., 2010). Both winter ice-fishing and open-water fishing are popular. The seasonal age/size composition of the harvest is influenced by the Walleye's age-specific migratory and movement patterns (Bowlby and Hoyle 2011). Most fishing pressure and Walleye harvest occurs in the upper/middle bay but there has been some increased importance in the lower bay more recently.

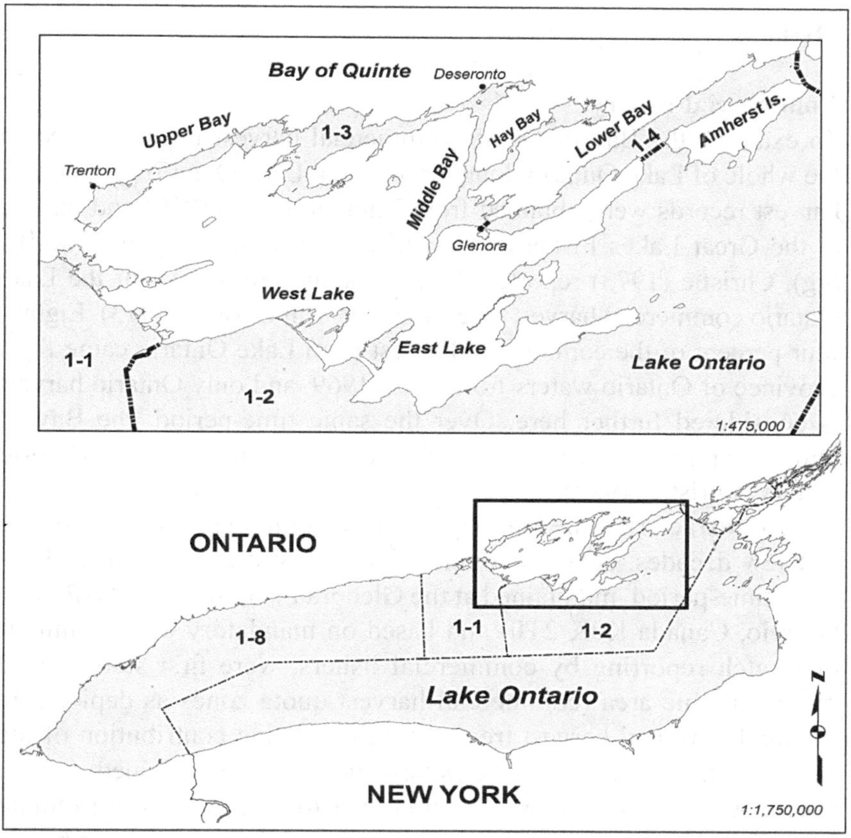

Figure 1. Map of Lake Ontario showing contemporary commercial fishing management units or "quota zones". The Bay of Quinte consists of quota zones 1-3 (upper Bay of Quinte) and 1-4 (lower Bay of Quinte). Lake Ontario proper consists of quota zones 1-1, 1-2, and 1-8.

An aboriginal fishery (spear fishing for spawning Walleye and gill netting primarily during spring and fall) also occurs in the Bay of Quinte but long-term monitoring data do not exist. This fishery is not considered further here.

The objective of this chapter is to characterize Bay of Quinte commercial and recreational fisheries in terms of their: 1) harvest trends over the last few decades, and 2) current status.

Methods

Commercial fishery

To examine the Bay of Quinte commercial harvest in the context of the whole of Lake Ontario, long term annual Lake Ontario commercial harvest records were obtained from Baldwin et al. (1979) and on-line at the Great Lakes Fisheries Commission web site (http://www.glfc.org). Christie (1973) reported the geographic breakdown of the Lake Ontario commercial harvest (see Appendix 1 in Christie, 1973). Eighty-four percent of the commercial harvest from Lake Ontario came from Province of Ontario waters from 1900-1969, and only Ontario harvest is considered further here. Over the same time-period, the Bay of Quinte commercial harvest contributed 34% of the total for Ontario waters (Christie, 1973).

To characterize the Bay of Quinte commercial fishery over the last few decades, detailed commercial harvest records for the 1993-2012 time-period, maintained at the Glenora Fisheries Station (Picton, Ontario, Canada K0K 2T0) and based on mandatory daily commercial catch reporting by commercial fishers, were first summarized by geographic area (commercial harvest quota zones as depicted in Figure 1). Annual harvest trends and the relative contribution of the Bay of Quinte to the total Lake Ontario harvest were examined. Annual species-specific harvest was summarized for the two Bay of Quinte commercial harvest quota zones: quota zones 1-3 (upper/middle bay of Quinte) and quota zone 1-4 (lower Bay of Quinte) to examine harvest trends in the context of the most important fish species. Bay of Quinte total commercial harvest for 2012 was reported by month to illustrate the seasonal pattern of commercial harvest. The number of commercial licences and the amount of licenced commercial gear in 2012 was also summarized.

Recreational fishery

The Bay of Quinte supports a recreational fishery based primarily on Walleye. There is an "open-water" fishery beginning on the first Saturday in May each year and lasting until "freeze-up" in December, and an ice fishery that occurs until Mar 1. The angling fishery is then closed for the rest of March and April until the first Saturday in May. The Bay of Quinte angling fishery occurs primarily from Trenton in the west to Glenora in the east. In recent years, a popular fishery for "trophy" sized Walleye has developed in the lower bay of Quinte during the fall of the open-water fishery. The two seasonal components of the fishery (open-water and ice) were regularly monitored by way of roving angler interviews in combination with instantaneous angling effort counts. Creel survey technicians used a boat (open-water) or snowmobile/ATV (ice) to approach and interview fishing parties at random. Fishing boat (open-water) or angler and ice hut (ice) counts were made by on-water creel survey technicians and by observers in a light aircraft for open-water and ice fisheries respectively. Sampling was stratified by geographic area, season, and day-type, and data were analyzed following the methods of Lester and Trippel (1985) to estimate total angling effort, catch rates, catch and harvest. Harvested Walleye were sampled for length and, in the case of the open-water fishery, a scale sample taken to determine age. Walleye weight was estimated based on the measured length. Open-water angling surveys were conducted in 1957-1959, 1976, 1979-1982, 1984-2006, 2008, and 2012. Ice angling surveys were conducted in 1982, 1984, 1986, 1988, 1989-2005, 2007, 2009 and 2013.

Open-water angling effort and harvest rate trends were compared graphically to Walleye population trends measured by assessment index gill nets set during summer in the Bay of Quinte. The gill net sampling occurred annually from 1958-2012 except for 1966. Gill net sampling methodology and Walleye abundance trends up to 2006 were described in Bowlby et al. (2010).

Age-class composition of Walleye harvested during the open-water angling fishery was reported separately for three major time-periods (1979-1984, 1985-1999, and 2000-2012). This was done to examine the impact of changes in Walleye growth (length-at-age), previously reported by Bowlby et al. (2010), on the age distribution of harvested Walleye.

Results

Commercial Fishery

The long-term trend in the Lake Ontario (Ontario waters) commercial harvest, summarized by decade and by cold- and warmwater fish species, is shown in Figure 2. The Lake Ontario fishery was dominated by coldwater species up until the 1950s and by warmwater species thereafter. Coldwater species included (in order of greatest to least harvest) ciscoes (four species of deepwater ciscoes and the shallow water Cisco or Lake Herring), Lake Whitefish, Lake Trout (*Salvelinus*

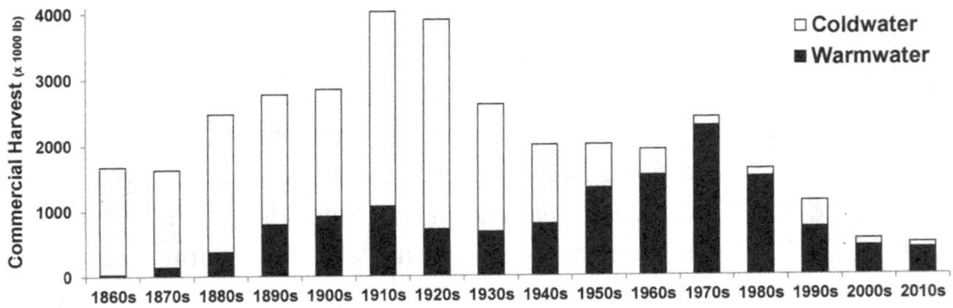

Figure 2. Commercial harvest trends (landed weight in lb) summarized by decade for the Canadian waters of Lake Ontario from the 1860s to 2012. Coldwater species include (in order of greatest to least harvest) Ciscos, Lake Whitefish, Lake Trout and Rainbow Smelt; warmwater species include Yellow Perch, bullheads and Channel Catfish, Carp, Northern Pike, American Eel, Blue Pike and Walleye, Freshwater Drum, Sunfish, *Morone* spp., Suckers and Redhorse, Lake Sturgeon, and Freshwater Drum.

namaycush) and Rainbow Smelt (*Osmerus mordax*); warmwater species include Yellow Perch, Bullheads and Channel Catfish (*Ictalurus punctatus*), Common Carp (*Cyprinus carpio*), Northern Pike (*Esox lucius*), American Eel, Blue Pike and Walleye, Freshwater Drum, Sunfish spp., suckers and redhorse, Lake Sturgeon (*Acipenser flavescens*), and Freshwater Drum.

There was a brief recovery of Lake Whitefish (coldwater species) harvest in the 1990s. Northern Pike, Yellow Perch, and Bullheads and Channel Catfish were the dominant warmwater species in the late 1800s and early 1900s while Yellow Perch, *Morone* spp. (mainly White Perch), American Eel, bullheads and Channel Catfish, carp, and sunfish all figured prominently in the 1970s commercial harvest. During the last decade, the major species in the entire commercial harvest were Yellow Perch and Sunfish.

Total annual commercial harvest (lb) by Lake Ontario commercial fishing quota zone from 1993-2012 is shown in Table 1. Commercial harvest from the Bay of Quinte figured prominently in the total Lake Ontario harvest, averaging 54% over the two decades including over 60% in the last ten years. Total commercial harvest declined about 50% over the time-period from about 1 million lb to about 500,000 lb. Most of the decline occurred in Lake Ontario proper; harvest declined much less in the eastern Lake Ontario embayment fisheries: Bay of Quinte, East Lake, and West Lake.

Annual species specific commercial harvest trends from the upper (quota zone 1-3) and lower (quota zone 1-4) Bay of Quinte are reported in Table 2. Species with highest harvest levels were Brown Bullhead, Yellow Perch, Sunfish and Lake Whitefish in the upper bay, and Yellow Perch, Lake Whitefish and Walleye in the lower bay. Harvest trend highlights included declines in Brown Bullhead and Lake Whitefish harvest and increase in sunfish harvest in the upper/middle bay, and the Lake Whitefish harvest decline in the lower bay. Also significant was the beginning of the Northern Pike harvest in 2006 (fall) and the decline followed by closure of the American Eel fishery in 2004. In most recent years the harvest consisted primarily of sunfish, Yellow Perch and Northern Pike in the upper/middle bay and Yellow Perch and Walleye in the lower bay.

The number of commercial licences and the amount of gear licenced (2012) are shown in Table 3. Note that no gill net gear is permitted in the upper Bay of Quinte. The Bay of Quinte commercial harvest is seasonal. Table 3 shows, by way of example, the 2012 commercial harvest, by month. In the upper/middle bay, commercial harvest is not permitted during July and August. Overall, highest harvest occurs during spring and fall.

Recreational Fishery

Open-water fishery

Bay of Quinte annual open-water angling fishery statistics, 1957-2012, are shown in Table 4. Over this time-period, angling effort, catch rates, and Walleye harvest increased to a peak in the early 1990s, declined steadily through to the early 2000s, and stabilized thereafter. Most (over 95%) of the angling effort targeted Walleye. Other species targeted by anglers included Yellow Perch and, particularly in recent years, Largemouth Bass. In recent years, Walleye targeted angling effort was about 200,000 angler hours with about 40,000 and 25,000 Walleye caught and harvested annually.

Trends in Walleye abundance (assessment gill nets), angling harvest rate, and angling effort are shown in Figure 3. Walleye abundance in assessment gill nets was highly correlated with angler catch rates: Angler catch rate (fish per hour) = 0.0159 * Walleye abundance (fish per gill net) + 0.1324; ($r^2 = 0.698$, $P < 0.001$). Walleye abundance in assessment gill nets was high in the late 1950s, declined to very low levels during the 1970s, increased dramatically to high levels in the 1980s and early 1990s, declined steadily through the early 2000s, and stabilized at

Table 1. Total annual commercial harvest (lb) by Lake Ontario commercial fishing quota zone, 1993-2012. Total value of all quota zones is also given, as is the proportion of total harvest (lb) attributed to the Bay of Quinte.

	Commercial Fishing Quota Zone									
Year	1-1 Central Lake Ontario	1-2 Eastern Lake Ontario	1-3 Upper Bay of Quinte	1-4 Lower Bay of Quinte	1-8 Western Lake Ontario	1-East Lake	1-West Lake	Total harvest (lb)	Total value	Proportion Bay of Quinte
1993	113,434	437,672	303,350	158,738	17,026	16,586	45,267	1,092,073	$ 735,224	42%
1994	107,244	84,963	381,194	177,020	13,293	20,364	35,215	1,219,293	$ 1,260,722	46%
1995	91,513	457,163	383,765	193,128	9,935	23,978	41,945	1,201,427	$ 1,316,218	48%
1996	62,290	630,455	346,725	179,374	4,245	15,605	28,927	1,267,621	$ 1,446,768	42%
1997	74,401	439,941	360,201	99,876	12,839	24,412	29,612	1,041,282	$ 964,005	44%
1998	77,450	349,513	385,051	143,894	9,960	25,006	47,021	1,037,895	$ 1,129,641	51%
1999	63,821	379,851	395,779	122,748	3,334	18,627	33,472	1,017,632	$ 1,117,883	51%
2000	58,722	374,672	340,177	131,687	6,952	15,934	29,688	957,832	$ 1,028,812	49%
2001	33,567	312,282	319,486	98,167	79,805	17,919	34,117	895,343	$ 908,797	47%
2002	9,198	214,145	276,224	68,564	34,567	16,030	35,088	653,816	$ 513,275	53%
2003	29,838	159,034	166,003	67,105	26,540	15,215	37,588	501,323	$ 363,987	46%
2004	20,160	112,884	213,965	58,466	2,479	20,655	40,127	468,736	$ 295,052	58%
2005	17,073	74,960	223,088	68,597	11,661	27,456	35,274	458,109	$ 361,849	64%
2006	14,252	184,878	229,046	137,212	14,351	15,423	17,331	612,492	$ 543,279	60%
2007	26,308	99,237	207,179	98,447	12,525	17,480	33,218	494,393	$ 464,467	62%
2008	11,454	107,172	182,402	72,834	63	13,849	15,545	403,319	$ 315,836	63%

Table 1. Continued

Year	Commercial Fishing Quota Zone: 1-1 Central Lake Ontario	1-2 Eastern Lake Ontario	1-3 Upper Bay of Quinte	1-4 Lower Bay of Quinte	1-8 Western Lake Ontario	1-East Lake	1-West Lake	Total harvest (lb)	Total value	Proportion Bay of Quinte
2009	7,172	104,387	179,989	94,171	-	15,384	30,066	431,169	$ 458,118	64%
2010	14,437	73,426	222,449	108,495	-	17,413	41,126	477,346	$ 551,014	69%
2011	18,401	94,741	234,828	103,393	-	17,299	40,541	509,203	$ 648,911	66%
2012	16,120	78,112	201,821	104,961	-	17,163	35,805	453,982	$ 530,309	68%

Table 2. Species-specific annual commercial harvest (lb) for quota zone 1-3 (Upper Bay of Quinte) and 1-4 (Lower Bay of Quinte), 1993-2012.

Year	Lake Whitefish	Cisco	Northern Pike	Suckers	Brown Bullhead	Eel	White Perch	Black Crappie	Sunfish	Yellow Perch	Walleye	Freshwater Drum	Other
	1-3 Upper Bay of Quinte												
1993	66,776	1,546	-	5,932	109,795	30,645	8,768	6,177	9,648	25,029	-	8,020	31,014
1994	66,041	1,794	-	8,608	150,469	23,827	19,811	3,057	17,829	53,807	-	10,529	25,422
1995	87,898	1,395	-	12,414	119,174	20,944	26,462	1,827	15,007	61,368	-	13,890	23,386
1996	91,848	2,369	-	10,571	111,997	19,069	8,715	3,673	14,066	44,521	123	15,977	23,796
1997	101,964	4,277	-	4,006	108,200	14,385	5,274	6,059	23,671	51,056	-	19,363	21,946
1998	86,487	4,562	-	5,077	140,073	6,198	3,705	8,995	48,665	40,674	43	24,418	16,154
1999	80,805	3,335	-	5,118	166,239	8,133	7,182	7,917	48,626	40,895	1	14,145	13,383
2000	63,069	1,056	-	6,776	127,287	9,640	7,776	5,994	44,307	39,054	-	19,050	16,168
2001	27,339	897	-	4,925	102,494	6,195	2,338	8,417	77,790	45,953	-	33,286	9,852
2002	33,724	817	-	3,439	102,247	2,783	1,863	4,800	72,612	30,515	-	15,859	7,565
2003	15,403	770	-	3,048	64,279	1,410	3,166	5,183	39,827	20,058	-	7,015	5,844
2004	14,179	482	-	10,633	79,350	4	8,339	5,136	42,790	22,701	-	20,602	9,749
2005	10,761	1,327	-	8,600	62,262	-	8,787	10,299	50,274	29,467	-	31,015	10,296

Table 2. Continued…

Year	Lake Whitefish	Cisco	Northern Pike	Suckers	Brown Bullhead	Eel	White Perch	Black Crappie	Sunfish	Yellow Perch	Walleye	Freshwa-ter Drum	Other
2006	5,194	1,132	2,783	5,627	64,992	-	8,273	12,081	42,648	58,964	-	15,725	11,628
2007	6,389	644	24,657	3,079	24,796	-	12,256	13,322	36,070	60,095	-	12,530	13,341
2008	5,143	2,247	9,912	3,039	26,276	-	12,870	15,587	37,463	45,315	-	14,495	10,055
2009	4,628	1,838	19,237	6,311	11,236	-	9,626	8,583	44,516	51,285	-	13,152	9,578
2010	4,890	1,845	24,784	5,235	7,085	-	11,085	7,660	81,905	55,709	-	11,975	10,276
2011	3,588	1,314	25,219	2,933	7,411	-	7,244	7,651	86,689	70,560	-	11,059	11,160
2012	5,218	1,463	22,551	14,744	6,450	-	11,141	5,944	62,047	48,660	-	14,975	8,628
	1-4 Lower Bay of Quinte												
1993	89,306	2,207	-	948	5,886	13,821	7,917	7	158	31,268	1,266	3,756	2,198
1994	77,568	2,726	-	706	1,391	11,813	29,657	34	277	42,510	3,419	5,724	1,195
1995	74,574	3,211	-	2,945	5,195	5,796	33,421	7	425	48,355	6,185	11,258	1,756
1996	74,504	2,672	-	3,363	6,057	6,838	18,491	695	455	37,994	9,158	16,520	2,627
1997	29,301	1,535	-	31	4,310	2,717	668	27	378	43,564	7,506	8,876	963
1998	46,265	458	-	411	7,123	849	800	36	1,112	71,763	10,138	4,211	728
1999	8,580	186	-	63	8,610	4,054	1,988	32	685	88,325	4,321	4,066	1,838
2000	13,531	509	-	10	5,551	1,614	1,368	3	412	94,866	6,001	5,475	2,347
2001	13,259	529	-	-	3,990	1,457	1,256	6	152	64,212	5,112	6,209	1,985

Table 2. Continued...

Year	Lake Whitefish	Cisco	Northern Pike	Suckers	Brown Bullhead	Eel	White Perch	Black Crappie	Sunfish	Yellow Perch	Walleye	Freshwater Drum	Other
2002	7,882	359	-	-	9,301	1,162	165	4	116	32,771	4,790	1,146	10,868
2003	5,754	199	-	47	5,188	1,527	2,521	-	26	37,292	3,979	2,472	8,100
2004	2,734	308	-	165	3,166	306	3,730	4	63	28,571	4,226	8,924	6,269
2005	519	203	-	48	1,547	-	5,183	5	65	44,902	5,373	9,090	1,662
2006	263	544	135	876	3,492	-	1,397	1	55	109,555	7,379	10,670	2,846
2007	2,096	340	2,713	1,213	1,905	-	4,483	13	142	67,583	10,452	5,282	2,226
2008	1,655	869	1,520	862	1,431	-	8,151	9	76	39,621	9,182	6,345	3,114
2009	1,792	752	2,330	961	1,031	-	8,805	25	469	48,370	15,920	11,846	1,871
2010	1,124	1,292	3,088	1,138	1,181	-	9,790	25	1,798	60,344	18,424	7,763	2,529
2011	645	1,204	4,673	2,152	1,037	-	1,728	22	601	58,155	20,499	10,140	2,537
2012	949	663	3,954	2,290	174	-	7,135	4	663	45,101	26,673	13,067	4,288

Table 3. Number of commercial licences, amount of gear licenced, and commercial harvest by month in the Bay of Quinte, 2012.

		1-3 Upper Bay of Quinte	1-4 Lower Bay of Quinte
Number of licences		28	19
Major gear types:			
	Hoop nets (#)	379	46
	Trap nets (#)	39	9
	Gill nets (m)	0	44,000
Harvest (lb):			
	Jan	295	355
	Feb	150	2,067
	Mar	4,279	7,517
	Apr	97,011	14,023
	May	43,065	8,632
	Jun	13,561	6,653
	Jul	n/a	10,551
	Aug	n/a	3,129
	Sep	25,888	17,704
	Oct	5,440	5,082
	Nov	15,700	6,136
	Dec	595	1,312

moderate levels thereafter. Angling surveys in the late 1950s indicated that angling effort was relatively low but harvest rate was high. Angling surveys were not conducted during most years in the 1960s and 1970s when Walleye abundance was low. Walleye harvest rates were high in the 1980s and early 1990s, especially in 1980 and 1981. Walleye harvest rate declined during the mid to late 1990s through to the early 2000s, increased slightly and stabilized at moderate levels. Walleye angling effort increased dramatically and steadily, and peaked in the mid-1990s, after Walleye abundance had declined. Angling effort declined rapidly after 1996 until about 2002 and then stabilized. After 2000, there was a modest increase in open-water angling effort targeted toward other species, primarily Largemouth Bass.

Table 4. Bay of Quinte open-water angling fishery statistics, 1957-2012, including angling effort (angler hours), both for all anglers and targeted Walleye anglers, Walleye catch and harvest rates (number of fish per hour), Walleye catch and harvest (number of fish), and the mean weight (kg) of harvested Walleye.

	All anglers	Walleye Anglers					
	Total effort	Effort	Catch rate	Harvest rate	Catch	Harvest	Mean weight (kg)
1957		128,040		0.299		38,318	0.638
1958		105,219		0.155		16,274	0.818
1959		67,000		0.254		17,037	0.963
1960						10,467	0.939
1961						22,117	0.596
1962						9,767	0.795
1963						2,466	1.422
1976		64,096		0.064		4,089	
1979		114,637		0.132		15,133	0.631
1980		321,388		0.598		192,305	0.464
1981		319,401		0.508		162,140	0.741
1982		382,306		0.236		90,182	1.030
1984		451,581		0.227		102,379	0.912
1985		442,717		0.263		116,415	0.859
1986		554,213		0.232		128,341	0.933
1987		589,163		0.172		101,092	0.756
1988		518,404	0.411	0.231	213,144	119,608	0.785

Table 4. Continued...

	All anglers	Walleye Anglers					
	Total effort	Effort	Catch rate	Harvest rate	Catch	Harvest	Mean weight (kg)
1989		466,008	0.512	0.290	238,549	135,151	0.760
1990		385,656	0.497	0.263	191,496	101,422	0.710
1991		634,101	0.543	0.302	344,156	191,785	0.789
1992		571,079	0.407	0.236	232,179	135,040	0.952
1993	644,477	637,401	0.417	0.227	265,551	144,476	0.912
1994	693,731	689,543	0.378	0.209	260,805	144,449	0.763
1995	519,276	512,054	0.320	0.189	163,875	96,631	0.710
1996	665,436	660,005	0.317	0.179	209,303	117,999	0.781
1997	544,476	539,276	0.250	0.154	134,672	82,821	0.747
1998	481,553	475,678	0.148	0.111	70,489	52,810	0.670
1999	379,012	374,128	0.127	0.090	47,562	33,575	0.958
2000	309,259	296,841	0.094	0.077	28,004	22,791	0.939
2001	247,537	222,052	0.182	0.126	40,512	28,037	0.916
2002	177,092	154,570	0.186	0.113	28,813	17,480	0.915
2003	219,684	194,169	0.344	0.178	66,706	34,543	0.637
2004	241,700	203,082	0.193	0.119	39,155	24,260	0.870
2005	225,385	205,933	0.204	0.125	42,031	25,757	0.693
2006	180,907	161,190	0.372	0.225	59,966	36,329	0.700

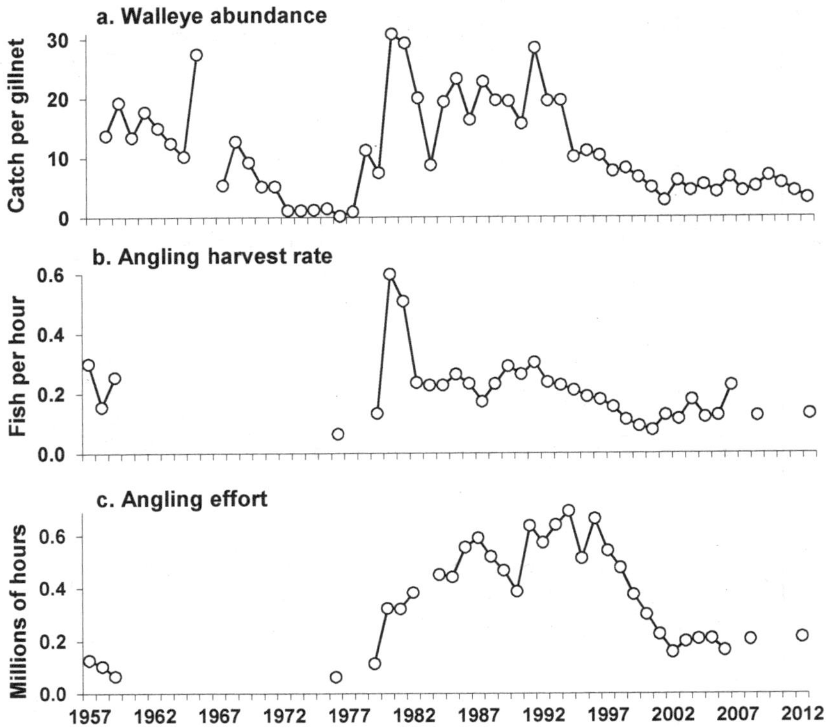

Figure 3. Bay of Quinte Walleye a) abundance in assessment index gillnets (summer); b) open-water angling harvest rate (number of fish per hour); and c) open-water angling effort (millions of hours), 1957-2012.

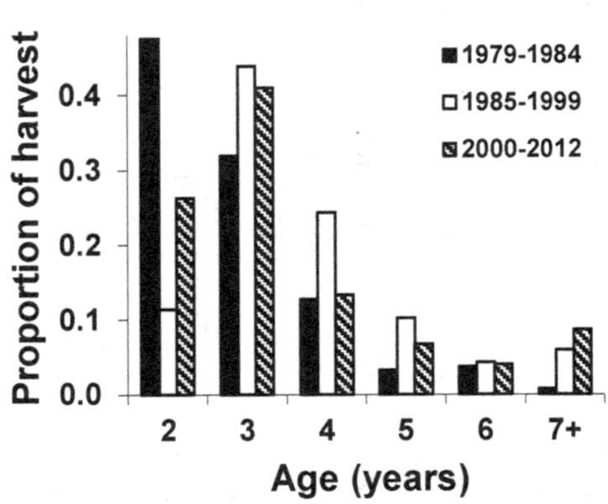

Figure 4. Age distribution of harvested Walleye angled during the open-water fishery for three time-periods indicated.

Most Walleye harvested during the open-water fishery were aged-2, 3 or 4 years. Most Walleye were aged-2 or 3 during earlier (1979-1984) and most recent (2000-2012) years but were aged 3 or 4 during intervening (1985-1999) years (Figure 4).

Ice fishery

Bay of Quinte annual ice angling fishery statistics, 1982-2012, are shown in Table 5. Over this time-period, angling effort and Walleye harvest increased to a peak in the early 1990s, declined steadily through to the early 2000s, and then increased. Angling catch and harvest rates were highly variable. In most recent years, Walleye targeted angling effort was about 128,000 angler hours with about 13,000 and 9,000 Walleye caught and harvested annually.

Discussion

Bay of Quinte fisheries harvest figured prominently relative to the total Lake Ontario harvest. From 1900-1969, the Bay of Quinte commercial harvest contributed 34% of the total for Ontario waters of Lake Ontario (Christie, 1973). Over the last two decades (1993-2012), the average contribution had increased to 54% including to more than 60% during the last ten years. The increase between the earlier time-period and more recently can largely be attributed to the decline in the abundance of coldwater species, in Lake Ontario proper, which included the extirpation of the deep water ciscoes and Lake Trout and the decline of Lake Whitefish and Cisco (Christie 1973). Currently, the only significant commercial fisheries occur for warm and cool water species in Lake Ontario's nearshore/embayment areas; primarily the Bay of Quinte, East Lake and West Lake.

The size of the Bay of Quinte recreational fishery also rivals that of the open Lake Ontario. Lake Ontario's native coldwater community, that once supported the commercial fishery, has largely been replaced by stocked salmon and trout and non-native prey fish species, Alewife and Rainbow Smelt. A large and popular recreational fishery developed around this contemporary species assemblage in the 1980s (Stewart et al., 2003). Annual catch of the two most popular species, Chinook Salmon (*Oncorhynchus tshawytscha*) and Rainbow Trout (*Oncorhynchus mykiss*) averaged about 44,000 and 20,000 fish, respectively, in Ontario waters over the last decade (Ontario Ministry of Natural Resources, 2013)

Table 5. Bay of Quinte ice angling fishery statistics, 1982-2013, including angling effort (angler hours), Walleye catch and harvest rates (number of fish per hour), Walleye catch and harvest (number of fish), and the mean weight (kg) of harvested Walleye.

	Walleye Anglers					
	Effort	Catch rate	Harvest rate	Catch	Harvest	Mean weight (kg
1982	80,129		0.103		8,223	1.209
1984	108,024		0.091		9,869	1.924
1986	143,960		0.165		23,768	2.272
1988	163,669		0.045		7,416	2.198
1989	175,119	0.145	0.109	25,458	19,147	1.738
1990	164,916					
1991	194,088	0.212	0.165	41,204	32,111	1.909
1992	327,546	0.172	0.132	56,494	43,343	1.388
1993	271,088	0.079	0.055	21,326	14,816	1.603
1994	300,049	0.104	0.029	31,060	8,557	2.239
1995	215,518	0.134	0.081	28,939	17,445	1.900
1996	392,602	0.149	0.053	58,468	20,972	1.563
1997	220,263	0.192	0.103	42,315	22,631	1.563
1998	117,602	0.095	0.052	11,167	6,089	2.327
1999	140,363	0.166	0.109	23,293	15,285	2.300
2000	139,047	0.072	0.066	9,949	9,240	2.359
2001	77,074	0.013	0.012	982	938	2.546
2002	37,129	0.070	0.066	2,601	2,468	2.358
2003	16,237	0.020	0.004	321	70	3.391
2004	79,767	0.105	0.051	8,413	4,075	1.668
2005	58,091	0.059	0.034	3,450	1,947	1.879
2007	99,368	0.176	0.114	17,480	11,313	1.008
2009	128,415	0.114	0.083	14,666	10,695	1.607

compared to 47,000 Walleye caught annually in the Bay of Quinte over the same time-period.

The Bay of Quinte's highly productive and diverse commercial and recreational fisheries may be viewed as complementary both in terms of maximizing potential benefits to humans and maintaining a balanced fish community trophic structure. During the last few decades the commercial fishery utilized a wide variety of species while the recreational fishery focused on Walleye. This complementary arrangement helped to ensure a balanced harvest pressure on the Bay of Quinte fish community structure as well as balanced the needs of resource users.

Walleye, the Bay of Quinte fish community's primary top piscivore, is the one fish species of major interest to both commercial and recreational fisheries. The Walleye population in the Bay of Quinte and eastern Lake Ontario is managed as a single large stock. Life history-specific movement and migration patterns between the bay and the lake are well understood (Payne, 1963; Bowlby and Hoyle, 2011). Walleye harvest by the commercial fishery is highly regulated and restricted. From fisheries management perspective, potential Walleye yield is largely allocated to the recreational fishery. No commercial Walleye is permitted in the upper and middle reaches of the bay (Trenton to Glenora). A relatively modest Walleye commercial quota is allocated in the lower Bay of Quinte and Lake Ontario with additional seasonal, gear, and fish-size restrictions. Here, the commercial harvest consists primarily of mature Walleye (age-4 years and older) that migrate from the upper/middle regions of the bay to the lower bay and eastern Lake Ontario for the summer months. Walleye harvest by the recreational fishery occurs primarily in the upper and middle reaches of the bay, and consists primarily of juvenile fish (age-2 and 3 years). Therefore, to the extent that the Walleye population is shared by both fisheries, the Walleye harvest is complementary in nature.

Major changes in total and species-specific harvest from the Bay of Quinte occurred over the last few decades. Harvest trends were related to changes in fish stock abundance. Hurley and Christie (1977) described the depreciated state of the fish community of the 1960s and 1970s, recognizing cultural eutrophication as the major driver. At that time small-bodied species tolerant of hyper-eutrophic conditions dominated, while top predators and, in the absence of submerged aquatic vegetation, species that thrive with sufficient aquatic vegetation were depressed. With low Walleye abundance the recreational fishery was also depressed. But the commercial fishery persisted and even thrived with the harvest consisting primarily of White Perch, Common Carp, Brown Bullhead and Channel Catfish, Yellow Perch and American Eel. The Walleye population recovered in the 1980s, in response to improved water quality management and an over-winter "die-off" of White Perch. This permitted a world-class recreational fishery to grow through to the mid-1990s, and peaking at a level of about 1 million angler hours of fishing effort and 150 thousand Walleye harvested annually. At the same time, Lake Whitefish populations resurged due to favourable climatic conditions and relaxation of a number factors (including exploitation, eutrophication, and predators) previously limiting their abundance (Hoyle, 2005). Despite a major reduction in the White Perch fishery, the commercial fishery continued to thrive.

A Dreissenid Mussel-induced regime shift in the Bay of Quinte ecosystem occurred after 1994 (Nichols et al., 2011). Increased water clarity following Dreissenid Mussel invasion led to the re-establishment of vast areas of submergent aquatic vegetation in the upper Bay of Quinte (Leisti et al., 2012). Pumpkinseed, Bluegill and Largemouth Bass abundance increased and Walleye abundance decreased. Yellow Perch and Brown Bullhead abundance increased initially but then declined (Hoyle et al. 2012). These fish abundance changes were also reflected in the fisheries harvest. Sunfish became much more important in the upper bay commercial harvest averaging greater than 50 thousand lb annually over the last ten years. Bullhead harvest was high (100-150 thousand lb annually) up until the early 2000s then declined to very low levels (less than 10 thousand lb in 2012). The Walleye recreational fishery (effort and harvest) declined after the mid-1990s then stabilized after 2002 at about 300 thousand hours of angling effort and a harvest of 40 thousand Walleye annually. Largemouth Bass angling effort increased in the post-dreissenid period and the Bay of Quinte became a popular location for competitive bass fishing tournaments (Kerr, 2012; 35 competitive fishing events were held on the Bay of Quinte in 2012).

Dreissenid Mussels also set the stage for Round Goby invasion (1999) and proliferation in the early 2000s. Round Goby, by being one of the few benthivorous fish species that exploit Dreissenid Mussels as prey, provides a new energy vector from the dreissenids to higher trophic levels (e.g. piscivores). Thus, Round Goby presence has the potential to significantly alter the Bay of Quinte foodweb. Round Goby soon became a significant component of the Walleye diet and indeed appeared in the diet of many of the bay's piscivores by 2003 (Hoyle et al., 2012). A corresponding increase in Walleye growth (Bowlby et al., 2010) may be linked to the diet change. Round Goby may impact other members of the fish community through competitive and other ecological interactions including species important to the fishery like Yellow Perch and the Centrarchids.

On-going changes outside the Bay of Quinte also impacted fish harvest in the bay. Lake Whitefish recruitment failed in the late-1990s after the loss of its primary diet item, the burrowing amphipod *Diporeia* from eastern Lake Ontario (Dermott, 2001; Hoyle et al., 2005). Commercial harvest of Lake Whitefish crashed in both the upper/middle (quota zone 1-3) and lower bay (quota zone 1-4) whitefish fisheries shortly thereafter. American Eel recruitment to the upper St. Lawrence/ Lake Ontario ecosystem decreased steadily after 1986 (Marcogliese and Casselman, 2009). Initially the commercial harvest remained high and stable but after the 1992 harvest of Eel in Lake Ontario, including the Bay of Quinte, declined dramatically due to the lack of recruitment. In 2004, the province of Ontario closed all commercial eel fisheries within its waters (Mathers and Stewart, 2009).

Northern Pike commercial harvest was significant from the Bay of Quinte prior to World War II (Christie, 1973). Pike harvest was discontinued when intensive commercial fishery management began in the early 1980s (Smith and Edwards, 2003) but was re-instated in the fall of 2006. Currently there is no quota restriction on pike harvest, and most of the annual harvest is taken during the early spring hoop and trap net fishery, which primarily targets Yellow Perch and Sunfish in the upper/middle reaches of the bay.

Summary and conclusions

Commercial and recreational fisheries in the Bay of Quinte, Lake Ontario were characterized in terms of their: 1) harvest trends over the last few decades, and 2) current status. The commercial fish harvest from the Bay of Quinte comprised a significant portion of that for all of Lake Ontario—60% over the last decade. Intensive management of the commercial fishery commenced in the early 1980s; individual license conditions were established that included harvest (species-specific quotas), season and gear restrictions. Commercial harvest trends in the years following reflected trends in fish population status. Highlights included collapse of Lake Whitefish fishery after the loss of *Diporeia* in the mid-1990s, increased prominence of sunfishes (Pumpkinseed and Bluegill) following Dreissenid Mussel invasion, closure of the

American Eel fishery in 2004, and re-establishment of Northern Pike harvest in 2006. Currently, the commercial harvest is comprised primarily of Yellow Perch and sunfishes and totals about 300,000 lb annually. The recreational fishery was focused on Walleye. Fishing effort and harvest trends reflected changes in Walleye abundance; resurgence in the 1980s, decline in the 1990s and stability in the 2000s. Size and age distribution of the Walleye harvest was a reflection of the Walleye's annual pattern of migratory behaviour and distribution pattern. Currently, the recreational fishery totals about 200,000 angler hours and 30,000 Walleye harvested during the open-water fishery and 100,000 angler hours and 10,000 Walleye harvested during the winter ice-fishery.

The Bay of Quinte continues to support important fisheries, even from a whole of Lake Ontario perspective. A feature of the commercial fishery is its capacity to adapt to changing fish species composition and abundance because of the multi-species nature of the fishery. The recreational fishery is dependent on Walleye, and fluctuates in size with Walleye abundance.

Acknowledgements

I wish to acknowledge the collectors and keepers of the commercial and recreational fisheries data at the Glenora Fisheries Station especially Kelly Sarley.

References

Baldwin, N. S., Saalfeld, R.W., Ross, M. A., Buettner, H. J., 1979. Commercial fish production in the Great Lakes 1867–1977. Great Lakes Fish. Comm. Tech. Rep. 3.

Bowlby, J.N., Hoyle, J.A., 2011. Distribution and movement of Bay of Quinte Walleye in relation to temperature, prey availability and dreissenid colonization. Aquat. Ecosyst. Health Mgmt. 14(1), 56–65.

Bowlby, J.N., Hoyle, J.A., Lantry, J.R., Morrison, B.J., 2010. The status of Walleye in Lake Ontario 1988–2006. In: Status of Walleye in the Great Lakes. Proceedings of the 2006 Symposium. Great Lakes Fish. Comm. Tech. Rep. 69.

COSEWIC, 2012. COSEWIC assessment and status report on the American Eel *Anguilla rostrata* in Canada. Committee on the Status of Endangered Wildlife in Canada. Ottawa. xii + 109 pp. (www.registrelep-sararegistry.gc.ca/default_e.cfm).

Christie, W. J., 1973. A review of the changes in the fish species composition of Lake Ontario. Great Lakes Fishery Commission, Technical Report 23.

Dermott, R. M., 2001. Sudden disappearance of the amphipod *Diporeia* from eastern Lake Ontario, 1993–1995. Journal of Great Lakes Res. 27, 423–433.

Hoyle, J.A., 2005. Status of lake whitefish (*Coregonus clupeaformis*) in Lake Ontario and the response to the disappearance of *Diporeia* spp. pp. 47–66. In: L.C. Mohr and T.F. Nalepa (Eds), Proceedings of a workshop on the dynamics of lake whitefish (*Coregonus clupeaformis*) and the amphipod *Diporeia* spp. in the Great Lakes. Great Lakes Fish. Comm. Tech. Rep. 66.

Hoyle, J.A., Hoyle, J.A., Bowlby, J.N., Morrison, B.J., 2008. Lake whitefish and walleye population responses to dreissenid mussel invasion in eastern Lake Ontario. Aquat. Ecosyst. Health Manage. 11, 403-411.

Hoyle, J.A., Bowlby, J.N., Brousseau, C.M., Johnson, T.B., Morrison, B.J., Randall, R.G., 2012. Fish community structure in the Bay of Quinte, Lake Ontario: The influence of nutrient levels and invasive species, Aquat. Ecosyst. Health Manage. 15 (4), 370-384.

Hurley, D.A., 1986. Fish populations in the Bay of Quinte, Lake Ontario, before and after phosphorus control. In: C.K. Minns, D.A. Hurley, K.H. Nichols (Eds.), Project Quinte: point source phosphorus control and ecosystem response in the Bay of Quinte, Can. Spec. Publi. Fish, Aquat. Sci., Lake Ontario 86.

Hurley, D.A., Christie, W.J., 1977. Depreciation of the warm water fish community in the Bay of Quinte, Lake Ontario. J. Fish. Res. Bd. Can. 34, 1849–1860.

Kerr, S. J., 2012. A survey of 2012 competitive fishing events in Ontario. Fisheries Policy Section, Biodiversity Branch. Ontario Ministry of Natural Resources. Peterborough, Ontario.

Leisti, K.E., Doka, S.E., Minns, C.K., 2012. Submerged aquatic vegetation in the Bay of Quinte: Response to decreased phosphorus loading and Zebra Mussel invasion. Aquat. Ecosyst. Health Manage. 15(4), 442–452.

Marcogliese, L.A., Casselman, J.M., 2009. Long-term trends in size and abundance of juvenile American Eels ascending the upper St. Lawrence River. In: J.M. Casselman and D.K. Cairns (Eds), *Eels at the Edge: Science Status, and Conservation Concerns*, pp. 191-205. American Fisheries Society, Symposium 58, Bethesda, Maryland.

Mathers, A., Stewart, T.J., 2009. Management of American Eels in Lake Ontario and the upper St. Lawrence River. In: J.M. Casselman and D.K. Cairns (Eds), *Eels at the Edge: Science, Status, and Conservation Concerns,* pp. 359-366. American Fisheries Society, Symposium 58, Bethesda, Maryland.

Nicholls, K.H., Hoyle, J.A., Johannsson, O.E., Dermott, R., 2011. A biological regime shift in the upper Bay of Quinte ecosystem (Lake Ontario) associated with the establishment of invasive dreissenid mussels. J. of Great Lakes Res. 37, 310-317.

Ontario Ministry of Natural Resources, 2013. Lake Ontario Fish Communities and Fisheries: 2009 Annual Report of the Lake Ontario Management Unit. Ontario Ministry of Natural Resources, Picton, Ontario, Canada. http://www.glfc.org/lakecom/loc/mgmt_unit/index.html

Payne, N. R., 1963. The life history of the yellow walleye (*Stizostedion vitreum*) (Mitchill), in the Bay of Quinte. University of Toronto, M.A. Thesis, Toronto, Ontario.

Smith, P.A., Edwards, P.A., 2003. Two decades of commercial fishery management on Lake Ontario, 1981-2001. Lake Ontario Fish Communities and Fisheries: 2002 Annual Report of the Lake Ontario Management Unit. Ontario Ministry of Natural Resources, Picton, Ontario, Canada. http://www.glfc.org/lakecom/loc/mgmt_unit/index.html.

Stewart, T.J., Bowlby, J.N., Rawson, M., Eckert, T.H., 2003. The recreational fishery for salmonids in Lake Ontario 1985-1995. In: M. Munawar (Ed), *State of Lake Ontario: Past, Present and Future*, pp. 517–537. Ecosystem World Monograph Series, Aquatic Ecosystem Health and Management Society, Burlington, ON.

Management of Great Lakes fisheries: Progressions and lessons

Charles K. Minns

Department of Ecology and Evolutionary Biology, University of Toronto, Toronto, Ontario, Canada

ken@minns.ca

Keywords: science, review

Introduction

This essay is concerned with the identification and assessment of progressions, sequences of progress, with respect to the science and management of large lake ecosystems. As context, I will begin with some quotations: George Bernard Shaw (Irish playwright, 1856–1950): "All progress is initiated by challenging current conceptions, and executed by supplanting existing institutions" — This is a reminder that changing our ideas is insufficient alone. We also need to change how our society operates if the benefits of the ideas are to flourish. Gaston Bachelard (French philosopher, 1884–1962): "The characteristic of scientific progress is our knowing that we did not know." — This statement speaks to the centrality of discovery and understanding to science. Thor Heyerdahl (Norwegian ethnologist, 1914–2002: "Progress is man's ability to complicate simplicity." — This last I disagree with and would reverse to read as, "Progress is man's ability to simplify complexity." I think that significant progress is usually marked by the integration of many accumulated events and observations into much simpler concepts which make it easier for people to articulate and defend actions. We have witnessed significant progressions and progress over the past 50 years in the science and management of the large lakes of the St. Lawrence-Great Lakes (SLGL). The purpose of this essay is to provide an overview of some of the progressions that have occurred with particular attention to their impact on the fisheries of the SLGL. Of necessity, a review of fisheries involves a view of ecosystems, since the former is entirely dependent on the latter.

This overview consists of four parts: (1) a brief contextual review at three nested spatial scales, SLGL, Lake Ontario and the Bay of Quinte; (2) a review of the main management activities for both ecosystems

Ecology of the Bay of Quinte: Health, Management and Global Implications
Editors: C.K. Minns, M. Munawar and M.A. Koops

Aquatic Ecosystem Health & Management Society, Canada

and fisheries; (3) a brief synopsis on the evolution of science; (4) a summing up in terms of lessons learned and future prospects.

The St. Lawrence-Great Lakes

Fishery resources include the fishes, the other biota they interact with, and the habitats they occupy. The historical sequences in the use and management of these resources may be considered as a series of interacting sequences of ecosystem change. These sequences can span from social, economic, institutional and landscape changes, through water quality, habitat supply, and climate changes, to biotic composition changes, introductions and extinctions, and species harvest changes. A selection of these sequences is examined for Figure 1. The Great Lakes Basin showing the lakes (white) and their drainage basins (shaded) and the position of the Bay of Quinte ecosystem within Lake Ontario's basin. SLGL which has been subjected to intense use over the last 200 years and studied intensely during the last 60 years. A nested set of ecosystems is considered as context: The SLGL, Lake Ontario and the Bay of Quinte (Figure 1). The history

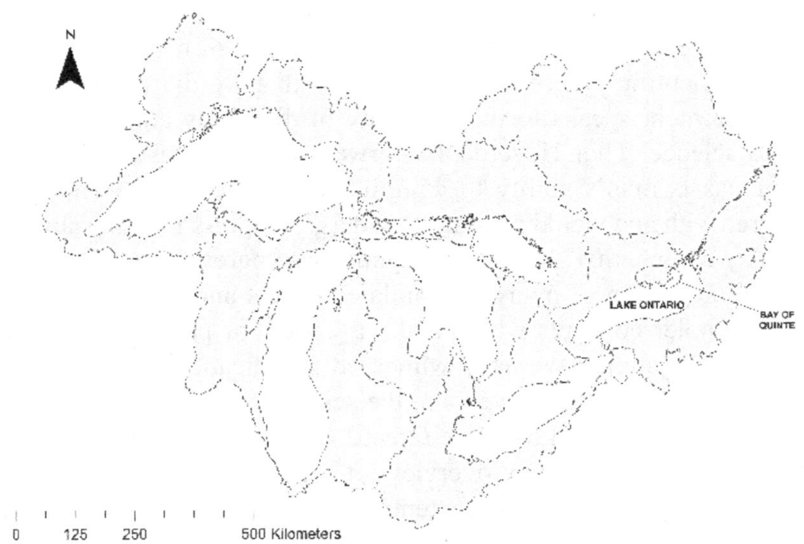

Figure 1. The Great Lakes Basin showing the lakes (white) and their drainage basins (shaded) and the position of the Bay of Quinte ecosystem within Lake Ontario's basin. of these ecosystems has been determined by the pat-tern of occupation and colonization by Europeans, exploitation of the natural resources, and industrialization (Table 1). In the 19th Century

the population grew steadily

Table 1. A brief selective history of the St. Lawrence-Great Lakes, Lake Ontario, and the Bay of Quinte, a nested set of ecosystems.

St. Lawrence-Great Lakes	Lake Ontario	Bay of Quinte
• The world's largest body of freshwater. • Major colonization route into North America. • Large fisheries from the 1800s on, heavily over-exploited and much habitat loss/damage due to negligent development. • A major hub of industrialization in the 19–20th Centuries. • 20th Century Sea Lamprey almost finished off salmonines and coregonines; many exotic salmonines introduced to create new recreational fisheries. • After eutrophication efforts began in the 1960s and 1970s, major remedial actions focused on 40+ areas from the 1980s on. • Home to 36 M (⅓ Canadian) in 2012.	• Deep oligotrophic lake. • Supported large Coregonine, Atlantic Salmon and Lake Trout fisheries. • Sea Lamprey and over-exploitation destroyed the cold-water fisheries. • Large-scale human development (eutrophication, habitat loss, etc.) with concentration of chemical industries led to serious contamination and toxicity problems mid 20th Century. • Major chemical load reductions begun in the 1970s–1980s. • Invasive species have caused many ecosystem changes. • Pacific salmon fisheries are sustained by stocking. Lake Trout rehabilitation has been very slow.	• Occupied by Empire Loyalists in the late 18th Century. Deforestation of watershed, many mill dams. Pioneer farming on sedimentary areas. Entrance to a major inland waterway. • Heavily eutrophied after the 1940s. Water levels regulated. • Major point source phosphorus controls implemented in the 1970s. • The Bay and the nearby lake outlet basin support much of the warm- and cool-water fisheries. • Walleye an important 20th Century species but likely a blip in an on-going successional sequence.

as convenient resources (wood, fish, wildlife, water, and minerals) were exploited and wild lands were turned to agriculture and emerging urban use. Shipping was important before the railways developed and urban centres grew up close to navigable waters. The open lakes had large coregonine and salmonine fishery resources while embayments like the Bay of Quinte supported percids, esocids and centrarchids. As population growth accelerated into the 20th Century so water quality problems (disease and eutrophication) developed at the urban centres. Industrialization was heavily dependent on water for power and for waste disposal thereby laying a path toward large-scale releases of chemicals and the emergence of bioaccumulating chemicals which caused health problems in biota at higher trophic levels. Fisheries were already heavily exploited by the end of the 19th Century and the problems intensified with the modernization of fishing technology into the 20th Century. Shipping canals and increased shipping from beyond the Great Lakes brought a living stream of invasive species alongside those deliberately introduced with the intent of enhancement the ecosystems, e.g. Common Carp was introduced as it was an important traditional food among the European-origin immigrant population. After the 1939–1945 World War, development activities and population size accelerated in the basin intensifying the environment impacts that arise when development is based on unsustainable assumptions. By the 1960s ecosystem damage had become widespread in the basin (Table 1).

In the Great Lakes as a whole, the fisheries were seriously depleted due to the combined effects of exploitation, Sea Lamprey parasitism, foodweb disruptions due to non-native species and ecosystem alterations, chemical contamination. In Lake Ontario, Lake Trout was almost extinct, coregonine stocks were seriously depleted, human-health threatening chemicals had contaminated much of the lake and its biota, and large urban centres had large eutrophication problems.

In the Bay of Quinte, eutrophication became a major issue from the 1940s on, the main percid fishery collapsed in the late 1950s due in part to eutrophication and invasive species, and large areas of macrophyte cover disappeared in the late 1950s.On all three spatial scales, considerable progress has been achieved since the 1970s particularly with respect to reducing or eliminating load-sustained and -driven stressors such as phosphorus (the primary cause of eutrophication) and many deleterious chemical substances notably health-threatening organic chemicals like PCBs, Dioxins, DDT, methyl-mercury and lead. All levels of government and industries combined to expand waste treatment facilities and, in many instances, to ban production of some chemicals entirely. Major fishery restoration efforts were based on controlling Sea Lamprey population primarily by using of a lampricide and on introduction stock (or load)-maintained populations of Pacific salmonines whilst longerterm efforts to restore native stocks proceeded. Unfortunately, efforts to limit load-driven stressors have only bought time as the underlying causes, human populations and their expanding resource consumption rates, have continued to increase. Over the same period, the accumulated losses of ecosystem goods and services resulting from permanent losses and deleterious alterations of ecosystem habitats (e.g. lake infilling and shore-line hardening) or from the foodweb disruptions of many successful invasive species (e.g. Dreissenid Mussels, Common Carp and European Common Reed) have only been partially offset with small-scale restoration efforts. The next section explores the history of management actions in more depth.

A brief management history

The management of the SLGL basin can be viewed by considering the history of two key institutions, the International Joint Commission and the Great Lakes Fishery Commission, and, to a lesser degree, the times before they were established. The early period may be characterized by the rapid colonization, development, and population growth in the century following the American Revolution. Population centres grew rapidly around the lake, fed by waves of immigration as the natural resource wealth (forests, mines, and water power) was exploited and the industrialization of Europe spilled across the Atlantic. Environment was little considered in this period and fisheries were viewed as limitless as utilitarian and pioneer thinking prevailed. Later in the 19th Century as urban centres and trade expanded, the lack of provision for sanitation led to disease problems, land clearance and hydrological disturbances released soils into watercourses and destroyed wetlands thereby impacting fisheries, and over-exploitation became widespread. As the consequences of unfettered development became apparent so transboundary issues arose and the need for dispute resolution mechanisms was recognized. In 1909, Canada and the United States signed the Boundary Waters Treaty (BWT) setting up the means to resolve conflicts and to cultivate action in areas of joint interest in the Great Lakes and other areas along the long Canada-U.S. border. A key instrument of that Treaty was the establishment of the International Joint Commission (IJC) which continues to this day to oversee many issues of concern to the two countries. The IJC oversaw numerous independent studies of issues seeking solutions. By the middle of the 20th Century, the impacts of eutrophication on lakes and rivers were obvious in many western countries. The eutrophication problem was attributable to three main sources: (i) excess nutrients in sewage waste released untreated by urban centres, (ii) nutrients released from soils either due to land disturbance or due to excess fertilizer use and (iii) the rise in the use of phosphate-based chemicals in cleaning compounds. After research clearly showed that phosphorus loading was the primary culprit in freshwaters, large investments were begun to reduce phosphorus loading. In the Great Lakes, following studies by the IJC, the two countries signed the first Great Lakes Water Quality Agreement (US State Dept., 1972) which is a permanent reference under the BWT and included agreed upon loading targets for each lake and country. Other water quality management issues were also addressed in the agreement. The GLWQA was renewed in 1978, revised in 1987 by Protocol with the guiding concept of the "Ecosystem Approach" being added. A further revision was negotiated beginning in January 2010, was signed on 7 September 2012, and entered into force on 12 February 2013.

Another significant binational Great Lakes institution, the Great Lakes Fisheries Commission (GLFC), was established in 1955 under the Great Lakes Fisheries Convention. Previously cross-border fishery disputes had been referred to the IJC. Continued declines in key fish stocks, particularly among salmonines and coregonines, were com-pounded by the devastating impacts of the Sea Lamprey in the 1930–1940s after 1919 improvements to the Welland Canal connecting Lakes Ontario and Erie allowed the parasite to penetrate into the upper lakes. The GLFC was charged with coordinating fisheries management and research where binational interests arose, arbitrating disputes, and, most importantly, implementing a program to control Sea Lamprey. The control program has been highly successful albeit that it will be required permanently as eradication is not deemed possible. Once Sea Lamprey controls began to have impact and because recovery of native Lake Trout and other fish stocks was expected to be a long-term challenge, the GLFC also coordinated efforts to promote new put-and-take fisheries especially for recreational fishers centred on Pacific salmonines like Coho and Chinook. As the activities of the GLFC and its partners evolved in the 1960–1970s, they became increasingly engaged in the activities in the water quality and ecosystem management arenas, recognizing the key role that healthy fish and fisheries have as symbols of success. The GLFC parties developed the Joint Strategic Great Lakes Fisheries Management Plan in 1981and revised in 1997 with four key elements: consensus decision-making, accountability, information sharing and, notably, an ecosystem-based approach to management. Both key institutions, the IJC and the GLFC, have largely operated as arms-length authorities where scientific and technical analyzes have been a key features coupled with decision-making shared among the representatives of the many parties involved for states, provinces and countries. Progressive science-based ideas on three fronts have guided the Great Lakes institutions and peoples to new ways of thinking about how manage the natural environment: (1) sustainable development (The Environment, Its People, and Their Economy); (2) ecosystem approach (balancing ecological, social, and economic concerns); (3) Fish Habitat (Habitat, Fishes, Fisheries). However, as a brief decadal synopsis of activities since the Second World War (Table 2) shows, complete realization of the ideals remains out of reach. Many able managers and scientists have come together in the shared institutions and for a, such as the long-standing annual meetings of the International Association of Great Lakes Research (IAGLR). However, many agencies have ultimately remained entrenched in their dated, home-based, siloed command-and-control management regimes (McLaughlin and Krantzberg, 2012). This has occurred despite the proliferation of a diversity of Non-Governmental Organizations (NGOs) lobbying for a growing range of ecosystem management concerns. While the sciences have evolved spawning new knowledge and understanding as a basis for wise choice of management actions, many of the institutions and interests of society have not evolved in parallel such that sustained progress can be hard to discern at present. In the future we might expect further reversals (Table 2). After the substantial investments begun in the 1970s and 1980s to rehabilitate the Great Lakes, overall federal and provincial contributions to public infrastructure investments have fallen far behind the levels needed to maintain past investments in Canada. As a result, municipal governments are increasingly resisting further investments to secure water quality gains and watershed protections. As suburban sprawl and population growth continue around the basin we should expect that the capacity of existing sewage treatment plants will be reached and new ones added without regard to the established limits to Point-source nutrient loading. These events will be overlain and exacerbated by the increasing impacts of climate change on the Great Lakes such as are already evident in the decreasing ice coverage in winter (Wang et al., 2012) and higher summer surface temperatures (Trumpickas et al.,2009). If the ideology-driven austerity programs in the public domain persist we may perhaps see a return of the sanitary diseases that were the scourge of 19th Century cities in the Great Lakes basin.

Table 2. A rough decadal synopsis of societal actions in the St. Lawrence Great Lakes Basin after the Second World War along with some forecasts for coming decades.

Decade	Highlights of Societal Activities
1950s	Piecemeal environmental clean-up actions amid rapid post-war economic expansion across the western hemisphere
1960s	Rising public concern and activism especially with regard to eutrophication and toxic chemicals; Beginning of the expansion of municipal waste management infrastructure
1970s	Major environmental clean-up activities going full tilt; Public demanding a greater say in decision-making; Growth of NGOs
1980s	Ecosystem-based management approaches flourish; Coordinate Remedial Action Plans initiated for 46 Areas of Concern (many embayments or large rivers draining into the lakes)
1990s	Institutional gridlock grows alongside greater public involvement; Magnitude of remedial actions needed is perceived along with the need for long-term planning; Realization that sustainable ecosystems require human activities to be bound
2000s	Long-term under-investment in public infrastructure along with opportunist tax cuts lead to environmental budget restraints; Public attention fatigue sets in as the array of issues demanding attention widens in the Internet Age
2010s	Economic crises have led to severe anti-Keynesian austerity measures in the public domain where environmental protection activity sits; A blinkered focus on non-renewable resource-based economic growth is fostering a push-back on environmental management objectives
...	*Possible Future*
2020s	*Old, seemingly solved problems like eutrophication begin to re-emerge*
2030s	*Climate change's many synergies will become more noticeable everywhere*
2040s	*Possibly, 19th Century problems like sanitary diseases will return in worse forms*

The evolution of ecosystem science

Sitting midway on a spatial scale between oceans and large marine coastal regions, where much of oceanography and fisheries science has evolved, and smaller inland lakes and streams, where much of limnology and water quality/habitat management has evolved, the development of SLGL's science has drawn heavily from both.

A small sample of recent studies across three spatial resolutions in the Great Lakes basin shows the rich benefits of the many progression sequences (Table 3). Hydro-acoustic assessment of distribution and abundance has expanded over the last 40 years from cumbersome racks of equipment with limited data management to compact integrated multi-sensor systems providing many alternate measures of fish and zooplankton abundance along with detailed pictures of habitat variables (bathymetry, substrates and vegetation). Radio-and acoustic-telemetry techniques have expanded from small boat operators chasing fleeting encounters of single fish to geo-referenced sono-buoy arrays passing back near-instantaneous 3-D positions for tens of fish, along with a widening spectrum of physiological indicators. Recognition of the central role of habitat has grown as intelligence about fish locations and movements has increased. The ability to simulate 3-D water movements and stratification cycles has been greatly enlarged thereby facilitating the ability to look at fish-habitat linkages more closely. Knowledge of the flows and pools of energy, nutrients and contaminants in the ecosystems and their foodwebs has grown as an ever-wider range of techniques, such as stable isotopes, bio-accumulating chemicals, mass balances and ecosystem models, yield insights into connections and dynamics. As in many biological fields, the rise of rapid automated sequencing of DNA and RNA strands has enhanced our appreciation of diversity in aquatic ecosystems, aiding in stock and species identification at all spatial scales. Amid the expansion of analytical techniques, formal consideration of uncertainty and risk is increasingly providing the basis for sounder management and decision-making. In the socio-economic arenas, sociologists' thoughtful analyses of institutional dynamics are emerging providing the basis for organizational improvements and ecological economists have been developing ways to assess the contribution of the many ecosystem goods and services to sustain-able human well-being in the Great Lakes and beyond. Backstopping the explosive growth in the many areas of ecosystem science has been the continuing rapid evolution of computer technology and the

Table 3. A brief selective overview of recent progress in Great Lakes ecosystem science with emphasis on fish and fisheries.

Topic	Great Lakes	Lake Ontario	Bay of Quinte
Hydro-acoustics	Annual surveys of pelagic stocks on 5 lakes (Rudstam et al., 2009)	Evaluation of Lake Trout stocking offshore (Lantry et al., 2011)	Macrophytes surveys every 3–5 years (Leisti et al., 2012)
Telemetry	Numerous studies with emphasis on Lake Trout, Walleye, Sturgeon (Landsman et al., 2011)	Nearshore grid along Toronto Waterfront examining residency and habitat selection of several species (S. Doka et al., Pers. Comm, DFO)	
Hydro-dynamics	3-D ELCOM models linking lake hydrodynamics to climate (Leon et al., 2005)	Long term trends in thermal regime (Huang et al., 2012)	3-D ELCOM modelling for Quinte and the outlet basin of Lake Ontario (ongoing studies Env Canada/Ont Min Env/Leon Boegman Queens)
Habitat	Process-driven models of population dynamics of fish tied to habitat supply (Hayes et al., 2009) Mapping aquatic biodiversity investment areas for conservation management (Morrison et al., 2001)	Basin scale habitat classification scheme (Minns and Wichert, 2005) Habitat supply limits on Northern Pike in Hamilton Harbour (Minns et al., 1996)	Long-term analysis of habitat supply of Walleye in relation to P control, temperature and water levels (Chu et al., 2004)
Foodweb	Geochemistry has contributed much to the paleolimnology (Meyers, 2003) Stable isotopes show effects of invasions on top predators (Vander Zanden et al., 1999)	Model of PCB transport and fate in foodwebs (Morrison et al., 2002) Sustainability of stocked salmonine fisheries (Jones et al., 1993)	Importance of the microbial foodweb in the pelagic zone of Quinte (Munawar et al., 2011)

Table 3. Continued....

Topic	Great Lakes	Lake Ontario	Bay of Quinte
Mass Balance	Hydrodynamic approach to P modelling in Lake Erie (Schwab et al., 2009)	Tools for predicting the movement of P from soils into the lake (Allaire et al., 2011)	Phosphorus management model (Minns and Moore, 2004) Fate of metal contaminants (Gandhi et al., 2011)
Ecosystem Modelling	Ecopath model of Lake Superior testing reasons for decline of Lake Herring (Cox and Kitchell, 2004)	Ecopath model of offshore foodweb pre- vs post-species invasion changes (Stewart and Sprules, 2011)	Ecopath model of upper BoQ covering pre- and post-P control and post-Zebra (Koops et al., 2009)
Genetics	Post-glacial dispersal of Walleye viz refugia (Stepien et al., 2009)	Diversity in Great Lakes bloaters viz. thier reintroduction into L. Ont. (Fave and Turgeon. 2008)	Genetic diversity of Lake Whitefish in BoQ relative to Lake Ontario viz. conservation (Bernard et al., 2009)
Uncertainty	Integrated pest management of Sea Lampreys (Christie and Goddard, 2003; Jones et al., 2009) Options for precautionary fisheries management (Fenichel et al., 2008)	Expected future declines in fisheries given environmental and cultural change (Rothlisberger et al., 2010)	Uncertainty in inputs to phosphorus budget (Arhonditis et al., Pers. Comm UTSC)
Socio-economics	Economic impact of nonidigenous species on ecological services (Rothlisberger et al., 2012) Analysis of information flows in social structure for GL fisheries management (Leonard et al., 2011) Historical development of fishery and environmental management practices (Bocking, 1997)	Pathological weakness of responsible management agencies (Mclaughlin and Krantzberg, 2012)	Analysis of the role of modelling and public consultation in the development of the Quinte remedial action plan (Stride et al., 1992) Communication of research and monitoring results to management (Berquist et al., 2012) Long-term studies prospectus (Minns et al., 2011)

Internet coupled with the similar expansion of statistical, analytical, and data management tools.

Lessons learned

Over the last 50 years, science methodology and analytical tools have grown dramatically, with the concentrated efforts in the Great Lakes Basin making significant contributions. This has fuelled a vastly increased understanding of the structural and functional dynamics of aquatic ecosystems, even as the array and strength of human-induced stressors continued, and continues to this day, to expand. These developments have led to the wide appreciation that successful management of human activities requires both an ecosystem based approach and a greater understanding of how humans make decisions, i.e. often irrationally based on instinctive fast-brain thinking rather than logically after careful consideration and analysis of many factors (Kahneman, 2011; Ariely, 2008). Society has become much more aware of and sensitized to its ecosystem management challenges due to the science-based actions to curb load-driven problems and the increased pressure from the many speaking up for conservation, restoration and protection of ecosystems (Table 3).

Unfortunately, the scientists and the understanding they have gathered are having a declining influence on public affairs as the demands of special interests continue to take precedence over communal needs, as the economy trumps the ecology (as usual). However, it is useful to distill scientific under-standing into a brief series of points.

Whole ecosystem view

We always need to use a wide-angle lens as there can be no externalities with ecosystems (Minns, 1999). For many, this has become the established norm in the Great Lakes basin (GLB) and particularly for fishers and fishery scientists and managers.

Keep monitoring

As with any complex systems, e.g. weather, stock market, and large-scale eco-systems, wise management depends on a spectrum of continuing intelligence inputs (Minns et al., 2011). GLB fisheries agencies have sup-ported an excellent long-term set of complementary programs while the water quality agencies have had mixed success.

Beware of the inertia of cumulative impacts
The accumulated impacts and pressures of past actions always weigh heavily on the present (Rees, 1995). While much clean-up progress has been achieved, in the GLB population and sub-urbanization have continued unchecked.

We are rehabilitating not restoring ecosystems
We cannot undo all the damage and changes and must settle for making what remains sustainable for the long term. In the GLB the irrevocable loss of much habitat, the less-than-pristine state of much of what remains, and the many established non-indigenous species preclude complete restoration of many ecosystem features.

Strive to involve all of society
In both the fisheries and ecosystem management arenas agencies in the GLB have greatly expanded their engagement with stakeholders and interested parties in recent decades although too often environmental advocacy groups just become special interests competing with the influence of the dominant capitalist interests.

Beware of shifting baselines
If society is inattentive to deteriorating ecosystem conditions the reference points are continually eroded. (Pinnegar and Engelhard, 2008). Increasing urbanization of the basin's population has inevitably meant most people have less direct contact with the natural ecosystems and hence less sensitivity to changing conditions.

Be actively adaptive
As the opportunities to run controlled experiments are very limited, society must be prepared to experiment and learn through evolving management actions, thereby discarding the traditional command and control approach (Walters and Holling, 1990).

Ecosystem-based management is forever
By trying to treat many ecosystem issues as externalities society often fails to recognize that continued sustainability of our life-supporting ecosystems will require eternal vigilance and interventions.

Summary and future prospects

Fishery resources include the fishes, the other biota they interact with, and the habitats they occupy. The historical sequences in the use and management of these resources may be considered as a series of interacting sequences of change. These sequences can span from social, economic, institutional and landscape changes, through water quality, habitat supply, and climate changes, to biotic composition changes, introductions and extinctions, and species harvest changes. A selection of these sequences is examined for the St. Lawrence-Great Lakes which has been subjected to intense development and study over the last 200 years. While the whole basin is considered, some detailed attention is given to Lake Ontario and, within it, the Bay of Quinte. Brief selective development histories are given for the three areas as context. The management history is outlined and critiqued.

While much progress has been achieved in cleaning up the load driven problems in the basin, little secure progress toward rehabilitation and sustainability has been achieved. In the current period of economic problems, governments, particularly Canada's, are undoing past ecosystem management progress. The development of St. Lawrence Great Lakes' ecosystem science has drawn heavily from both oceanography and limnology. A brief, selective overview of several progressions in ecosystem science illustrates how knowledge and understanding of this ecosystem has expanded over the last 60 years, providing an improved basis for management action.

As with use and management, the science of fishery resource management has consisted of many historical progressions. The many sequences in the St. Lawrence-Great Lakes management and science histories lend support for recognition of (i) the importance of taking an ecosystem approach to renewable resource management, (ii) the value of adaptive management practices and, particularly, (iii) the vital complementary roles of long-term monitoring and mathematical modelling. We have progressively developed many of the scientific tools needed to manage large lake eco-systems but society as a whole still lacks the collective persistence and will to implement measures needed to secure the earth's natural capital (the environment), to ensure equity and health for all people, and to ensure that our economic activities are sustainable. Based on the guiding principle at the core of the 1986 Canadian fish habitat management policy. "No net loss of the productive capacity of fish habitats," it is possible to articulate a trio of guiding principles for freshwater, and other natural, ecosystems:

- No net loss of the natural productive capacity of ecosystems.
- No net loss of the biological diversity of ecosystems.
- No net loss of the potentially utilizable productivity of natural ecosystems.

The first two principles address the need to con-serve and protect natural capital while the third addresses the need for inter-generational equity. The productive capacity is the full set of inherent structural and functional properties of ecosystems that collectively support the ability of biota to be productive. The naturally evolved diversity of species, communities and ecosystems provides the basis for maximizing the productive potential of ecosystems. The potentially utilizable productivity must be able to meet the aggregate human use of ecosystem goods and services sustainably. Many countries and peoples began to aspire to some form of these guiding principles in the 1960s and 1970s but these aspirations have not been realized as yet, as the power and influence of rampant capitalism have come to dominate most public discourse since the 1990s. The recent failures at Copenhagen, December 2009 (UN Conference of the Parties (15th) on the Framework Convention on Climate Change) and RioC20, June 2012 (UN Conference on Sustainable Development) have been disheartening with many countries, including Canada and the United Kingdom, rolling back many of the environment protection rules established in the 1970s in order to facilitate unfettered development, especially of non-renewable resources. In 2012 the Canadian government passed legislation that significantly reduced federal protection of fish and their habitat to facilitate its priority for economic development particularly with respect to non-renewable resource extraction (Favaro et al., 2012). The new Fisheries Act only covers existing fisheries, reducing the spatial and ecological scope of the protections, and significantly raised the threshold with respect to what constitutes harm; the new regulations implementing the changes have yet to be released but the "no net loss" guiding principle, will no longer be followed. The changes to the Canadian Fisheries Act were part of a wider attack on environmental protections in Canada (Gibson, 2012). As the full consequences of cumulative ecosystem negligence and destruction are realized in the coming decades, people will turn again toward implementation of the three "no net loss" principles as the only viable path to truly sustainable existence. Meanwhile, I will have to be comforted by the words of St. Francis of Assisi (Italian friar and preacher; 1181/2-1226), the patron saint of nature:

"True progress quietly and persistently moves along without notice."

Acknowledgements

Thanks to the organizing committee for the "State of Lake Vannern Ecosystem" symposium for the invitation to participate as a keynote. Special thanks to Dr. Goran Dave for his tireless assistance for my wife and me during our visit to Sweden. Thanks to Mohi Munawar and his staff at AEHMS for his support and encouragement on this project.

References

Allaire, S. E., van Bochove, E., Denault, J.-T., Dadfar, H., Theriault, G., Charles, A. and De Jong, R., 2011. Preferential pathways of phosphorus movement from agriculturall and to water bodies in the Canadian Great Lakes basin: A predictive tool. Can. J. Soil Sci. 91, 361–374.

Ariely, D., 2008. *Predictably irrational: the hidden forces that shape our decisions.* Harper-Collins, New York.

Bernard, A.M., Ferguson, M. M., Noakes, D.L.G., Morrison, B.G., Wilson, C.C., 2009. How different is different? Defining management and conservation units for a problematic exploited species. Can. J. Fish. Aquat. Sci. 66(9), 1617–1630.

Berquist, M.K., Campbell, L.M., Whitelaw, G.S., Millard, S.E., 2012. Communicating research findings and monitoring data in support of management: A case study of the Bay of Quinte Remedial Action Plan. Aquat. Ecosyst. Health Mgmt. 15(4), 473–483.

Bocking, S., 1997. Fishing in the inland seas: Great Lakes research, fisheries management, and environmental policy. Env. History 2, 52–73.

Christie, G.C., Goddard, C.I., 2003. Sea Lamprey International Symposium (SLIS II): Advances in the integrated management of sea lamprey in the Great Lakes. Journal of Great Lakes Research 29 (Supplement 1), 1–14.

Chu, C., Minns, C.K., Moore, J.E., and Millard, E.S., 2004. Impact of oligotrophication, temperature, and water level son walleye habitat in the Bay of Quinte, Lake Ontario. Trans. Amer. Fish. Soc. 133, 868–879.

Cox, S.P., Kitchell, J.F., 2004. Lake Superior Ecosystem, 1929-1998: Simulating Alternative Hypotheses for Recruitment Failure of Lake Herring (*Coregonus artedi*). Bulletin of Marine Science, 74(3), 671–683.

Favaro, B., Reynolds, J.D., C^ote, I.M., 2012. Canada's weakening aquatic protection. Science 337, 154.Fave, M-J., Turgeon, J., 2008. Patterns of genetic diversity in Great Lakes bloaters (*Coregonus hoyi*) with a view to reintroduction in Lake Ontario. Conservation Genetics 9(2), 281–293.

Fenichel, E.P., Tsao, J. I., Jones, M.L., Hickling, G.J., 2008. Real options for precautionary fisheries management. Fish and Fisheries 9, 1–17.

Gandhi, N., Diamond, M. L., Razavi, R., Bhavsar, S.P., Hodge, E.M., 2011. A modeling assessment of contaminant fate in the Bay of Quinte, Lake Ontario: Part 1. Metals. Aquat. Ecosyst Health Mgmt. 14(1), 85–93.

Gibson, R.B., 2012. In full retreat: the Canadian government's new environmental assessment law undoes decades of progress. Impact Assessment and Project Appraisal 30(3), 179–188.

Hayes, D., Jones, M.L., Lester, N., Chu, C., Doka, S.E., Netto, J., Stockwell, J., Thompson, B., Minns, C.K., Shuter, B.J., Collins, N.C., 2009. Linking fish population dynamics habitat conditions: insights from the application of a process-oriented approach to several Great Lakes species. Rev Fish Biol Fisheries 19, 295–312. (DOI 10.1007/ s11160-009-9103-8).

Huang, A., Rao, Y., Zhang, W., 2012: On Recent Trends in Atmospheric and Limnological Variables in Lake Ontario. J. Climate 25, 5807–5816. doi: http://dx.doi.org/10.1175/JCLI-D-11-00495.1

Jones, M.L., Koonce, J.F., O'Gorman, R., 1993. Sustainability of hatchery dependent salmonine fisheries in Lake Ontario: the conflict between predator demand and prey supply. Transactions of the American Fisheries Society. 122, 1002–1018.

Jones, M.L., Irwin, B.J., Hansen, G. J. A., Dawson, H.A., Treble, A.J., Liu, W., Dai, W., Bence, J.R., 2009. An operating model for the integrated pest management of the Great Lakes Sea Lampreys. The Open Fish Sci. J. 2, 59–73.

Kahneman, D., 2011.Thinking, fast and slow. Farrar, Strausand Giroux, New York.

Koops, M.A., Dermott, R. M., Leisti, K.E., Johannsson, O.E., Millard, E.S., Minns, C.K., Munawar, M., Nicholls, K.H., Hoyle, J.A., 2009. The Bay of Quinte: a model for large lake ecosystem management. Verh. Int. Ver. Limnol.30, 7, 1024–1029. (SIL Proceedings).

Landsman, S.J., Nguyen, V.M., Gutowsky, L.F.G., Gobin, J., Cook, K.V., Binder, T.R., Lower, N., McLaughlin, R.L., Cooke, S.J., 2011. Fish movement and migration studies in the Laurentian Great Lakes: research trends and knowledge gaps. J. Great Lakes Res. 37, 365–370.

Lantry, B.F., O'Gorman, R., Strang, T.G., Lantry, J.R., Connerton, M.J., Schaner, T., 2011. Evaluation of offshore stocking of Lake Trout in Lake Ontario. N. Amer. J. Fish. Manage. 31, 671–682.

Leisti, K.E., Doka, S.E., Minns, C.K., 2012. Submerged Aquatic Vegetation in the Bay of Quinte: Response to decreased phosphorous loading and Zebra Mussel invasion. Aquat. Ecosyst. Health Mgmt 15(4), 442–452.

Leon, L.F., Lam, D., Schertzer, W., Swayne, D., 2005. Lake and climate models linkage: a 3-D hydrodynamic contribution. Advances in Geosciences 4, 57–62.

Leonard, N.J., Taylor, W.M., Goddard, C.I., Frank, K.A., Krause, A.E., Schecter, M.G., 2011. Information flow within the social network structure of a Joint Strategic Plan for Management of Great Lakes Fisheries. North Amer. J. Fish. Manage. 31, 629–655.

McLaughlin, C., Krantzberg, G., 2012. An appraisal of management pathologies in the Great Lakes. Sci. Total Environ. 416, 40–47.

Meyers, P.A., 2003. Applications of organic geochemistry to paleolimnological reconstructions: a summary of examples from the Laurentian Great Lakes. Organic Geochem. 34, 261–289.

Minns, C.K., 1999. The ecosystem pyramid and the means forattaining ecological sustainability: an essay in honour of Jack Christie. Aquatic Ecosyst. Health Mgmt. 2(3), 209–221.

Minns, C.K., Moore, J.E., 2004. Modelling phosphorus management in the Bay of Quinte, Lake Ontario in the past,1972 to 2001, and in the future. Can. Manuscr. Rep. Fish. Aquat. Sci. 2695.

Minns, C.K., Wichert, G.A. 2005. A framework for defining fishhabitat domains in Lake Ontario and its drainage. J. Great Lakes Res. 31(Suppl.1), 6–27.

Minns, C.K., Munawar, M., Koops, M.A., Millard, E.S., 2011.Long-term Ecosystem Studies in the Bay of Quinte, Lake Ontario 1972-2008: A Prospectus. Aquat. Ecosyst. Health Mgmt. 14 (1), 3–8.

Minns, C.K., Randall, R.G., Moore, J.E., Cairns, V.W., 1996. Amodel simulating the impact of habitat supply limits on northern pike, *Esox lucius*, in Hamilton Harbour, Lake Ontario.

Morrison, H.A., Minns, C.K., Koonce, J.F., 2001. A methodology for identifying and classifying aquatic biodiversity investment areas: Application in the Great Lakes basin. Aquat. Ecosyst. Health and Mgmt. 4(1), 1–12.

Morrison, H.A., Whittle, D. M., Haffner, D., 2002. A comparison of the transport and fate of polychlorinated biphenyl congeners in three Great Lakes food webs. Environ. Toxi-col. Chem. 21, 683–692.

Munawar, M., Fitzpatrick, M., Niblock, H., Lorimer, J., 2011.The relative importance of autotrophic and heterotrophicmicrobial communities in the planktonic food web of the Bay of Quinte, Lake Ontario 2000–2007. Aquat. Ecosyst. Health Mgmt. 14(1P), 21–32.

Pinnegar, J.K. and Engelhard, G.H., 2008. The 'shifting baseline' phenomenon: a global perspective. Rev. Fish. Biol. Fisheries 18, 1–16.Rees, W., 1995. Cumulative environmental assessment and global change. Environ. Impact Assess. Rev. 15, 295–309.

Rothlisberger, J.D., Lodge, D.M., Cooke, R.M., Finnoff, D.C.,2010. Future declines of the binational Great Lakes fisheries: the importance of environmental and cultural change. Front. Ecol. Environ. 8, 239–244.

Rothlisberger, J.D., Finnoff, D.C., Cooke, R.M., Lodge, D.M., 2012. Ship-borne nonindigenous species diminish Great Lakes ecosystem services. Ecosystems 15, 462–476.

Rudstam, L. G., Parker-Stetter, S. L., Sullivan, P. J., and Warner, D. M., 2009. Towards a standard operating procedure for fishery acoustic surveys in the Laurentian Great Lakes, North America. – ICES Journal of Marine Science 66, 1391–1397.

Schwab, D.J., Beletsky, D., DePinto, J., Dolan, D.M., 2009. Ahydrodynamic approach to modelling phosphorus distribution in Lake Erie. J. Great Lakes Res. 35, 50–60.

Stepien, C.A., Murphy, D.J., Louthier, R.N., Sepulveda-Villet., O.J., Haponski, A.E., 2009. Signatures of vicariance, post-glacial dispersal and spawning phiopatry: population genetics of the walleye, *Sander vitreus*. Molecular Ecol. 18, 3411–3428.

Stewart, T.J., Sprules, W.G., 2011. Carbon-based balanced trophic structure and flows in the offshore Lake Ontario foodweb before (1987–1991) and after (2001–2005) invasion-induced ecosystem change. Ecol. Modelling 222(3), 692–708.

Stride, F., German, M., Hurley, D.A., Millard, E.S., Minns, C.K., Nicholls, K.H., Owen, G.E., Poulton, D.A., and de Geus, N., 1992. An overview of the modelling and public consultation processes used to develop the bay of Quinte Remedial Action Plan. In: J.H. Hartig, J.H. and M.A. Zarull(Eds.), Under RAPs: toward grassroots ecological democracy in the Great Lakes Basin, pp. 161–183. University of Michigan Press, Ann Arbor, Michigan.

Trumpickas, J., Shuter, B.J., Minns, C.K., 2009. Forecasting impacts of climate change on Great Lakes surface water temperatures. J. Great Lakes Res. 35(3), 454–463. U.S. State Dept., 1972. Great Lakes Water Quality Agreement with Annexes and Texts and Terms of Reference, Between the United States of America and Canada, signed at Ottawa, 1972 April 15. Washington, D.C.

Vander Zanden, M.J., Casselman, J.M., Rasmussen, J.B., 1999.Stable isotope evidence for the food web consequences of species invasions in lakes. Nature 410, 464–467. Walters, C.J. and Holling, C.S., 1990. Large scale management experiments and learning by doing. Ecology 71, 2060–2068.

Wang, J., Bai, X., Hu, H., Clites, A., Colton, M., Lofgren, B.,2012. Temporal and spatial variability of Great Lakes ice cover, 1973–2010. J. Climate 25, 1318–1329.

The above article is a reprinted version of the following previously published article:

Minns, C.K., 2014.
Management of Great Lakes fisheries: Progressions and lessons.
Aquatic Ecosystem Health & Management (2014) 17 (4): 382–393.
https://doi.org/10.1080/14634988.2014.967163

A comparison of approaches for integrated management in coastal marine areas of Canada with the historical approach used in the Great Lakes (Bay of Quinte)

R. G. Randall, M. A. Koops[1], and C. K. Minns[2*]

[1]Fisheries and Oceans Canada, 867 Lakeshore Road, Burlington, ON L7S 1A1
[2] Retired
**Corresponding author: ken@minns.ca*

Keywords: conservation objectives, coastal zone, ecosystem-based management, freshwater, valued ecosystem component, knowledge transfer

Introduction

Canada's Oceans Act (1997) authorized Fisheries and Oceans Canada (DFO) to lead the development of a national oceans management strategy, to be guided by the principles of sustainable development and integrated management. Although the Oceans Act does not apply to the Great Lakes, there is a close parallel between activities in marine areas, initiated recently to support integrated management, and activities in the Great Lakes, developed historically and currently ongoing. In the Great Lakes, there is a long history of ecosystem-based management that has shifted from managing degraded areas (IJC, 1987) to the integrated management of valued ecosystem components (Minns et al., 1986; Minns, 1995, 1999). Past and existing management frameworks in the Great Lakes are consistent with the oceans framework. The goal of this paper is to compare frameworks and similarities between the historical ecosystem approach for management (EAM) of aquatic resources at the Bay of Quinte, Lake Ontario, with the more recent process developed for marine areas in Canada. Through this comparison, we demonstrate that strong communication of EAM processes between freshwater and marine ecosystems are mutually beneficial for advancing integrated management. Additionally, the novel science needs required to support the EAM are summarized to emphasize that they apply equally to freshwater and marine areas. The three

Ecology of the Bay of Quinte: Health, Management and Global Implications
Editors: C.K. Minns, M. Munawar and M.A. Koops

Aquatic Ecosystem Health & Management Society, Canada

overarching conservation goals of maintaining productivity, preserving biodiversity and protecting habitat (Gavaris, 2009) are common to both freshwater and marine ecosystems.

Comparison of frameworks in oceans and freshwater

The geographic scale of marine ecosystems is large relative to the Great Lakes. Under the Ocean Action Plan (OAP) (DFO, 2005), five Large Ocean Management Areas (LOMAs) were identified: the Pacific North Coast, Beaufort Sea, Gulf of St. Lawrence, Eastern Scotian Shelf and Placentia Bay/Grand Banks (DFO, 2007a). Taken together, the Great Lakes have an area similar to the Gulf of St. Lawrence, the smallest LOMA. For each LOMA, Ecosystem Overview and Assessment Reports (EOAR) were prepared as part of an integrated management framework (Table 1). As components of the EOAR, four candidate conservation priorities were considered, and all are relevant to this comparison: (1) Ecologically and Biologically Significant Areas (EBSA); (2) degraded areas; (3) Ecologically Significant Species (ESS); and (4) depleted or rare species. Each of these four conservation priorities are further discussed below. If coastal areas were included in the LOMA, coastal EBSAs were identified. The geographic scale (area and length of coast-line) of Quinte is comparable to marine EBSAs. In addition to the four conservation priorities, integrated management in the Bay of Quinte is compared and contrasted to the oceans framework in the context of (5) science-based Conservation Objectives,

Table 1. Components of the Oceans Action Plan framework (DFO, 2005, 2007a, 2008).

Large Ocean Management Area (LOMA)
Ecosystem Overview and Assessment Report (EOAR)
Candidate Conservation Priorities:
Ecologically and Biologically Significant Areas (EBSA)
Ecologically Significant Species (ESS)
Depleted or Rare Species
Degraded Areas
Conservation Objectives (CO)
(with indicators and reference points)
↓
Integrated Management Planning

(6) the use of ecosystem indicators for deter-mining status in the context of reference points, and finally (7) science support. To demonstrate the similarities in ecosystem approaches between ocean and freshwater management, DFOs oceans framework and criteria are introduced and defined initially for each conservation component, followed by a discussion of the relevance and application to the Bay of Quinte.

Ecologically and biologically significant areas

The identification of significant areas is under-taken jointly and collaboratively by community, scientific and EBSA evaluation workshops (DFO, 2004), demonstrating an integrated approach from the outset of the Oceans Action Plan (DFO, 2005). Once identified, the Ecologically and Biologically Significant Areas (EBSAs) receive enhanced protection while allowing sustainable activities. EBSAs are smaller areas nested within Large Ocean Management Areas; for example, the Beaufort Sea LOMA includes 10 EBSAs (DFO, 2007b). Knowledge of both the characteristics of the particular area, and a process-based understanding of these characteristics in terms of ecosystem structure and function are used as criteria to identify EBSAs (DFO, 2004). Functions are related to the life history of fishes, including reproduction, rearing, feeding, and migration, which together determine growth, survival and production rates. Significance of an area, based on ecosystem function and structure, is judged in the context of three major dimensions, uniqueness, aggregation and fitness consequences, along with two additional considerations, resilience and naturalness. An example of a significant ecological function would be a fish spawning area that is high for uniqueness (site used by many species),- aggregation (high proportion of a population, or number of species, use the area), and fitness con-sequences (reproductive habitat can greatly affect the productivity of species). Significant structural features would be extensive macrophyte beds that demonstrate uniqueness (absent elsewhere), aggregation (density and variety of species), fitness consequences (provide greater population fecundity resilience (perennial species, temporal and spatialstability) and naturalness (native aquatic plants).

Based on the EBSA definition, the Bay of Quinte meets the criteria of a significant area, and it has received enhanced management and protection within Lake Ontario for a number of years (Minns et al.,1986). In terms of ecosystem function, the bay is known to be an important spawning and nursery habitat for coastal wetland and offshore species in Lake Ontario. Historically, outcroppings and shoals of limestone (Sherman Fall Unit, Makatewis Island shoal, Telegraph Narrows and others) were preferred spawning surfaces for Lake Whitefish *Coregonis clupeaformis* (Hart, 1930; cited by Scott and Crossman, 1973 and Sly, 1986). Spawning conditions and seasonal thermal habitats were also important for successful hatch and initial feeding for Walleye (*Sander vitreus*), Yellow Perch (*Perca flavescens*) and other fishes. Extensive coastal wetlands, dominated by emergent and submerged aquatic macrophytes, alternate with rocky areas along the diverse shoreline of the Bay of Quinte. From 1979 aerial photographs, Crowder and Bristow (1986) estimated the area of submerged and floating-leaved macrophytes to be 627 ha and the area of emergent wetlands to be 4074 ha. Area of plant cover would have been greater in years before the upper Bay became eutrophic (below). With the exception of the Bay of Quinte, extensive coastal wetlands are rare along the north shore of Lake Ontario. Macrophyte beds provide structural habitat for spawning, nursery and refugia for numerous phytophilic fishes. The Quinte macrophyte beds would meet all the criteria of a significant ecosystem structure – unique (lack of beds elsewhere), aggregation (phytophilic fishes), fitness consequence (reproduction and survival), resilience in terms of temporal and spatial stability and high diversity of predominantly native macrophytes.

Degraded area

For integrated ocean management planning, identifying and managing degraded areas was identified as a candidate conservation issue. Degraded areas are defined as "areas where environmental quality has been impacted by human activities to a state where the natural structure and function of the ecosystem in the area is severely diminished" (DFO, 2007a). Criteria and guidelines for identifying degraded areas in marine areas have not yet been established (DFO, 2008). However, a workshop process for doing so has been described, involving many levels of government, and knowledge transfer from the Great Lakes and international marine organizations that deal with environmental quality (IMO, MarPOL, ICES and others). The Bay of Quinte was designated (listed) by the International Joint Commission (IJC) in 1986 as one of 43 degraded coastal areas in the Great Lakes that required Remedial Action Plans (IJC, 1987). Designation of Quinte as an Area of Concern was based on a number of use impairments (listing criteria; IJC, 1991), some of which were related to fishes and fish habitat

(Table 2). A long-standing comprehensive ecosystem-based remediation plan for the Bay Quinte has resulted in a significant improvement in the aquatic community and habitat (Minns et al., 1986). Ecologically Significant Species and community properties (ESS). In the context of the OAP, "significant" indicates that the "species or community property has controlling influence over key aspects of ecosystem structure and function" (DFO, 2006a). Four types of significant species or properties were identified: (1) species with a crucial trophic-dynamic role (e.g. for age species and predators); (2) species that provide three-dimensional structure to biodiversity and productivity (e.g. corals, plants); (3) above the species level, aggregate ecosystem properties that are essential for maintaining ecosystem structure and function (e.g. size based properties, or frequency distribution of biomass across species) and (4) species or species groups that if abundant pose a threat to ecosystem structure and function (e.g. invasive or toxic species). The latter type requires enhanced management to control their abundance and distribution, rather than enhanced protection. Additionally, rarity and sensitivity/recoverability affect the application of the four criteria. Definitions, diagnostic features, and considerations for use of these criteria and species groups are provided by DFO (2006a). The diverse habitat in the Bay of Quinte sup-ports a rich fish fauna, with 65 species representing 45 genera and 23 families (Hurley, 1986). Many of the Quinte fishes are warm water species, but cool and cold water species also use the area as a migration corridor for spawning (coregonids, salmonids) and as seasonal habitat for feeding and refugia. All four of the ESS types of significant species or properties apply to Quinte. First, top predators including Walleye function in a crucial trophodyamic role. Hurley et al. (1986) cited evidence that the fish community changed when piscivore biomass was less than 20% of the total fish biomass. Walleye was identified as a key top predator (significant species) in the Bay of Quinte, but population abundance has varied significantly (Hurley, 1986) and habitat quality has declined recently (Chu et al., 2004). The relative abundance of piscivores is a key metric in the Index of Biotic Integrity (Minns et al., 1994), which is used as a delisting criteria (indicator) for Quinte (Brousseau et al., 2011). Second, the role of macrophytes in providing 3-D habitat structure was highlighted previously in the discussion of significant areas (EBSA). Third, a size-spectrum model of aggregate biomass for Quinte showed reduced transfer of production from plankton to fishes (particularly large bodied fishes) with increased eutrophication (Hurley et al., 1986). Finally, invasive species posed a threat to ecosystem structure and function historically, and continue to be a primary threat to the Quinte ecosystem (Table 3).

Depleted species

Within the context of Large Ocean Management Areas, depleted or rare species are 'species that are both currently at a very low abundance, and usually were much more abundant at some time in the past (DFO, 2007a). Because of their status, depleted species warrant risk-averse management to ensure their survival and recovery. Depleted species include COSEWIC listed species-at-risk, and species identified by DFO that are below a Critical-Cautious boundary in the Precautionary Approach framework (DFO, 2006b). Nine fishes that inhabit the Bay of Quinte are listed by COSEWIC as being Extirpated (1 species), Endangered (1), Threatened (2) or Special Concern (5) (Table 4). Historically, Atlantic Salmon were known to inhabit the Trent, Moira and Salmon Rivers (COSEWIC, 2006a), but these populations disappeared by the late 1800s (a re-introduction program for Lake Ontario Atlantic Salmon is now being conducted). A shoal-spawning population of Lake Sturgeon (*Acipenser fulvescens*) was known at Amherst Island (COSEWIC, 2006b), but the current status is unknown (shoal-spawning Lake Sturgeon are rare). Habitat degradation or fragmentation was listed as threats for most of these at-risk species; historically, Atlantic Salmon, Lake Sturgeon and American Eel (*Anguilla rostrata*) were also fished. Additional (non-COSEWIC) depleted fishes in the Bay of Quinte have been identified by the Ontario Ministry of Natural Resources, the responsible agency for fisheries management in Ontario (Hurley, 1986; Casselman et al., 1999). Restoring top piscivores is a goal of the Quinte Remedial Action Plan.

Table 2. Examples of use impairments and delisting targets (IJC, 1991) related to fish population, fish habitat and lower trophic status in the Bay of Quinte (BQRAP, 1993; German and Stride, 1993; Johanson and McNevin, 2007). The Bay of Quinte was designated an Area of Concern (degraded area) in 1986 (IJC, 1987).

Use Impairment	Delisting targets	Measures, indicators and reference points
Degradation of fish populations	Demonstrate that key fish and wildlife species—including Walleye, Bass and Pike—are present in numbers consistent with a stable, diverse and healthy aquatic ecosystem	Demonstrate that Walleye, Bass and Pike populations are all self-reproducing and that none of these species dominates the fish community
		For littoral fish communities, maintain an IBI[1] score >60
Degradation of benthos	Using 1977–84 as the “base-time period,” demonstrate a positive trend and change in the benthic macroinvertebrate community structure of the Bay of Quinte.	Measure the presence and increasing abundance of pollution sensitive benthic macroinvertebrate species in the Bay of Quinte that were absent in earlier community surveys
Eutrophication and undesirable algae	Demonstrate a 25% reduction in total phosphorus concentrations in the upper Bay of Quinte	Reduce the average concentration of total phosphorus from 40 $\mu g\ l^{-1}$ to 30 $\mu g\ l^{-1}$
	Demonstrate a 40% reduction in algae levels in the upper Bay of Quinte	Demonstrate a change in total phytoplankton biomass such that the average pre-1990 algae density of 7.0–8.0 $mm^3\ l^{-1}$ to is reduced to 4.0–5.0 $mm^3\ l^{-1}$

Table 2 Continued...

Use Impairment	Delisting targets	Measures, indicators and reference points
	Using 1986 to 1994 as a "base year period," increase area of submerged aquatic macrophytes	Measure positive changes in distribution and areal coverage of submerged macrophytes
Degradation of phytoplankton and zooplankton communities	Demonstrate that the upper Bay of Quinte phytoplankton and zooplankton community structure does not diverge significantly from an unimpaired Great Lakes/control site of comparable physical and chemical characteristics	Associated with the Bay of Quinte RAP phosphorus reduction targets for the upper Bay of Quinte, demonstrate that chlorophyll *a* concentrations are about 12 ug l^{-1} in the upper Bay of Quinte without the presence of Zebra Mussels (with Zebra Mussel, a further 2–3 ug l^{-1} chlorophyll *a* reduction should be anticipated)
Loss of fish habitat	To the greatest extent possible, protect and restore fish habitat in the Bay of Quinte	Complete a Bay of Quinte Fish Habitat Management Strategy

[1] IBI = Index of Biotic Integrity (Minns et al., 1994)

Table 3. Non-native and invasive species in the Bay of Quinte (Johanson and McNevin 2007). Species intentionally introduced to sustain recreational fishing were excluded. Year of introduction and potential impacts to the conservation of productivity, biodiversity and habitat are indicated.

			Impacts to Conservation Objectives			
Species	**Common name**	**Year**	**Productivity**	**Diversity**	**Habitat**	**References**
Alosa pseudoharengus	Alewife	1873	√	√		Scott and Crossman, 1998
Osmerus mordax	Rainbow Smelt	1931	√			Scott and Crossman, 1998
Dorosoma cepedianum	Gizzard Shad	unknown	√			Johanson and McNevin, 2007
Cyprinus carpio	Common Carp	1896	√	√	√	Scott and Crossman, 1998
Carrassius auratus	Goldfish	<1878				Mills et al., 1993
Morone americana	White Perch	1950	√			Scott and Christie 1963; Minns and Hurley, 1986
Neogobius melanostomus	Round Goby	1999	√	√		Dietrich et al., 2006; Hoyle et al., 2011
Dreissena spp.	Zebra Mussels	1993	√	√	√	Hoyle et al., in press; Koops (pers. comm.)
Cercopagis pengoi	Fishhook Waterflea	1998	√	√		Kipp and Ricciardi, 2010
Hemimysis anomala	Bloody Red Shrimp	2009				Koops (pers. comm.)

Table 4. Depleted or extirpated fish species of the Bay of Quinte watersheds and coastal areas (modified from Johanson and McNevin, 2007).

Species	Common name	COSEWIC Status	Comments and threats (F = fishing; H = habitat perturbations)
Salmo salar	Atlantic Salmon	Extirpated	Lake Ontario population; F, H
Notropus anogenus	Pugnose Shiner	Endangered	West Lake population[1]; H
Percina copelandi	Channel Darter	Threatened	H
Acipenser fulvescens	Lake Sturgeon	Threatened	F, H
Anguilla rostrata	American Eel	Special concern	F, H
Notropis bifrenatus	Bridle Shiner	Special concern	H
Esox americanus vermiculatus	Grass Pickerel	Special concern	H
Myoxocephalus thompsonii	Deepwater Sculpin	Special concern	H
Moxostoma carinatum	River Redhorse	Special concern	H

[1]first recorded in 2009 by electrofishing (C. Brousseau, pers. comm.)

Conservation objectives

An important component of the assessment and advisory framework for ecosystem-based advice on integrated management of human activities is the development of explicit criteria-based Conservation Objectives (COs) (DFO, 2008). Under the OAP, COs are identified by workshop consensus and peer-review to address the four types of conservation priorities (EBSA, ESS, depleted or rare species, and degraded areas). Each priority, if it applies to a specific LOMA, has separate COs collectively prioritized based on risk assessment (DFO, 2007a; DFO, 2008). For LOMAs, the COs are identified on relatively large spatial scales. However, the objectives can be scaled down depending on either the scale of the human activity being managed or the ecosystem property being conserved. COs are the main product of Ecosystem Overview and Assessment Reports (EOAR), and represent the central science input to integrated management planning (Table 1).The DPSIR framework (United Nations Environmental Program: UNEP/GRID-Arendal,

2002), which identifies Drivers (social-economic forces driving human activities); Pressures (type and intensity of perturbations; stresses on ecosystems); State (status of ecosystem structure and function); Impacts (consequence of pressures on state; e.g. loss of biodiversity); and Responses (governance to mitigate impacts), has been used as a default conceptual framework and starting point for developing COs, but this framework is not meant to be prescriptive. Using the DPSIR concept, COs for both Pressures (e.g. fishing pressure) and State (e.g. spawning biomass/population status) properties are needed to protect ecosystem structure and function. An important qualification is that "the specificity in phrasing of a Conservation Objective should be sufficient. . . to select appropriate indicator(s) and reference points" (DFO, 2008). Ecosystem indicators are discussed in the next section.

The EOAR and CO approach was developed for marine systems, but it provides a reasonable starting point for an ecosystem approach to conservation and sustainable use in freshwater ecosystems as well. As noted, COs can be scaled to an appropriate spatial scale for coastal systems. For the Bay of Quinte, listing/delisting criteria (IJC, 1991) already exist to address the degraded area conservation issue (Table 2).

Ecosystem indicators

Ecosystem indicators are tools for management of aquatic ecosystems that can be used to assess the status of the ecosystem, its sub-components and the impacts of human activities (Rice and Sutton 2007). An indicator provides the currency by which a specific aspect of ecosystem status is measured relative to ecologically based reference points or benchmarks. For evaluating and selecting indicators, the client needs are paramount. Two roles of indicators are audit function (determine status/trends relative to reference points) and control function (to support decision making and management actions). Ecosystem indicators are chosen after COs are finalized and prioritized. Four types of ecosystem indicators were identified by Rice and Sutton (2007): (1) compliance indicators (are rules being followed?); (2) limit indicators (are COs being respected for properties essential to ecosystem structure and function); (3) target indicators (are social, economic, cultural, pol-icy aspirations being met?); and (4) status and trend indicators (what is the status of the ecosystem and its component properties?). All types of indicators are relevant to integrated management planning.

For the Bay of Quinte, several delisting criteria and targets were established as part of the Remedial Action Plan (Table 2). Designation of the Bay of Quinte as a degraded area is analogous to discussion of Degraded Areas in marine ecosystems (DFO, 2008). In the Great Lakes, indicators to assess targets have been both qualitative and quantitative (Table 2), but there has been increased emphasis on establishing quantitative science-based targets for rehabilitation (Hartig et al., 1997; Johanson and McNevin, 2007). Qualitative targets for Areas of Concern may lead to uncertainty in evaluating ecosystem status if the endpoints are not well dejfined. Examples of quantitative targets for Quinte that have been instructive are the phosphorus concentration target (TP $\leq 30 \mu g l-1$) and the Index of Biotic Integrity (IBI) scores (>60). Applying the OAP terminology to Quinte targets (Table 2) would identify total phosphorus concentration and IBI as indicators of water quality and littoral fish components of ecosystem status, respectively, and the quantitative targets are the desired reference points.

Science needs

The Ecosystem Research Initiative program was implemented by DFO to provide science support for integrated management. Seven ERIs were established in Canada, mainly in marine ecosystems: Newfoundland Shelf, Gulf of Maine, Northumberland Strait, Lower St. Lawrence Estuary, Lake Ontario, Beaufort Sea Shelf, and the Strait of Georgia. Each ERI was geographically distinct with different primary stressors affecting their ecosystems, including fishing (Gulf of Maine) oil and gas exploration (Beaufort) and human activities in watershed and coastal areas (Northumberland Strait). The ERI program was to serve as a pilot for DFO's ecosystem-based approach by focusing on regional research priorities. General research themes for all ERIs included understanding ecosystem processes, understanding the impacts of climate variability, and developing tools for ecosystem-based management. Lake Ontario, the only freshwater ERI, was chosen as a pilot for a number of reasons: coastal habitat is affected by multiple stressors (urban development, water quality, invasive species, water level management), a long history of applying an ecosystem approach for management, and a long-term (> 35 years), multi-trophic data set for the Bay of Quinte (Koops et al., 2009). The Lake Ontario ERI focuses spatially on the coastal zone (Figures 1 and 2). The goal of the Lake Ontario ERI is to build on research and partnerships in Lake Ontario to (i) evaluate how the nearshore coastal areas contribute to the function of the whole lake ecosystem, (ii) assess the sensitivity of the nearshore to cumulative impacts from multiple stressors, and (iii) predict how the nearshore will respond to projected future conditions. This is being accomplished through field studies to fill information gaps about the function of the nearshore and its linkages to the offshore, evaluation of ecosystem indicators that assess cumulative impacts of multiple stressors, and modelling to synthesize data and explore scenarios to assess the sensitivity and responses of the nearshore. The Lake Ontario Ecosystem Research Initiative will lead to a quantification of the productivity of coastal areas, relative to offshore, and the services that coastal areas provide to the Lake Ontario ecosystem.

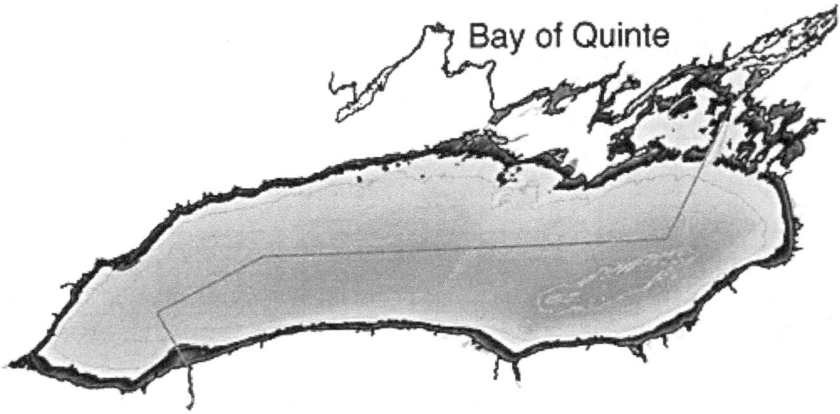

Figure 1. Lake Ontario and the Bay of Quinte. The area of the near shore zone (0–20 m) is 3025 km^2 (shaded area), and the surface area of the whole lake is about 20,000 km^2. The working hypothesis of the Lake Ontario Ecosystem Research Initiative is that community fish production is proportionately higher (per unit area) in the coastal zone than in the pelagic zone, because of the high productivity and biodiversity provided by the unique habitat (coastal wetlands and ecotone between watersheds/land and lake).

Discussion and recommendations

The science-based frameworks for integrated management in freshwater and marine ecosystems are similar in that three overarching conservation objectives (Gavaris, 2009) apply: maintaining productivity, conserving diversity and protecting habitat. A key role for science to support area based management is to develop specific and explicit conservation objectives and indicators as tools to guide integrated management.

Our comparison of this process between marine (OAP) and freshwater ecosystems (with the Bay of Quinte as a case study) confirms that the process is similar and communication between scientists from both ecosystems is mutually beneficial for advancing the science needed to support integrated management. Given the improvements in water quality and fish habitat, the Bay of Quinte is expected to be 'delisted' by the IJC as a degraded area in the foreseeable future. Despite the potential change in ecosystem status, continuation of enhanced management of human activities at Quinte should remain a priority. The development and adoption of additional conservation objectives, moving beyond the goals of restoration of a degraded area, would explicitly recognize the

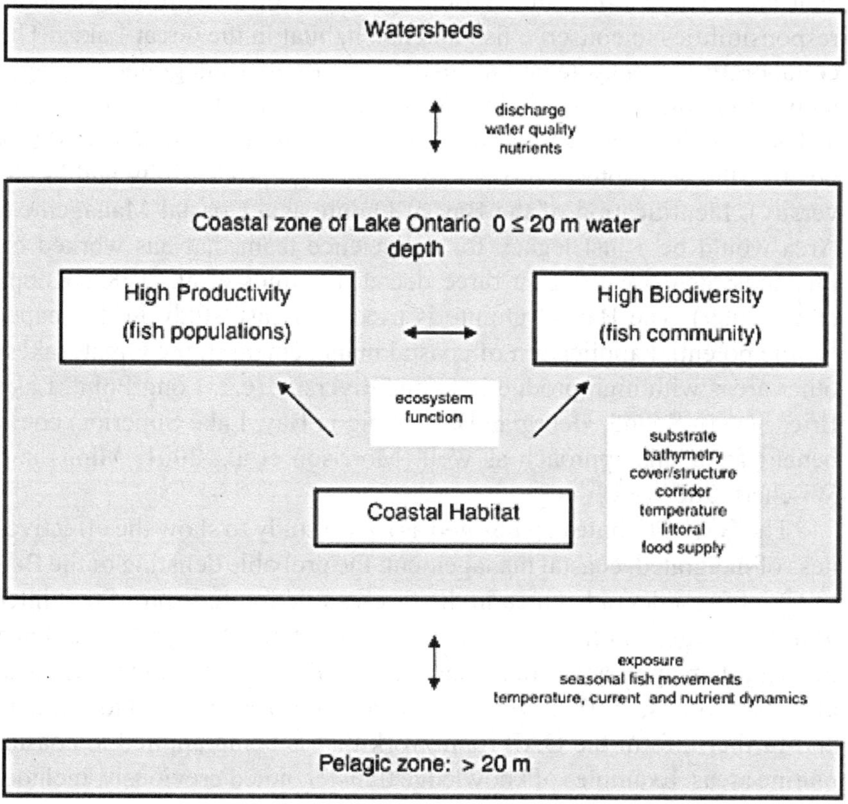

Figure 2. Conceptual schematic of the Lake Ontario Ecosystem Research Initiative showing the links between the coastal zone (0 to 20 m depth) and the watersheds and the pelagic zone. Stressors that cumulatively affect the near shore habitat are aquatic invasive species (AIS), water quality degradation, habitat loss and perturbations, fishing, water level changes, and climate change. The three generic conservation themes for the coastal zone are to

maintain productivity, conserve biodiversity and protect habitat. The long-term data set and history of ecosystem science in the Bay of Quinte will be used to anchor this coastal study.

Bay of Quinte as an ecologically significant area with unique habitat that supports high fish diversity and productivity. To continue to manage and protect this significant area, we recommend that the Bay of Quinte be identified as a Great Lakes Coastal Management Area, using science based conservation objectives to inform and justify this designation. The Oceans Act does not apply to the Great Lakes, so governance and funding sources for a coastal management area would need to be developed. Designation of the Bay of Quinte as a significant area would fulfill international, federal, provincial and regional responsibilities to conserve fish and fish habitat in the Great Lakes. The collaborative agency teams needed for continued integrated management of Quinte are already in place. Significantly, this would result in a change in the conservation focus from a negative (degraded area) to a positive theme (protection of an area with high productivity and biodiversity). Identification of the Bay of Quinte as a Coastal Management Area would be a just legacy for the science team that has worked on this ecosystem for the past three decades (Minns et al., 1986; Koops et al., 2009). The Bay of Quinte is used as a case study in this paper for the potential application of coastal management in the Great Lakes; other areas with high productivity and diversity (e.g. Long Point, Lake Erie; Severn Sound, Georgian Bay; Nipigon Bay, Lake Superior) could benefit from this approach as well (Morrison et al., 2001; Minns and Wischert, 2005).

The Bay of Quinte can be cited as a case study to show the effectiveness of integrated, coastal management: the probable delisting of the Bay of Quinte as a degraded area in the foreseeable future is proof that integrated management has been effective. Long-term (>35 yrs) experience of ecosystem approaches to management at the Bay of Quinte (Minns et al., 1986; Minns, 1995, 1999; Koops et al., 2009) can be used to evaluate and further inform the OAP framework that is being applied to coastal marine areas. Examples of knowledge transfer, noted previously, include: extrapolation of criteria for identifying degraded areas from the Great Lakes to marine systems, criteria and usefulness of identifying aggregate significant species (biomass size spectra, goals for trophic structure, including the role of top predators) (Hurley et al., 1986), impacts and management challenges associated with invasive species (Mills et al., 1993), the efficacy of quantitative versus qualitative ecosystem indicators

(Hartig et al., 1997), and the challenge of identifying and adjusting science-based indicators for ecosystems undergoing constant change. Hypotheses and methods that are difficult to investigate or test in open, large marine systems may be tractable in the Great Lakes or elsewhere in freshwater ecosystems.

Summary and conclusions

Canadian approaches for ecosystem-based management in coastal marine areas, prompted by Canada's Oceans Act, are similar to and consistent with the long-standing, integrated management of the Bay of Quinte in Lake Ontario. The similarities include the criteria for the identification of ecologically significant areas and habitat, degraded areas, ecologically significant species, depleted or rare species, conservation objectives and ecological indicators to assess the status of ecosystem components. The rationale for integrated management in the Bay of Quinte is compared to marine areas, using the terminology that has been adopted for ocean area-based management in Canada. The three overarching goals of maintaining productivity, preserving biodiversity and protecting habitat are common to both freshwater and marine ecosystems. Communication of science priorities is mutually beneficial for advancing integrated management in both ecosystem types. Identifying the Bay of Quinte as a coastal management area by Great Lakes agencies, using elements of the oceans framework, would demonstrate a transition from a negative (degraded area) to a positive ecosystem status (essential habitat with high productivity and biodiversity).

From a science perspective, knowledge transfer between freshwater and marine ecosystems is paramount. Area-based management, the identification of designated areas that require enhanced protection, will increase in the future in both marine (Rodwell et al., 2003) and freshwater ecosystems (Morrison et al., 2001). Demands for science to address knowledge gaps will also increase. Specific and novel research is needed to: develop methods and tools, including GIS, for measuring and integrating cumulative effects across managed human activities that impact fish and fish habitat; spatial and individual-based models linking habitat to populations and communities, including mesohabitat-dependent fish movements (Hayes et al., 2009); functional links between habitat and populations; habitat area-dependent survival (MacKenzie et al., 2003); science-based methods for defining geographic boundaries for coastal management areas; and methods for tracking the import/export of nutrients and biota across these boundaries. The hypothesis that coastal zones contribute disproportionately to total fish production needs to be addressed in both freshwater and marine areas, as the general principles of ecosystem structure and function in the spatial context of the land/water interface apply to both ecosystems.

Acknowledgements

The Lake Ontario Ecosystem Research Initiative, funded by Fisheries and Oceans Canada, motivated this comparison of science-based conservation objectives for integrated management in marine and freshwater ecosystems. This paper benefited from the comments of two anonymous reviewers.

References

BQRAP (Bay of Quinte Remedial Action Plan), 1993. Time to act. Stage 2 Report, Ontario Canada. Prepared by the BQRAP Coordinating Committee, Trenton, ON. Brousseau, C. M., Randall, R. G., Hoyle, J. A., Minns, C. K., 2011. Fish community indices of ecosystem health: How does the Bay of Quinte compare to other coastal sites in Lake Ontario? Aquatic Ecosystem Health and Management 14(1),75–84. Casselman, J. M., Scott, K. A., Brown, D. M., Robinson, C.J., 1999. Changes in the relative abundance, variability and stability of fish assemblages of eastern Lake Ontario and the Bay of Quinte – the value of long-term community sampling. Aquat. Ecosys. Health and Manage. 2(3), 255–269.

Chu, C. Minns, C. K., Moore, J. E., Millard, E. S., 2004. Impact of oligotrophication, temperature, and water levels on walleye habitat in the Bay of Quinte, Lake Ontario. Trans. Amer. Fish. Soc. 133, 868–879.

COSEWIC, 2006a. COSEWIC assessment and status report on the Atlantic salmon *Salmo salar* (Lake Ontario population) in Canada. Committee on the Status of Endangered Wildlife in Canada. Ottawa.

COSEWIC, 2006b. COSEWIC assessment and update status report on the lake sturgeon *Acipenser fulvescens* in Canada. Committee on the status of Endangered Wildlife in Canada. Ottawa, ON.

Crowder, A. A., Bristow, J. M., 1986. Aquatic macrophytes in the Bay of Quinte, 1972–82. p. 114–127. In: C.K. Minns, D.A. Hurley, and K.H. Nichols (Ed.), Project Quinte: point-sources phosphorus control and ecosystem response in the Bay of Quinte, Lake Ontario. Can. Spec. Publ. Fish. Aquat. Sci. 86.

DFO, 2004. Identification of Ecological and Biological Significant Areas. DFO Can. Sci. Advis. Sec. Ecosystem Status Rep. 2004/006.DFO, 2005. Canada's Oceans Action Plan For Present and Future Generations. Communications Branch, Ottawa.

DFO, 2006a. Identification of ecologically significant species and community properties. DFO Can. Sci. Advis. Sec. Sci. Advis. Rep. 2006/041.

DFO, 2006b. A harvest strategy compliant with the Precautionary Approach. DFO Can. Sci. Advis. Sec. Sci. Advis. Rep.2006/023.

DFO, 2007a. Guidance document on identifying conservation priorities and phrasing conservation objectives for Large Ocean Management areas. Can. Sci. Advis. Sec. Sci. Advis. Rep. 2007/010.

DFO, 2007b. Beaufort Sea Large Ocean Management Area. Ecosystem Overview and Assessment Report. Winnipeg, Manitoba.

DFO, 2008. Further guidance on the formulation, prioritization and use of conservation objectives in an ecosystem approach to integrated management of human activities in aquatic ecosystems. DFO Can. Sci. Advis. Sec. Sci. Advis. Rep.2008/029.

Dietrich, J. P., Morrison, B. J., Hoyle, J. A., 2006. Alternative ecological pathways in the eastern Lake Ontario food web—round goby in the diet of lake trout. J. Great Lakes Res. 32,395–400.

Gavaris, S., 2009. Fisheries management planning and support for strategic and tactical decisions in an ecosystem approach context. Fisheries Research 100, 6–14.

German, M., Stride, F., 2003. Bay of Quinte RAP Monitoring and Delisting Strategy. IBU Assessment Statements 2003. Bay of Quinte RAP Restoration Council, Trenton, ON.

Hart, J. L., 1930. The spawning and early life history of the whitefish *Coregonus clupeaformis* (Mitchell) in the Bay of Quinte, Ontario. Contri. Can. Biol. Fish. VI(20), 427–444.

Hartig, J. H., Zarull, M. A., Reynoldson, T. B., Mikol, G., Harris, V. A., Randall, R. G., Cairns, V. W., 1997. Quantifying targets for rehabilitating degraded areas of the Great Lakes. Envirn. Manage. 21, 713–723.

Hurley, D. A., Christie, W. J., Minns, C. K., Millard, E. S., Cooley, J. M., Johnson, M. G., Nicholls, K. H., Robinson, G. W., Owen, G. E., Sly, P. G., Geiling, W. T., Crowder, A. A., 1986.Trophic structure and interactions in the Bay of Quinte, Lake Ontario, before and after point-source phosphorus control,. In: C.K. Minns, D.A. Hurley, and K.H. Nicholls (Ed.), Project Quinte: point-source phosphorus control and ecosystem response in the Bay of Quinte, Lake Ontario, p. 259–270. Can. Spec. Publ. Fish. Aquat. Sci. 86.

International Joint Commission (IJC), 1987. Protocol for the Great Lakes Water Quality Agreement. Annex2: Great Lakes Areas of Concern and Remedial Action Plans. International Joint Commission, Windsor, ON (www.ijc.org/en/activities/consultations/glwqa/guide33.htm)

IJC, 1991. Commission approves list/delist criteria for Great Lakes Areas of Concern. www.ijc.org/rel/focus/listdelist/

Johanson, P., McNevin. B., 2007. A fish habitat management plan for the Bay of Quinte. Bay of Quinte Remedial Action Plan, Trenton, ON.

Kipp, R. M., Ricciardi, A., 2010. *Hemimysis anomala.* USGS Nonindigenous Aquatic Species Database, Gainesville, FL.(http://nas.er.usgs.gov/queries/GreatLakes/SpeciesInfo.asp)

Koops, M. A., Dermott, R. M., Leisti, K. E., Johannsson, O. E., Millard, E. S., Minns, C. K., Munawar, M., Nicholls, K. H., Hoyle, J. A., 2009. The Bay of Quinte: a model for large lake ecosystem management. Verh. Internat. Verein. Limnol. 30,1024–1029.

MacKenzie, B. R., Myers, R. A., Bowen, K. G., 2003. Spawner-recruit relationships and fish stock carrying capacity in aquatic ecosystems. Mar. Ecol. Prog. Ser. 248, 209–220.

Mills, E. L., Leach, J. H., Carlton, J. T., Secor, C. L., 1993. Exotic species in the Great Lakes: a history of biotic crises and anthropogenic introductions. J. Great Lakes Res. 19, 1–54.

Minns, C. K., 1995. Approaches to assessing and managing cumulative ecosystem change, with the Bay of Quinte as a case study. J. Aquat. Ecosystem Health 4, 1–24. Minns, C. K., 1999. The ecosystem pyramid and the means for attaining ecological sustainability: an essay in honour of Jack Christie. Aquatic Ecosystem Health and Management 2(2),209–221.

Minns, C. K., Hurley, D. A., 1986. Population dynamics and production of white perch, *Morone americana,* in the Bay of Quinte, Lake Ontario. In: C.K. Minns, D.A. Hurley, K.H. Nicholls (Ed.), Project Quinte: point-source phosphorus control and ecosystem response in the Bay of Quinte, Lake Ontario, p.215–223. Can. Spec. Publ. Fish. Aquat. Sci. 86.

Minns, C. K., Wischert, G. A., 2005. A framework for defining fish habitat domains in Lake Ontario and its drainage. J. Great Lakes Res. 31(Suppl. 1), 6–27.

Minns, C. K., Hurley, D. A., Nicholls, K. H., 1986. Project Quinte: point-source phosphorus control and ecosystem response in the Bay of Quinte, Lake Ontario. Can. Spec. Publ. Fish. Aquat. Sci. 86.

Minns, C. K., Cairns, V. W., Randall, R. G., Moore, J. E., 1994. An Index of Biotic Integrity (IBI) for fish assemblages in the littoral zone of Great Lakes' Areas of Concern. Can. J. Fish. Aquat. Sci. 51, 1804–1822.

Morrison, H., Minns, C. K., Koonce. J. F., 2001. A methodology or identifying aquatic biodiversity investment areas: application in the Great Lakes basin. Aquatic Ecosystem Health and Management 4(1), 1–12.

Rice, J., Sutton, V., 2007. National Science Workshop: Ecosystem Indicators. Can. Sci. Advis. Sec. Proc. Ser. 2007/045.

Scott, W. B., Christie, W. J., 1963. The invasion of the lower Great Lakes by the white perch, *Roccus americanus* (Gmelin). J. Fish. Res. Board Can. 20, 1189–1195.

Scott, W. B., Crossman, E. J., 1998. Freshwater Fishes of Canada. Galt House Publications, Oakville, ON.

Sly, P., 1986. Review of postglacial environmental changes and cultural impacts in the Bay of Quinte. In: C.K. Minns, D.A. Hurley, and K.H. Nichols (Ed.), Project Quinte: point-sources phosphorus control and ecosystem response in the Bay of Quinte, Lake Ontario, pp. 7–26. Can. Spec. Publ. Fish. Aquat. Sci. 86.

UNEP/GRID-Arendal. 2002. The DPSIR Framework. UNEP/GRID Maps and Graphics library (http://maps.grida.no/go/graphic/thedpsirframework)

The above article is a reprinted version of the following previously published article:

Randall, R.G.; Koops; M.A., C. K. Minns, C.K., 2011.
A comparison of approaches for integrated management in coastal marine areas of Canada with the historical approach used in the Great Lakes (Bay of Quinte)
Aquatic Ecosystem Health & Management (2011) 14 (1): 104–113.
https://doi.org/10.1080/14634988.2011.549440

Assessing ecosystem health impairments using a battery of ecological indicators: Bay of Quinte, Lake Ontario example

M. Munawar[1] M. Fitzpatrick,[1] I. F. Munawar,[2] H. Niblock,[1] and D. Kane[3,4]

[1]Fisheries and Oceans Canada, 867 Lakeshore Road, Burlington, ON L7S 1A1,
[2]Plankton Canada, Burlington, ON
[3]Defiance College, Defiance, OH.
[4]Present address: Heidelberg University, Tiffin, OH
**Corresponding author: mohiuddin.munawar@dfo-mpo.gc.ca*

Keywords: Beneficial Use Impairment, Area of Concern, Great Lakes, eutrophication, remediation, stressed ecosystems

Introduction

The North American Great Lakes are an immense global aquatic resource, covering 245,000 km^2 and accounting for an estimated 20% of the global supply of freshwater (Beeton, 1984). The lakes are shared by Canada and the United States, are home to over 30 million people in both countries and have been estimated to contribute more than $2 trillion U.S. to the regional and global economy (Environment Canada and the United States Environmental Protection Agency, 1995). The Great Lakes have also been subject to multiple anthropogenic stressors including chemical pollution, eutrophication, exotic species, overfishing, loss of biodiversity and climate change (Vollenweider et al., 1974; Mills et al., 1993; Kerr and Ryder, 1997; Trumpickas et al., 2009). Managing these multiple anthropogenic stressors in the Great Lakes has required international cooperation in addition to a shared commitment to maintain the ecosystemic integrity of this enormous global resource.

Concern over eutrophication in the Great Lakes became widespread in the 1950s and 1960s with the problem being most evident in the densely populated coastal zones of Lakes Erie and Ontario, as well as the southern shores of Lake Michigan. External phosphorus loadings, primarily from detergents and agricultural wastes, were identified as the probable cause of widespread algal blooms associated with eutrophication (Vollenweider, 1968; Vallentyne, 1974; Schindler, 1978). The governments of Canada and the United States signed the Great Lakes

Ecology of the Bay of Quinte: Health, Management and Global Implications
Editors: C.K. Minns, M. Munawar and M.A. Koops

Aquatic Ecosystem Health & Management Society, Canada

Water Quality Agreement (GLWQA) in 1972 which committed both governments "to restore and maintain the chemical, physical and biological integrity of the Great Lakes basin ecosystem" (Article II: IJC, 1988) using science based management.

The relationship between phosphorus loadings, primary production and algal standing crop (chlorophyll-*a*) was modelled by Vollenweider et al. (1974) and served as the basis for establishing targets for phosphorous load reductions into the Great Lakes under the auspices of the GLWQA. The governments of Canada and the United States agreed to meet these targets beginning in 1978 by banning phosphates in soaps and detergents and by improving municipal wastewater treatment (IJC, 1988). Success in alleviating eutrophic conditions in the lower Great Lakes, especially Lake Erie, demonstrated that the ecosystem approach to management enshrined in the GLWQA–that put sound science ahead of national interests–was a viable model for the rehabilitation of shared aquatic resources (Burns, 1985; Vallentyne, 1993).

Phosphorous loadings were not the only issue affecting the health of the Great Lakes. Severe pollution and degradation of various coastal regions, embayments, harbours, river mouths and tributaries of the Great Lakes led to the designation of 42 Areas of Concern (AOC) (IJC, 1989). AOC were selected based on a criterion of having at least 1 of 14 Beneficial Use Impairments (BUIs), although multiple impairments are the norm (Hartig and Zarull, 1992). BUIs included: "degraded fish and wildlife populations," "fish tumours or other deformities," "degradation of benthos," "eutrophication or undesirable algae" and "degradation of phytoplankton and zooplankton populations," to name a few. Although symptomatic in nature, the established list of BUIs was representative of environmental degradation at all trophic levels in keeping with the ecosystem approach adopted by the GLWQA.

Areas of Concern were intended to serve as focal points for remediation efforts, but some controversy has arisen because objective, quantitative and scientifically defensible parameters for beneficial use impairments were never fully defined (Krantzberg, 2004; George and Boyd, 2007). Researchers and managers are faced with the question of what constitutes an impaired beneficial use and at what point is it no longer impaired? These are known as "listing" and "delisting" criteria. There is an obvious need for researchers to develop and deploy relevant ecological indicators in order for managers to objectively define impairments of ecosystem health and to measure potential recovery.

The Bay of Quinte is one such AOC located on the northeastern shore of Lake Ontario, with 10 Beneficial Use Impairments including: "Eutrophication or undesirable algae" and "degradation of phytoplankton and zooplankton communities" (BQ RAP, 1993). Fisheries and Oceans Canada in partnership with the Ontario Ministry of Natural Resources, Ontario Ministry of the Environment and Environment Canada began monitoring the bay in 1972 as part of the initial efforts to control eutrophication in the Great Lakes which included regular measurements of nutrients, chlorophyll-*a*, phytoplankton and zooplankton. Because of this effort, an enormous, long term (almost 40 years) data set was developed which in itself provides unique opportunities for addressing ecological questions in a robust manner.

The current article employs a "battery of tests" strategy to assess the status of beneficial use impairments (BUIs) of "eutrophication or undesirable algae" and "degradation of phytoplankton and zooplankton communities" in the Bay of Quinte. This strategy integrates structural (chlorophyll-*a*, phytoplankton biomass and taxonomy, zooplankton biomass and taxonomy), functional (primary production) and chemical (phosphorous loadings) parameters and applies them to established ecosystem health models and scales which include:

- The trophic ladder (phytoplankton biomass) of Munawar and Munawar (1982)
- A qualitative assessment of phytoplankton composition (Hutchinson, 1967)
- Vollenweider's eutrophication models (Vollenweider et al., 1974)
- The Planktonic Index of Biotic Integrity (P-IBI) (Kane et al., 2009)

Collectively, these metrics incorporate a large body of the long term monitoring data from Project Quinte (2000–2008). Our goal is to assess the current state of BUIs in the Bay of Quinte and offer scientific guidance towards the development of delisting criteria that are founded upon sound ecosystem-based science.

Materials and Methods

Study area and sampling program

The Bay of Quinte is a large Z – shaped embayment (254 km^2) that is also relatively shallow (4–8 m), but is much deeper at the interface with Lake Ontario. One shallow station in the upper bay (Belleville–5m) was sampled bi-weekly from May 1st to October 31st in each year (Figure 1) from 2000 to 2008. Integrated epilimnetic water was collected and subsamples were drawn for primary production, chlorophyll-*a*, and phytoplankton. Zooplankton was sampled from discrete depths.

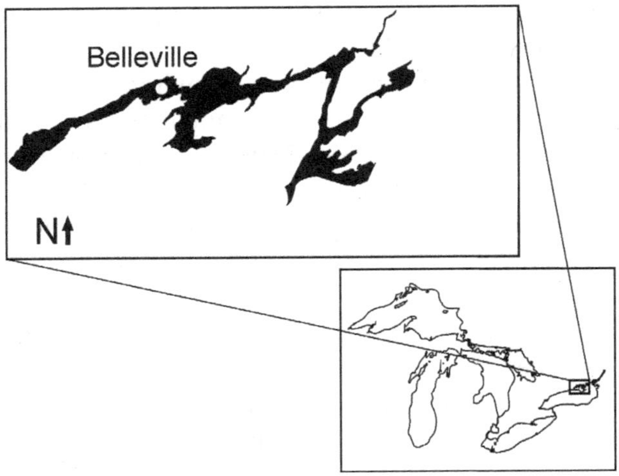

Figure 1. Map of the Bay of Quinte, Lake Ontario including sampling location.

Chemical, structural and functional measurements

Point-source phosphorous loadings were compiled by P. Kintsler, Ontario Ministry of the Environment, Kingston, Ontario (pers. comm.). Chlorophyll-*a* concentrations were determined by filtering up to 1 litre of water through Whatman GF/C filters followed by cold acetone pigment extraction and spectrophotometric analysis (Strickland and Parsons, 1968).

Phytoplankton samples were fixed immediately with Lugol's iodine. Identification and enumeration followed the Utermöhl (1958) inverted microscope technique as described in Nicholls et al. (2002) which are broadly compatible with the established techniques of Munawar et al. (1987). All species names were verified with either the Integrated Taxonomic Information Service (ITIS) (http://www.itis.gov) or algae BASE (Guiry and Guiry, 2012) prior to publication. The listing on ITIS was given preference.

The zooplankton data used in this article were obtained from Bowen and Johannsson (2011). Samples were collected using a 41 litre Schindler–Patalas trap fitted with a 64 μm mesh. Three depths were sampled at Belleville (1 m, 2 m and 3 m) and preserved immediately with 4% sugared formalin. A single composite sample was constructed for each station-date by combining 50% of the sample from each depth. Biomass of each taxon was determined in each sample from length–weight regressions.

Annual primary production was estimated by ^{14}carbon uptake. Samples from each cruise were spiked with $Na^{14}CO_3$ and incubated for 2–4 h at irradiance levels ranging from 25–1300 µmol m^{-2} s^{-1}. Following incubation, acidification and bubbling was used to remove excess ^{14}C and radioactivity was determined by liquid scintillation counting (Millard et al., 1999). Seasonal areal primary production (SAPP) was then estimated according to the model of Fee (1990). The original estimates of SAPP were provided by E.S. Millard and M. Frank (M. Burley) of Fisheries and Oceans Canada, Burlington, Ontario (pers. comm.). These estimates were increased by 10% in order to account for winter primary production (Vollenweider et al., 1974; Fitzpatrick et al., 2007).

Ecosystem health indices

The trophic ladder (Munawar and Munawar, 1982) defines trophic state as a function of phytoplankton biomass along a linear scale, where the lowest levels (<1 g m^{-2}) being indicative of ultra-oligotrophic conditions and the highest levels (>5 g m^{-2}) indicative of hyper eutrophic conditions. Phytoplankton species composition is also used to make assumptions about trophic state based on well established patterns in community structure (Hutchinson, 1967).

Vollenwieder's eutrophication models (Vollenweider et al., 1974)

$A = 420[10\,P^{0.6}(9 + 10\,P^{0.6})^{-1}]$,

where

A = Annual Primary Production (g C m^{-2} y^{-1})

P = Annual Phosphorus Loadings (g m^{-2} y^{-1})

$A = 420[1.15C^{1.33}(9 + 1.15C^{1.33})^{-1}]$,

where

A = Annual Primary Production (g C m^{-2} y^{-1})

C = Annual Mean Chlorophyll-*a* (µg l^{-1})

Planktonic index of biotic integrity (P-IBI) (Kane et al., 2009)

$$\text{P-IBI} = \frac{1}{B}\sum_{k=1}^{B}\frac{1}{S}\sum_{j=1}^{S}\frac{1}{M}(EA_{jk} + CB_{jk} + RJ_{jk} + LM_{jk} + RA_{jk} + ZB_{jk})$$

where:

EAjk = June biomass of edible algae taxa metric score,

CBjk = June % *Microcystis*, *Anabaena*, and *Aphanizomenon* of total phytoplankton biomass metric score,

RJjk = June zooplankton ratio (Calanoida/ (Cladocera + Cyclo poida)) metric score,

LMjk = July *Limnocalanus macrurus* density metric score, *RAjk* = August zooplankton ratio (Calanoida/ (Cladocera + Cyclo poida)) metric score,

ZBjk = August crustacean zooplankton biomass metric score, *M* = number of metrics,

S = number of sites (within a basin), and

B = number of basins.

Results and discussion

The earliest efforts at remediation in the Bay of Quinte included reductions in Point-source phosphorus loads from >200 kg d^{-1} in 1972 to <70 kg d^{-1} in 1978 (Minns et al., 1986) with reductions continuing to the present. Following the Bay of Quinte's designation as an Area of Concern, target levels were established for total phosphorus concentrations (30 μg l^{-1}), phytoplankton biomass (4–5 g m^{-3}) and chlorophyll-*a* (12–15 μg l^{-1}) (BQ RAP, 1993). Current research suggests that while these targets have been achieved, the bay is still eutrophic and the remediation targets are not robust enough to affect a change in trophic state (Munawar et al., 2011).

In previous work, we recommended that a "battery of tests" strategy be adopted to assess Beneficial Use Impairments using a variety of structural and functional measurements (Munawar et al., 2011). The battery of tests can be applied to test relevant impairments with multiple indicators being available to add confidence to the results. These tests range from simple structural indicators which would include estimation of phytoplankton biomass and its species composition based on microscopic analysis and functional/experimental assessments of primary productivity. Also included in the battery are integrated mathematical models that combine multiple structural, functional and chemical parameters. We chose a subset of the complete battery for consideration in this article.

Trophic ladder/phytoplankton biomass

The trophic ladder approach of Munawar and Munawar (1982) uses the structural measurement of total phytoplankton biomass to define a trophic gradient. Ultra-oligotrophic conditions at one extreme are defined as occurring where phytoplankton biomass is below 1 g m^{-3} and at the other extreme hyper-eutrophic conditions are defined when biomass exceeds 5 g m^{-3}. According to this scale, the Bay of Quinte could be classified as either eutrophic (>3 g m^{-3} of phytoplankton) or hypereutrophic (>5 g m^{-3}) from 2001–2008, with only a single year (2000) observed where the biomass was low enough to result in mesotrophic conditions (Figure 2). Phytoplankton biomass has also been used as a Remedial Action Plan target and we have previously noted that based on the trophic ladder approach, achieving the interim RAP target of 4–5 g m^{-3} will still result in a eutrophic embayment (Munawar et al., 2011). While not presented here, long term data shows that annual mean phytoplankton biomass never fell below the threshold of 3 g m^{-3} from 1972–1999 (Nicholls and Carney, 2011) and indicates that eutrophic and hyper eutrophic conditions were the norm both before and after the imposition of phosphorus controls.

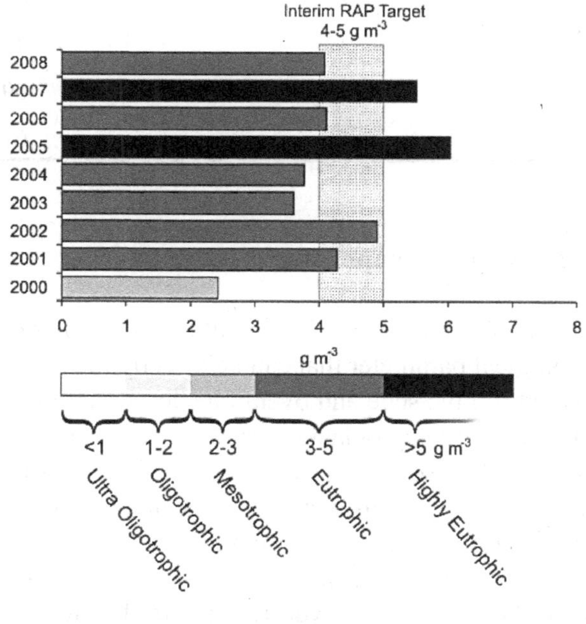

Figure 2. Phytoplankton biomass index (trophic ladder) of Munawar and Munawar (1982) applied to the Bay of Quinte, Lake Ontario.

Actually, our results are rather understated, since phytoplankton biomass was calculated as the weighted mean of 13 observations from May to October, whereas the original scale was based on single observations, so this model could be applied to each individual sampling date. For example, in 2000 when mean biomass was observed to be at its lowest historical level, eutrophy was observed on 3 (of 13) sampling dates (Table 1). At the other extreme, in 2007 eutrophy was observed on 9 sampling dates. Phytoplankton biomass suggests a consistent long term pattern of eutrophication in the upper Bay of Quinte.

Table 1. Frequency of eutrophic conditions (phytoplankton biomass >3 g m^{-3}) in the Bay of Quinte.

Year	# of observations	% of observations
2000	3	23.1
2001	7	53.8
2002	7	53.8
2003	7	53.8
2004	5	38.5
2005	8	61.5
2006	6	46.2
2007	9	69.2
2008	8	61.5

Phytoplankton species composition

The species composition of the phytoplankton is another example of a structural ecological parameter that can be directly measured and is also instructive about trophic state and overall health. It is impossible to give a thorough treatment to phytoplankton species composition for all of our observations (n = 104), so we have selected two examples from 2006 for intensive analysis. As previous work has indicated, 2006 was the second consecutive year of recurring late summer algal blooms (Munawar et al., 2011; Nicholls and Carney, 2011) and we selected two dates where eutrophic conditions were observed (i.e. phytoplankton biomass >3 g m^{-3}). We chose 26 July, the first occurrence of eutrophic conditions that year and 19 September, the last sampling date of the summer.

The species composition and relative abundance are shown in Table 2. The July bloom was dominated by species of Cyanophyta, namely *Anabaena spiroides*, *A. circinalis*, *A. planktonica* and *Micros cystis aeruginosa* which are all strongly associated with eutrophic conditions (Munawar and Munawar, 1996; Hutchinson, 1967). Centric diatoms (e.g. *Cyclotella* sp.) and phytoflagellates (*Cryptomonas caudata, C. erosa, C. ovata* and *Rhodomonas minuta*) were also abundant. The late summer bloom was dominated by blue-greens including *Lyngbya limnetica*, *Anabaena spiroides*, *A. circinalis* and *Microscystis aeruginosa*. A similar assortment of centric diatoms and phytoflagellates to the July sample was also found in September. The presence of nitrogen– fixing species of *Anabaena* for example is characteristic of nutrient enriched habitats (Hutchinson, 1967; Munawar and Munawar, 1982; Stockner and Shortreed, 1988). On the other hand, *Lyngbya limnetica* (observed here in late summer) tend to out-compete other groups of phytoplankton for nitrogen when there is ample phosphorus and there fore thrive in turbid, light limited environments (Reynolds, 1984; Scheffer et al., 1997; Salmaso, 2000). The composition of the phytoplankton community provides hard evidence that the observed blooms are the consequence of external phosphorus loadings and that cultural eutrophication is a continuing threat to the Bay of Quinte ecosystem.

The presence of algal toxins, including microcystins and anatoxins, has become an increasingly important issue in the Great Lakes and other watersheds and is closely related to eutrophication. In Areas of Concern, however, the specific beneficial use impairment is often considered to be taste and odour (Watson et al., 2008), although Microcystin levels have been used as an indicator of phytoplankton health in other AOC (Irvine and Murphy, 2009; Gillett and Steinman, 2011). Many of the dominant species observed during both blooms are known to produce Microcystin and other algal toxins. For example, *Anabaena spiroides*, *A. planktonica*, *Micocystis aeroginosa* and *A. flos-aquae* were the major toxin producers observed in July and *A. spiroides*, *Lyngbya diguettii*, *A. planktonica* and *Aphanizomenon flos-aquae* were found in September. While the occurrence of these species does not imply the production of algal toxins, it does present an obvious risk. More work is needed to link phytoplankton taxonomy with toxin production and other physiological properties including photosynthesis and respiration rates in order to better understand the hazard posed by algal blooms.

Table 2. List of Phytoplankton species observed at Belleville during eutrophic conditions on 25 July 2006 and on 19 September 2006. Species are ranked from highest to lowest in terms of relative abundance.

25 July 2006		19 September 2006	
Species	Rank	Species	Rank
Cyanophyta		**Cyanophyta**	
Anabaena spiroides Klebahn	xxxxxxx	*Lyngbya limnetica* Lemmermann	xxxxxxxx
A. circinalis (Kütz.) Rabenhorst	xxxxxx	*Anabaena spiroides*	xxxx
A. planctonica Brunnthaler	xxxxx	*A. circinalis*	xxxx
Microcystis aeruginosa (Kütz.) Lemmerman	xxxxx	*Microcystis aeruginosa*	xxxx
Diatomeae		*L. diguetii*	xxx
Cyclotella sp.	xxx	*A. planctonica*	xxx
Cryptophyceae		*Aphanizomenon flos-aquae* (Ralphs) Ex	xxx
Cryptomonas caudata Schiller	xxx	Bornet and Flah.	
C. erosa Ehrenberg	xxx	*Microcystis pulverea incerta* (Lemm.)	xxx
Cryptomonas sp.	xxx	Crow	
Rhodomonas minuta Skuja	xxx	*Gomphosphaeria lacustris* Chodat	xxx
Cyanophyta		**Diatomeae**	
Anabaena flos-aquae (Lyngb.) De Brebisson	xx	*Cyclotella sp.*	xxx
Anabaena sp.	xx	**Cryptophyceae**	
Chlorophyta		*Cryptomonas ovata*	xxx
Chroococcus limneticus Lemmermann	xx	*C. caudata*	xxx
Gomphosphaeria lacustris Chodat	xx	*Rhodomonas minuta*	xxx

Table 2. Continued...

25 July 2006		19 September 2006	
Species	Rank	Species	Rank
Dictyosphaerium pulchellum Wood	xx	*Raphidiopsis sp.*	xx
Scenedesmus quadricauda (Turp.) de Brebisson	xx	**Chlorophyta**	
		Coelastrum sphaericum Naegeli	xx
Sphaerocystis schroeteri Chodat	xx	*Gloeocystis gigas*	xx
Diatomeae		*Crucigenia fenestrata* (Schm.)	xx
Fragilaria crotonensis Kitton	xx	*Scenedesmus quadricauda*	xx
Aulacoseira granulata (Ehr.) Ralfs (syn. *Melosira granulata*)	xx	*Oocystis lacustris* Chodat	xx
		Dictyosphaerium pulchellum	xx
A. islandica (O. Mu¨ ll) Somonsen (syn. *M. islandica*)	xx	*Chlamydomonas globosa*	xx
		Gemellicystis neglecta Teiling	xx
Cryptophyceae		**Chrysophyceae**	
Cryptomonas ovata Ehrenberg	xx	*Mallomonas spp.*	xx
Dinophyceae		**Diatomeae**	
Peridinium aciculiferum Lemmermann	xx	*Aulacoseira islandica*	xx
P. Cunningtonii (Lemm.) Lemmermann	xx	**Cryptophyceae**	
Cyanophyta		*Cryptomonas erosa*	xx
Aphanocapsa elachista West and West	x	*C. tetrapyrenoidifera*	xx

Table 2. Continued...

25 July 2006		19 September 2006	
Species	Rank	Species	Rank
Oscillatoria limnetica Lemmermann	x	*C. tenuis* Pascher	xx
Chlorophyta		*Katablepharis ovalis* Skuja	xx
Chlamydomonas globosa Snow	x	**Cyanophyta**	
Closterium parvulum Naeg.	x	*Anabaena flos-aquae* (Lyng.) Breb.	x
Crucigenia quadrata Morren	x	*A. wisconsinense* Prescott	x
Gloeocystis gigas (Ku¨ tz.) Lagerheim	x	*A. affinis* Lemmermann	x
Pediastrum boryanum (Turp.) Meneghini	x	*Oscillatoria limnetica* Lemmermann	x
P. duplex Meyen	x	**Chlorophyta**	
P. simplex (Meyen) Lemm.	x	*Closterium parvulum*	x
Staurastrum paradoxum Meyer	x	*Pediastrum duplex*	x
		Cryptophyceae	
		Cryptomonas Marssonii Skuja	x
		Dinophyceae	
		Peridinium aciculiferum	x
		P. Cunningtonii	x
		Gymnodinium varians Maskell	x

Vollenweider's eutrophication models

Phosphorus loadings were identified early on as the probable cause of eutrophication in the Bay of Quinte (Johnson and Owen, 1971) and a phosphorus abatement program was implemented beginning in 1978. Currently, Point-source phosphorous loads are in the range of 9–22 kg d^{-1} (equivalent to 0.01 – 0.03 g m^{-2}; Table 3) and generally fall below the recommended cap of 15 kg d^{-1} (Minns and Moore, 2004). Vollen weider's eutrophication models (Vollenweider et al., 1974) were developed to predict the relationship between phosphorus loadings, primary production and algal standing crop and provide a scale for classifying lakes and embayments based on any of the 3 parameters. With these models, managers could calculate the magnitude of reductions in external phosphorus loads needed to alleviate eutrophic conditions. It is worth noting that Vollenweider's models provided the scientific basis for the phosphorus abatement provisions of the Great Lakes Water Quality Agreement (Bruce, 2011). Although the models were intended to be applied on a lakewide basis, recent work has demonstrated that they can also be applied to smaller embayments, providing a valuable tool for assessing AOC (Munawar and Fitzpatrick, 2011).

The question remains as to whether or not the models are suitable for a shallow, largely riverine embayment, such as the Bay of Quinte, with relatively high flushing rates and long phosphorous retention times. The models were developed from surveys of over 150 temperate water sheds which varied greatly in their physical, biological and chemical characteristics and included riverine systems (Vollenweider, 1968; Janus and Vollenweider, 1981). Importantly, the models include critical P loading estimates for watersheds as shallow as 5 m (mean depth), as well as much deeper lakes (Vollenweider, 1968). P retention in the Bay of Quinte (0.55 g m^{-2} y^{-1}; Johnson and Owen, 1971) was in the middle of the range (0.11–1.08 g m^{-2} y^{-1}) of the lakes surveyed by Vollenweider.
Similarly, flushing rates in the Bay of Quinte (2.6–26.8 y^{-1}; Minns et al., 2004) were well within the range of 0.26–62.7 y^{-1} observed in various southern Ontario watersheds to which the models were applied (Vollenweider and Dillon, 1974).

As a practical matter, high flushing rates have the effect of diluting the total phosphorous (TP) concentration and thereby reducing primary production and algal stranding crop, whereas high P retention would have the effect of increasing the TP concentration and result in increased primary production and a larger standing crop (Vollenweider and Dillon,

Table 3. Point-source phosphorous loadings, annual primary production and annual mean chlorophyll *a* observed in the Bay of Quinte and predicted levels of primary production and chlorophyll *a* from Vollenweider's models at observed phosphorus loadings.

	Phosphorous loadings		Observed		Predicted	
			Annual primary production	Mean chlorophyll *a*	Annual primary production	Mean chlorophyll *a*
	$g\ m^{-2}$	$kg\ d^{-1}$	$g\ C\ m^{-2}$	$\mu g\ l^{-1}$	$g\ C\ m^{-2}$	$\mu g\ l^{-1}$
2000	0.018	12.4	315.7	11.7	37.9	0.8
2001	0.015	10.1	297.0	15.7	33.9	0.8
2002	0.018	12.6	343.2	12.9	38.2	0.8
2003	0.032	22.2	424.6	13.3	51.9	1.1
2004	0.019	13.2	310.2	14.1	39.2	0.9
2005	0.016	11.1	365.2	25.8	35.7	0.8
2006	0.027	18.4	335.5	18.4	47.0	1.0
2007	0.012	8.6	326.7	16.2	31.0	0.7
2008	0.013	9.1	281.6	10.8	32.0	0.7

1974; Schindler, 1978). In terms of applying Vollenweider's model to the Bay of Quinte as a tool for predicting P loads, we would expect that the dilution effect of high flushing rates will be mitigated by the high P retention capacity and thus not hinder our estimates.

In the Bay of Quinte, Vollenweider's models (Figure 3) show that: (1) annual primary production is higher than predicted from phosphorus loadings and (2) eutrophic conditions persist in the bay based on primary production and chlorophyll-*a* concentrations. The annual primary production observed (282–425 g m^{-2}) was much higher than predicted by phosphorus loadings (0.01–0.03 g m^{-2}) as was the algal standing crop (11–26 μg l^{-1} of chlorophyll-*a*). The reason for this disparity is that only Point-sources of phosphorus were considered, whereas the model was calibrated to include both point and non-Point-sources. In fact, Point-source phosphorus loads, according to the model, account for only 10–14% of the annual primary production and 3–8% of the algal standing crop (Table 3). By implication, the models suggest that non-Point-sources (tributaries, storm sewer overflows, etc.) contribute significantly to the total phosphorous loadings in the bay and are a major factor behind the persistent eutrophic conditions. In fact, Minns et al. (2004) estimated tributary loadings to the Bay of Quinte to be >600 kg d^{-1} – roughly 40x greater than Point-source loads–which would account for a significant amount of the observed primary production.

The predicted values of chlorophyll-*a* and primary production from Point-source loads using Vollenweider's models (Table 3) suggest that the bay would be oligotrophic if there were no other sources of phosphorus. As past management plans have focused on reductions in Point-source loads, it is apparent that future management strategies must consider ways of reducing non-Point-sources of phosphorus. The use of Vollenweider's models indicates that the Beneficial Use Impairment of eutrophication is likely to continue into the future without reductions in non-point phosphorus loadings.

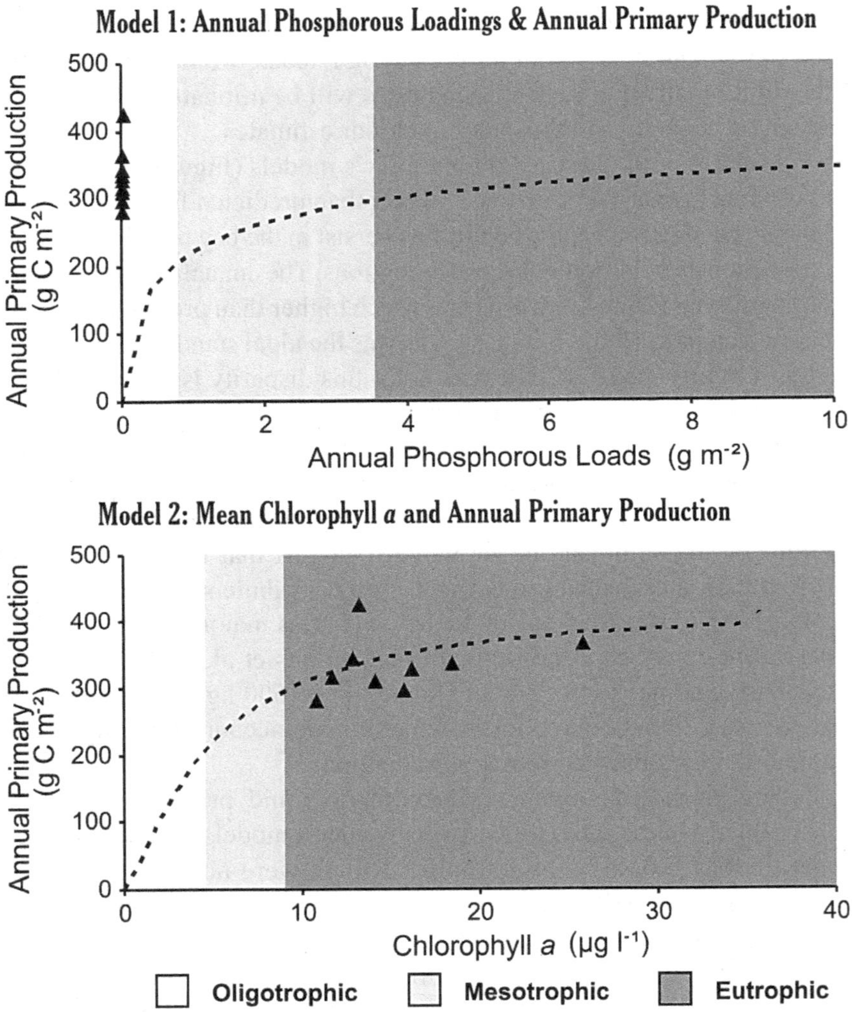

Figure 3. Vollenweider's eutrophication models applied to the Bay of Quinte, Lake Ontario. Model 1 shows the relationship between annual phosphorus loadings and primary production. Only Point-source phosphorus loads for the Bay of Quinte are plotted while the model also accounts for non-Point-sources. Model 2 shows the relationship between annual mean chlorophyll-*a* and annual primary production.

Planktonic index of biotic integrity

Degradation of phytoplankton and zooplankton communities is also listed as a Beneficial Use Impairment for the Bay of Quinte, so metrics which assess the relative health of both organisms are also needed. One such metric is the Planktonic Index of Biotic Integrity (Kane et al., 2009) which considers the biomass and composition of phytoplankton and zooplankton populations however the index does not directly consider the impacts of predators. The P-IBI was originally developed for another ecosystem in the Great Lakes basin, Lake Erie, which shares some major stressors with the Bay of Quinte including eutrophication and invasive species (Johnson and Owen, 1971; Vollenweider et al., 1974; Nicholls et al., 2002). From 2000–2008, we found that P-IBI scores in the Bay of Quinte ranged from 1.67–3.57 (Figure 4) with scores of less than 2 considered hyper-eutrophic, 2–3 eutrophic and >3–4 mesotrophic. In 6 of 9 years, the bay was eutrophic according to the P-IBI, mesotrophic for two of the years and hyper-eutrophic during one year. Overall, the P-IBI model indicates the continuance of eutrophication in the bay some 25–30 years after the implementation of Point-source phosphorous controls.

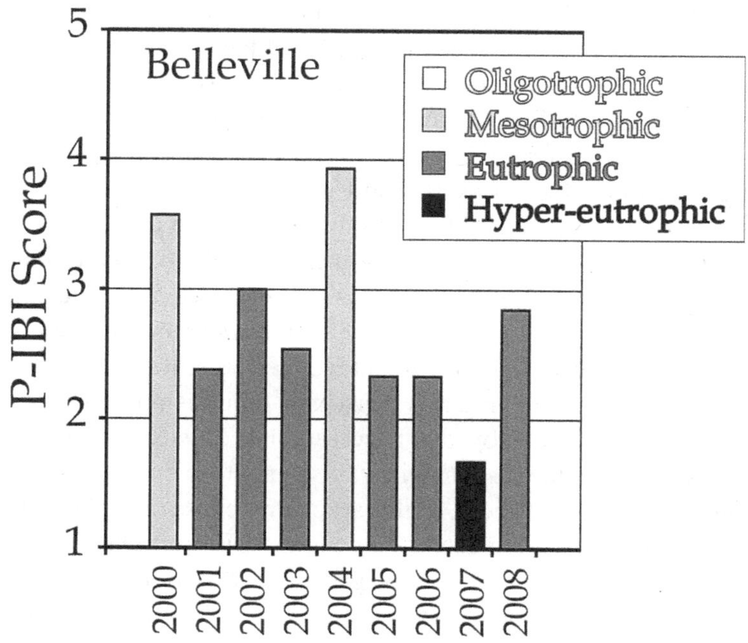

Figure 4. Planktonic index of biotic integrity (Kane et al., 2009) applied to the Bay of Quinte, Lake Ontario.

While the P-IBI model explicitly addresses eutrophication, it is built on the implicit linkages between phytoplankton and zooplankton communities and the results can therefore be useful in addressing the related BUI of "degradation of phytoplankton and zooplankton communities." This is because some of the driving factors behind low P-IBI scores include a high proportion of Cyanophyta relative to other phytoplankton (i.e. blue-green algal blooms), a relatively small biomass of edible phytoplankton (which would include centric diatoms and smaller phytoflagellates) and a reduced zooplankton population, especially among Cladocerans. These findings are consistent with our recently published work on the Bay of Quinte (Munawar et al., 2011) which concluded that organic carbon was likely being sequestered by other micro-organisms (i.e. heterotrophic nanoflagellates) indicating a potential lack of energy transfer from phytoplankton to zooplankton resulting in an impaired zooplankton community.

In defining the zooplankton impairment, Johannsson and Nicholls (2002) noted that increases in the relative densities of *Bosmina longirostris*, *Daphnia pulicaria*, total calanoid copepods, littoral cladocerans, *Leptodiaptomus minutus* and *Diacyclops thomasi* would be expected before the zooplankton community could be considered unimpaired. Furthermore, Bowen and Johannsson (2011) have observed that while zooplankton biomass did increase in the early 1980s following nutrient load reductions, significant reductions in zooplankton biomass followed the establishment of Dreissenid Mussels in the 1990s—particularly among Cladocerans, Cyclopoids and Calanoids—suggesting continued impairment in the zooplankton community. However, more work is needed to address the current state of the zooplankton impairment. The concept of Beneficial Use Impairments was intended to provide robust, multi-trophic criteria for assessing ecosystem health however a consensus as to what constitutes impairment and recovery is still required (George and Boyd, 2007). Achieving this consensus requires proven ecological indicators that can be deployed as necessary in ecosystems facing anthropogenic stress. It is the responsibility of scientists to choose appropriate indicators to meet these needs. In this article, we chose our battery of tests in order to make optimum use of the long term Project Quinte database and included phosphorous loadings, primary production, chlorophyll-*a*, phytoplankton and zooplankton in the various metrics. We focused the battery of tests on measurements that address eutrophication in terms of phosphorus, algal standing crop and primary production; factors that were prevalent in the 1970s and that later contributed to the Bay of Quinte's listing as an AOC. To a lesser extent, we addressed potential bottom up impacts on the zooplankton community with the P-IBI model.

We did not, however, consider the synergistic impacts of other known stressors such as Zebra Mussels on the base of the foodweb nor did we consider top down stressors on the zooplankton community. We simply used an established set of criteria that clearly define eutrophication to determine the current state of the ecosystem.

Summary and conclusions

Large freshwater and marine ecosystems suffer from a variety of anthropogenic stressors which include eutrophication, chemical contamination, coastal degradation and overexploitation of fisheries to name only a few. Attempts at remediation are often confounded by the multitude of local, regional, national and international governments and agencies that exercise jurisdiction over smaller parts of these ecosystems. In the North American Great Lakes, there exists a (nearly) 40 year track record for international cooperation in managing anthropogenic stressors that emphasizes sound ecosystem based science. Among these efforts was the designation of 42 severely polluted coastal regions as Areas of Concern (AOC) which were deemed to have at least 1 of 14 possible Beneficial Use Impairments. The Bay of Quinte, Lake Ontario, is one AOC with 10 listed impairments. We used a “battery of tests” strategy to assess the health of the bay with respect to the impairments of “eutrophication or undesirable algae” and “degradation of phytoplankton and zooplankton communities” in the bay. This strategy integrates structural, functional and chemical parameters into established ecosystem health models. The results of the battery of tests showed continued eutrophication of the bay and not coincidentally, continued degradation of the phytoplankton community. We also found that Point-sources of phosphorous do not account for all of the (pelagic) primary production observed in the bay and suggest that non-Point-sources of phosphorous contribute significantly to eutrophication. Our results further suggest that the battery of tests strategy is a sensitive science-based tool for assessing ecosystem health. These tests could also be applied to the evaluation of ecosystem health in other Great Lakes AOC as well as large lakes and marine environments where cultural eutrophication is a problem.

In Figure 5, we summarized the results of the battery of tests with respect to the assessment of trophic state. Phytoplankton taxonomy was not included in this summary because it is only available on a limited basis. Nonetheless, our ecological indicators showed good agreement in concluding eutrophic or hyper eutrophic conditions in 7 of the 9 years studied. In 2000, 2 of 3 indicators (trophic ladder and P-IBI) suggested mesotrophic conditions, while Vollenweider's models suggested eutrophic conditions. Similarly in 2004, the P-IBI model suggested mesotrophic conditions, while the other 2 models showed eutrophic conditions. Although Vollenweider's models do not distinguish between eutrophic and hyper-eutrophic conditions, it would not be unreasonable to assume that higher levels of chlorophyll-*a* and primary production were in fact indicative of hyper-eutrophic conditions. We conclude from the weight of evidence (24/27) that eutrophic conditions persist in the Bay of Quinte roughly 25 years after reductions in Point-source phosphorus loadings were first introduced.

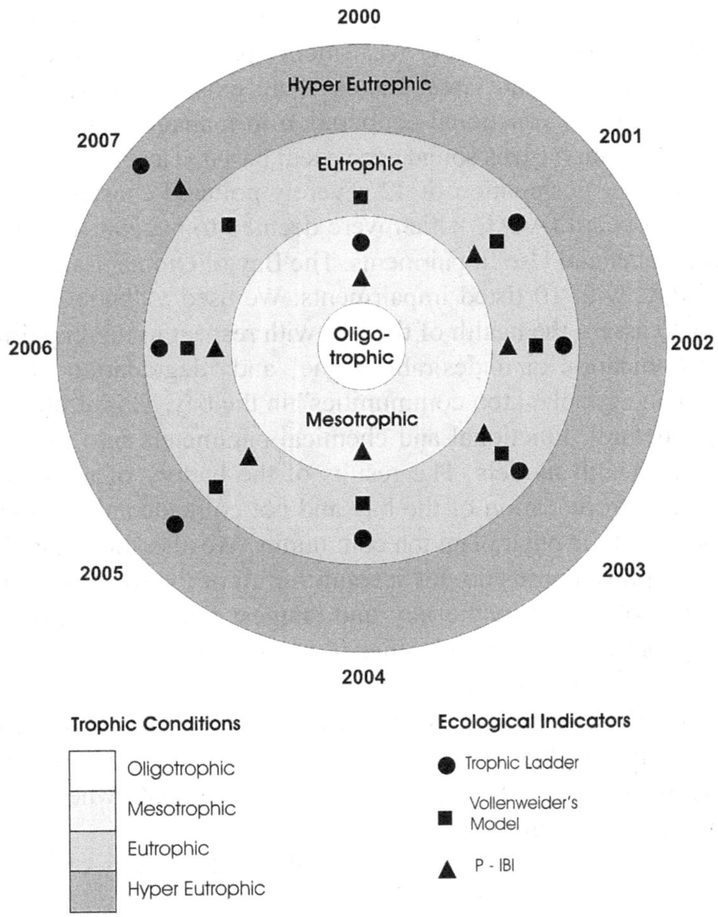

Figure 5. Results of the battery of tests deployed in this article. Phytoplank-

ton taxonomy was excluded from this comparison since it was only applied in 2005.

With respect to Beneficial Use Impairments in the Bay of Quinte, specifically "eutrophication or undesirable algae" and "degradation of phytoplankton and zooplankton communities," we conclude from the battery of tests that the Bay of Quinte remains eutrophic and that the phytoplankton community remains in a state of degradation. It is also apparent that future phosphorus abatement strategies will need to address the impact of non-Point-sources on primary production and algal standing crop. There is also evidence from the P-IBI model that the zooplankton community is also impaired, but our efforts were directed towards assessing the phytoplankton impairment. The battery of tests is composed of a mixture of structural, functional and chemical ecological indicators that are applicable to assessing the impact of multiple stressors in both marine and freshwater systems. The experiences and lessons learned from the Laurentian Great Lakes could be effectively applied towards assessing the health of the remaining Great Lakes Areas of Concern, as well as other freshwater and marine environments across the world, which have been impacted by cultural eutrophication.

Acknowledgements

We thank Drs. J. H. Leach, E. L. Mills and C. K. Minns, for their constructive reviews of the manuscript. A special thank you is due to J. Lorimer for the graphics and to S. Blunt (AEHMS) for technical editing. Finally, we gratefully acknowledge the hard work and perseverance of the DFO staff whose dedication to Project Quinte have made this research possible including: E. S. Millard, M. Frank (M. Burley), J. Gerlofsma, C. Timmins, R. Bonnell and A. Bedford.

References

Bay of Quinte Remedial Action Plan (BQ RAP), 1993. Time to Act. The Bay of Quinte Remedial Action Plan Stage 2 Report. Bay of Quinte Remedial Action Plan (RAP) Coordinating Committee, Ontario Ministry of the Environment, Kingston, ON.

Beeton, A.M., 1984. The world's great lakes. J. Great Lakes Res. 10(2), 106–113.

Bowen, K.L. and Johannsson, O.E., 2011. Changes in zooplankton biomass in the Bay of Quinte with the arrival of the mussels, *Dreissena polymorpha* and *D. rostiformis*, and the predatory cladocerans, *Cercopagis pengoi*: 1975 to 2008. Aquat. Ecosyst. Health Mgmt. 14(1), 44–55.

Bruce, J.P., 2011. Dr. Richard Vollenweider and the Great Lakes. Aquat. Ecosyst. Health Mgmt. 14(2), 186–187.

Burns, N.M., 1985. *Erie: the lake that survived.* Rowman & Allanheld, N.J.

Environment Canada and the United States Environmental Protection Agency, 1995. *The Great Lakes: An environmental atlas and resource book, 3rd Edition.* http:// www. epa.gov/glnpo/atlas/index.html. Last accessed: November 9, 2011.

Fee, E.J., 1990. Computer programs for calculating in situ phytoplankton photosynthesis. Can. Tech. Rep. Fish. Aquat. Sci. No. 1740.

Fitzpatrick, M.A.J., Munawar, M., Leach, J.H., and Haffner, G.D., 2007. Factors regulating primary production and phytoplankton dynamics in western Lake Erie. Fundam. Appl. Limnol. (Archiv. Hydrobiol.) 169, 137–152.

George, T.K., Boyd, D., 2007. Limitations on the development of quantitative monitoring plans to track progress of Beneficial Use Impairment restoration at Great Lakes Areas of Concern. J. Great Lakes Res. 33, 686–692.

Gillett, N.D., Steinman, A.D., 2011. An analysis of long-term phytoplankton dynamics in Muskegon Lake, a Great Lakes Area of Concern. J. Great Lakes Res. 37, 335–342. Guiry, M.D., Guiry, G.M., 2012. *AlgaeBase. World-wide electronic publication.* National University of Ireland, Galway. http://www.algaebase.org; searched on 23 January 2012

Hartig, J.H. and Zarull, M.A., 1992. Towards defining aquatic ecosystem health for the Great Lakes. J. Aquat. Ecosyst. Health 1, 97–107.

Hutchinson, G.E., 1967. *A treatise on limnology. Vol. 2.* John Wiley & Sons, New York.

International Joint Commission (IJC), 1988. Revised Great Lakes Water Quality Agreement of 1978. Agreement, with Annexes and Terms of Reference, between the United States and Canada signed at Ottawa, November 22, 1978 and Phosphorus Load Reduction Supplement signed October 7, 1983 as amended by protocol signed October 7, 1983, as amended by protocol signed November 18, 1987.

International Joint Commission (IJC), 1989. Revised Great Lakes Water Quality Agreement of 1978 between the United States and Canada, as amended by protocol, signed November 18, 1987.

Irvine, K.N., Murphy, T.P., 2009. Assessment of eutrophication and phytoplankton community impairment in the Buffalo River Area of Concern. J. Great Lakes Res. 35, 83–93.

Janus, L.L., Vollenweider, R.A., 1981. Summary Report: The OECD Cooperative Programme on Eutrophication, Canadian Contribution. IWD Series No. 131. Environment Canada, Burlington, ON.

Johannsson, O.E., Nicholls, K.H., 2002. Bay of Quinte Remedial Action Plan Assessment of the State of Impairment of Beneficial Uses: II. Zooplankton. Prepared for the Department of the Environment, Ontario Region. Bay of Quinte Remedial Action Plan, Kingston, ON.

Johnson, M.G. and Owen, G.E., 1971. Nutrients and nutrient budgets in the Bay of Quinte, Lake Ontario. J. Water Pollut. Control Fed. 43, 836–853.

Kane, D.D., Gordon, S.I., Munawar, M., Charlton, M.N., Culver, D.A., 2009. The Planktonic Index of Biotic Integrity (P-IBI): an approach for assessing lake ecosystem health. Ecol. Ind. 9, 1234–1247

Kerr, S.R., Ryder, R.A., 1997. The Laurentian Great Lakes experience: a prognosis for the fisheries of Atlantic Canada. Can. J. Fish. Aquat. Sci. 54, 1190–1197.

Krantzberg. G., 2004. Science must inform Great Lakes policy. J. Great Lakes. Res. 30, 573–574.

Millard, E.S., Fee, E.J., Myles, D.D., Dahl, J.A., 1999. Comparison of phytoplankton photosynthesis methodology in Lakes Erie, Ontario, the Bay of Quinte and the Northwest Ontario Lake Size Series. In: M. Munawar, T. Edsall, I.F. Munawar (Eds.), *State of Lake Erie (SOLE)—Past, Present and Future*, pp. 441–468. Backhuys, Leiden, the Netherlands.

Mills, E.L., Leach, J.H., Carlton, J.T., Secor, C.S., 1993. Exotic species in the Great Lakes: a history of biotic crises and anthropogenic introductions. J. Great Lakes Res. 19(1), 1–54.

Minns, C.K., Moore, J.E., 2004. Modelling phosphorus management in the Bay of Quinte, Lake Ontario in the past, 1972 to 2001, and in the future. Can. Manuscr. Rep. Fish. Aquat. Sci. No. 2695.

Minns, C.K., Owen, G.E., Johnson, M.G., 1986. Nutrient loads and budgets in the Bay of Quinte, Lake Ontario, 1951–81. In: C.K. Minns, D.A. Hurley, K.H. Nicholls. (Eds), (*Project Quinte: point-source phosphorus control and ecosystem response in the Bay of Quinte, Lake Ontario*. Can. Spec. Publ. Fish. Aquat. Sci. 86, 59–76.

Minns, C.K., Moore, J.E., Seifried, K.E., 2004. Nutrient loads and budgets in the Bay of Quinte, Lake Ontario, 1972 to 2001. Can. Manuscr. Rep. Fish. Aquat. Sci. No. 2694. Munawar, M., Fitzpatrick, M., 2011. The application of Vollenweider's eutrophication models for assessing ecosystem health: Hamilton Harbour (Lake Ontario) example. Aquat. Ecosyst. Health Mgmt. 14(2), 204–208.

Munawar, M., Munawar, I.F., 1982. Phycological studies in Lakes Ontario, Erie, Huron and Superior. Can. J. Bot. 60, 1837–1858.

Munawar, M., I.F. Munawar, I.F., 1996. *Phytoplankton dynamics in the North American Great Lakes Vol. I: Lakes Ontario, Erie and St. Clair*. Ecovision World Monograph Series. SPB Academic Publishing, Amsterdam.

Munawar, M., Munawar, I.F., McCarthy, L., 1987. Phytoplankton ecology of large eutrophic and oligotrophic lakes of North America: Lakes Ontario and Superior. Arch. Hydrobiol. Bieh. Ergebn. Limnol. 25, 51–96.

Munawar, M., Fitzpatrick, M., Niblock, H., Lorimer, J., 2011. The relative importance of autotrophic and heterotrophic microbial communities in the planktonic food web of the Bay of Quinte, Lake Ontario 2000–2007. Aquat. Ecosyst. Health Mgmt. 14(1), 21–32.

Nicholls, K.H., Carney, E.C., 2011. The phytoplankton of the Bay of Quinte, 1972–2008: Point-source phosphorous loading control, dreissenid mussel establishment, and a proposed community reference. Aquat. Ecosyst. Health Mgmt. 14(1), 33–43.

Nicholls, K.H., Heintsch, L, Carney, E., 2002. Univariate step-trend and multivariate assessments of the apparent effects of P loading reductions and zebra mussels on the phytoplankton of the Bay of Quinte, Lake Ontario. J. Great Lakes Res. 28(1), 15–31.

Reynolds, C.S., 1984. *The ecology of freshwater phytoplankton*. Cambridge University Press, Cambridge.

Salmaso, N., 2000. Factors affecting the seasonality and distribution of cyanobacteria and chlorophytes: a case study from the large lakes south of the Alps, with special reference to Lake Garda. Hydrobiol. 438, 43–63.

Scheffer, M., Rinaldi, S., Gragnani, A., Mur, L.R, van Nes, E.H., 1997. On the dominance of filamentous cyanobacteria in shallow, turbid lakes. Ecology 78(1), 272–282. Schindler, D.W., 1978. Factors regulating phytoplankton production and standing crop in the world's freshwaters. Limnol. Oceanogr. 23(3), 478–486.

Strickland, J.D.H., Parsons, T.R., 1968. *A practical handbook of seawater analysis*. Bull. Fish. Res. Board Can. Ottawa, ON. Stockner, J.G. and Shortreed, K.S. 1988. Response of *Anabaena* and *Synechococcus* to manipulation of nitrogen: phosphorous ratios in a lake fertilization experiment. Limnol. Oceanogr. 33 (6, part 1), 1348–1361.

Trumpickas, J., Shuter, B.J., Minns, C.K., 2009. Forecasting impacts of climate change on Great Lakes surface water temperatures. J. Great Lakes Res. 35, 454–463.

Utermöhl, H., 1958. Zur vervolkommnung der quantitativen phytoplankton-methodik. (The improvement of quantitative phytoplankton methodology. In German.) Mitt. Internat. Verein. Limnol. 9, 1–38.

Vallentyne, J.R., 1974. The algal bowl: lakes and man. Government of Canada, Misc. Spec. Publ. 22.

Vallentyne, J.R., 1993. Biospheric foundations of the ecosystem approach to environ mental management. J. Aquat. Ecosyst. Health 2(1), 9–14.

Vollenweider, R.A., 1968. Scientific fundamentals of the eutrophication of lakes and flowing waters, with particular reference to nitrogen and phosphorus as factors in eutrophication. Organization for Economic Co-operation and Development, Paris, France.

Vollenweider, R.A., Dillon, P.J., 1974. The application of the phosphorous loading concept to eutrophication research. National Research Council of Canada, Publication No. 13690. Environment Canada, Burlington, ON.

Vollenweider, R.A., Munawar, M., Stadelmann, P., 1974. A comparative review of phytoplankton and primary production in the Laurentian Great Lakes. J. Fish. Res. Board Can. 31, 739–762.

Watson, S.B., Rydal. J., and Boyer, G.L., 2008. Taste and odour cyanobacterial toxins: impairment, prediction, and management in the Great Lakes. Can. J. Fish. Aquat. Sci. 65, 1779–1796.

The above article is a reprinted version of the following previously published article:

Assessing ecosystem health impairments using a battery of ecological indicators: Bay of Quinte, Lake Ontario example
M. Munawar; M. Fitzpatrick; I. F. Munawar; H. Niblock; D. Kane
Aquatic Ecosystem Health & Management (2012) 15 (4): 430–441.
https://doi.org/10.1080/14634988.2012.731367

Fish community indices of ecosystem health: How does the Bay of Quinte compare to other coastal sites in Lake Ontario?

C. M. Brousseau,[1]* R. G. Randall,[1] J. A. Hoyle[3] and C. K. Minns[1,2]

[1]*Fisheries and Oceans Canada, 867 Lakeshore Road, Burlington, ON L7S 1A1*
[2]*University of Toronto, 25 Willcocks Street, Toronto, ON M5S 3B2*
[3]*Ontario Ministry of Natural Resources and Forestry, 41 Hatchery Lane, Picton, ON K0K 2T0*
**Corresponding author: Christine.brousseau@dfo-mpo.gc.ca*

Keywords: Index of Biotic Integrity, Great Lakes, nearshore habitat, species richness, electrofishing

Introduction

The Index of Biotic Integrity (IBI) was developed as an indicator of ecosystem health by Karr (1981) for streams in North America. Until then, physical-chemical water quality monitoring was used to assess the degree of environmental degradation in running waters. Fish were chosen as the indicator group to develop the IBI for two reasons: (1) the fish community could be accurately and rapidly assessed and (2) fish act as integrators of both the biotic (foodweb) and abiotic (habitat and water quality) environment (Karr, 1981). IBI's are widely used in North America to assess impairment, guide management and monitor fish habitat restoration in aquatic ecosystems (Barbour et al., 1999; Fausch et al., 1990; Fayram and Mitro, 2008; Karr, 1991; Wilson and Xenopoulos, 2008). Karr's IBI has also been modified to assess impairments at a landscape scale (Meador et al., 2008; Weijters et al., 2009) and for lacustrine habitats that include coastal wetlands (Seilheimer and Chow-Fraser, 2006) and littoral areas of the Great Lakes (Bhagat et al., 2007).

Minns et al. (1994) adapted Karr's IBI for the littoral zone in the Great Lakes (GLIBI) to be used as a quantitative tool for assessing aquatic ecosystem health status. Data used to develop the GLIBI were collected by boat electrofishing in three Great Lakes Areas of Concern (AOC): Hamilton Harbour, Bay of Quinte and Severn Sound. Electrofishing was selected as a tool that can sample a large proportion of the fish community (e.g.

Ecology of the Bay of Quinte: Health, Management and Global Implications
Editors: C.K. Minns, M. Munawar and M.A. Koops

Aquatic Ecosystem Health & Management Society, Canada

species and size) and a wide range of habitats (Portt et al., 2006). Minns et al. (1994) used reference "conditions" from within the AOCs where, environmental conditions ranged from degraded to relatively healthy, assuming that sites at the upper bounds of the GLIBI were healthy and comparable to reference "locations" outside the AOCs. The GLIBI was adopted as a tool for establishing quantitative delisting targets by Remedial Action Plan (RAP) teams in the three AOCs from which it was derived.

The nearshore electrofishing program in the Bay of Quinte (BoQ) and other areas in Lake Ontario has been ongoing for >20 years (1988–2009). Data has been gathered for a number of research programs including the Great Lakes Action Plan (GLAP), productive capacity of nearshore habitats (Randall and Minns, 2002), Aquatic Invasive Species (AIS) monitoring and the Ontario Ministry of Natural Resources' (OMNR) nearshore trap net assessment program. The objectives of this paper were to use these historic data to: (1) test the assumption that the electrofishing data from the upper BoQ included sites comparable to reference locations outside of the AOC and also to compare trends from the OMNR's trap net dataset at the same locations; (2) examine temporal and spatial trends in the littoral fish community of the BoQ and reference locations, using both electrofishing and trap net data, in light of recent changes in water quality and clarity; and (3) compare fish community attributes (IBI, species richness) at locations in Lake Ontario that were outside the BoQ and immediate vicinity, using both electrofishing and trap net data. The three objectives were complementary, testing the hypothesis that GLIBI values at locations in the upper BoQ were similar to or exceeded values at reference locations elsewhere in Lake Ontario.

Methods

Study area

Boat electrofishing (EF) surveys were conducted by Fisheries and Oceans Canada (DFO) between 1990 and 2009 at coastal habitats in three locations of the upper BoQ (Trenton, Belleville and Big Bay) and 12 other locations in Lake Ontario following a standardized protocol (Brousseau et al., 2005). EF data were collected from two AOCs (BoQ and Hamilton Harbour), from reference locations in the vicinity of the BoQ AOC (West Lake, Black River mouth, Prince Edward Bay), at five other Lake Ontario locations (Presquille to Port Dalhousie) and Long Point, Lake Erie (Figure 1). These locations represented a variety of habitats (Table 1) that were sampled sporadically between 1998 and 2009.

Trap net data were collected by the OMNR following the Nearshore Community Index Netting Surveys (NSCIN) protocol (Haxton and Punt, 2004). Trap net surveys were conducted between 2001 and 2009 at six locations in the upper BoQ, middle BoQ (Hay Bay) and other Lake Ontario embayments (Figure 1).

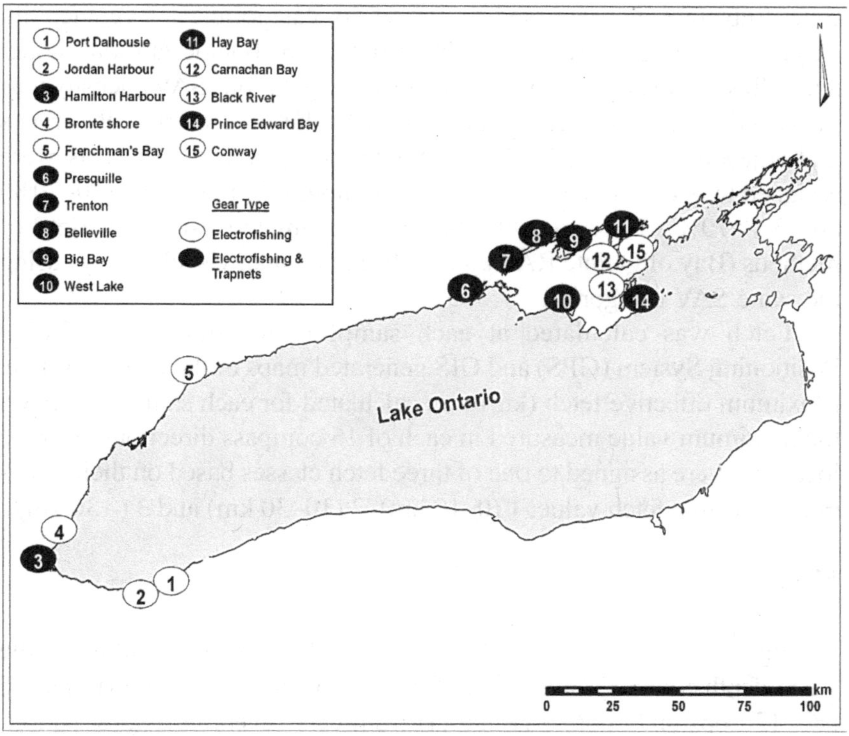

Figure 1. Sampling locations in the Bay of Quinte (Trenton, Belleville, Big Bay, Hay Bay, Carnachan Bay and Conway), reference sites in Prince Edward County (Black River, Prince Edward Bay, West Lake) and other locations in Lake Ontario. Data from Long Point, western Lake Erie (not shown) were also included in the comparison.

Habitat

Coastal habitats surveyed included embayments, wetlands, harbours, river mouths and exposed shoreline. Individual sampling sites were categorized by three physical habitat attributes: substrate size, macrophyte density and fetch. Habitat data were collected at 100 m transects prior to electrofishing (Brousseau et al., 2005) and at an area of 100 m adjacent to either side of a trap net (Haxton and Punt, 2004). Substrate size was determined by visual observation (cobbles or larger) or Ponar grab samples (smaller substrates) taken at each transect end or on either side of a trap net. A median substrate category was calculated for each location based on observations from the August-September sampling period over time. Substrate was assigned to one of six categories: 1 (organic, silt, clay), 2 (sand), 3 (gravel), 4 (cobble/rubble), 5 (boulder/armour stone) and 6 (bedrock/shale). Submerged aquatic vegetation (SAV) density was measured visually as percent bottom cover during the transect surveys and adjacent to trap nets in late summer or early fall. Transects were assigned to one of four SAV categories: 0 (none), 1 (1–19%), 2 (20–70%) and 3 (>70%). Prior to 1999, SAV was estimated using a variety of methods (Bay of Quinte RAP Plan, 1993) other than visual surveys using the same SAV categories.

Fetch was calculated at each sampling location using Global Positioning System (GPS) and GIS generated maps of the lake shoreline. Maximum effective fetch (km) was calculated for each sample based on the maximum value measured in each of 16 compass directions. Sample locations were assigned to one of three fetch classes based on their maximum effective fetch value: 1 (0–10 km), 2 (10–30 km) and 3 (>30 km).

Fish

During EF, fish were collected along 100 m line transects following the 1.5 m depth contour using a 6.1 m Smith-Root electrofishing boat (pulsed DC, 120 pps and about 8 A output) from early evening to after sunset. For each species, individuals were identified and up to 20 measured (fork length (mm) and wet weight (g)); after 20, fish were counted and bulk weighed (Brousseau et al., 2005). For each transect, GLIBI scores were calculated following the methodology of Minns et al. (1994), both including and excluding offshore species (e.g. Alewife). In addition to the IBI scores, total biomass (kg), total numbers, piscivore biomass (%), and the relative number of native, centrarchid and non-native species were used to assess the fish community at sampling locations. Piscivores were classified taxonomically in the GLIBI as defined in Minns et al. (1994).

Fish were collected from trap nets following the NSCIN protocol (1.83 trap nets, gap depth 2.0–2.5 m, 24 hour set) at a number of random nearshore sampling locations in Lake Ontario between 2001 and 2009

(Hoyle et al., in press). Sampling occurred from mid-August until the end of September with the exception of West Lake (2001) which was surveyed in October. For this paper, trap net data collected from the upper BoQ (Trenton, Belleville and Big Bay), middle BoQ (upper and lower Hay Bay), West Lake, Prince Edward Bay and Hamilton Harbour were used because of their proximity to electrofishing sites and because they included both AOC and reference locations.

Statistics

For statistical analysis, different subsets of data were chosen depending on one of three study objectives. For each objective, sample size varied dependent upon year and location.

Objective 1: Are there reference conditions in the upper BoQ? EF data from the upper BoQ were compared to locations in the middle BoQ and reference locations in Prince Edward Bay, the Black River mouth and West Lake. Statistical analyses were used to test for differences among locations for the dependent variables IBI, adjusted IBI (IBI*-exclusion of offshore species), biomass, numbers, piscivore biomass (%) and the relative number of native, centrarchid and non-native species. For GLIBI scores and catch data among locations, a one-way analysis of variance (ANOVA, SYSTAT 10.2) by year was used on square root transformed IBI and IBI* values and log (x + 1) transformed catch data (biomass and numbers) to achieve a normal distribution (Minns et al., 1994). The Bonferroni multiple comparisons test was used to determine which pairs of means differed significantly. For other GLIBI metrics, the non-parametric Kruskal-Wallis analysis of variance or the Mann Whitney U-test (MWU) was used. All test statistics were considered significant at $\alpha < 0.05$. For trap nets, data from individual nets were used to calculate average numbers, piscivore biomass (%) and species richness values by location and year. Statistical tests used were the same as for electrofishing (ANOVA, MWU).

Objective 2: For the temporal analyses, datasets were examined at three upper Bay locations both individually and collectively, using the dependent variables from Objective 1. For spatial analyses, datasets from locations in the middle BoQ and West Lake were examined.

Objective 3: IBI values calculated from electrofishing datasets at 12 locations were examined to determine spatial variability across Lake Ontario (Figure 1). Data were displayed from east to west and included two AOCs, BoQ and Hamilton Harbour, 10 non-AOC areas and one reference location from Lake Erie (Figure 3). In addition, species richness and piscivore biomass (%) from six of these locations were compared, using data from both electrofishing and trap nets, to determine if catch trends were similar for both gears.

Results

Habitat

Substrate classes varied among locations from silt to boulders (Table 1). Both Prince Edward Bay and Bronte shore had coarse substrate, predominantly boulders, which was different from other Lake Ontario locations examined in this study. SAV density was highest in coastal wetland areas (upper BoQ, West Lake) at eastern Lake Ontario sites (Figure 1) and Lake Erie (Table 1). SAV density was lowest at coastal areas with higher fetch values and harbours. Two-thirds of the Lake Ontario sites had low fetch values (Table 1).

Fish

Objective 1

Electrofishing: Fish assemblage data from locations sampled in the upper BoQ had average IBI values and other fish community variables that generally fell within the range of values from sites in the middle BoQ and at reference locations (Table 1), except where values in the upper BoQ were higher than elsewhere.

For 1999, fish catch data from the upper BoQ was compared to datasets from the middle BoQ, Black River and West Lake. No significant differences were found between the upper BoQ and other locations for IBI, numbers, piscivore biomass, native, centrarchid and non-native species. However, IBI* varied among locations ($F_{3,89} = 3.14$, $p = 0.029$), with West Lake having a significantly higher score than the upper BoQ ($p = 0.055$, Bonferroni) but not the middle BoQ or Black River. Cyprinid species richness was significantly higher at West Lake ($KW_{3,93} = 14.86$, $p = 0.002$) than at the upper BoQ to account for the higher IBI* score in West Lake. No other significant differences in species richness or percent piscivore biomass were found between the two locations (Table S1). Biomass was significantly lower at West Lake than at the other two locations ($F_{3,89} = 3.68$, $p = 0.015$).

For 2001, catch data from Big Bay in the upper BoQ was compared to data from the mouth of the Black River. There were no significant

Table 1. Physical habitat and fish community values for electrofishing and trap net locations surveyed between 1990 and 2009. Median habitat categories are indicated with min and max values in parenthesis. Samples (n) used to classify habitat were not equal among variables and were year dependent. Prior to 1999, habitat data was sampled sporadically. For fish data, the number (n) of samples indicates years of sampling.

Location	Habitat Type	n	Substrate class[1]	n	Macrophyte class[2]	n	Fetch class[3]	n	IBI	Native species	Cen-trarchid species	Piscivore biomass (%)
ELECTROFISHING												
Lake Ontario												
Upper Bay of Quinte	EMB, AOC	151	2 (1-4)	165	2 (1-3)	165	1 (1-1)	5	75 (63-85)	8 (7-9)	3 (2-4)	35 (28-41)
Middle Bay of Quinte	EMB, AOC	42	1 (1-3)	59	2 (2-3)	59	1 (1-1)	3	74 (62-83)	8 (5-9)	3 (1-4)	36 (30-43)
Lower Bay of Quinte	EMB, AOC	6	3 (1-5)	12	1 (0-3)	12	1 (1-1)	1	51 (51-51)	4 (4-4)	1 (1-1)	33 (33-33)
Black River	CW; RM	49	1 (1-6)	49	3 (0-3)	49	2 (1-2)	4	72 (68-77)	8 (7-9)	3 (2-3)	18 (11-38)
Prince Edward Bay	EXE	57	5 (4-5)	58	0 (0-1)	58	2 (1-2)	5	63 (61-65)	5 (4-5)	2 (2-2)	36 (17-50)
West Lake	CW	132	1 (1-4)	132	2 (1-3)	132	1 (1-2)	6	76 (70-80)	9 (7-10)	4 (3-5)	38 (22-45)
Presquille	CW	40	1 (1-1)	40	3 (3-3)	40	1 (1-1)	1	68 (68-68)	6 (6-6)	2 (2-2)	23 (23-23)
Frenchman's Bay	EMB	10	1 (1-2)	10	1 (0-1)	10	1 (1-3)	1	40 (40-40)	4 (4-4)	1 (1-1)	6 (6-6)
Bronte shore	ES	95	5 (1-6)	95	0 (0-2)	95	3 (1-3)	5	33 (20-40)	2 (1-3)	0 (0-1)	8 (0-20)
Hamilton Harbour	HAR, AOC	413	2 (1-5)	413	1 (0-3)	786	1 (1-1)	11	37 (24-50)	1(1-5)	1 (0-2)	7 (2-13)
Jordan Harbour	EMB	25	1 (1-5)	25	0 (0-2)	25	1 (1-3)	2	45 (42-50)	4 (3-5)	1 (0-1)	5 (0-10)
Port Dalhousie	HAR, EXS	46	3 (2-5)	46	0 (0-3)	46	3 (1-3)	3	53 (50-57)	3 (3-4)	1 (1-2)	23 (4-40)

Table 1. Continued…

Location	Habitat Type	n	Substrate class[1]	n	Macrophyte class[2]	n	Fetch class[3]	n	IBI	Native species	Centrarchid species	Piscivore biomass (%)
ELECTROFISHING												
Lake Erie												
Long Point	CW	21	1 (1-1)	21	3 (1-3)	21	2 (1-2)	1	78 (78-78)	9 (9-9)	4 (4-4)	50 (50-50)
TRAP NETS												
Lake Ontario												
Upper Bay of Quinte	EMB, AOC	92	3 (1-6)	92	1 (0-3)	92	1 (1-1)	5		9 (9-10)	4 (4-4)	25 (15-32)
Middle Bay of Quinte	EMB, AOC	28	3 (1-6)	28	1 (0-2)	28	1 (1-1)	4		9 (8-12)	4 (2-4)	26 (16-38)
Prince Edward Bay	EXE	34	5 (1-6)	34	0 (0-3)	30	2 (1-2)	1		7 (7-7)	3 (3-3)	36 (36-36)
West Lake	CW	30	1 (1-5)	30	1 (0-2)	30	1 (1-1)	2		9 (8-12)	4 (4-5)	31 (19-42)
Presquille	CW	24	2 (2-2)	24	2 (1-3)	30	1 (1-1)	1		7 (7-7)	3 (3-3)	27 (27-27)
Hamilton Harbour	HAR, AOC	43	2 (1-5)	43	1 (0-2)	43	1 (1-1)	2		7 (7-7)	2 (2-2)	9 (7-10)

[1] Substrate class: 1-(organic, silt, clay), 2-(sand), 3-(gravel), 4-(cobble/rubble), 5-(boulder/armour stone) and 6-(bedrock, shale).
[2] Macrophyte class: 0 (none), 1 (1–19%), 2 (20–70%) and 3 (>70%).
[3] Fetch class: 1 (0–10 km), 2 (10–30 km), 3 (>30 km).
AOC – Area of Concern, CW – Coastal wetland, EXE–Exposed embayment, EXS–Exposed shore, EMB–Embayment; HAR–Harbor, RM–River mouth

differences between the two locations for IBI, IBI*, biomass, numbers, native or non-native species. However, both centrarchid species richness ($MWU_{1,22}$ = 18.00, p = 0.007) and piscivore biomass ($MWU_{1,22}$ = 14.86, p = 0.011) were significantly higher at the BoQ.

In 2007 and 2009, no significant differences in the IBI scores, species richness or piscivore biomass (%) were found between the upper BoQ and West Lake (Table S1). Biomass (kg) was significantly higher at the upper Bay than at West Lake or the middle BoQ ($F_{2,50}$ = 9.23, p < 0.001). The mean number of fish caught per transect was significantly lower at West Lake than at the two BoQ locations ($F_{2,90}$ = 5.12, p = 0.009).

Trap nets: In 2001 and 2007, average catch data from the upper BoQ was pooled and compared to data from West Lake. In 2001, native species richness ($MWU_{1,22}$ = 83.00, p = 0.009) and average numbers (ANOVA, $F_{1,20}$ = 13.08, p = 0.002) were the only variables that were significantly different (i.e. higher) at the upper BoQ than at West Lake. In 2007, no significant differences in fish community variables were found between the two locations (Table S1).

In 2004 and 2009, catch data from the upper BoQ was compared to data from two locations in the middle BoQ. In 2004, there was a significant difference in centrarchid species richness ($KW_{2,30}$ = 7.69, p = 0.021) between the upper and middle bays; average values were highest at upper Hay Bay (4.0) and the upper BoQ (3.9). In 2009, non-native species richness was significantly different among locations ($KW_{2,30}$ = 9.02, p = 0.011); average numbers of non-native species were lowest at the upper BoQ (0.7) and highest at upper Hay Bay (1.7). Numbers of fish were not different among areas in 2004 but were significantly higher in 2009 at upper Hay Bay than at the upper BoQ ($F_{2,27}$ = 3.58, p = 0.042).

In 2008, data from Presquille Bay was compared to data from the upper BoQ. Both centrarchid species richness ($MWU_{1,29}$ = 47.00, p = 0.011) and numbers ($F_{1,27}$ = 5.24, p = 0.030) were significantly higher at the upper BoQ than at Presquille.

Objective 2

IBI, biomass and number of fish: IBI and IBI* scores in the upper BoQ increased significantly over time ($F_{3,166}$ = 12.11, p < 0.001); average IBI scores were significantly higher in the years from 1999 to 2009 than in 1990 (Figure 2). A significant increase in IBI scores at Belleville and Big Bay drove the overall increase in the mean upper BoQ scores (Figure 2). At Trenton, middle BoQ and West Lake, no significant changes in IBI scores were detected (Figure 2) however, IBI scores in the middle Bay increased from 70 (1992) to 83 (2009). Similarly, the adjusted IBI scores were significantly higher in the latter three upper BoQ surveys (66–70) than in 1990 (46). In middle BoQ, average IBI* in 2009 (83) was significantly higher than scores from 1992 (56) and 1999 (73). No differences in IBI scores were detected at West Lake (1998 to 2009).

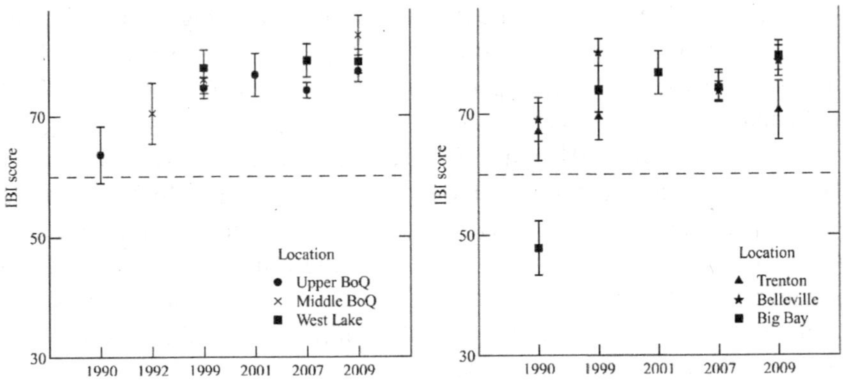

Figure 2. Average (±SE) Index of Biotic Integrity (IBI) scores between 1990 and 2009 for the upper Bay of Quinte (BoQ), middle BoQ and West Lake (left) and individual locations sampled in the upper BoQ (right). The horizontal reference line indicates the lower limit of the IBI delisting target in the BoQ (Brousseau and Randall, 2009).

Fish biomass in electrofishing catches varied significantly in the upper BoQ: biomass (kg transect^{-1}) was highest in 1990 (8.9) and lowest in 1999 (3.6) ($F_{3,166} = 8.36$, $p < 0.001$). Biomass decreased by more than 50% from 1990 to 1999 but then increased to former levels by 2009. A similar pattern was found at Trenton, Belleville and Big Bay. No significant difference in annual biomass was detected at the middle Bay or West Lake. Average fish catch in the middle BoQ was significantly higher in 2009 (83) than 1992 (34) ($F_{2,21} = 5.73$, $p = 0.010$). In contrast, average catch did not differ significantly among years in either the upper BoQ or West Lake.

Species richness and piscivore biomass: Between 1990 and 2009, significant changes in centrarchid and non-native species richness occurred at both the upper BoQ and other locations (Table S2). Native species richness increased significantly in the upper BoQ, from 6.9 to 9.0 species in 1990 and 2009, respectively ($KW_{3,170} = 13.44$, $p = 0.004$). The increase in native species in the upper BoQ was driven by a significant increase at Big Bay which almost doubled between 1990 (5.3) and 2009 (9.8). No significant changes in native species richness were noted in the middle BoQ or West Lake. The increase in native species richness at the upper Bay was correlated with an increase in the number of centrarchid species ($KW_{3,170}$ = 67.84, $p < 0.001$). Centrarchid species also increased significantly at the middle BoQ and West Lake (Table S2).

Non-native species richness in the upper BoQ decreased between 1990 and 1999 but then increased again in 2007 and 2009 ($KW_{3,170}$ = 54.05, $p < 0.001$). No temporal changes in non-native species richness were detected in the middle BoQ but the number of non-native species in West Lake was higher in 2007 than in other years (Table S2). Piscivore biomass (%) in the upper BoQ, middle BoQ and West Lake did not change over time (Table S2); however, the composition of the piscivore community did. In 1990, Walleye was the key contributor to piscivore biomass in upper BoQ EF catches, but has since been replaced as top predator by Largemouth Bass ($F_{3,166}$ = 22.76, $p < 0.001$) (Figure S1).

Objective 3

GLIBI scores were higher at the east end of the lake (upper BoQ to Presquille) and Long Point, Lake Erie than at other locations in Lake Ontario (Frenchman's Bay to Port Dalhousie) relating to habitat and exposure (Table 1, Figure 3). Both SAV density and fetch were correlated with location (Table 1). Exploratory analysis (ANCOVA), with distance as the covariate, indicated that both SAV density and fetch affected fish catches, but there were significant interactions between factors and locations. Results were also confounded by the differences in the macrophyte beds as spatial extent is generally restricted at the west and southwest parts of the lake behind breakwalls and in marinas. Further analysis of the habitat effects was beyond the scope of this study.

Both EF and trap net datasets showed similar trends in piscivore biomass (CPUE), native and centrarchid species richness (Figure 4) and non-native species richness at six locations in Lake Ontario. Piscivore biomass and centrarchid species richness were higher in the BoQ and at reference locations in the vicinity (e.g. West Lake) than in Hamilton Harbour or elsewhere in Lake Ontario, in catches for both gears. In contrast, non-native species richness was lower at the BoQ and reference locations than in Hamilton Harbour. From EF catches, total native species richness was lower at Hamilton Harbour than at BoQ and reference locations; this trend was also seen in the trap net catches but the differences among locations were not as great (Figure 4).

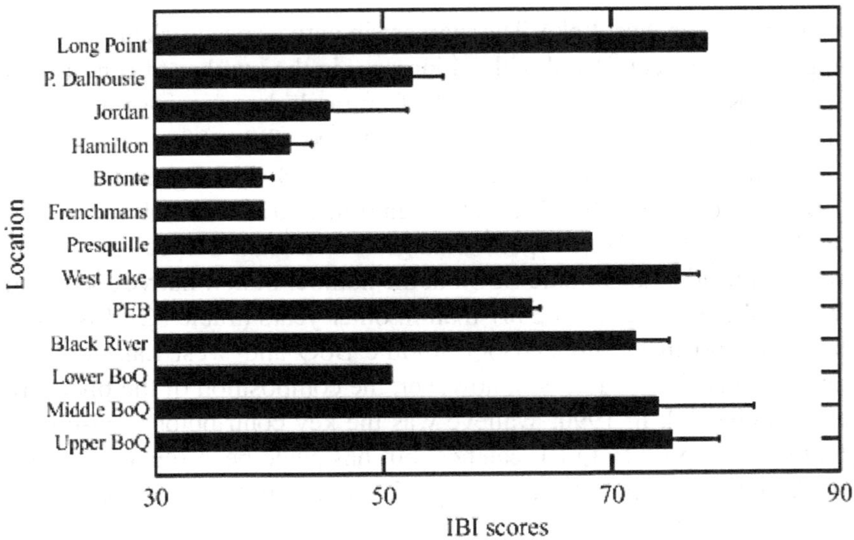

Figure 3. Comparison of average IBI scores (±SE) at locations in Lake Ontario and Long Point, Lake Erie. Sampling locations included two Great Lakes' Areas of Concern: Bay of Quinte (BoQ: Upper, Middle and Lower) and Hamilton, and 9 other non-AOC locations including Prince Edward Bay (PEB).

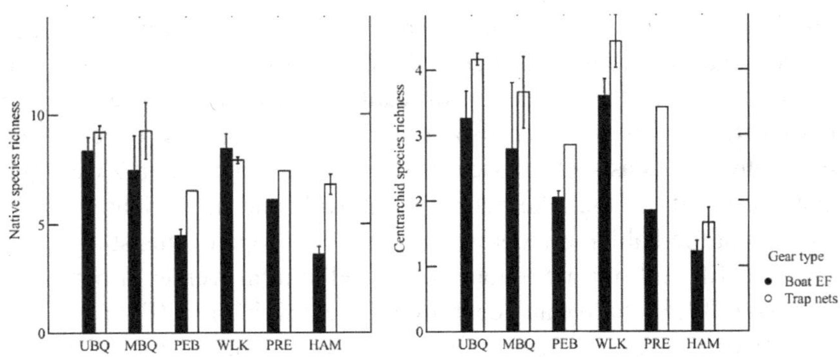

Figure 4. The average number (±SE) of centrarchid species (left) and total native species (right) per unit of sampling effort at six locations in Lake Ontario, to show` consistency in catches trends between gear types. Sampling effort was defined as an electrofishing boat transect (100 m) or one 24-hour trap net set. Data used to generate the average and standard error was collected between 1999 and 2009. Locations are as follows: Upper Bay (UBQ), Middle Bay (MBQ), Prince Edward Bay (PEB), West Lake (WLK), Presquille (PRE) and Hamilton (HAM).

Discussion

GLIBI values from the upper Bay of Quinte were high in comparison to values from reference locations in the vicinity and elsewhere in Lake Ontario. Results of three objectives supported the hypothesis that coastal areas in the BoQ, although an Area of Concern, included relatively healthy habitat and good reference conditions compared to fish habitat at reference locations. Locations in the upper BoQ had GLIBI scores, diversity metrics and total catches that fell within the same range or exceeded values from reference locations with similar habitat in the vicinity (e.g. West Lake) or elsewhere in Lake Ontario (Presquille Bay) and Lake Erie (Long Point). Observed temporal trends in the fish catch from the upper BoQ, for the post-phosphorus control and post-Zebra Mussel invasion periods (Nicholls et al., 2004), were also comparable to trends from the middle BoQ and reference locations. Spatial trends in the electrofishing data were consistent with NSCIN trap net data. Coastal areas surveyed in Lake Ontario outside BoQ and immediate vicinity (e.g. West Lake) had lower GLIBI values, but many of the sites had dissimilar habitats (e.g. exposure, SAV) which confounded the comparison with Quinte. Other fish community metrics (e.g. centrarchid species richness) showed similar trends, with values at the BoQ comparable to other reference locations with similar habitat. Taken together, the results of this study supported the assumption that the upper BoQ, although within the boundary of an AOC, included healthy habitat and fish assemblages (Minns et al., 1994).

Identifying reference locations in Lake Ontario is a challenge due to the extensive anthropogenic impacts on the nearshore zone (e.g. loss of coastal wetlands; Whillans, 1979). Identifying reference locations has also been a challenge for researchers globally (Magalhaes et al., 2008; Akre Matzen and Berge, 2008). During their study, Hurley and Christie (1977) identified West Lake as a reference location for the upper BoQ due to its similarity in morphometry and limnology but lack of man-made impacts. Between 1969 and 1976, Hurley and Christie (1977) compared catches of fish in West Lake to those from two BoQ locations (Trenton and Hay Bay); they found significant differences in the fish assemblages. One key difference was the percentage of piscivores in the catch between the two areas; at the time, piscivore biomass was 22.5% in West Lake compared to 9.0% and 6.7% in Trenton and Hay Bay, respectively (Hurley and Christie, 1977). The proportion of centrarchids in the catch was also higher at West Lake than Quinte.

Hurley and Christie (1977) suggested that the Quinte nearshore fish community would not be restored until a significant piscivore population was established. Our results and a more extensive analysis of temporal change in the BoQ fish community (Hoyle et al., in press) both suggest fish assemblages in West Lake and the BoQ are now more similar, with comparable contributions of piscivores, centrarchids and other native fishes.

The IBI was designed to integrate the effects of four main factors: water quality, physical habitat supply, piscivore abundance and non-native fishes (Minns et al., 1994). All four factors have had a positive effect on the IBI scores in the upper BoQ since the initial 1990 surveys. GLIBI scores in the upper BoQ increased significantly between 1990 and 1999 and have since remained constant. Improvements in water clarity due to upgrades in wastewater treatment (1977–1978) and Dreissenid Mussel effects (mid-1990s) led to a significant increase in SAV in the upper BoQ (Nicholls et al., 2004; Leisti et al., 2006). An increase in macrophytes led to an increase in phytophilic fishes, centrarchids in particular, which use vegetated habitats for cover, feeding and reproduction. The increase in centrarchids led to a significant increase in native species and a shift in the piscivore assemblage of the upper BoQ. Walleye was the key predator of the 1990 littoral fish community, but based on EF data from this study, Largemouth Bass have replaced Walleye as top predator (Figure S1). According to trap net catches, Walleye are still the key predator of the nearshore fishery in the upper BoQ, but are not at the same population level that was seen in the late 1980s to early 1990s (Hoyle et al., in press) or prior to the late 1960s (Hurley and Christie, 1977). A significant decline in offshore and hyper-eutrophic species (e.g. Alewife and Gizzard Shad) that were abundant in the BoQ prior to phosphorus control and Zebra Mussel invasion has also occurred (Brousseau and Randall, 2009; Hoyle et al., in press). Significant declines in non-native species metrics between 1990 and 1999 were mainly due to a decrease in Alewife and Common Carp in the nearshore (Brousseau and Randall, 2009), but since, non-native species richness has increased due to the introduction of a new (i.e. Round Goby) and the resurgence of an old (i.e. White Perch) invader. White Perch have also increased in the offshore fish community (Hoyle et al., in press).

Spatially and temporally, there have been differences in the IBI scores and other metrics at the three upper BoQ locations. The nearshore fish community has changed significantly at Belleville and Big Bay (specifically, an increase in native species richness and piscivore biomass) but not at Trenton Harbour (Brousseau and Randall, 2009). The reason for the disparity among locations in the upper Bay can in part be explained by SAV. Historically, Trenton Harbour has had higher SAV density and bed extent than Belleville or Big Bay (Leisti et al., 2006) which would have supported a larger number of species than areas with less extensive macrophyte beds. SAV density at Belleville and Big Bay increased significantly post-dreissenid invasion to levels similar to those found at Trenton pre-dreissenids (Leisti et al., 2006). Since 1999, there has been no significant change in IBI scores at the three locations.

Lake-wide trends in our fish surveys support the contention that the upper and middle BoQ included good reference conditions when compared to reference locations and other embayments in Lake Ontario. Both electrofishing and trap net surveys were consistent in supporting the delisting of the degraded fish population Beneficial Use Impairment (BQRAP, 2003) for the BoQ. Both gears indicated differences in the fish community at different habitats and levels of degradation. Fish assemblages in the upper BoQ were similar to those found at reference locations independent of gear type (Hoyle et al., in press). Recent surveys to support both GLAP and AIS programs has shown that the combination of nearshore electrofishing (active) and NSCIN trap nets (passive) offers a more complete picture of the nearshore fishery (C. Brousseau, person. comm.) than either gear by itself. Lake Ontario wide comparisons are difficult as SAV beds at the west end of the lake are limited spatially; most SAV beds are found behind breakwalls or in harbours where habitat patch size is limited.

Summary and conclusions

The Index of Biotic Integrity (Minns et al., 1994) was used to evaluate ecosystem health in the Bay of Quinte, Lake Ontario. Despite being classified an Area of Concern (AOC) in 1985, Index of Biotic Integrity (IBI) values at Bay of Quinte fell within the range or exceeded values from reference locations elsewhere in Lakes Ontario and Erie. Fish survey data collected from 1988–2009 in the Bay of Quinte and elsewhere indicated that (1) the Bay of Quinte has relatively healthy fish habitat despite the AOC designation, (2) Bay of Quinte IBI scores increased significantly between 1990 and 1999 due to changes in relative species richness and (3) differences in fish communities were correlated with physical habitat attributes at survey locations. Data from both nearshore electrofishing and trap net surveys confirmed that the Bay of Quinte supports a highly productive and diverse fish community.

In future, the Great Lakes IBI will likely continue to be used as an indicator of ecosystem health. To improve its usefulness, two revisions are proposed: (1) revisit and update the species classifications identified in the original IBI and (2) calibrate the upper limits to IBI based on habitat conditions (e.g. exposed versus non-exposed shoreline reaches). Recent literature from river ecosystems indicate that habitat variables at the landscape scale such as catchment land use (e.g. urban, agricultural), percent riparian land cover or other measures of human disturbance (e.g. total impervious area) have proven useful in explaining differences in biotic integrity (Akre Matzen and Berge, 2008; Weijters et al., 2009). Landscape or region scale variables may also help explain differences in the nearshore fish community. In addition, GLIBI should be compared and tested against other measures including, multivariate statistics (Kennard et al., 2006; Novotny et al., 2009) and indices (Barbour et al., 1999; Seilheimer and Chow-Fraser, 2006) to examine and explain differences in aquatic communities.

Acknowledgements

We thank the various field crews that have assisted in collecting data over the years: Carolyn Bakelaar for generating the fetch data and Andrew Doolittle for preparing the map. Funding was provided by the Great Lakes Action Plan Environment Canada, DFO Science, and the Lake Ontario Management Unit (OMNR).

References

Akre Matzen, D., Berge, H. B., 2008. Assessing Small-Stream Biotic Integrity Using Fish Assemblages across an Urban Landscape in the Puget Sound Lowlands of Western Washington. Trans. Am. Fish. Soc. 137, 677–689.

Barbour, M. T., Gerritsen, J., Snyder, B. D., Stribling, J. B., 1999. *Rapid bioassessment protocols for use in streams and wadeable rivers: periphyton, benthic macroinvertebrates, and fish, Second Edition.* EPA841-B-99/002. United States Environmental Protection Agency, Washington, DC.

Bay of Quinte RAP Plan, 1993. Time to Act. The Bay of Quinte Remedial Action Plan Stage 2 Report. Bay of Quinte RAP Coordinating Committee, September, 1993. Ontario Ministry of the Environment, Kingston, Ontario.

Bhagat, Y., Ciborowski, J. J. H., Johnson, L. B., Uzarski, D. G., Burton, T. M., Timmermans, S. T. A., Cooper, M. J., 2007. Testing a fish index of biotic integrity for responses to different stressors in Great Lakes coastal wetlands. J. Great Lakes Res. 33(3), 224–235.

Brousseau, C. M., Randall, R. G., 2009. Long-Term Trends in the Littoral Fish Assemblages in the Upper Bay of Quinte-2007. In: Project Quinte Annual Report 2007. Bay of Quinte RAP Restoration Council/Project Quinte. Monitoring Report #18. pp. 113–130. Bay of Quinte RAP. Kingston, Ontario, Canada.

Brousseau, C. M., Randall, R. G., Clark, M. G., 2005. Protocol for boat electrofishing in near shore areas of the lower Great Lakes: transect and point survey methods for collecting fish and habitat data, 1988 to 2002. Fisheries and Oceans Canada, Burlington, Ontario.

Fausch, K.D., Lyons, J., Karr, J.R., Angermeier, P.L., 1990. Fish communities as indices of environmental degradation. In: S.M. Adams (Ed.), *Biological indicators of stress in fish*, pp. 123-144. American Fisheries Society, Symposium 8, Bethesda, Maryland.

Fayram, A. H., Mitro, M. G., 2008. Relationships between Reach-Scale Habitat Variables and Biotic Integrity Score, Brook Trout Density, and Brown Trout Density in Wisconsin Streams. North America. Fisheries 28, 1601–1608.

Haxton, T., Punt, K., 2004. Abundance and Demography of Channel Catfish in a Large Northern River. North America. Fisheries 24, 1440–1446.

Hurley, D. A., Christie, W. J., 1977. Depreciation of the warmwater fish community in the Bay of Quinte, Lake Ontario. J. Fish. Res. Board Can. 38, 1849–1860.

Karr, J. R., 1981. Assessment of Biotic Integrity Using Fish Communities. Fisheries 6, 21–26.

Karr, J. R., 1991. Biological integrity: a long neglected aspect of water resource management. Ecological Applications 1, 66–84.

Kennard, M. J., Harch, B. D., Pusey, B. J., Arthington, A. H., 2006. Accurately defining the reference condition for summary biotic metrics: a comparison of four approaches. Hydrobiologia 572, 151–170.

Leisti, K. E., Millard, E. S., Minns, C. K., 2006. *Assessment of submergent macrophytes in the Bay of Quinte, Lake Ontario, August 2004, including historical context.* Fisheries and Oceans Canada, Burlington, Ontario.

Magalhaes, M. F., Ramalho, C. E., Collares-Pereira, M. J., 2008. Assessing biotic integrity in a Mediterranean watershed: development and evaluation of a fish-based index. Fish. Manage. Ecol. 15, 273–289.

Meador, M. R., Whittier, T. R., Goldstein, R. M., Hughes, R. M., Peck, D. V., 2008. Evaluation of an Index of Biotic Integrity Approach used to assess biological condition in Western U.S. streams and rivers at varying spatial scales. Trans. Am. Fish. Soc. 137, 13–22.

Minns, C. K., Cairns, V. W., Randall, R. G., Moore, J. E., 1994. An Index of Biotic Integrity (IBI) for fish assemblages in the littoral zone of Great Lakes Area of Concern. Can. J. Aquat. Sci. 51, 1804–1822.

Nicholls, K. H., Heintsch, L., Carney, E. C., 2004. A multivariate approach for evaluating progress towards phytoplankton community restoration targets: Examples from eutrophication and acidification case histories. Aquat. Ecosyst. Health Mgmt. 7(1), 15–30.

Novotny, V., Bedoya, D., Virani, H., Manolakos, E., 2009. Linking indices of biotic integrity to environmental and land use variables: multimetric clustering and predictive models. Water Sci. and Tech. 59(1), 1–8.

Portt, C. B., Coker, G. A., Ming, D. L., Randall, R. G., 2006. *A review of fish sampling methods commonly used in Canadian freshwater habitats*. Fisheries and Oceans Canada, Burlington, Ontario.

Randall, R. G., Minns, C. K. 2002. Comparison of a Habitat Productivity Index (HPI) and an Index of Biotic Integrity (IBI) for the measuring the productive capacity of fish habitat in nearshore areas of the Great Lakes. J. Great Lakes Res. 28(2), 240–255.

Seilheimer, T. S., Chow-Fraser, P., 2006. Development and use of the Wetland Fish Index to assess the quality of coastal wetlands in the Laurentian Great Lakes. Can. J. Fish. Aquat. Sci. 63, 354–366.

Weijters, M. J., Janse, J. H., Alkemade, R., Verhoeven, J. T. A., 2009. Quantifying the effect of catchment land use and water nutrient concentrations on freshwater river and stream biodiversity. Aquatic Conserv: Mar. Freshw. Ecosyst. 19, 104–112.

Whillans, T. H., 1979. Historic transformations of fish communities in three Great Lakes bays. J. Great Lakes Res. 5(2), 195–215.

Wilson, H. F., Xenopoulos, M. A., 2008. Landscape influences on stream fish assemblages across spatial scales in a northern Great Plains ecoregion. Can. J. Fish. Aquat. Sci. 65(2), 245– 257.

The above article is a reprinted version of the following previously published article:

Brousseau, C.M., Randall, R.G., Hoyle, J.A., Minns, C.K., 2011. Fish community indices of ecosystem health: How does the Bay of Quinte compare to other coastal sites in Lake Ontario?

Aquatic Ecosystem Health & Management (2011) 14 (1): 75–84.

https://doi.org/10.1080/14634988.2011.5473311

Supplementary material

Table S1. Results of a non-parametric one-way analysis (Mann Whitney U test) for electrofishing and trap net data surveys conducted in the upper Bay of Quinte and West Lake for the dependent variables centrarchid species richness, non-native species richness and piscivore biomass (%) in West Lake by year.

Year	Category	**Electro-fishing**			**Trap nets**		
		df	Test statistic	P	df	Test statistic	P
1999	Centrarchid species	1,58	371.00	0.843			
	Non-native species	1,58	414.00	0.090			
	Piscivore biomass (%)	1,58	311.00	0.410			
2001	Centrarchid species				1,22	36.00	0.344
	Non-native species				1,22	43.00	0.665
	Piscivore biomass (%)				1,22	50.00	0.879
2007	Centrarchid species	1,45	109.00	0.114	1,34	140.50	0.900
	Non-native species	1,45	151.50	0.113	1,34	166.00	0.399
	Piscivore biomass (%)	1,45	136.00	0.461	1,34	125.50	0.513
2009	Centrarchid species	1,47	166.00	0.245			
	Non-native species	1,47	266.00	0.081			
	Piscivore biomass (%)	1,47	237.00	0.510			

Table S2. Results from non-parametric one-way analyses of variance for temporal differences (year) in nearshore electrofishing data for the dependent variables centrarchid species richness, non-native species richness and piscivore biomass (%) by location. Upper Bay of Quinte transect locations were compared to those from reference locations in middle Bay of Quinte and West Lake. Sampling years varied by location.

Location	Fish category	df	K-W test statistic	P
Upper Bay of Quinte	Centrarchid species	3,170	67.84	<0.001
	Non-native species		54.05	<0.001
	Piscivore biomass (%)		3.87	0.424
Middle Bay of Quinte	Centrarchid species	2,24	10.31	0.006
	Non-native species		3.49	0.175
	Piscivore biomass (%)		0.15	0.927
West Lake	Centrarchid species	2,37	9.30	0.010
	Non-native species		8.68	0.013
	Piscivore biomass (%)		0.29	0.865

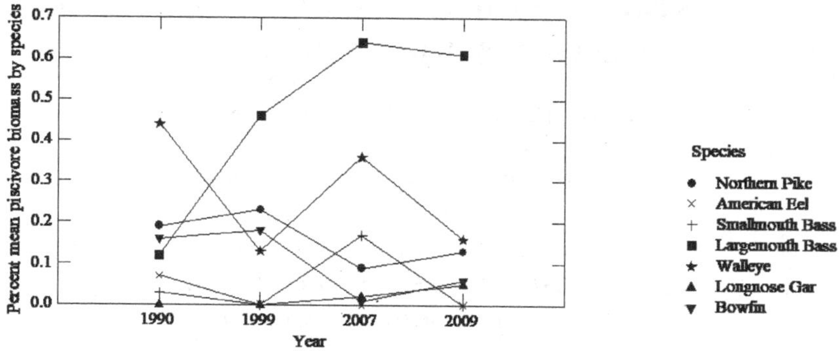

Figure S1. Trends in the piscivore biomass at nearshore electrofishing locations in the upper Bay of Quinte, showing the increase in catch of Largemouth Bass.

An integrated approach to identifying ecosystem recovery targets: Application to the Bay of Quinte

E.A. Blukacz-Richards[1,2] and M.A. Koops[1,*]

[1]*Fisheries and Oceans, 867 Lakeshore Road, Burlington, ON L7S 1A1*
[2]*Environment and Climate Change Canada, 867 Lakeshore Road, Burlington, ON L7S 1A1*
**Corresponding author: marten.koops@dfo-mpo.gc.ca*

Keywords: Areas of Concern, Ecopath, mass-balance, Beneficial Use Impairment, Remedial Action Plans, restoration targets

Introduction

Ecological restoration, whether it is targeting individual species or whole ecosystems, needs to be considered within the context of the interacting components of the ecosystem. An ecosystem approach to management has received a lot of attention in the marine fisheries literature (Browman and Stergiou, 2004; Hall and Mainprize, 2004; Francis et al., 2007). However, an ecosystem approach is equally applicable, and has a longer history, in the management of freshwater ecosystems (Koops et al., 2009). Most authors, when describing an ecosystem approach, include the identification of clear, explicitly stated, long-term objectives and integration across multiple levels. When managing ecosystem restoration, it is common to explicitly identify targets at multiple trophic levels. While the literature on ecological restoration recognizes the importance of foodweb interactions (Peterson and Lipcius, 2003; Vander Zanden et al., 2003; Ryder and Miller, 2005; Lake et al., 2007), it is rare to use a modelling framework to set or evaluate targets at multiple trophic levels.

The advantage to using a modelling framework to evaluate ecosystem recovery targets is that it provides the opportunity to ensure that targets at multiple trophic levels are compatible. The modelling framework employed must depend on the context and goals to be achieved. A common framework within which recovery targets are evaluated will also help when there are multiple human uses being served by the ecosystem or restoration goals. When there are multiple interests, it may not be possible to achieve optimal outcomes for all users, but there are

Ecology of the Bay of Quinte: Health, Management and Global Implications
Editors: C.K. Minns, M. Munawar and M.A. Koops

Aquatic Ecosystem Health & Management Society, Canada

tools that can be employed to find acceptable trade-offs (McLaughlin et al., 2013). Here we consider ecosystem recovery targets at multiple trophic levels. These targets set desirable outcomes for the foodweb of a single ecosystem, but have not been evaluated to determine if it is possible to achieve these targets within the same foodweb. In 1985, the Great Lakes Water Quality Board of the International Joint Commission (IJC) identified degraded areas in the Great Lakes Basin, as Areas of Concern (AOCs). These were areas where beneficial uses of water were impaired resulting in ecosystem effects such as presence of undesirable algae and degraded communities of fish, benthic invertebrates, and zooplankton. Other contributing factors include excessive phosphorus loadings, persistent toxic contaminants, bacteriological contamination, and degraded coastal fish and wildlife habitat. To remediate these areas, the IJC recommended that stakeholders implement a Remedial Action Plan (RAP) with specific targets and timelines for each AOC.

In 1987, the Bay of Quinte was listed in the Great Lakes Water Quality Agreement (GLWQA) as an AOC. The Bay of Quinte has been one of the most intensively studied ecosystems on the Great Lakes since 1972. The presence of the Ontario Ministry of Natural Resources (OMNR), Lake Ontario Fisheries Assessment Unit on the bay and ongoing nutrient and lower trophic level monitoring (Fisheries and Oceans Canada [DFO] and Ontario Ministry of the Environment [OME]) has ensured a consistent and abundant data set on biomass, species composition, and production from phytoplankton to fish.

Historically, the Bay of Quinte was one of the most productive fisheries in the Great Lakes (Johnson and Hurley, 1986) which has been influenced by cultural eutrophication (1950s and 1960s), phosphorus control (1980s to 1990s), and invasive species (e.g. Zebra Mussel (*Dreissena polymorpha*) invaded in 1994) resulting in rapid foodweb changes. Ecopath models, which combine traditional mass-balance with population dynamics, have been widely used to examine foodweb changes (Stewart and Sprules, 2011). Such models have been used as fishery management tools (Pauly et al., 2000) and for analyses of basic ecosystem properties (Vasconcellos et al., 1997). Ecopath could serve as a modelling framework to determine if foodweb recovery targets are mutually compatible or mutually exclusive.

The Bay of Quinte RAP has set restoration targets corresponding to the beneficial use impairments, as identified in the GLQWA. Previous assessments have focused on one BUI at a time. However, many BUIs relate to foodweb components and are not independent of each other. To evaluate the compatibility of these foodweb targets, we used output from a multi-trophic Ecopath model to determine if the delisting targets set by the Bay of Quinte RAP are internally consistent and realistically achievable. The ultimate goal for the RAP team is to delist the Bay of Quinte as an AOC by achieving its targets. This study provides a quantitative framework for evaluating whether the targets can all be achieved and to guide adaptive management actions.

Materials and Methods

Ecopath and RAP delisting targets

The Bay of Quinte (44°09′ N, 77°15′ W) is situated on the northeast margin of Lake Ontario and forms a Z-shaped body of water between the mainland to the north and the Prince Edward County peninsula to the south and west (Figure 1). The Bay of Quinte has an area of 254 km^2 and it is composed of three bays (upper, middle and lower). We used an Ecopath model developed by Koops et al. (2006) that focused on the upper bay, representing just over 50% (133.4 km^2) of the whole bay, based on input data from the post-Zebra Mussel invasion period (1994–2000). The model, including the age structure for key fish species, was composed of 47 trophic groups (Table 1). We modified the Koops et al. (2006) model so we could evaluate four BUIs and a total of seven targets across multiple trophic levels (Table 2).

Cormorants were used to assess the degradation of wildlife populations. A total of 22 groups of fishes, in some cases age classes (>1 to 5 years) were included (Table 1), were used to assess fish community health. The final fish biomass output from Ecopath was used to estimate Minns et al.'s (1994) index of biotic integrity (IBI) and compared to the delisting target (Table 2). The benthic community in Ecopath was divided into pollution sensitive and insensitive species to assess changes in biomass and community structure (Table 2). Delisting targets for zooplankton and phytoplankton populations were addressed separately. Zooplankton were divided into herbivores, carnivores, and invasive species, while phytoplankton were categorized as edible, non-edible, and nuisance fractions (USEPA, Planktonic state indicators; Kane et al., 2008). The remaining targets were assessed with weighted statistical models (Table 3) because it was not possible to stabilize the variation in the error terms with data transformations. The weights were estimated as the inverse of the absolute value of the residuals (Blukacz et al., 2010; Neter et al., 1996).

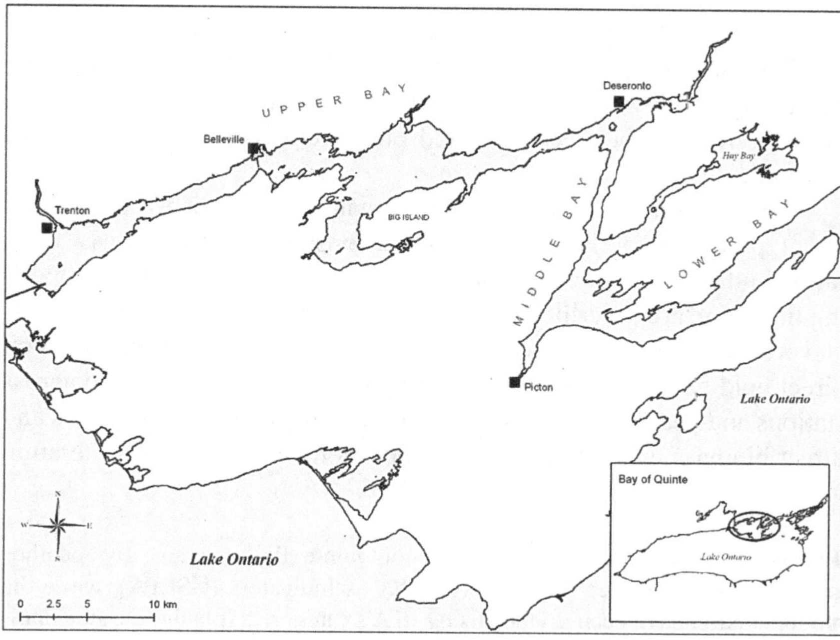

Figure 1. The Bay of Quinte is situated on the northeast margin of Lake Ontario, forming a Z-shaped body of water. It is composed of three bays (upper, middle and lower). The model focused on the upper bay.

Ecopath approach and input data

Ecopath is a mass-balance approach to modelling ecosystems based on biomass (B), production (P/B), consumption (Q/B) and diet composition (DC; Christensen and Walters 2004). The mass-balance for each group in the Bay of Quinte model was:

$$B_i \times \left(\frac{P}{B}\right)_i - \sum_{j=1}^{n} B_j \times \left(\frac{Q}{B}\right)_j \times DC_{ij} - B_i \times \left(\frac{P}{B}\right)_i \times (1 - EE_i) = 0$$

where B_i is the biomass of group *i*, $(P/B)_i$ is the production to biomass ratio, $(Q/B)_j$ is the predator's consumption to biomass ratio, DC_{ij} is the proportion of prey *i* in predator *j*'s diet, EE_i is the ecotrophic efficiency (proportion lost to predation), and n is the number of groups (Table 1). These linear equations were set up and balanced for each group by iteratively varying B until EE < 1 for each group. In a balanced model the EEs must be between zero and one (inclusive) and values above one indicate a prey deficit.

For each group, Ecopath requires mandatory input for biomass (t km^{-2}), production to biomass (P/B) and consumption to biomass (Q/B) ratios, and a diet composition matrix. These estimates, including supporting literature, are described in the online appendix for Blukacz-Richards and Koops (2012). The biomass values for each group were based on direct field observations and, when required, data were averaged across stations and years. Q/B ratios were estimated using literature values and direct biomass estimates. All diet proportions were based on literature values.

Table 1. Trophic groups used in the Bay of Quinte. Pollution sensitive benthos are based on categories from Invertebrates as Indicators (USEPA), while the phytoplankton have been divided using EPA's categories (planktonic state indicators). YOY: young of the year fish; DOC: dissolved organic carbon.

Category	Number of Groups	Trophic Groups
Birds	1	Cormorants
Fish	22	Piscivores, Walleye (age >5, age 4, age 3, age 2, age 1 and YOY), Smallmouth Bass, Alewife, Yellow Perch (age >1, YOY), White Perch (age >1, YOY), panfish (age >1, YOY), invertivores, planktivores, Trout-Perch, Freshwater Drum, Round Goby, Common Carp, Gizzard Shad
Benthos	8	Pollution-sensitive (insects, bivalves, amphiopods, gastropods), pollution-insensitive (oligochaetes, chironomids, isopods, dreissinids) and other benthos (no insects)
Zooplankton	5	Rotifers, *Cercopagis*, predatory cladocerans, copepods, herbivorous zooplankton
Primary Producers	8	Phytoplankton (inedible, microcystis, anabena, aphanixomenon, remaining edible), macrophytes, periphyton, epiphyton
Detrital Pools	3	Sedimented detritus, pelagic detritus, DOC
Total	47	

Balancing Ecopath

To determine if the Bay of Quinte delisting targets were achievable, we quantified the consequences of bottom-up-control by gradually reducing phytoplankton biomass until the Ecopath model became unbalanced (i.e. EE's exceeded one). Total phytoplankton biomass was reduced by increments of 5% in proportion to each group. The unbalanced model was then re-balanced using the Ecopath (version 5.0 Beta) auto-balance routine (Kavanagh et al., 2004). For each run, the auto-balance routine randomly selected biomass values that fell within zero to 100% of the initial biomass inputs. All biomass values were randomly varied, except for aquatic vegetation (macrophytes, epiphytes, and periphytes) and all of the phytoplankton groups (inedible, microcystis, anabena, aphanixomenon, and remaining edible). As recommended by Stewart and Sprules (2011), the auto-balanced procedure was repeated 25 times where each replicate was based on up to 10,000 iterations, ensuring that a mass-balanced solution was reached. The replicates were compared with the baseline balanced model through estimates of percent differences for each group.

Results

The total biomass in the baseline model was composed of 58% benthic invertebrates (55% represented by Zebra Mussel), 22% aquatic vegetation, 12% detritus, ~2% phytoplankton, ~1% zooplankton, and the remaining 5% was fish and cormorants. In the baseline balanced model, the total phytoplankton biomass was 9.96 (t km^{-2}) and the model remained balanced until phytoplankton biomass was reduced by 27% (7.23 t km^{-2}). This reduction resulted in EEs for remaining edible algae that exceeded one. Percent differences between the original baseline balanced model and the 25 auto-balanced replicates (at reduced phytoplankton biomass) were computed by guild (cormorants, piscivores, planktivores, benthivores, benthic invertebrates, and zooplankton) and for each species-group. On average, benthic invertebrates had the highest biomass loss (54%), followed by cormorants which showed a biomass loss of 13%, while the remaining guilds had gains in biomass ranging from 12% to 25% (Figure 2A). On average, Walleye age one to four and young of the year (YOY) panfish showed a loss in biomass while all the remaining fishes showed a gain in biomass (Figure 2B–D). For the invertebrate groups, on average, Zebra Mussel had the highest biomass loss followed by chironomids, all the remaining groups showed an increase in biomass ranging from 0.35% to 57% (Figure 2E and F).

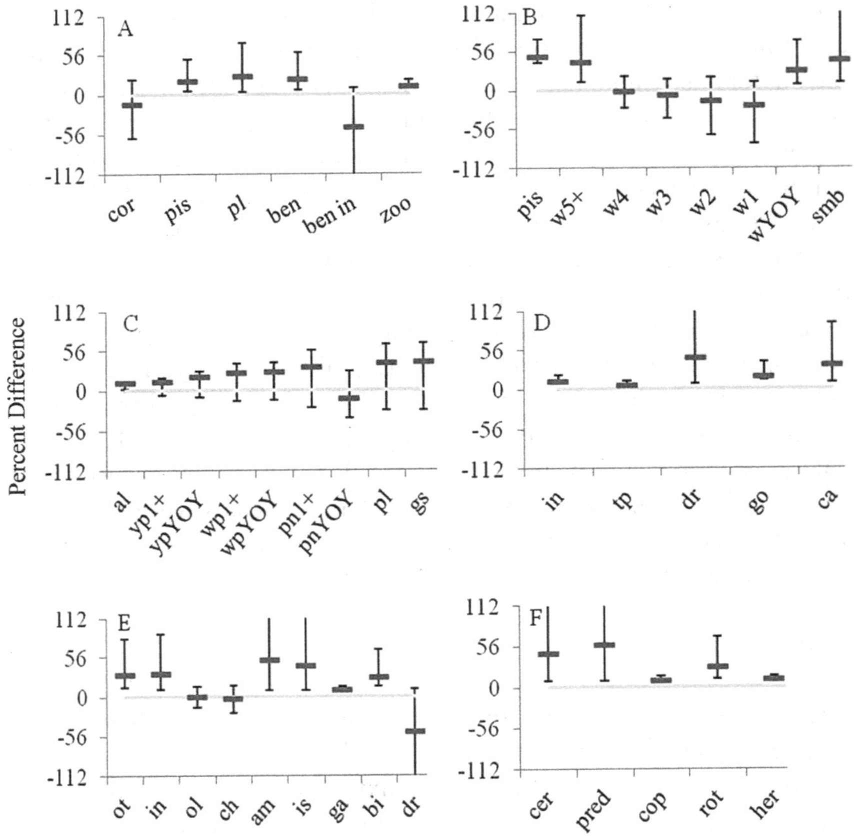

Figure 2. Averaged percent differences computed between the baseline-model and the auto-balanced replicates in response to a simulated 27% reduction in phytoplankton shown by guild (A) and by group (B-F). Abbreviations start from panel (A): Cormorants (cor), Piscivores (pis), Planktivores (pl), Benthivores (ben), Benthic Invertebrates (ben in), Zooplankton (zoo); (B): Piscivores (pis), Walleye (w) with age indicated, Smallmouth Bass (sm); (C): Alewife (al), Yellow Perch (yp) with age indicated, White Perch (wp) with age indicated, Panfish (pn) with age indicated, Gizzard Shad (gs); (D): invertivores (in), Trout-Perch (tp), Freshwater Drum (dr), Round Goby (go), Common Carp (ca); (E): other Benthos insects not included (ot), Insects (in), Oligochaetes (ol), Chironomids (ch), Amphipods (am), Isopods (is), Gastropods (ga), Bivalves (bi), Dreissinids (dr); (F): Cercopagis (cer), Predatory Cladocerans (pred), Copepods (cop), Rotifers (rot), and Herbivorous Zooplankton (her). The gray horizontal lines represent zero percent difference.

The BUIs, the corresponding targets, and the targets derived from the averaged auto-balanced Ecopath output are shown in Table 2 where all the reductions were estimated as percent differences from baseline. The fish population RAP target was exceeded (IBI = 86), indicating an excellent fish community. The piscivore biomass represented 15–22% of the total fish biomass (average: 18%) which fell short of the 20% RAP target. The biomass increase of pollution sensitive benthic invertebrates ranged from 10% to 52% according to the Ecopath responses, while pollution tolerant groups such as Chironomids and Zebra Mussels showed a decrease in biomass (Figure 2E). Chlorophyll-*a* concentration and total phosphorous were back-calculated (using relationships defined in Table 3) and both of these estimates fell within the RAP targets representing reduction from baseline of 23% and 36%, respectively (Table 2). The 27% reduction in phytoplankton biomass was applied across all algal species, and represents a reduction to just less than the RAP algae target.

Discussion

Our Ecopath model simultaneously addressed seven RAP targets and four BUIs dealing with communities of fish, benthic invertebrates, and phytoplankton. We quantified the trophic consequences of bottom-up control by gradually reducing phytoplankton biomass until the model became unbalanced (27% decrease); after that the auto-balance routine was used to identify the range of foodweb outcomes that could be supported. These replicated mass-balanced solutions represent robust estimates of trophic dynamics in the Bay of Quinte (1995–2002), where, on average, total biomass decreased by 26% and ranged between 50% and 105% of the baseline total. In the baseline model, 55% of the biomass was composed of Zebra Mussel and in the mass- balanced replicates Zebra Mussel biomass was consistently lower (maximum decrease of 93%). This is not a prediction that Zebra Mussel biomass will decline under a reduced phytoplankton scenario, but does suggest that a reduction in Zebra Mussel biomass was required to accommodate all the predatory demands on phytoplankton. In the Bay of Quinte, Millard and Burley (2004) attributed Zebra Mussel grazing to lower phytoplankton biomass and increased water clarity. This is also consistent with Koops et al.'s (2006) estimation that 99% of phytoplankton production in the Bay of Quinte was being consumed post-Dreissenid invasion. Zebra Mussel grazing has been implicated in declines of chlorophyll concentrations in the Hudson River (Caraco et al.,

Table 2. The Bay of Quinte, beneficial use impairments (BUIs) and the delisting targets from the RAP. The responses from Ecopath were estimated from the averaged 25 auto-balanced replicates and all reductions were estimated as percent differences from the baseline. NA are BUIs that were not assessed with the Ecopath model.

BUI	RAP Targets	Ecopath Responses to Simulated Phytoplankton Reduction
3. Degraded fish populations	Index of Biotic Integrity (IBI) ≥ 60 (Minns et al., 1994).	IBI = 86
	Proportion of piscivore biomass (PPB) in the fish community of 20% or greater	15-22% (average: 18%) of the fish biomass represented by piscivores.
6. Degraded Benthic Invertebrates	Presence and increasing abundance of pollution sensitive benthic invertebrates species.	Average gain in biomass for pollution sensitive benthic invertebrates: (1) 10% for Gastropods, (2) 33% for Insects, and (3) 52% for Amphiopods.
	Increased species and community diversity.	
8. Eutrophication or undesirable algae	25% reduction in total phosphorus (TP) concentrations from 40 μg l^{-1} to 30 μg l^{-1}	23% reduction with TP = 24.28 μg l^{-1}
	40% reduction in algal level with phytoplankton biomass of 4-5 g m^{-3} for the May to October period.	27% reduction with phytoplankton biomass of 7.24 t km^{-2} (annual) = 3.8 g m^{-3} (seasonal).
11. Degraded aesthetics	Seasonal mean Secchi depth of 1.3 m.	NA
13. Degraded phytoplankton populations	Decrease in nuisance and eutrophic indicator phytoplankton species.	27% reduction for all algal species.
	Chlorophyll-*a* concentrations of 9-10 μg l^{-1}	36% reduction with Chlorophyll-*a* = 9.10 μg l^{-1}.
14. Loss of fish and wildlife habitat	20% increase in area of submerged aquatic macrophytes.	NA

Table 3. Statistical models used to assess total phosphorus (TP) and chlorophyll (Chl) concentration. Phyto-B refers to phytoplankton biomass.

Model	P-values and R^2
Phyto-B = 2.48 + 0.523 *Chl	P < 0.0001, 82.7%
Phyto-B = −1.77 + 0.371*TP	P = 0.017, 52.5%

2006) and lower primary production in Lake Erie (Madenjian, 1995). However, Stewart and Sprules' (2011) Lake Ontario offshore Ecopath model suggests that the observed decline in primary production (2001–2005) is unlikely to be the result of Zebra Mussel grazing because it accounted for less than 1% of the total phytoplankton grazing. Our model represented a snapshot of the upper bay where Zebra Mussel were relatively common and their grazing accounted for a significant proportion of the total phytoplankton grazing (Project Quinte Annual Report, 2005; Koops et al., 2006). These results show that the trophic dynamics in the upper Bay of Quinte are distinct from the Lake Ontario dynamics quantified by Stewart and Sprules' (2011) offshore Lake Ontario model, suggesting that spatially distinct dynamics should be considered in future modelling efforts. The bathymetry of the Bay of Quinte changes from the upper to the lower bay. With the lower bay being relatively deep, the dynamics in the lower bay may be different from those observed in our model of the upper bay. Future modelling efforts could consider an approach that models the spatial differences in the Bay of Quinte coupled with an offshore model of Lake Ontario to provide further insight into whole lake production dynamics.

The 27% decrease in phytoplankton biomass (bottom-up control) did not result in proportionate declines in the upper trophic levels (i.e. zooplankton) (Figure 2) and individual species-groups responded differently. On average, Zebra Mussels had the highest biomass loss followed by chironomids, while the remaining groups showed an average increase in biomass with some balanced solutions showing a decrease (e.g. herbivorous zooplankton). For the vertebrate groups, YOY panfish, Walleye (age one to four), and cormorants all showed an average decrease in biomass. These results show that none of the species-groups responded in a linear manner, suggesting that future Ecopath models should thoroughly investigate mass-balance solutions using different bottom-up and top-down controls. For example, we only decreased phytoplankton groups, however, as was observed earlier in the Bay of Quinte (Koops et al., 2009; Minns et al., 2011), decreases in phosphorus and algal biomass led to increases in macrophytes and benthic algae. This creates the possibility for alternative energy pathways, the implications of which could be explored with future modelling efforts.

Summary and conclusions

In 1985, the International Joint Commission identified the Bay of Quinte as an Area of Concern due to its degraded ecosystem. A Remediation Action Plan was established with delisting targets including the goals of decreasing phosphorous loading and restoring the upper (fish and wildlife) and lower (phytoplankton, zooplankton, and benthic invertebrates) trophic levels. We examined the consistency among seven Remedial Action Plan targets using Ecopath, a mass-balance model, for the upper Bay of Quinte for the post-Zebra Mussel (*Dreissena polymorpha*) invasion period (1995–2002). We quantified the trophic consequences of bottom-up-control by gradually reducing the phytoplankton biomass until the Ecopath model became unbalanced (27% reduction). Replicate (n = 25) mass-balance solutions consistently showed that reductions in Zebra Mussel biomass were necessary to achieve mass-balance. This bottom-up control met the nutrient (total phosphorus), fish, benthic invertebrates, and phytoplankton population RAP targets. Based on these consistent results, it is recommended that future modelling efforts examine the effects of further phytoplankton biomass reductions.

Our modelling predicted that there is broad compatibility among the fish, benthic invertebrate, and phytoplankton population targets set out by the RAP team (Table 2), and that achieving these targets is consistent with bottom-up control of the foodweb. The one exception was that the piscivore proportion of total fish biomass may not be met under a reduced phytoplankton biomass scenario. The model was unbalanced, and there were foodweb perturbations, when phytoplankton biomass was reduced below the RAP algae target. Given that phytoplankton biomass has been observed to vary by as much as 50% (Project Quinte Annual Report, 2005), we recommend that future modelling efforts further examine the impact of reduced phytoplankton biomass on the other RAP delisting targets. As a snapshot of the foodweb, Ecopath models do not evaluate the impact of inter-annual variation, which may affect the dynamics of aquatic ecosystems and may affect the capacity of a foodweb to support the desired fish community. This study provides a scientific framework for evaluating the compatibility of ecological restoration targets, and demonstrates that the foodweb targets set by the Bay of Quinte RAP are mutually compatible. While the targets evaluated are specific to the Bay of Quinte, the approach is broadly applicable and could be used to help set or evaluate restoration targets in degraded ecosystems (e.g. Hamilton Harbour; Hossain et al., 2012).

Acknowledgements

We thank the Quinte-Oneida Comparative Modelling Project for the use of the Ecopath model they developed for the upper Bay of Quinte. We also thank two anonymous reviewers for their helpful comments. Funding for this work was provided through the Great Lakes Action Plan (GLAP).

References

Blukacz-Richards, E.A., Koops, M.A., 2012. An integrated approach to identifying ecosystem recovery targets: application to the Bay of Quinte. Aquatic Ecosystem Health Management 15, 464-472.

Blukacz, E.A., Koops, M.A., Sutton, T.M., Arts, T.M., Fitzsimons, J.D., Muir, A.M., Claramunt, R.M., Johnson, T.B., Kinnunen, R.E., Ebener, M.P., Suski, C., Burness, G., 2010. Linking Lake Whitefish (*Coregonus clupeaformis*) condition with male gamete quality and quantity. J. Great Lakes Res. 36, 78–83.

Browman, H.I., Stergiou, K.I. (Eds.), 2004. Perspectives on ecosystem-based approaches to the management of marine resources. Mar. Ecol. Prog. Ser. 274, 269–303.

Caraco, N.F., Cole, J.J., Strayer, D.L., 2006. Top-down control from the bottom: regulation of eutrophication in a large river by benthic grazing. Limnol. Oceanogr. 51, 664–670.

Christensen, V., Walters, C.J., 2004. Ecopath with Ecosim: methods, capabilities and limitations. Ecol. Model. 172, 109–139. Francis, R.C., Hixon, M.A., Clark, M.E., Murawski, S.A., Ralston, S., 2007. Ten commandments for ecosystem-based fisheries scientists. Fisheries 32, 217–233.

Francis, R.C., Hixon, M.A., Clark, M.E., Murawski, S.A., Ralston, S., 2007. Ten commandments for ecosystem-based fisheries scientists. Fisheries 32, 217-233.

Hall, S.J., Mainprize, B., 2004. Towards ecosystem-based fisheries management. Fish Fisheries 5, 1–20. Invertebrates as Indicators. USEPA. http://www.epa.gov/bioiweb1/html/invertebrate.html. Accessed 25 August 2012.

Hossain, M., Arhonditsis, G.B., Koops, M.A., Minns, C.K., 2012. Towards the development of an ecosystem model for the Hamilton Harbour, Ontario, Canada. J. Great Lakes Res. 38, 628-642.

Johnson, M.G., Hurley, D.A., 1986. Overview of Project Quinte – 1972–82. In: Project Quinte: point source phosphorus control and ecosystem response in the Bay of Quinte, Lake Ontario. Can. Spec. Publ. Fish Aquat. Sci. 85, 1–6.

Kane, D.D., Gordon, S.I., Munawar, M., Charlton, M.N., Culver, D.A., 2008. A planktonic index of biotic integrity (P-IBI) for Lake Erie: A new technique for checking the pulse of Lake Erie. In: M. Munawar and R. Heath (Eds.), *Checking the Pulse of Lake Erie*. pp. 347–367. Ecovision World Monograph Series, Aquatic Ecosystem Health and Management Society, Burlington, ON.

Kavanagh, P., Newlands, N., Christensen, V., Pauly, D., 2004. Automated parameter optimization for Ecopath ecosystem models. Ecol. Modell. 172, 141–149.

Koops, M.A., Irwin, B.J., MacNeil, J.E., Millard, E.S., Mills, E.L., 2006. Comparative ecosystem modelling of the ecosystem impacts of exotic invertebrates and productivity changes on fisheries in the Bay of Quinte and Oneida Lake. Great Lakes Fishery Commission Project Completion Report, Ann Arbor, MI.

Koops, M.A., Dermott, R.M., Leisti, K.E., Johannsson, O.E., Millard, E.S., Minns, C.K., Munawar, M., Nicholls, K.H., Hoyle, J.A. 2009. The Bay of Quinte: a model for large lake ecosystem management. Verh. Internat. Verein. Limnol. 30, 1024–1029.

Lake, P.S., Bond, N., Reich, P., 2007. Linking ecological theory with stream restoration. Freshw. Biol. 52, 597–615.

Madenjian, C.P., 1995. Removal of algae by the Zebra Mussel (*Dreissena polymorpha*) population in western Lake Erie: a bioenergetics approach. Can. J. Fish. Aquat. Sci. 52, 381–390.

McLaughlin, R.L., Smyth, E.R.B., Castro-Santos, T., Jones, M.L., Koops, M.A., Pratt, T.C., Vélez-Espino, L.A. 2013. Unintended consequences and trade-offs of fish passage. Fish Fisheries DOI: 10.1111/faf.12003.

Millard, S., Burley, M., 2004. Photosynthesis, Chlorophyll-*a* and Light Extinction. In: Monitoring Report #13. Project Quinte Annual Report 2002, pp. 21–38. Bay of Quinte Remedial Action Plan, Kingston, ON, Canada.

Minns, C.K., Cairns, V.W., Randall, R.G., Moore, J.E., 1994. An index of biotic integrity (IBI) for fish assemblages in the littoral zone of Great Lakes' Areas of Concern. Can. J. Fish. Aquat. Sci. 51, 1804–1822.

Minns, C.K., Munawar, M., Koops, M.A., Millard, E.S. 2011. Long-term ecosystem studies in the Bay of Quinte, Lake Ontario, 1972–2008: a prospectus. Aquatic Ecosystem Health & Management 14(1), 3–8.

Neter, J., Kutner, M.H., Nachtsheim, C.J., Wasserman, W., 1996. *Applied Linear Statistical Models, Fourth Ed.* McGraw Hill, New York.

Pauly, D., Christensen, V., Walters, C., 2000. Ecopath, Ecosim, and Ecospace as tools for evaluating ecosystem impact of fisheries. ICES J. Mar. Sci. 57, 697–706.

Peterson, C.H., Lipcius, R.N., 2003. Conceptual progress towards predicting quantitative ecosystem benefits of ecological restorations. Mar. Ecol. Prog. Ser. 264, 297–307.

Planktonic state indicators. USEPA. http://www.epa.gov/med/grosseile_site/indicators/plankton.html. Accessed 25 August 2012.

Project Quinte Annual Report, 2005. Bay of Quinte Remedial Action Plan Monitoring Report 16. Kingston, ON, Canada.

Ryder, D.S., Miller, W., 2005. Setting goals and measuring success: linking patterns and processes in stream restoration. Hydrobiologia 552, 147–158.

Stewart, T.J., Sprules, W.G., 2011. Carbon-based balanced trophic structure and flows in the offshore Lake Ontario food-web before (1987–1991) and after (2001–2005) invasion-induced ecosystem change. Ecol. Model. 222, 692–708.

Vander Zanden, M.J., Chandra, S., Allen, B.C., Reuter, J.E., Goldman, C.R., 2003. Historical foodweb structure and restoration of native aquatic communities in the Lake Tahoe (California-Nevada) basin. Ecosystems 6, 274–288.

Vasconcellos, M., Mackinson, S., Sloman, K., Pauly, D., 1997. The stability of trophic mass-balance models of marine ecosystems: a comparative analysis. Ecol. Model. 100, 125–134.

The above article is a reprinted version of the following previously published article:

Blukacz-Richards, E.A., Koops, M.A., 2012.
An integrated approach to identifying ecosystem recovery targets: Application to the Bay of Quinte.
Aquatic Ecosystem Health & Management (2012) 15 (4): 464–472.
https://doi.org/10.1080/14634988.2012.738997

Integration of best management practices in the Bay of Quinte watershed with the phosphorus dynamics in the receiving waterbody: What do the models predict?

George B. Arhonditsis,[1]* Dong-Kyun Kim,[1] Yuko Shimoda,[1] Weitao Zhang,[1] Sue Watson,[2] Shan Mugalingam,[3] Maria Dittrich,[1] Kristin Geater,[4] Christine McClure,[5] Bryon Keene,[5] Andrew Morley,[6] Agnes Richards,[2] Tanya Long,[7] Yerubandi R. Rao,[2] and Rimi Kalinauskas[4]

[1]*Department of Physical and Environmental Sciences, University of Toronto, Toronto, ON M1C 1A4*

[2]*Environment Canada, National Water Research Institute, Burlington, ON L7S 1A1*

[3]*Lower Trent Conservation, Trenton, ON K8V 5P4*

[4]*Environment Canada, Great Lakes Areas of Concern Section, North York, ON M3H 5T4*

[5]*Quinte Conservation, Belleville, ON K8N 4Z2*

[6]*Kingston Regional Office, Ontario Ministry of the Environment, Kingston, ON K7M 8S5, Canada*

[7]*Environmental Monitoring and Reporting Branch, Ontario Ministry of the Environment, Toronto, ON M9P 3V6*

**Corresponding author: george.arhonditsis@utoronto.ca.*

Keywords: eutrophication, phosphorus modelling, internal loading, watershed management, water quality criteria, sediment dynamics

Introduction

A rigorous analysis of decision problems in water quality management requires specification of (i) a finite set of alternative management actions, including any constraints on their use; (ii) an objective function for evaluating the alternative management strategies; (iii) predictive models of system dynamics formulated in terms of quantities relevant to management objectives; and (iv) a monitoring program to follow system evolution and responses to management (Walters, 1986). The objective function specifies the value of alternative management actions and usually accounts for social benefits and costs, as well as conditional constraints. The predictive models must realistically reproduce the relevant behaviors of aquatic ecosystems, which often involve complex non-linear

Ecology of the Bay of Quinte: Health, Management and Global Implications
Editors: C.K. Minns, M. Munawar and M.A. Koops

Aquatic Ecosystem Health & Management Society, Canada

dynamics and are characterized by spatial, temporal, and organizational heterogeneity (Arhonditsis et al., 2007). Perhaps, even greater challenges are posed by the uncertainty in the predictions of management outcomes. This uncertainty may stem from incomplete control of management actions, errors in measurement and sampling, environmental variability, or incomplete knowledge of system behaviour. Failure to recognize and account for these sources of uncertainty can severely affect management performance, and in some cases, has led to catastrophic environmental and economic losses (Reckhow et al., 2005). The currently popular notion of adaptive management implementation is specifically designed to accommodate the dynamics of uncertainty in both the environmental policy making process as well as model development (Dorazio and Johnson, 2003).

The Bay of Quinte, a Z-shaped embayment at the northeastern end of Lake Ontario, represents a characteristic case, where scientific uncertainty underlies the current management efforts to integrate environmental concerns with socioeconomic values. This uncertainty exists despite one of the longest continuing research and monitoring programs of any site in the Great Lakes. The Bay of Quinte has a long history of eutrophication problems resulting in frequent and spatially extensive algal blooms, predominance of toxic cyanobacteria, and hypolimnetic oxygen depletion (Minns et al., 2011). Thus, it was one of the 43 degraded sites around the Great Lakes designated by the International Joint Commission as Areas of Concern (AOCs) in 1986[1]. Following the AOC designation, the Bay of Quinte Remedial Action Plan (RAP) was formulated through a wide variety of government, private sector, and community participants to provide the framework for actions aimed at restoring the system. In this regard, all the restoration efforts in the Bay were founded upon an "ecosystem" management approach, which was selected as a pragmatic means for

---

1 Areas of Concern are sites on the Great Lakes system where environmental quality is significantly degraded and beneficial uses are impaired. In the Bay of Quinte, there are 9 beneficial use impairments: Degraded Fish and Wildlife Populations, Degradation of Benthos, Eutrophication or Undesirable Algae, Restrictions on Drinking Water Consumption or Taste and Odor Problems, Beach Closings, Degradation of Aesthetics, Degradation of Phytoplankton and Zooplankton Populations, Loss of Fish and Wildlife Habitat, Restrictions on Fish and Wildlife Consumption (impaired for fish consumption), while further assessment is required for the beneficial use "Fish Tumors or Other Deformities."

addressing the effects of a suite of tightly intertwined stressors related to eutrophication. One of the critical questions arising from the management efforts in the Bay of Quinte area is to identify the optimal water quality criteria that can impartially represent the prevailing conditions in the system. Recent empirical evidence suggested that the historical target of a seasonal average TP concentration of 30 μg l^{-1} was not representative of the actual temporal variability in the system nor did it reflect the water quality conditions (e.g. toxic algal blooms) in inshore areas of high public exposure, such as beaches (Kim et al., 2013; Zhang et al., 2013). In this regard, Zhang et al. (2013) asserted that one of the contemporary challenges in the Bay of Quinte is the establishment of eutrophication delisting objectives that will be framed upon: (i) a finer temporal scale, such as monthly or even daily snapshots from samplings that cover the entire growing season; (ii) the need to accommodate the considerable spatial variability in the system; (iii) the importance of identifying a numerical value that is both scientifically sound and achievable; (iv) the pragmatic view that the criteria setting process should explicitly accommodate the natural variability of the system or the substantial uncertainty characterizing the existing exogenous loading estimates by permitting a realistic frequency of goal violations; and (v) the fact that all management decisions should effectively balance between environmental concerns and socioeconomic values. This balance reflects how all citizens, businesses, industries, and the resource management agencies view the potential of the Bay of Quinte to be a symbol of an environmentally sustainable community.

In this study, we present a modelling analysis of the management practices that would lead to significant reduction of phosphorus export from the Bay of Quinte watershed and an evaluation of the overall uncertainty associated with the assessment of the Beneficial Use Impairment Eutrophication and Undesirable Algae. Our analysis is based on a network of models that connects the watershed processes with the dynamics of the Bay of Quinte, using the SPARROW non-linear regression strategy along with a mathematical model that aims to reproduce the phosphorus dynamics in the system. First, we present an exploratory analysis of the landscape characteristics, land cover patterns, and phosphorus export spatial variability. We then examine the impact of various exogenous phosphorus loading sources relative to the role of dreissenids and macrophytes on the eutrophication processes in the receiving waterbody. Our study undertakes an analysis of management scenarios in the Bay of Quinte watershed and subsequently attempts to shed light on critical aspects of the system functioning that invite further investigation and will likely determine our predictive capacity to assess compliance with the most sensible delisting water quality targets.

Materials and methods

Watershed modelling

Prior to the SPARROW application, we used a novel classification technique, Self-Organizing Maps (SOM), to gain insights into the physiographical features and land-use patterns in the Bay of Quinte watershed, and to subsequently associate them with the phosphorus non-Point-source loading. SOM is a neural network technique known to provide a powerful means for extracting information from complex multi-dimensional data and map it onto a reduced dimensional space (Kohonen, 1997). Being particularly convenient in visual understanding, SOM has been widely applied to pattern recognition of various learning problems (Chon et al., 1996). In the present application, we used eighteen classification variables, such as the landscape slope, saturated soil hydraulic conductivity, soil bulk density, and areal fractions for different land use types (lakes, ponds, alvars, bogs, coniferous swamps, deciduous swamps, fens, marshes, deciduous forests, coniferous forests, cutovers, mining areas, urban lands, pastures, and croplands) in 73 gauged and 137 ungauged subwatersheds. Thus, a total of 210 spatial units were distributed on 2-dimensional hexagonal maps, and then clustered in different groups according to their similarities. We used the SOM Toolbox developed by Helsinki University of Technology (Vesanto et al., 2000).

The SPARROW model was implemented to estimate nutrient loads, yields, and deliveries at landscape and regional scales (McMahon et al., 2003). The model uses a hybrid empirical/process-based approach designed to be applied to a network of water quality monitoring stations. SPARROW consists of a two-level hierarchical spatial structure. Watersheds are first divided into subwatersheds, each of which drains to a water quality monitoring station, and then each subwatershed is disaggregated into reach catchments draining to a particular stream segment (Schwarz et al., 2006). SPARROW considers two basic processes: (i) the source processes, represented by export coefficients, accounting for the constituent mobilization; (ii) the sink processes introduced by delivery factors, predicting how landscape attributes modulate the delivery of the mobilized constituent to streams, and attenuation coefficients, predicting the amount of the delivered constituent remaining in transit per length of stream or per reservoir (see parameter definitions in Table 1). In particular, the governing SPARROW equation can be expressed as:

Table 1. Posteriors estimates of all the SPARROW parameters.

Parameter	Unit	Mean	S.D.	Parameter description
Delivery coefficient (α)	hr cm^{-1}	0.172	0.063	It takes approximately 10 min for P to be transmitted for 1 cm.
TP export coefficient from forest areas (β_1)	tons km^{-2} yr^{-1}	0.018	0.006	18 kg of *TP* per km^2 are released from forested areas on an annual basis.
TP export coefficient from pasture areas (β_2)	tons km^{-2} yr^{-1}	0.040	0.020	40 kg of *TP* per km^2 are released from pasture areas on an annual basis.
TP export coefficient from urban areas (β_3)	tons km^{-2} yr^{-1}	0.126	0.087	126 kg of *TP* per km^2 are released from urban areas on an annual basis.
TP export coefficient from septic tanks (β_4)	tons $tank^{-1}$ yr^{-1}	0.001	0.001	1 kg of *TP* is released from septic tanks on an annual basis.
TP export coefficient from wheat-cultivated areas (β_{5wheat})	tons km^{-2} yr^{-1}	0.076	0.032	76 kg of *TP* per km^2 are released from wheat-cultivated areas on an annual basis.
TP export coefficient from oat-cultivated areas (β_{5oat})	tons km^{-2} yr^{-1}	0.127	0.065	127 kg of *TP* per km^2 are released from oat-cultivated areas on an annual basis.
TP export coefficient from corn-cultivated areas (β_{5corn})	tons km^{-2} yr^{-1}	0.044	0.026	44 kg of *TP* per km^2 are released from corn-cultivated areas on an annual basis.
TP export coefficient from alfalfa-cultivated areas ($\beta_{5alfalfa}$)	tons km^{-2} yr^{-1}	0.030	0.011	30 kg of *TP* per km^2 are released from alfalfa-cultivated areas on an annual basis.
TP export coefficient from fallow land areas ($\beta_{5fallow}$)	tons km^{-2} yr^{-1}	0.072	0.038	72 kg of *TP* per km^2 are released from fallow lands on an annual basis.
1st order settling rate in lakes (kr)	m yr^{-1}	4.980	1.551	*TP* settles with an average rate of 4.98 m per year in the lakes.
1st order attenuation rate for small streams (ks_1)	km^{-1}	0.037	0.022	3.7% of *TP* per kilometer is attenuated in small streams.
1st order attenuation rate for large streams (ks_2)	km^{-1}	0.011	0.007	1.1% of *TP* per kilometer is attenuated in large streams.

$$MAL_i = \left\{ \sum_{n=1}^{N} \sum_{j=1}^{J_i} \beta_n S_{n,j} e^{(-\alpha Z_j)} H_{i,j}^{S} H_{i,j}^{R} \right\} \varepsilon_i$$

where the subscripts i and j refer to subwatersheds and reach catchments, respectively; MALi refers to the mean annual total phosphorus load measured at the monitoring station of the subwatershed i in metric tonnes per year; n, N refers to the source index, N is the total number of sources (diffuse and Point-sources) and n is an index for each source; Ji refers to the number of reaches in subwatershed i; βn refers to the estimated source coefficient for source n with units metric tonnes per unit area (km^2) and year; Sn,j refers to the quantity (area in km^2 or number of septic tanks) of source n in reach j. In the Bay of Quinte application, we explicitly considered the export of phosphorus from forested areas, pastures, urban sites, septic tanks as well as five crop types, i.e. wheat, oat, corn, alfalfa and fallow land areas. The parameter α refers to the vector of land to water delivery coefficients, Zj is a vector of the land-surface characteristics associated with drainage to reach j. Exploratory analysis prior to the SPARROW application showed that the vertical hydraulic conductivity was the optimal surrogate variable to characterize Zj in the Bay of Quinte watershed. represents the fraction of nutrient mass originating in reach j remaining at the monitoring station of the subwatershed i as a function of first order loss processes in streams; refers to the fraction of nutrient mass originating in reach j remaining at station i as a function of first order loss processes in lakes and reservoirs; and ϵi refers to a random multiplicative error term assumed to be independently and identically distributed across all subwatersheds.

First order loss processes in streams include loss to sediments and biota, and are expressed as:

$$H_{i,j}^{S} = \exp(-k_s L_{i,j})$$

where ks refers to the first order loss coefficient for streams (km^{-1}), and Li,j refers to the stream length in kilometers between reach i and station j. Our study explicitly characterizes the attenuation processes in low and high flow streams, using the flow level of 1 $m^3 s^{-1}$ as a cutoff point. First order loss processes operating in lakes and reservoirs are limited to loss to sediment, which is expressed as:

$$H_{i,j}^{R} = \prod_{l} \exp(-k_r q_l^{-1})$$

where l refers to any lakes or reservoirs between reach i and station j, kr. refers to the first order loss coefficient or settling velocity (m year^{-1}), ql refers to the aerial hydraulic loading of the lake/reservoir (m year^{-1}).

Bayesian parameter estimation was also used to address several critical issues related to the SPARROW applications, such the uncertainty of calibration data, the importance of informative prior parameter distributions in assisting model calibration, and the implications of the covariance of model parameters on the inference drawn and the posterior patterns derived (see Equation 12 in Wellen et al., 2014). The phosphorus loading estimates in our calibration dataset were based on total phosphorus concentrations and flow data from 73 and 48 monitoring sites in the Bay of Quinte watershed, respectively. The development of a regression model ($r^2 > 0.94$) that connected the annual average flows with the corresponding catchment areas enabled the calculation of the annual TP loads in sites, where the flow data were not available. Further information about the methodological protocol used to derive the loading estimates and the associated errors can be found in Wellen et al. (2014) and Kim et al. (2014a,b submitted).

Phosphorus modelling for the receiving waterbody

The basis of the present study is the TP mass-balance model for the Bay of Quinte, originally developed by Minns and Moore (2004) and recently modified by Kim et al. (2013). Improvements in the mechanistic foundation of that model involved the explicit representation of macrophyte dynamics; the role of dreissenids in the system; improved spatial resolution; and several processes related to the fate and transport of phosphorus in the sediments along with the interplay between water column and sediments, such as particulate sedimentation dependent upon the standing algal biomass, sediment resuspension, sorption/desorption in the sediment particles, and organic matter decomposition. Schematic illustration of the modelled processes in the receiving waterbody is provided in Figure 1, while the basic conceptual design, the mathematical formulations, and the forcing functions used are described in detail in Kim et al. (2013). The TP concentrations in the calibration dataset were generated as time-weighted monthly means of data provided by the Project Quinte members (2011) in their annual monitoring report. Monthly mean nutrient concentrations for the Lake Ontario outlet basin were also provided by the Department of Fisheries and Oceans' Long-term Bio-monitoring (Bioindex) Program (Johannsson et al., 1998).

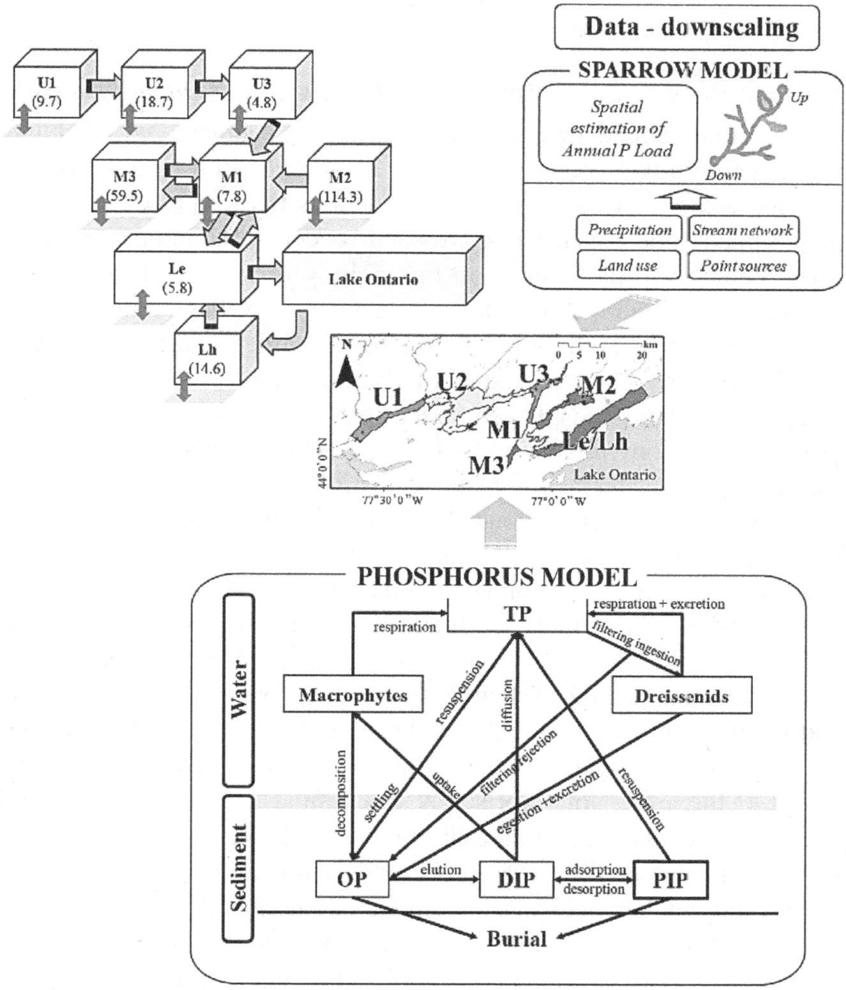

Figure 1. Conceptual diagram of the integrated phosphorus modelling framework for the Bay of Quinte. The spatial segmentation of the model for the receiving waterbody consists of the following compartments: (U1) the segment that extends from the mouth of Trent River until the city of Belleville; (U2) the segment that begins from the mouth of Moira River and comprises the Big Bay, Muscote Bay, and North Point Bay; (U3) the area influenced by the inflows of Napanee River, extending until the outlet of Hay Bay. In the middle Bay, there are three segments corresponding to the main stem (M1) and the two adjacent embayments: Hay Bay (M2), and Picton Bay (M3). The lower segment of the Bay, representing the transitional area to Lake Ontario, was separated into the epilimnetic (Le) and hypolimnetic (Lh) compartments. Numbers in parentheses correspond to the average flushing rate of each segment.

Results and discussion

Land use patterns and phosphorus trends in the Bay of Quinte watershed

The Canadian Shield separates the northern and southern Bay of Quinte watershed into two seemingly distinct sub-basins. In the northern part, forested areas with coniferous and deciduous trees represent the most predominant land cover, while croplands occupy a large portion of the lower catchment, particularly in the vicinity of lakes and the Bay of Quinte itself (Figure 2a). Lakes, pastures, and wetlands are sparsely distributed in the southern part of the basin, while the urban landscape mainly corresponds to the cities of Peterborough, Trenton, Belleville and other smaller towns. Based on the spatial heterogeneity of the eighteen classification variables, SOM delineated six spatial clusters in the Bay of Quinte watershed with fairly distinct land-use patterns (Figure 2b). Coniferous and deciduous coverage along with pastures and croplands dominate the landscape in cluster 1. It is also interesting to note that different types of wetlands, such as fen (≈10%), coniferous swamp (≈8%), and alvar (≈0.4%) have their highest areal fraction values in the same cluster. In cluster 2, the average landscape slope is steep and the soil bulk density is high. The areal fractions of forests as well as mining and logging sites are also high. In cluster 3, most of the subwatersheds are located in the vicinity of the Bay of Quinte, where crops occupy ≈75% of the area. Not surprisingly, the annual TP yield per area and average TP concentrations are the highest (528 kg km^{-2} yr^{-1} and 103 μg l^{-1}) in these same regions. In cluster 4, soil hydraulic conductivity is significantly higher, deciduous swamp are more abundant relative to the rest watershed, cropland coverage is the second highest (≈41%), and thus the net TP export is high. In cluster 5, urban land represents ≈74% of the land-use coverage and net TP export and yield are the second highest (3.72 tonnes yr^{-1} and 209 kg km^{-2} yr^{-1}), which is further accentuated by the increased Point-source loading (2.44 tonnes yr^{-1}). In cluster 6, pasture and cropland approximately correspond to 60% of the area, and these subwatersheds are mainly located adjacent to the Bay of Quinte.

The goodness-of-fit between the observed and predicted TP loading values from SPARROW model was excellent in the logarithmic scale (r^2 > 0.95), although there were four sites with error greater than 10 tonnes yr^{-1} when the SPARROW predictions were back-transformed in the original scale (Figure 3). The posterior parameter values offered interesting insights (and/or testable hypotheses) into the patterns of phosphorus export and delivery in the Bay of Quinte watershed (Table 1).

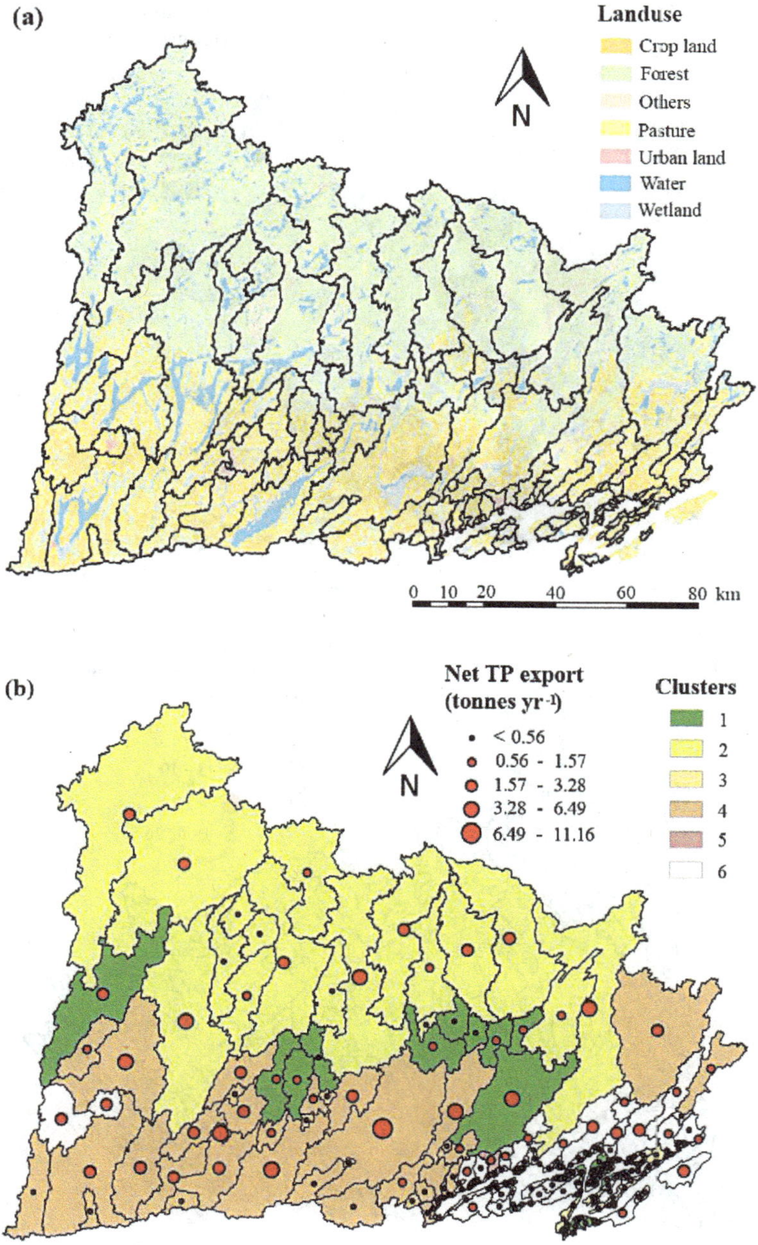

Figure 2. Map of Bay of Quinte watershed: (a) land use types and (b) classification based on artificial neural networks and associated phosphorus export per subwatershed.

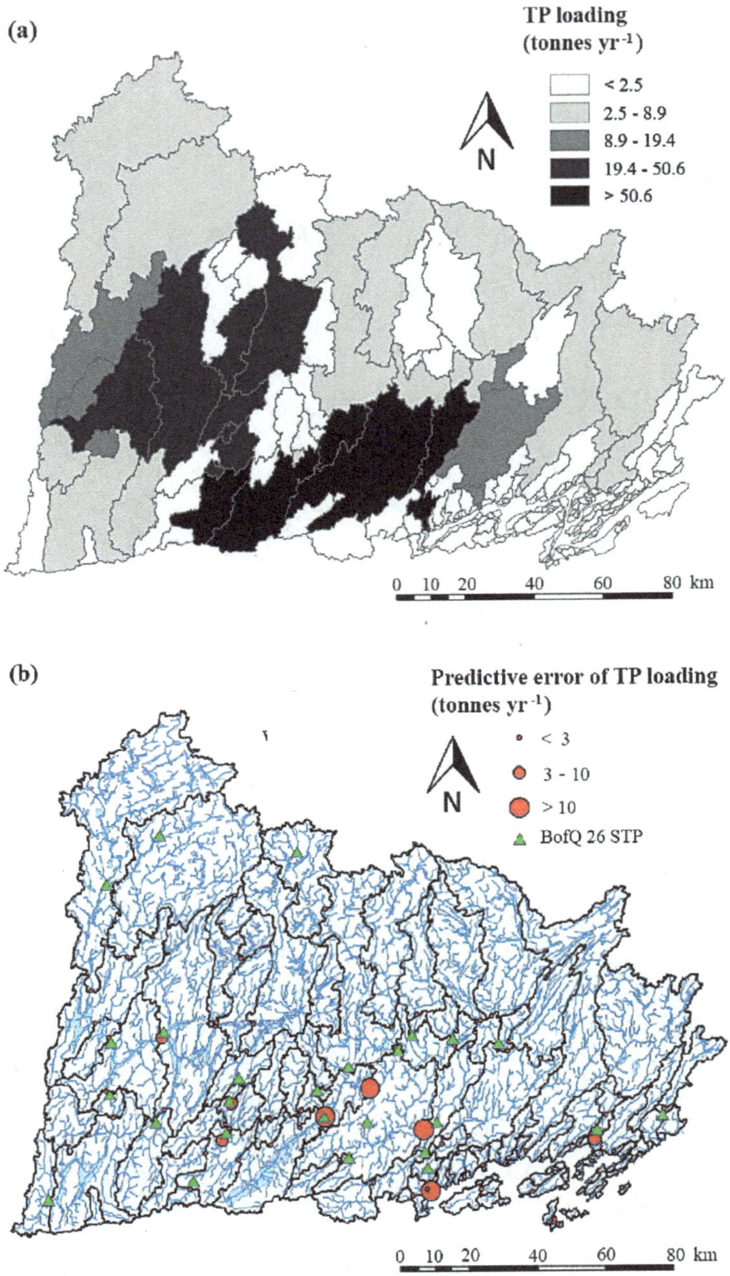

Figure 3. (a) Mean TP loading predictions and (b) associated errors of the SPARROW model application in the Bay of Quinte watershed. Triangles (green) indicate the locations of 26 sewage treatment plants.

The main findings from the SPARROW modelling exercise were as follows: (i) urban areas are characterized by a fairly high areal phosphorus export with a mean estimate of 126 kg of TP per km^2 on an annual basis; (ii) the contribution of phosphorus from agricultural land uses can vary considerably among the various crop types (30–127 TP kg per km^2), but is generally lower than the impact of urban sites. This finding appears to contradict the popular notion that rates nutrient export of urban lands below that of agricultural lands due to lower anthropogenic nutrient subsidies, such as fertilizer implementation (Moore, 2004; Soldat and Petrovic, 2008; Soldat et al., 2009). Nonetheless, other studies in the region of Southern Ontario have found urban total phosphorus export rates to be comparable (or even higher) than agricultural total phosphorus export rates (Winter and Duthie, 2000; Wellen et al., 2014); (iii) the crop-specific export coefficient values were on par with those typically reported in the literature (Harmel et al., 2008), although the confounding effects of the uncertainty associated with the delineation of the areal coverage of the various crop types may be responsible for some counterintuitive results (e.g. $\beta 5oat > \beta 5corn$); (iv) the attenuation rate in low flow streams (3.7% of TP per kilometer) appears to be distinctly greater than in those with high flow (1.1% of TP per kilometer), reflecting a tighter coupling between the streambed and water column in the former streams that allows the biotic (uptake) and abiotic (settling) processes to exert control on the nutrient load en route to the receiving waterbody (Basu et al., 2011); and (v) fallow areas are responsible for approximately 70 kg of TP per km^2 on an annual basis.

Interesting temporal trends were also found about the relative contribution of the different point and non-Point-sources to the total phosphorus loading in the Bay of Quinte (Figure 4). The inflows from Trent River represented less than 50% of the total loading discharged into the system before the reduction of phosphorus in detergents and the upgrades at the local wastewater treatment plants in the 1970s, but the relative contribution of the Trent River to total loading has gradually increased up to 70% over the last decade. By contrast, the proportion of phosphorus loading associated with the Point-sources was reduced from 21% to 3%, while the relative contribution of the rest of the tributaries has remained essentially unaltered over the time period examined. We also calculated the net loading[2] from both point and non-Point-sources to weigh the different

2 The net loading accommodates the idea that two equal total loads with opposite pairs of flow and concentration, high flow with low concentration

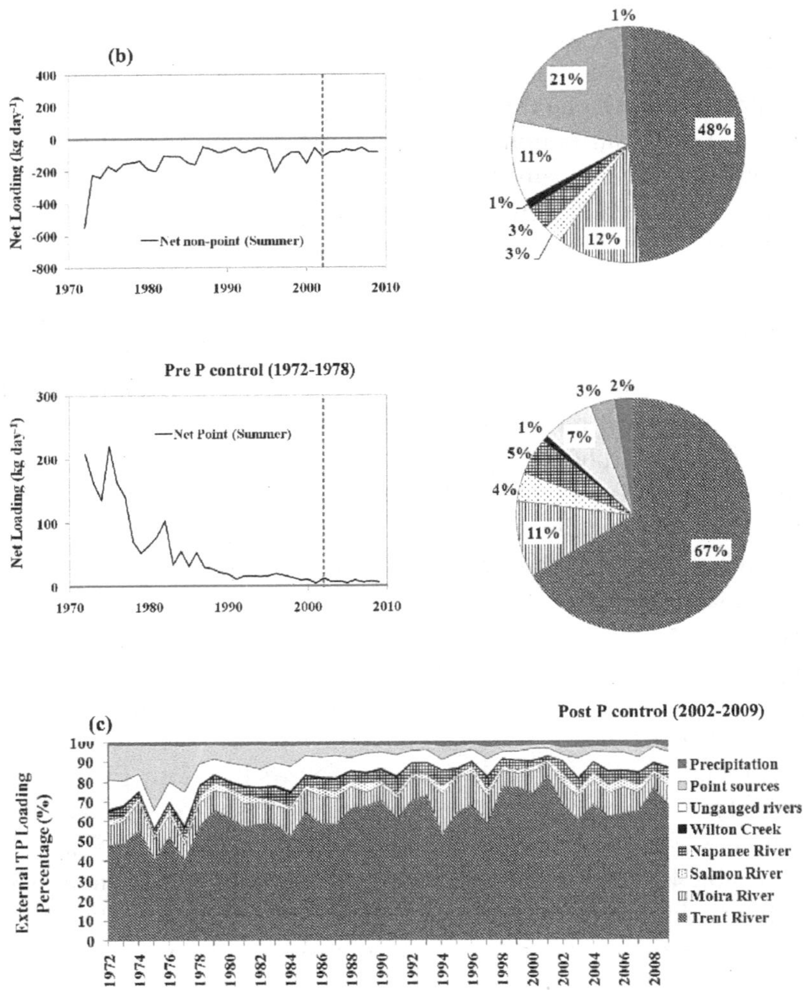

Figure 4. Temporal trends of exogenous TP loading in the Bay of Quinte during the study period (1972–2009): (a) net non-point (upper panel) and point (lower panel) sources of phosphorus loadings into the Upper Bay (= Flowsexogenous × [TPexogenous − TPUpper Bay]); (b) comparison between the pre- (1972–1978) and post-control (2002–2009) periods; (c) relative contribution from the different loading sources over time (modified from Kim et al., 2013).

---

or low flow with high concentration, could potentially have a very different effect on the trophic state of the system, and is simply calculated by multiplying the inflows from a particular source with the difference between the inflow and outflow concentrations (Johnson and Owen, 1971).

transport of phosphorus due to the variability in the corresponding flow regimes (Johnson and Owen, 1971; Minns et al., 2004). During the earlier years of the study period, our analysis showed that the computed total seasonal P loads would always indicate the river loads were greater than Point-source loading, sometimes much more so when river flows are higher in the upper Bay of Quinte. However, the net loading values clearly suggest that the Point-sources always have a positive sign, while the river net loads were mostly negative. Net river loads were negative when Point-source loads were higher, because the Point-source inputs elevated TP concentrations in the system far above the river levels. Hence, when Point-source loads were high, river inputs actually displaced some of the excess loading. In recent years, both Point-source loadings and ambient concentrations in the Bay have significantly declined; the river net loads still remain negative, but their absolute values have decreased. The impact of the point loading on the prevailing conditions in the Bay of Quinte has been reduced. It is important to note though that the ambient TP concentrations towards the end of summer (August-September) still rise to levels well above the 30 $\mu g\ l^{-1}$ target. Because this period coincides with the time when the river flows decline to their lowest annual values, we believe that this recurring pattern is indicative of the role of internal nutrient sources (e.g. sediment reflux) in the system.

Phosphorus dynamics in the Bay of Quinte: Evaluation of the role of exogenous loading relative to internal recycling

The SPARROW model was subsequently integrated with the Kim et al. (2013) phosphorus model. The downscaling of the annual nutrient loading predictions to daily inputs was based on a Monte Carlo algorithm, which was designed to generate the probability distributions of the annual average TP concentrations in the different tributaries, while accounting for the SPARROW structural and parametric uncertainty as well as the observed flow variability. The ratios between simulated and long-term average TP concentrations (2002–2012) were then used as (stochastic) weights for adjusting the tributary-specific rating curves (Zhang et al., 2013), and thus used to obtain daily estimates of tributary phosphorus loads from the contemporaneous flows. The model was then calibrated to match the observed monthly TP patterns in the upper, middle, and lower segments of the Bay during the 2002–2009 period (Kim et al., 2013). The model demonstrated satisfactory ability to fit the monthly TP levels in the Bay of Quinte, and was able to reproduce the end-of-summer increase of the ambient TP levels in the upper segment, even in years (e.g. 2005) when the corresponding concentrations were greater than 60 μg l^{-1}. The model also faithfully depicted the spatial gradients in the system, with distinctly higher TP levels in the upper segment relative to those experienced in the middle/lower Bay (Kim et al., 2013, 2014c submitted).

We also used the model to draw inference on the spatial variability of the various external and internal TP flux rates in the Bay of Quinte (Figure 5 and Table 2). The net TP contributions (sources or sinks) represent the mass of phosphorus associated with the various compartments (water column, sediments, macrophytes, dreissenids) throughout the growing season (May-October) averaged over the 2002–2009 period. In the U1 segment, the phosphorus budget is predominantly driven by the external sources (phosphorus loading: 158.6 kg day^{-1}) and sinks

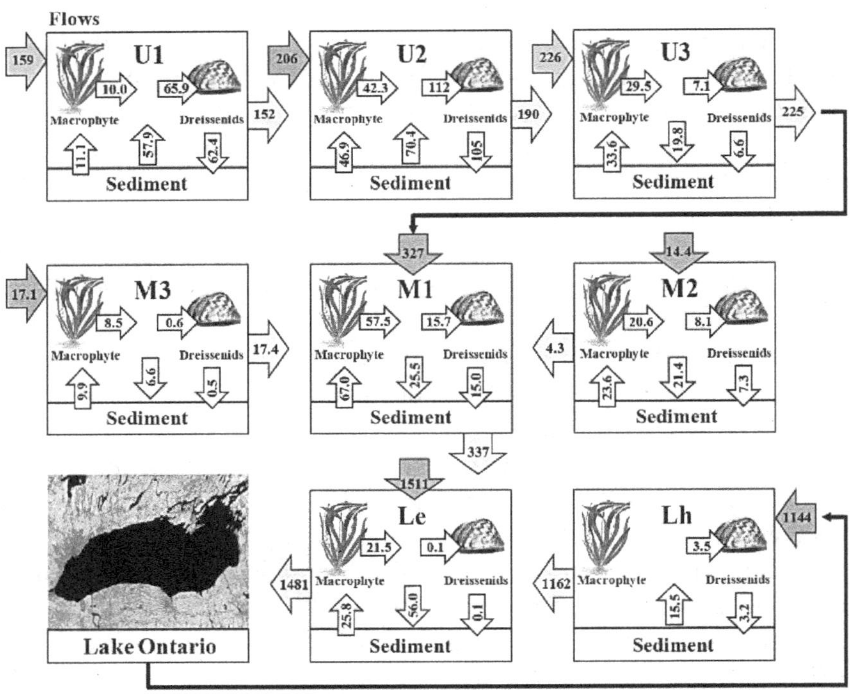

Figure 5. Spatial variability of the various external and internal TP flux rates (kg day^{-1}) in the Bay of Quinte. Arrow directions indicate the net contribution (sources or sinks) of the various compartments (water column, sediments, macrophytes, dreissenids). Dark gray arrows show the TP inflows in a spatial segment, while the light-gray arrows depict the corresponding outflows.

(outflows: 152.2 kg day^{-1}). Our model suggests that the sediments (resuspension and diffusion from the sediments to water column minus particle settling) act as a net source of phosphorus in this segment (57.9 kg day^{-1}). Dreissenids subtract approximately 65.9 kg day^{-1} from the water column (particle filtration minus respiration) and subsequently deposit 62.4 kg day^{-1} via their excretion and particle rejection. Likewise, the U2 segment receives 205.8 kg day^{-1} from exogenous sources, including the upstream

Table 2. Estimation of TP flux rates from different components of the Bay of Quinte model.

	Net TP flux (kg day^{-1})						
Site	*Water–Sed*	*Water–Macro*	*Water–Zebra*	*Sed–Zebra*	*Sed – Macro*	*Water in*	*Water out*
			Present loading conditions				
U1	57.9	10.0	65.9	62.4	11.1	158.6	152.2
U2	70.4	42.3	111.8	105.5	46.9	205.8	190.0
U3	-19.8	29.5	7.1	6.6	33.6	226.3	225.5
M1	-25.5	57.5	15.7	15.0	67.0	326.9	337.4
M2	-21.4	<u>20.6</u>	<u>8.1</u>	<u>7.3</u>	<u>23.6</u>	14.4	4.3
M3	-6.6	8.5	0.6	0.5	9.9	17.1	17.4
Le	-56.0	21.5	0.1	0.1	25.8	1510.6	1480.8
Lh	15.5	0.0	3.5	3.2	0.0	1144.0	1162.5
			Reduced loading conditions				
U1	63.6	10.0	58.1	55.4	11.1	130.3	137.3
U2	76.2	42.3	105.3	99.6	46.9	185.8	180.5
U3	-17.8	29.5	6.5	6.1	33.6	209.9	211.4
M1	-24.0	57.5	14.8	14.2	67.0	309.5	321.3
M2	-19.9	20.6	7.3	6.6	23.6	12.3	4.0
M3	-6.3	8.5	0.5	0.5	9.9	16.2	16.6
Le	-54.7	21.5	0.1	0.1	25.8	1487.7	1457.0
Lh	15.7	0.0	3.4	3.2	0.0	1142.0	1158.7

Water–Sed: resuspension and diffusion from the sediments to water column − particle settling; *Water–Macro*: macrophyte respiration; *Water–Zebra*: respiration and 0.4 × excretion of dreissenids − particle filtration of dreissenids; *Sed–Zebra*: 0.6 × excretion, egestion and rejection from dreissenids; *Sed–Macro*: macrophyte mortality − macrophyte intake from sediment; *Water in*: upstream inflow and external loading; *Water out:* downstream outflow.

inflows, and transports downstream 190.0 kg day^{-1}. The net contribution of the sediments accounts for 70.4 kg day^{-1}, while dreissenids on average reduce the ambient TP levels by 111.8 kg day^{-1}. The main differences between the two segments in the upper Bay are the TP fluxes related to macrophyte P intake from the sediments and respiration that can reach the levels of 46.9 and 42.3 kg day^{-1} relative to the fluxes of 11.1 and 10.0 kg day^{-1} in the U1 segment. Likewise, the macrophyte intake from the sediments minus the amount of P regererenated from the decomposition of the dead plant tissues varies between 35–65 kg day^{-1} in segments U3 and M1, while the subsequent release of their metabolic by-products is approximately responsible for 19–26 kg day^{-1}. Interestingly, the settling of particulate P dominates over the resuspension and diffusion from the sediments to the water column with the corresponding net fluxes ranging between 25–35 kg day^{-1}. In Hay Bay (M2), the fluxes mediated by the macrophytes and dreissenids primarily modulate the TP dynamics and the same pattern appears to hold true in Picton Bay (M3). In the lower Bay of Quinte (Le and Lh), the model postulates a significant pathway (>1100 kg P day^{-1}) through which the inflowing water masses from Lake Ontario well up from the hypolimnion to the epilimnion and are subsequently exported from the system. In the same area, the internal biotic sources (macrophytes) similarly represent an important vector of phosphorus transport.

Relative to the present loading conditions, we evaluated a number of management scenarios intended to reduce the exogenous levels of phosphorus entering the system. For the sake of brevity, we here present the predicted TP levels along with the broader implications for ecosystem functioning of a scenario related to the reduction of the contribution of the agricultural land uses, urban storm water, and Point-sources by 20%, 50%, and 60%, respectively (Figure 6). The latter reduction aimed to evaluate the effects of a "phosphorus load cap" that stipulates a total phosphorus effluent concentration of 0.1 mg l^{-1} in the local sewage treatment plants. Importantly, after the consideration of the deposition of biogenic material (e.g. production of pseudo-feces by dreissenids, decomposition of the dead macrophyte tissues), the simulated scenario does not reach the "threshold load" for rapid system recovery at which the flux downward equals the flux upward from anoxic sediments (Table 2; see net TP exchange between water column and sediments in segments U3, M1, M2, M3). In fact, the sediment reflux in the inner segments of the upper Bay (U1 and U2) is predicted to increase under the reduced nutrient loading scenario, as

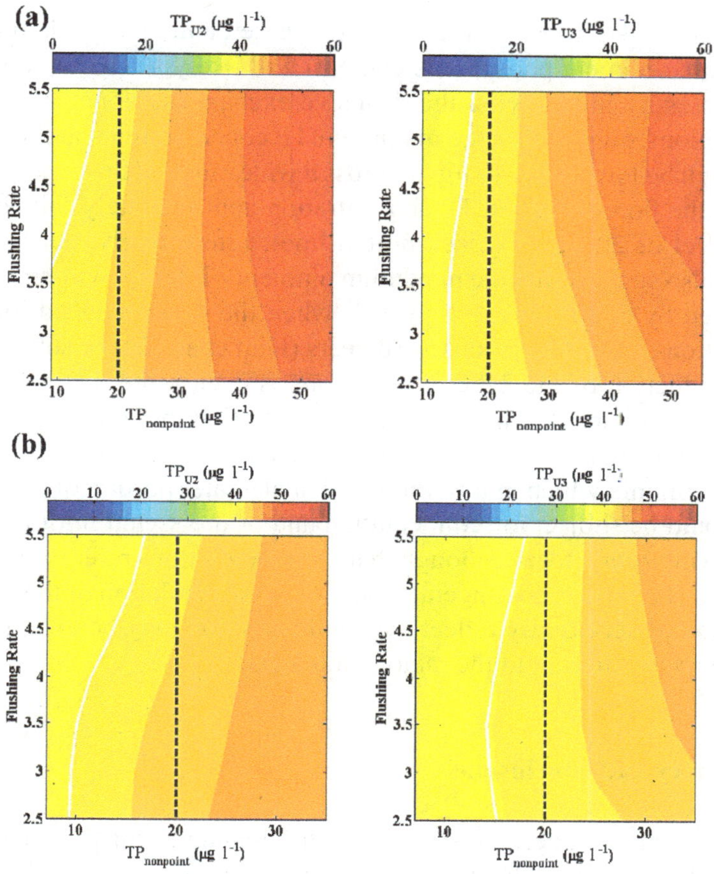

Figure 6. Simulated maximum TP concentrations during the growing season (May-October) in the Bay of Quinte. Upper (a) panels refer to the predictions associated with the reference environmental conditions; and lower (b) panels represent the predictions of a TP loading reduction scenario (60% Point-sources, 20% non-Point-sources, and 50% urban storm water). The first eleven years (2002–2012) were based on real meteorological and nutrient loading conditions, while the final (12th) year was forced with a wide range of combinations of TP riverine concentrations and flows that were generated from the mean (± error) predictions of the SPARROW model. The white contour line corresponds to the proposed targeted level of 40 μg TP l^{-1}. The flushing rates express the frequency (number of times) of water renewal in the upper Bay during the growing season. The black dotted line represents a threshold level of 20 μg l^{-1} for the flow-weighted TP concentration in all the major tributaries in the upper Bay of Quinte.

the increased vertical gradient (i.e. decreased ambient levels relative to stable sediment phosphorus content) will likely shape the transient dynamics by intensifying the diffusive exchanges. Nonetheless, our predictions suggest that the maximum TP concentrations in the upper Bay will be lower than 40 μg l^{-1} across a wide range of flow conditions, when the flow-weighted TP concentration approximately falls below the level of 20 μg l^{-1} (black dotted line). Interestingly, our simulations also suggest that the maximum ambient TP levels will be further reduced by 5–20% (or 1–5 μg l^{-1}), when the same reduced loading conditions prevail for about 5–10 years (Kim et al., 2014c submitted). The reduction of the TP concentrations is even more distinct, if we consider the fact that the increased sequestration into the algal cells is likely to decrease the dissolved to particulate phosphorus ratio in the water column, which in turn implies that a greater portion of phosphorus could be subject to sedimentation and/or dreissenid infiltration. It is also important to note though that these predictions refer to spatially average TP values, contingent upon the coarse resolution of the model, and do not necessarily reflect the conditions that may be experienced in the areas adjacent to the major tributaries.

Summary and conclusions

We present a modelling analysis of the management practices that could lead to significant reduction of phosphorus export from the Bay of Quinte watershed and an evaluation of the overall uncertainty associated with the assessment of the Beneficial Use Impairment Eutrophication and Undesirable Algae. Our work highlights the internal recycling as one of the key drivers of phosphorus dynamics in the Bay. The flow from the Trent River is the predominant driver of the upper Bay dynamics until the main stem of the middle area however, the sediments in the same segment release a significant amount of phosphorus and the corresponding fluxes are likely amplified by the macrophyte and dreissenid activity. From a management standpoint, the presence of a significant positive feedback loop in the upper Bay suggests that the anticipated benefits of additional reductions of the exogenous point and non-point loading may not be realized within a reasonable time frame (i.e. 5 – 10 years). Our analysis of nutrient loading scenarios shows that the restoration pace of the Bay could be slow, even if the riverine total phosphorus concentrations reach levels significantly lower than their contemporary values, <25 μg TP l^{-1}. We believe that the on-going management decisions, monitoring, and modelling should also explicitly consider the role and broader ramifications of internal phosphorus loading into the system. The anticipated lessons from such a multi-faceted exercise are a unique aspect of the Bay of Quinte ecosystem because of the long history of research and monitoring data. This study can produce transferable knowledge to other systems worldwide, experiencing similar hysteresis patterns associated with internal nutrient loading.

Significant progress has been made with the reduction of both point and non-point loading discharges in the Bay of Quinte over the last four decades. However, Kim et al. (2013) recently underscored the challenges of the current management efforts to further improve the water quality conditions, as the sediments in the upper middle Bay area act as a net source of phosphorus and the corresponding fluxes are likely magnified by the macrophyte and dreissenid activity. The presence of a feedback loop in the system suggests that the emergence of hysteresis patterns may compromise the efficiency of additional external loading reductions (Scheffer et al., 2001; Minns et al., 2004; Gudimov et al., 2011). While the Bay of Quinte morphology minimizes the likelihood of regime shifts to occur (Seifried, 2002), the contribution of internal nutrient loading could significantly protract the realization of an improved state (Zhang et al., 2013). Similar to Kim et al.'s (2013) projections, the present modelling analysis provides evidence that the restoration pace of the Bay could be slow, even if the riverine average TP concentrations reach levels significantly lower than their current values, <25 μg TP l^{-1}.

Because of the skepticism arising from the projected limited system response to the nutrient loading scenarios examined, we caution that there is considerable uncertainty regarding the characterization of the phosphorus cycle and the assumptions made about the sedimentation rates or the nutrient subsidies from dreissenids, macrophytes, and sediment diagenesis. In particular, Manning (1996) reported that a significant portion of the riverine fine-grained assemblage of phosphate-oxides-organic matter is transported through the upper Bay and deposits downstream in the middle and lower segments. As a result, his estimates of the phosphorus sedimentation rate were significantly lower than those assigned by Minns et al. (2004) and Kim et al. (2013) to reproduce the observed TP concentrations in the water column. Nonetheless, counter to the Minns et al.'s (2004) characterization of the sediments, Kim et al. (2013) (and the present model) assumed a significantly lower permanent loss of phosphorus mass from the system through the sediment burial. This feature along with the explicit consideration of the recycling mediated by dreissenids and macrophytes resulted in a strong feedback loop in our modelling exercise relative to the Manning's (1996) and Minns et al.'s (2004) estimates that appears to profoundly modulate the capacity of the management scenarios examined to shape the dynamics of phosphorus in the Bay. In this regard, one of the compelling research questions in the system involves the delineation of the sediment TP profiles and the improvement of our quantitative understanding of the mechanisms of phosphorus mobilization in the sediments. The on-going determination of the TP fractionation in the sediment profiles and the characterization of the diagenesis processes are likely to shed light on the plausibility of the hypotheses proposed from our modelling work (Dittrich et al., unpublished data).

There are two major differences between the system conceptualization postulated by Kim et al. (2013) and the present study. First, counter to the Kim et al. (2013) calibration strategy, we downplay here the direct release of phosphorus in the water column due to macrophyte respiration. Instead, greater weight is placed on the indirect pathway through which the dead plant tissues are subject to microbial decomposition in the sediments and phosphorus is subsequently released into the water column through diffusive exchanges in the sediment-water column interface. The latter conduit can be an important source of nutrients, as the macrophytes demonstrate high capacity of luxury uptake and thus tend to accumulate nutrients in concentrations higher than their physiological requirements (Bini et al., 2010). Second, the SPARROW application showed the use of the Wilton Creek equivalents[3] approach likely underestimates the levels of phosphorus export from the ungauged catchments (Kim et al., 2014b submitted), and thus rendered support to our concerns that the role of the nutrient recycling may have been overstated somewhat in our earlier work (Kim et al., 2013). This finding is particularly relevant to the middle sections of the system, such as the Hay Bay area (M2), where the nutrient regeneration processes were originally predicted to overwhelmingly dominate the TP budget (Kim et al., 2013). The greater reliance of inshore waters in the upper-middle Bay of Quinte upon external nutrient subsidies attests to the benefits of

3 The Wilton Creek equivalent approach, originally introduced Minns et al. (1986), assumes that the daily load from an un-gauged subwatershed equals the daily load from Wilton Creek weighted by the ratio of the area of the ungauged site to the Wilton Creek area.

optimizing the management practices in the corresponding catchments. In particular, both our SOM and SPARROW analysis highlighted the urban areas adjacent to the Bay as major phosphorus contributors per unit area, and thus a more effective stormwater management and Point-source loading control could ameliorate the prevailing local water quality conditions. Interestingly, our analysis still predicts a co-dependence between macrophyte and dreissenid activity in Hay Bay (M2) (Kim et al., 2013; see also bold numbers in Table 2), suggesting that the magnitude of the macrophyte "nutrient pump effect" and the associated stimulation of algal growth can potentially induce significant variations in the amount of particles filtered/ingested by the dreissenids as well as in their excretion/pseudofeces production rates (Vanderploeg et al., 2001).

In a recent paper, Zhang et al. (2013) argued against the historical delisting criterion of a seasonal average TP concentration lower than 30 $\mu g\ l^{-1}$, asserting that it is neither a reflection of the considerable intra-annual variability in the upper Bay nor representative of the water quality conditions in nearshore areas of high public exposure (e.g. beaches). It is would seem very unlikely that a single-value water quality standard monitored in a few offshore sampling stations can capture the entire range of dynamics in the system (e.g. the extremes seen in the nearshore sites) or the magnitude of the end-of-summer TP peaks. Responding to these skeptical views, the Bay of Quinte RAP considers adopting the pragmatic stance that the delisting objectives should revolve around extreme (and not average) values of variables of management interest and must explicitly accommodate all the sources of uncertainty (insufficient information, lack of knowledge, and natural variability) by permitting a realistic frequency of standard violations. Namely, the critical threshold level was set at a value of 40 μg TP l^{-1}, which cannot be exceeded more than 10–15% in both time and space. Under the assumption that the TP concentrations in the Bay of Quinte follow a log-normal distribution and that TP values $<15\ \mu g\ l^{-1}$ are likely to occur only 10% of the time during the growing season, then 10–15% exceedances of the 40 μg TP l^{-1} level are approximately equivalent to a targeted seasonal average of 25–28 μg TP l^{-1}. Thus, the replacement of the historical paradigm (binary assessment) with a probabilistic approach to water quality criteria does not intend to make the delisting of AOCs easier, but rather it offers a more comprehensive method to track the prevailing conditions in the Bay.

Bearing in mind that the TP targeted levels merely represent a "means to an end" and not "the end itself", the actual question that the local stakeholders should ponder is to what extent the anticipated benefits from a more efficient external phosphorus loading control could also be capitalized as a significant decrease of the algal bloom frequency? With respect to the total phytoplankton biovolume, Nicholls et al. (2002) showed that it declined after the control of phosphorus in the 1970s, but did not change significantly after the establishment of dreissenids in the system. Moreover, Nicholls and Carney (2011) asserted that the arrival of Dreissenid Mussels may be associated with both desirable (e.g. Aphanizomenon and Oscillatoria decline) and undesirable (e.g. Microcystis increase) changes in the integrity of the Bay of Quinte ecosystem. Importantly, the post-Dreissena increase of the cyanophyte Microcystis has had significant implications for the aesthetics and other beneficial uses in the Bay, through the formation of "scums" on the water surface (Jacoby et al., 2000) as well as due to the fact that some strains of Microcystis are toxin producers (Vanderploeg et al., 2001). Some of these structural changes in the phytoplankton community composition could stem directly from the feeding selectivity of dreissenids or indirectly from the improvements in the transparency of the water column, but the role of the feedback loop associated with their nutrient recycling activity could conceivably be another important factor (Shimoda et al., 2016). On positive note, according to the predictions of a non-linear quantile regression model (Carvalho et al., 2013), the current average TP concentrations (30–40 μg l^{-1}) represent the area where the algal biovolume vs TP relationship is characterized by a steep slope and thus any further improvements in the ambient nutrient levels are likely to induce more favourable quantitative and qualitative changes in phytoplankton (Figure 7).

Nonetheless, existing empirical evidence from the system suggests that a complex interplay among physical, chemical, and biological factors may drive the spatiotemporal abundance and composition patterns of the algal assemblages in the Bay of Quinte. Using a segmented regression model with a stochastic breakpoint of microcystin concentrations estimated at 0.54 μg l^{-1}, Kelly et al. (2019) demonstrated that the environmental conditions associated with increased ambient toxin levels are chlorophyll-*a* concentrations >7 μg l^{-1}, water temperature >20 °C, ammonium concentrations <40 μg l^{-1}, total phosphorus concentrations >25 μg l^{-1}, and wind speeds <37 km h^{-1}. Thus, in a system like the Bay of Quinte, where the severity of eutrophication phenomena is driven by both external and internal loading, there will inevitably be some uncertainty in the overall assessment of the Beneficial Use Impairment Eutrophication and Undesirable Algae. It is our belief though that the recent shift in the focus of the Bay of Quinte RAP activities (management decisions,

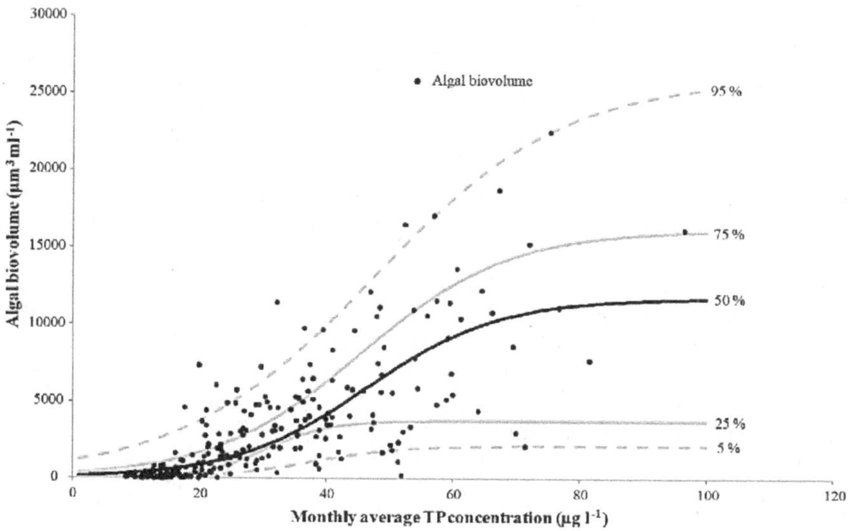

Figure 7. Quantile regression model for total phytoplankton biovolume against monthly average TP concentration in the Bay of Quinte.

monitoring, and modelling) on extreme states (>40 μg TP l^{-1}, toxic algal blooms) offers a more sensible way to evaluate the actual progress over time (Arhonditsis et al., 2019). Recognizing that it is practically impossible to eliminate these events in the foreseeable future, the robust assessment of their frequency of occurrence and the effective communication of the actual trends to the end-users (stakeholders and public) should be essential steps of the local management efforts.

Funding

This project was undertaken with the financial support of the Lower Trent Region Conservation Authority provided through the Bay of Quinte Remedial Action Plan Restoration Council. Aspects of the project were also supported by the Ontario Ministry of Research and Innovation (Early Researcher Award granted to George B. Arhonditsis).

References

Arhonditsis, G. B., Qian, S.S., Stow, C. A., Lamon, E. C., Reckhow, K. H., 2007. Eutrophication risk assessment using Bayesian calibration of process-based models: Application to a mesotrophic lake. Ecol. Model. 208, 215–229.

Arhonditsis, G.B., Neumann, A., Shimoda, Y., Javed, A., Blukacz-Richards, A., Mugalingam S. 2019. When can we declare a success? A Bayesian framework to assess the recovery rate of impaired freshwater ecosystems. Environ. Int. 130, 104821.

Basu, N. B., Rao, P. S. C., Thompson, S.E., Loukinova, N.V., Donner, S. D., Ye, S., Sivapalan, M., 2011. Spatiotemporal averaging of in-stream solute removal dynamics. Water Resour. Res. 47, W00J06.

Bini, L., Thomaz, S., Carvalho, P., 2010. Limnological effects of Egeria najas Planchon (Hydrocharitaceae) in the arms of Itaipu Reservoir (Brazil, Paraguay). Limnology 11, 39–47.

Carvalho, L., McDonald, C., de Hoyos C., Mischke, U., Phillips, G., Borics, G., Poikane, S., Skjelbred, B., Solheim, A.L., Van Wichelen, J:, Cardoso, A. C., 2013. Sustaining recreational quality of European lakes: minimizing the health risks from algal blooms through phosphorus control. Journal of Applied Ecology 50, 315–323.

Chon, T.-S., Park, Y.-S., Moon, K. H., Cha, E. Y., 1996. Patterning communities by using an artificial neural network. Ecol. Model. 90, 69–78.

Dorazio, R. M., Johnson, F. A., 2003. Bayesian inference and decision theory: a framework for decision making in natural resource management. Ecol. Appl. 13, 556–563.

Gudimov, A., Ramin, M., Labencki, T., Wellen, C., Shelar, M., Shimoda, Y., Boyd, D., Arhonditsis, G.B., 2011. Predicting the response of Hamilton Harbour to the nutrient loading reductions: A modeling analysis of the "ecological unknowns." J. Great Lakes Res. 37, 494–506.

Harmel, D., Qian, S., Reckhow, K., Casebolt, P., 2008. The MANAGE database: nutrient load and site characteristic updates and runoff concentration data. J. Envion. Qual. 37, 2403–2406.

Jacoby, J. M., Collier, D. C., Welch, E. B., Hardy, F. J., Crayton, M., 2000. Environmental factors associated with a toxic bloom of Microcystis aeruginosa. Can. J. Fish. Aquat. Sci. 57, 231–240.

Johannsson, O. E., Millard, E. S., Ralph, K. M., Myles, D. D., Graham, D. M., Taylor, W.D., Giles, B. G., and Allen, R. E., 1998. The changing pelagia of Lake Ontario (1981 to 1995): A report of the DFO Long-Term Biomonitoring (Bioindex) Program. Can. Tech. Rep. Fish. Aquat. Sci. 2243.

Johnson, M. G., Owen, G.E., 1971. Nutrients and nutrient budgets in the Bay of Quinte, Lake Ontario. Journal WPCF 43, 836–853.

Kelly, N.E. Javed, A., Shimoda, Y., Zastepa, A., Watson, S., Mugalingam, S., Arhonditsis, G.B. 2019. A Bayesian risk assessment framework for Microcystin violations of drinking water and recreational standards in the Bay of Quinte, Lake Ontario, Canada. Water Res. 162, 288-301.

Kim, D.-K., Zhang, W., Rao, Y., Watson, S., Mugalingam, S., Labencki, T., Dittrich, M., Morley, A., Arhonditsis, G. B., 2013. Improving the representation of internal nutrient recycling with phosphorus mass balance models: A case study in the Bay of Quinte, Ontario, Canada. Ecol. Model. 256, 53–68.

Kohonen, T., 1997. *Self-Organizing Maps*. Springer, New York.

Manning, P.G., 1996. Bioavailability of riverine, sewage plant, and sediment phosphorus in the Bay of Quinte, Lake Ontario. Canadian Mineralogist 34, 667–675.

McMahon, G., Alexander, R.B., Qian, S., 2003. Support of total maximum daily load programs using spatially referenced regression models. J. Water Resour. Plan. Manage. 129, 315–329.

Minns, C. K., Moore, J. E., 2004. Modelling Phosphorus Management in the Bay of Quinte, Lake Ontario in the Past, 1972 to 2001, and in the Future, Canadian Manuscript Report of Fisheries and Aquatic Sciences No. 2695. Burlington, Ontario.

Minns, C. K., Moore, J. E., Seifried, K. E., 2004. Nutrient loads and budgets in the Bay of Quinte, Lake Ontario, 1972 to 2001. Burlington, Ontario: Great Lakes Laboratory for Fisheries and Aquatic Sciences, Bayfield Institute

Minns, C. K., Moore, J. E., Doka, S. E., St. John, M.A., 2011. Temporal trends and spatial patterns in the temperature and oxygen regimes in the Bay of Quinte, Lake Ontario, 1972–2008. Aquat. Ecosyst. Health Mgmt. 14(1), 9–20.

Moore, R. B., 2004. Estimation of total nitrogen and phosphorus in New England streams using spatially referenced regression models. New Hampshire: US Department of the Interior, US Geological Survey.

Nicholls, K. H., Carney, E.C., 2011. The phytoplankton of the Bay of Quinte, 1972–2008: point-source phosphorus loading control, Dreissenid Mussel establishment, and a proposed community reference. Aquat. Ecosyst. Health Mgmt. 14(2), 33–43.

Nicholls, K.H., Heintsch, L., Carney, E., 2002. Univariate step-trend and multivariate assessments of the apparent effects of P loading reductions and zebra mussels on the phytoplankton of the Bay of Quinte Lake Ontario. J. Gt. Lakes Res. 28, 15–31.

Project Quinte members, 2011. Project Quinte Annual Report 2009. Kingston, Ontario, Canada.

Reckhow, K. H., Arhonditsis, G. B., Kenney, M. A., Hauser, L., Tribo, J., Wu, C., Elcock, K.J., Steinberg, L.J., Stow, C. A., McBride S. J., 2005. A predictive approach to nutrient criteria. Environ. Sci. Technol. 39, 2913–2919.

Scheffer, M., Straile, D., van Nes, E. H., Hosper, H., 2001. Climatic warming causes regime shifts in lake food webs. Limnol. Oceanogr. 46, 1780–1783.

Schwarz, G. E., Hoos, A. B., Alexander, R. B., Smith, R. A., 2006. The SPARROW surface water-quality model: Theory, application, and user documentation: US Dept. of the Interior, US Geological Survey.

Seifried, K. E., 2002. Submerged macrophytes in the Bay of Quinte: 1972 to 2000, Department of Zoology. University of Toronto, University of Toronto.

Shimoda, Y., Watson, S., Palmer, M.E., Koops, M., Mugalingam, S., Morley, A., Arhonditsis, G.B., 2016. Delineation of the role of nutrient variability and dreissenids (Mollusca, Bivalvia) on phytoplankton dynamics in the Bay of Quinte, Ontario, Canada. Harmful Algae 55, 121-136.

Soldat, D. J., Petrovic, A. M., 2008. The fate and transport of phosphorus in Turfgrass Ecosystems. Crop Sci. 48, 2051–2065.

Soldat, D., Petrovic, A. M., Ketterings, Q., 2009. Effect of Soil Phosphorus Levels on Phosphorus Runoff Concentrations from Turfgrass. Water Air Soil Pollut 199, 33–44.

Vanderploeg, H. A., Liebig, J. R., Carmichael, W. W., Agy, M. A., Johengen, T. H., Fahnenstiel, G. L., Nalepa, T. F., 2001. Zebra Mussel (Dreissena polymorpha) selective filtration promoted toxic Microcystis blooms in Saginaw Bay (Lake Huron) and Lake Erie. Can. J. Fish. Aquat. Sci. 58, 1208–1221.

Vesanto, J., Himberg, J., Alhoniemi, E., Parhankangas, J., 2000. SOM Toolbox for Matlab 5.

Walters, C. J., 1986. *Adaptive Management of Renewable Resources: Biological Resource Management*. Macmillan, New York.

Wellen, C., Arhonditsis, G. B., Long, T., Boyd, D., 2014. Accommodating environmental thresholds and extreme events in hydrological models: a Bayesian approach. J. Great Lakes Res. 40 (Supplement 3), 102–116.

Winter, J. G., Duthie, H.C., 2000. Export coefficient modeling to assess phosphorus loading in an urban watershed. JAWRA 36, 1053–1061.

Zhang, W., Kim, D.-K., Rao, Y., Watson, S., Mugalingam, S., Labencki, T., Dittrich, M., Morley, A., Arhonditsis, G. B., 2013. Can simple phosphorus mass balance models guide management decision?: A case study in the Bay of Quinte, Ontario, Canada. Ecol. Model. 257, 66–79.

The above article is a reprinted version of the following previously published article:

Integration of best management practices in the Bay of Quinte watershed with the phosphorus dynamics in the receiving waterbody: What do the models predict?
Arhonditsis, G.B., Kim, D-K., Shimoda, Y.; Zhang, W., Sue Watson, S., Mugalingam, S., Dittrich, M., Geater, K., McClure, C., Bryon Keene, B., Morley, A., Richards, A., Long, T.; Rao, Y.R., Kalinauskas, R., 2016.
Aquatic Ecosystem Health & Management (2016) 19 (1), 1–18.
https://doi.org/10.1080/14634988.2016.1130566

Genetic fingerprints reveal geographically and seasonally variable algal virus communities in the Bay of Quinte

Steven M. Short[1,2*], Ankita Virdi[2], Cindy M. Short[2], and Robin M. Rozon[1]

[1]*Department of Ecology and Evolutionary Biology, University of Toronto, Toronto, ON*

[2]*Department of Biology, University of Toronto Mississauga, Mississauga, ON*

**Corresponding author : steven.short@utoronto.ca*

Keywords: *Phycodnaviridae*, algal virus, DNA polymerase, *polB*, PCR, DGGE

Introduction

The Bay of Quinte is scientifically unique as it has been the subject of a particularly intense, long-term ecosystem study (Minns et al., 2011). Unfortunately, the impetus to implement such an intensive study includes the facts that more than 40 years ago it was recognized as having suffered severe cultural eutrophication with numerous "Beneficial Use Impairments". The bay was eventually classified as a Great Lakes "Area of Concern", and it became the target of a "Remedial Action Plan" that focused on controlling phosphorus loading (Munawar et al., 2011; Nicholls, 2012). Following the implementation of a phosphorus abatement program in the late 1970s that targeted Point-sources such as sewage treatment plants, P concentrations in the upper bay were reduced by as much as 38% (Nicholls et al., 2002). However, despite these reductions targets for total P (30 $\mu g\ l^{-1}$) have not been met consistently in recent years, and eutrophication continues resulting in several beneficial use impairments (Munawar et al., 2012; Nicholls, 2012).

Reductions of P loading in the bay have had a statistically significant effect on phytoplankton communities and declines in the biovolume of total phytoplankton as well as certain species of diatoms, dinoflagellates, cryptophytes, green algae, and cyanobacteria were observed soon after the implementation of P control (Nicholls et al., 2002). After this initial phytoplankton decline following P control, phytoplankton rebounded in the mid-1980s to pre-P control levels. Later, in the

Ecology of the Bay of Quinte: Health, Management and Global Implications
Editors: C.K. Minns, M. Munawar and M.A. Koops

Aquatic Ecosystem Health & Management Society, Canada

mid-1990s, Zebra Mussels (*Dreissena* spp.), invaded the bay leading to additional changes in phytoplankton community composition and biomass. Most notably, the potentially toxic cyanobacteria *Microcystis* dramatically increased in biomass (Nicholls et al., 2002). Thus, scientists monitoring the Bay of Quinte over the last few decades have been presented with a unique opportunity to observe dramatic ecosystem changes first hand, and the perturbations that have challenged the embayment and conservation efforts alike have resulted in numerous lessons in aquatic ecosystem science and management.

An important area of scientific progress that has spun off of work on the Bay of Quinte has been the development of statistical tools to detect ecological regime shifts, or "relatively sudden changes between contrasting, persistent states" (Nicholls, 2011). First, multivariate statistical tests were used to demonstrate that three distinct phytoplankton communities existed: before P control, post-P control up to the Dreissena invasion, and post-invasion (Nicholls et al., 2002). Later, additional multivariate analytical tools were used to determine if community regime shifts could be reliably detected in a 37 year record and also demonstrated significant impacts of P reductions and the Dreissena invasion (Nicholls et al., 2011). This work led to the development of community reference that has been proposed as a way to relate ongoing changes in phytoplankton to 'improving' versus 'deteriorating' conditions in the bay (Nicholls et al., 2004; Nicholls and Carney, 2011). Despite these excellent contributions, forecasting phytoplankton regime shifts has remained elusive suggesting that either that the phytoplankton dynamics in the Bay of Quinte are intractably complex, or that the system as a whole is not entirely understood.

Viruses are a potentially important group of microorganisms that have not received any attention in Bay of Quinte research until very recently. This is not surprising considering that until recently, freshwater viruses received much less attention than marine viruses and freshwater viral ecology is a relatively young discipline within the aquatic sciences (Middelboe et al., 2008). Although the bulk of the literature on aquatic viruses has focused on marine environments, microbial ecologists have learned that viruses are important components in all aquatic environments that influence both primary and secondary production, carbon and nutrient flows, species succession and community dynamics, and even gene flow and evolutionary processes (for reviews see: Brussaard, 2004; Fuhrman, 1999; Jacquet et al., 2010; Short, 2012; Suttle, 2007; Van Etten et al., 1991; Wilhelm and Suttle, 1999; Wommack and Colwell,

2000). Currently, only two reports of viruses in the Bay of Quinte exist. The first focused on phosphorus bioavailability and microbial process in Lake Ontario, but it included an abundance estimate of approximately 2.09 x 10^7 virus particles ml^{-1} in the surface waters of the bay (Gouvea et al., 2006). More recently, Rozon and Short (2013), examined the dynamics of several different phytoplankton viruses in the bay and observed complex patterns of seasonality and spatial distribution. Despite the paucity of information on viruses in the Bay of Quinte, these two reports alone suggest that viruses are abundant and ecologically relevant components of the bay's microbial community and lower foodweb.

Hence, the overarching goal of this research was to obtain a broad perspective on the seasonal and spatial dynamics of a group of related viruses that infect different species of eukaryotic algae, specifically dsDNA viruses belonging to the family *Phycodnaviridae*. To that end, the study was designed to address the hypothesis that the phycodnavirus community is an active component of the bay's microbiota that is both seasonally and spatially variable. Since different types of viruses cannot be discriminated microscopically, a gene-based molecular approach was used to generate 'fingerprints' of phycodnavirus communities in samples from three locations in the bay collected from May to October, 2011 (Fig. 1). More specifically, the polymerase chain reaction

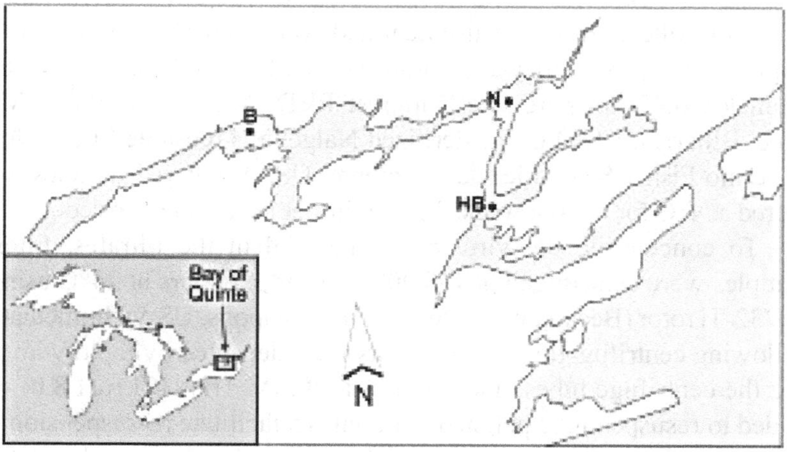

Figure 1. Map of the Bay of Quinte. The inset map shows the North American Great Lakes, with the Bay of Quinte on Lake Ontario. The approximate locations of the bi-weekly stations sampled during 2011 are indicated with the labels B (Belleville), HB (Hay Bay), and N (Napanee). Map image courtesy of Dr. Munawar, DFO.

(PCR) was used to amplify *Phycodnaviridae* DNA polymerase (*polB*) gene fragments and denaturing gradient gel electrophoresis (DGGE) was used to generate a genetic banding pattern, or 'fingerprint', characterizing the virus community in each sample. This approach has been used previously to compare and contrast algal virus communities in a variety of marine and freshwater environments (Clasen and Suttle, 2009; Short and Suttle, 2002; 2003; Zhong and Jacquet, 2014) and remains one of the most cost effective methods to compare diverse microbial communities in environmental samples.

Methodology

Sample collection and processing

As part of an ongoing monitoring program conducted by Fisheries and Oceans Canada researchers at the Great Lakes Laboratory of Fisheries and Aquatic Sciences (GLLFAS), 12 depth-integrated epilimnetic samples were collected approximately bi-weekly from May 25th to October 25th, 2011 from three Project Quinte index stations including the upper bay station Belleville (44° 09' 13.201" N, 77° 20' 44.002" W) and the middle bay stations Napanee (44° 09' 13.201" N, 77° 20' 44.002" W) and Hay Bay (44° 05' 35.999" N, 77° 04' 18.001" W). Within 12 hours of collection the depth-integrated water sample was mixed and approximately 250 ml was subsampled and filtered through 47 mm diameter, 0.45 µm pore-size Durapore PVDF membrane filters (Milli pore, Billerica, USA) using sterilized Nalgene™ reusable filter holders (Thermo Fisher Scientific Inc., Waltham, USA). Filtered samples were stored at 4°C for up to 4 weeks before further processing was conducted.

To concentrate the virus-sized material in the filtrates, filtered samples were centrifuged at 118,000 X g for 3.5 hours at 24°C using a SW32-Ti rotor (Beckham Coulter, Inc., Indianapolis, USA). Immediately following centrifugation, supernatants were decanted by rapidly inverting the centrifuge tubes, and 300 µl of 10 mM Tris·HCl (pH 8.0) was added to resuspend the pelleted material. To facilitate re-suspension the tubes were left overnight at 4°C to allow the pelleted material to soften, and the following day pellets were mechanically resuspended using a vortex mixer, transferred to a screw-cap microcentrifuge tube, and were stored at -20°C. To prepare the virus-size fraction pellets as templates for PCR, 50 µl aliquots of the resuspended pellets were subjected to a freeze-thaw treatment as previously described (Short and Short, 2008).

Molecular characterization of virus communities

To specifically amplify *Phycodnaviridae polB* genes from each sample, PCR was conducted using the algal-virus specific primers AVS1 and AVS2 (Chen and Suttle, 1995). As previously described, to enhance the yields of amplified material and facilitate DGGE resolution of individual amplified gene fragments PCR was conducted in two stages where excised bands from the first-round reaction were used as templates in a second-round reaction with a modified version of the forward primer that had an additional 30 nucleotide GC-clamp added (Short and Suttle, 2002; Clasen and Suttle, 2009; Zhong et al. 2013; Zhong and Jacquet, 2014). For the first-round reactions, 5 µl of the freeze-thaw treated virus-size fraction pellets were added to reaction mixtures containing 20 and 30 pmol of AVS1 and AVS2, respectively, 1.5 mM $MgCl_2$, 10 mM each dNTP, and 0.625 U of Platinum® *Taq* DNA polymerase (Life Technologies, Carlsbad, CA, USA) with the manufacturer supplied reaction buffer. Negative control reactions were set up using 5 µl of H_20 substituted as template. Thermal cycling was conducted with an initial denaturation step at 95°C for 2 min, followed by 36 cycles of 95°C for 30 s, 50°C for 1 min, and 72°C for 1 min, with a final extension step at 72°C for 30 min using a Bio-Rad C1000™ thermal cycler (Bio-Rad Laboratories, Inc., Hercules, USA). Approximately 40 µl of each PCR reaction was electrophoresed in 1.5 % LE agarose (Promega, Madison, USA) gels in 1X TAE (40 mM Tris·HCl, 20 mM acetic acid, 1.0 mM EDTA, pH 7.6) for 75 min at 5 V cm^{-1}. Agarose gels were stained in a 0.5 µg ml^{-1} ethidium bromide solution and were imaged using a Molecular Imager® ChemiDoc™ XRS system (Bio-Rad Laboratories).

After electrophoresis and imaging was complete, a disposable Pasteur pipette was used to excise agarose plugs from bands (i.e. DNA fragments) in each sample corresponding to the expected ~700 bp *polB* PCR product. When excising bands, amplified fragments were exposed as little UV radiation as possible by placing only the lane of interest over the lit part of the transilluminator, and the transilluminator was turned off between subsequent excisions. Excised agarose plugs were placed in 100 µl of sterile Tris buffer (10 mM Tris·HCl, pH 8.5) and were heated to 65°C for 20 min to elute DNA fragments. Five microlitres of the eluant was used as template for a second-stage of PCR that was identical to the first-stage PCR described above except 5 pmol of the primer AVS1-GC (5' - CGC GCC CCG GGC GGC GCG GGC GCA CGG GGC GAR GGI GCI ACI GTI YTI GAY GC - 3') was used in place of AVS1, the amount of AVS2 was reduced to 20 pmol, the total volume of the reactions were 100 μl, and the number of PCR cycles was reduced to 30. Negative controls were included in each set of reactions as described above. To ensure enough volume was available for the subsequent DGGE steps, two independent second-stage PCR reactions were carried out for each sample, and these were pooled prior to electrophoresis. To visualize the second-stage PCR products, 10 μl of the pooled samples were electrophoresed in agarose gels as described above for the first stage reactions.

Prior to DGGE, the relative DNA concentration in each pooled second-stage PCR was estimated via gel densitometry using Quantity One 1-D Analysis Software (Bio-Rad Laboratories), and the amount of DNA loaded into the DGGE gel was normalized such that all lanes in a particular gel were loaded with approximately the same amount of DNA and the maximum volume loaded for any sample was 30 μl. DGGE was conducted using 35 to 50% denaturing gradients (where 100% denaturant is 7 M urea and 40 % formamide) cast in 6% polyacrylamide gels prepared from a 40% 37.5:1 Acrylamide: Bis-Acrylamide solution (Bio-Rad Laboratories). To provide reference markers in the gels, at least two wells in each gel were loaded with 10 μl of Fermentas GeneRuler™ 100 bp DNA ladder (Thermo Fisher Scientific Inc., Waltham, USA). Electrophoresis was carried out for 15 h in 1.0 X TAE at 105 volts and a constant temperature of 60°C using the Cipher DGGE Electrophoresis System (C.B.S. Scientific Company Inc., Del Mar, USA). Following electrophoresis, gels were soaked for 1 h in a 1X SYBR® Gold (Life Technologies, Carlsbad, USA) nucleic acid gel stain solution prepared with 1X TAE buffer. Gels were imaged using a Molecular Imager® ChemiDoc™ XRS system (Bio-Rad Laboratories).

Analytical approach

To compare and contrast phycodnavirus communities in the Bay of Quinte, four separate DGGE analyses were conducted during this study: one for each of the three sample locations, and one containing samples from each sample site collected at four different times of the year. Since each DGGE lane was loaded with DNA amplified from a single environmental sample, lane banding patterns in individual DGGE gels were compared pairwise to produce lane similarity matrices and create dendro grams of sample similarities. First, after all lanes in a given gel were identified and background was subtracted using the rolling disk method, densitometry profiles were created for each lane using Quantity One 1-D Analysis Software (Bio-Rad Laboratories). The data from each lane profile (i.e. pixel intensities at > 620 discrete intervals from the top to the bottom of the gel lanes) was exported into Microsoft Excel 2007 (Micro soft Corporation, Redmond, USA) and a matrix of pairwise distances was created using the formula $1 - r$, where r is the Pearson product-moment correlation coefficient. The matrix was then imported into MEGA 5 (Tamura et al., 2011), and a dendrogram of pairwise distances was generated using unweighted pair group method with arithmetic mean (UPGMA) as the clustering method. All graphics were drawn using Adobe Illustrator CS5 (Adobe Systems, Inc., San Jose, USA).

Results

All first-stage PCR reactions produced visible bands of approximately 700 bp corresponding to the expected fragment size for *Phycodnaviridae polB* amplicons. Likewise, second-stage PCR reactions with the AVS1-GC primer produced the expected size fragments and densitometry of the agarose gels demonstrated that the DNA concentrations of the PCR products were very similar (i.e. concentration differences between samples were always less than 2-fold). DGGE analysis of the second-stage PCR reactions produced distinct banding patterns for each sample with numerous discrete bands demonstrating the presence of diverse Phycodnavirus communities throughout the Bay of Quinte. Visual inspection of the gels loaded with PCR products from a single site revealed that some bands appeared and disappeared over time, and the intensity of some bands changed markedly throughout the year. Additionally, in every gel there were some bands that are present in nearly every sample as well as some that appear transiently in only one or a few samples (Fig. 2). For each of the three sampling sites there appears to be subtle seasonal pattern of band richness, with certain samples producing the greatest band richness. For example, amplification of Belleville samples from the late summer and early autumn (i.e. July, August and September) produced the greatest number of bands, whereas samples from June and October produced the fewest (Fig. 2A). Amplification of Napanee samples from July also produced the greatest number of bands, yet the sample collected on October 25 was similarly rich while the August samples were relatively depauperate (Fig. 2B). For Hay Bay, samples from August appear to have the least number of discernable bands while samples from the other months sampled (May, June, July, September and October) were all richer with respect to the number of bands (Fig. 2C).

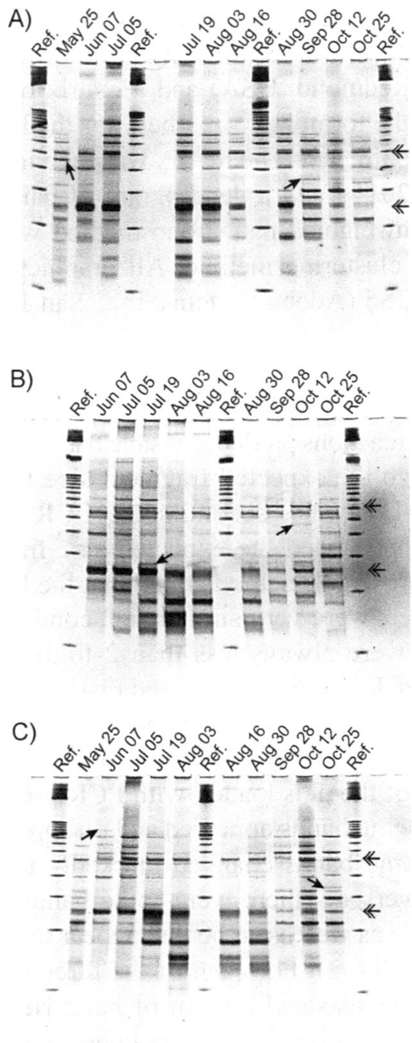

Figure 2. Fingerprint analysis of algal viruses at each station. Algal virus gene fragments were amplified from A) Belleville, B) Napanee, and C) Hay Bay water samples on the dates indicated above each lane. As examples related to comments in the text, the single headed arrows indicate bands that appear in only one or a few samples, whereas double-headed arrows indicated the position of bands that appear in almost all samples. Lanes labelled as Ref. were loaded with a 100 bp DNA marker as a reference marker for the gel. See text for further details.

Comparing dendrograms of lane similarities to the gel images reveals that cluster analysis, in general, reflected general trends that can be picked out by visual inspection of banding patterns. For example, in the gel of the Belleville samples, the banding patterns that appeared most similar, like those from October 12 and 25th or July 19th and August 3rd, formed neighboring branches in the corresponding dendrogram (Figs. 2A & 3A). Similar correspondence of banding patterns and dendrogram branching can also be discerned for the Napanee samples from August 3rd and 6th, or the Hay Bay samples from October 12th and 25th, for example (Figs. 2 & 3). Broadly speaking, dendrograms of bi-weekly samples from the same locations within the Bay of Quinte all demonstrated that samples from the same time of year produced the most similar finger prints (i.e. samples from the same seasons clustered), but no two samples in any gel were identical (Fig. 3).

The gel loaded with samples from each location at several times of the year (specifically, June 7th, July 19th, August 30th, and October 12th) also produced patterns demonstrating the seasonality inherent in these virus fingerprints (Fig. 4A); generally, with respect to the number of bands, samples from July produced the richest patterns and this signal was reflected in the dendrogram of fingerprint similarities (Fig. 4B). For example, the samples from all stations collected on June 7th clustered together as a group as did the samples from July 19th. In contrast, samples from the late fall in the Bay of Quinte (i.e. October 12th) were more distinct as obvious differences in the fingerprint patters were apparent for each station. This observation is also reflected in the cluster analysis since the samples from October 12 have the longest branch lengths in the dendrogram indicating that they were the least similar samples of the entire set (Fig. 4B). Another interesting observation from this fingerprint analysis was that at different times of the year, different stations were most similar to each other. For example, on July 19th fingerprints generated from the Napanee and Belleville stations were most similar, yet on August 30th, the Napanee fingerprint was much more similar to the fingerprint from Hay Bay (Fig. 4B).

Discussion

To examine Bay of Quinte algal virus diversity and dynamics, the universal PCR primers AVS1 and AVS2 that target conserved virus genes were used to amplify Phycodnavirus *polB* gene fragments from various water samples. Because PCR with these universal primers can

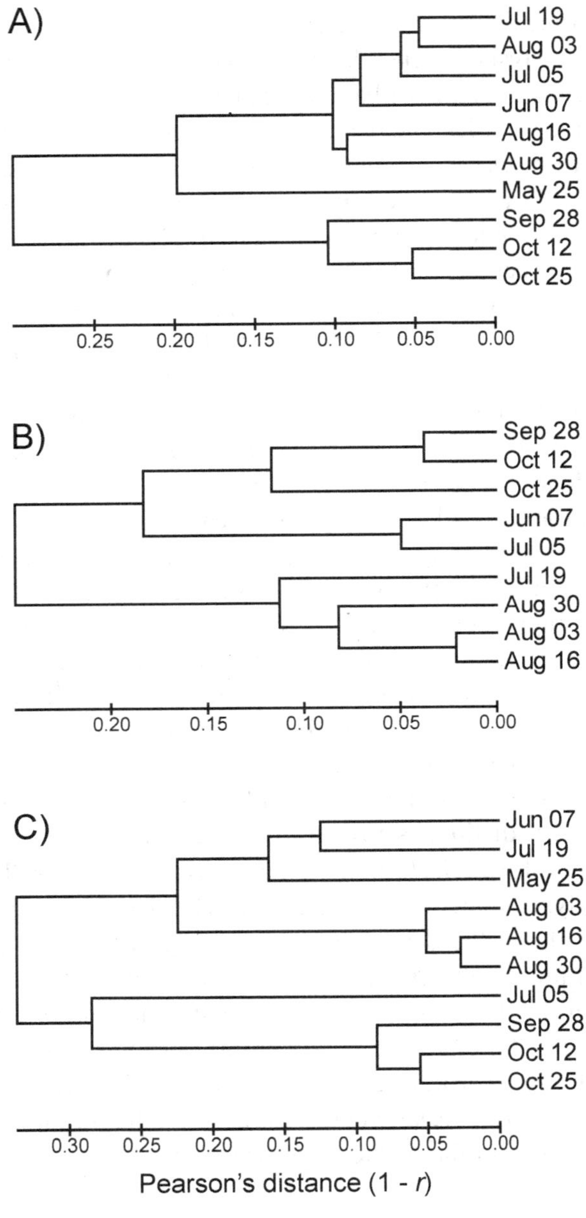

Figure 3. Cluster analysis of fingerprint similarities. The Dendrograms are based on UPGMA cluster analyses of pair-wise correlations of DGGE densitometry profiles and show sample to sample distances for algal-virus gene fragments amplified via PCR from A) Belleville, B) Napanee, and C) Hay Bay.

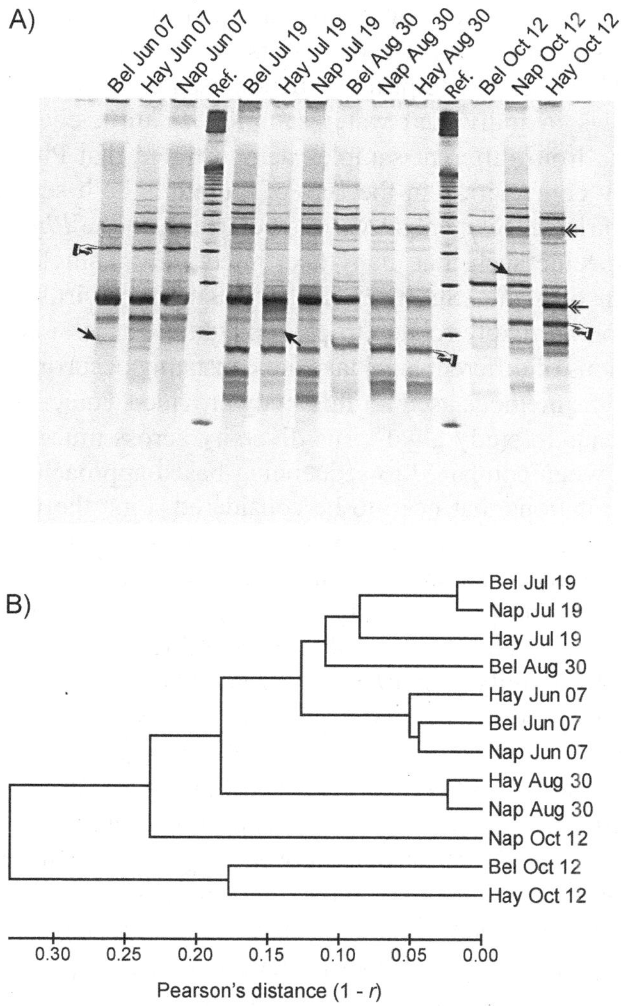

Figure 4. Fingerprint and cluster analysis of algal viruses sampled at four different dates. Samples are labeled with the sampling location and date of collection, and locations abbreviated as Bel, Nap, and Hay correspond to Bellevile, Napanee, and Hay Bay, respectively. A) DGGE analysis. As examples related to comments in the text, the single headed arrows indi cate bands that appear in only one or a few samples, double-headed arrows indicated the position of bands that appear in almost all samples, and the hand pointer indicates the positions of bands that were most predominant at certain times of the year. Lanes labelled as Ref. were loaded with a 100 bp DNA marker as a reference marker for the gel. B) Dendrogram of UPGMA cluster analysis of the fingerprints shown in A.

amplify gene fragments from many different viruses simultaneously, DGGE was used to separate fragments derived from different virus taxa and generate a fingerprint, or banding pattern, of the algal virus communities in individual water samples. In turn, comparisons of fingerprints from different samples demonstrated that Phycodnavirus community composition in the Bay of Quinte is both seasonally and spatially variable. Since algal viruses belonging to the *Phycodnavidae* are host specific such that many taxa infect only a single species, or even strain of algae (Short, 2012), fluctuations in virus community composition are likely related to phytoplankton succession and imply that infections of different algal taxa are constantly occurring in the bay.

While the methods used for this study provided a quick and cost-effective means to study algal virus diversity across multiple samples, especially when compared to sequencing-based approaches, there are distinct limitations that need to be considered. First the PCR method described here is not truly universal as it cannot capture the complete diversity of viruses that infect eukaryotic algae, nor can it even capture all phycodnavirus diversity (Short, 2012). As other studies of fresh water algal viruses have shown, the AVS1 and AVS2 primers appear to preferentially amplify gene fragments from phycodnaviruses that infect green algae, especially those that are most closely related to the genus *Prasinovirus* (cultivated viruses that infect prasinophyte algae) (Clasen and Suttle, 2009; Short and Short, 2008; Short et al., 2011; Zhong and Jacquet, 2013). Unfortunately, due to the limited number phycodnavi ruses that have been cultivated, it is not entirely clear which algal species or even genera are infected by the diverse group of viruses that can be detected with AVS1 and AVS2. Nevertheless, considering that almost all of the environmental phycodnavirus sequences obtained from freshwater environments form clusters nested among the virus genera *Chlorovirus* and *Prasinovirus*, it is very likely that genes amplified with AVS1 and AVS2 are derived from viruses that infect green algae (i.e. Chloro phytes, or in earlier literature Chlorophyceae). Chlorophyte ecology in the Bay of Quinte is particularly interesting since, following each of the major perturbances the bay has experienced over the last few decades (i.e. P control and *Dreissena* establishment), this group has declined with respect to its contribution to total algal biomass (Nicholls et al., 2002). Importantly, relative to the current status of the Bay of Quinte, increased Chlorophyte biomass is an aspect of the target, or reference phytoplankton community that was defined for ongoing comparisons of changes in phytoplankton community composition (Nicholls et al., 2004;

Nicholls and Carney, 2011). Hence, this study of algal virus community dynamics focusing on viruses of Chlorophyte algae is relevant not only from the perspective of increasing knowledge of phytoplankton ecology, but perhaps even for future monitoring and remediation efforts. Indeed, this work demonstrates that there is a previously unrecognized, additional layer of complexity in the ecology of primary producers in the Bay of Quinte. By acting as highly selective top-down control agents of phytoplankton (Short, 2012), viruses could confound bottom-up based remediation efforts (e.g. nutrient abatement) leading to unexpected or even undesirable results. Clearly, much is left to learn about the role of viruses in algal ecology, but this first glimpse into algal virus communities highlights the fact they are dynamic and potentially important players in the Bay of Quinte's lower foodweb.

Another important technical aspect of this work that needs to be considered is that the fingerprinting approach cannot be considered quantitative with respect to gene abundances. Differing band intensities between and even within samples cannot be interpreted as arising from differences in target gene abundance alone since they can also be due to sample to sample variations in PCR efficiency, differences in the amount of DNA loaded each lane, or even differing amplification efficiency of different fragments within the same sample; since the PCR primers used were highly degenerate, different fragments targeted with these primers can amplify with differing efficiency. Additionally, some studies have demonstrated that individual bands in a DGGE gel do not necessarily correspond to single gene sequences since multiple sequences can be obtained from a single DGGE band (e.g. Sekiguchi et al., 2001). Thus, DGGE banding patterns must be interpreted conservatively to avoid overstating the results of the analysis. Despite these caveats, DGGE fingerprints are reproducible and can be used to discern changes in aquatic virus communities (Sandaa et al., 2010; Short and Suttle, 2002). While it is tempting to use band presence/absence and intensity (i.e. abundance) data for Bray-Curtis statistical comparisons, that approach was not employed here. Rather, to minimize analytical artefacts stemming from erroneous interpretations of band intensities and to avoid issues associated with discriminating bands within a lane or matching bands across lanes, statistical comparisons of sample finger prints were based on lane densitometry profiles instead of band presence/ absence data. Pair-wise comparisons of each lane in each gel were used to generate matrices for cluster analyses that provide easily visualized representations of the overall similarities and differences between algal virus communities, or more specifically, algal virus gene fragments amplified from discrete water samples. As a final comment on technical issues related to DGGE, it is important to note that the chemical gradients in DGGE gels can vary from gel to gel such that gel to gel comparisons of fingerprints can be problematic. To avoid this issue, only within-gel comparisons were conducted for this study.

Broadly speaking, the results of this study highlight the seasonality and patchiness of some algal viruses in the Bay of Quinte. A striking result that is apparent in all of the DGGE gels (Figs. 2 & 4A) is that some bands (i.e. virus gene fragments) appeared in almost every sample while others were more transient and appeared in only one or a few samples. While the transient bands were almost certainly derived from gene fragments amplified from transient viruses, persistent bands may be dismissed as DGGE artifacts since, as noted above, different gene fragments can localize to the same position in DGGE so that bands at the same position across all lanes might not represent the same virus. That said, previous studies based on DGGE or even quantitative PCR have provided growing evidence that some phytoplankton viruses are persistent and seasonally stable and could represent seed bank populations, while other are more ephemeral and may even be allocthonous (Hewson et al., 2012; Rozon and Short, 2013; Short and Suttle, 2003; Short and Short, 2009; Short et al., 2011). Although it appears that intensity of the persistent bands is variable across samples, it is unclear if this variability actually reflects variations in the abundance of a single virus in the environment, or if it stemmed from the DGGE artifacts noted above. Whatever the case, this study of algal virus communities in the Bay of Quinte confirms and extends the findings of a previous study based on quantitative analysis of individual virus taxa; algal virus dynamics are complex and some viruses may stably coexist with their hosts, while others cycle through multiple boom and bust cycles each growing season (Rozon and Short, 2013). This observation of complex virus dynamics implies that the ecology of their hosts are similarly complex and raises questions about the predictability of phytoplankton community dynamics considering that even simple models of phytoplankton competition can yield chaotic results (Huisman and Weissing, 1999, 2001).

At the temporal and geographic scale of sampling conducted for this study (e.g. bi-weekly samples collected from stations at least 10 km apart), every sample produced a unique banding pattern suggesting that the virus community in every sample was also unique. Furthermore, marked changes in the fingerprints can be observed at certain times of the year. For example, at the upper bay stations, Belleville and Napanee, many bands appeared in the July 5th sample that were not present in the preceding sample (Fig. 2A & B) suggesting that new infections that gave rise to newly detectable viruses and an apparent increase in algal virus richness occurred sometime after June 7th. In the mid-bay station, Hay Bay, a similar transition with an apparent increase in virus richness was observed in the samples collected after August 30th (Fig. 2C). However, besides the obvious transitions in virus community structure, and despite subtle differences in all fingerprints, samples from the same time of year were generally more similar to each other. This autocorrelation of algal virus community composition demonstrates that algal virus communities in the Bay of Quinte undergo seasonal succession. In all likelihood, the succession of algal virus and their hosts are intimately linked.

Another interesting observation is that the sampling stations that were most similar with respect to algal virus fingerprints differed at different times of the year. For example, fingerprints from samples collected on July 19th from Napanee and Belleville that clustered with very low Pearson's distance values and were distinct from Hay Bay, whereas on August 30th it was the samples from Napanee and Hay Bay clustered most closely and were distinct from Belleville (Fig. 4B). This data corroborates a previous Mantel test showing that geographic distance was positively correlated to the abundances of several different phyto plankton viruses at the Belleville, Napanee, and Hay Bay stations high lighting the patchiness of phytoplankton viruses in the bay (Rozon and Short, 2013). On the other hand, the fingerprints from samples collected on June 7th were all relatively similar suggesting that the algal virus community was relatively homogenous throughout the upper and mid reaches of the bay (Fig. 4B). Although factors driving seasonal contrasts of algal virus community patchiness are likely very complex, it is notable that these patterns of sample similarity coincide with at least one other relevant biological parameter. Although chlorophyll-*a* concentrations were very similar among these sampling stations at each of these times of year, differences in Chlorophyte biomass, the presumed hosts for the viruses fingerprinted, can be related to the fingerprint clustering. On June 7th when all fingerprints were equally similar, Chlorophyte biomass was also similar at Belleville, Napanee, and Hay Bay with values of 196.1, 179.3, and 176.2 mg m^{-3}, respectively (M. Munawar, unpubl. data). In contrast and following the pattern of virus fingerprint similarities, on July 19th Belleville and Napanee were most similar with Chlorophyte biomass values of 219.8 and 187.6 mg m^{-3} compared to the Hay Bay value of 424.2 mg m^{-3}. On August 30th, the Napanee and Hay Bay Chlorophyte biomass was 239.6 and 302.8 mg m^{-3}, respectively, whereas Belleville was 407.9 mg m^{-3} (M. Munawar, unpubl. data). Thus, while it is premature to conclude that algal virus community composition was directly linked to the patterns observed for biomass, or visa versa, it is reassuring that these intuitively related biological parameters corresponded.

Summary and conclusions

To examine the genetic richness and dynamics of algal virus communities in the Bay of Quinte, the Polymerase Chain Reaction (PCR) and Denaturing Gradient Gel Electrophoresis (DGGE) were used to amplify and 'fingerprint' Phycodnavirus DNA polymerase gene (*polB*) fragments from depth-integrated, epilimnetic samples collected at three stations throughout 2011. Fingerprint analysis demonstrated that the community in every sample was unique, despite the fact that some individual virus gene fragments were persistent and homogenously distributed throughout the bay. In contrast, other gene fragments were ephemeral and only appeared at certain locations, or at certain times of the year. Cluster analysis of the fingerprint patterns demonstrated the seasonality of the bay's algal virus communities. Samples from the same time of year collected at each individual station tended to be most similar, and samples from all three stations collected at four different dates spanning the sampling period also tended to cluster by date. Considering that the molecular methods used in this study are biased towards viruses that infect Chlorophyte algae, it was notable that clustering patterns of algal virus community structure coincided with patterns of Chlorophyte biomass. Together, the results of this study demonstrate that viral infections of phytoplankton are constantly occurring in the bay and highlight the potential importance of these parasitic microorganisms in the Bay of Quinte's phytoplankton ecology.

This first study of a broad group of phytoplankton viruses in the Bay of Quinte confirmed previous observations that some algal viruses are environmentally persistent while others are much more ephemeral (e.g. Hewson et al., 2012; Rozon and Short, 2013; Short and Suttle, 2003; Short and Short, 2009; Short et al., 2011), and also provided evidence for the patchy spatial and seasonal distributions of algal viruses in the bay. An important distinction of this study from the previous work on

Bay of Quinte phytoplankton viruses is that it demonstrated for the first time that numerous algal viruses can be homogenously distributed throughout the bay at certain times of the year. Perhaps even more important is the fact that every sample collected during this study generated a unique fingerprint. Even if the differences between some samples were very subtle, this observation suggests that algal virus communities in the Bay of Quinte are highly dynamic. All together, the data reported here demonstrate that viral infections are continuously occurring in the bay. Hence, to achieve a complete understanding of the bay's algal ecology, host-parasite interactions must be considered. How, or if, this information will influence future management decisions and remediation criteria remains to be seen, but it is clear that viruses are a vital component of the Bay of Quinte's lower foodweb.

Acknowledgements

Special thanks to Dr. Mohi Munawar, Mark Fitzpatrick, and Heather Niblock and colleagues at Fisheries and Oceans Canada and the Ontario Ministry of the Environment for sample collection. This research was supported in part by an NSERC Undergraduate Student Research Award given to A. Virdi, and Canadian Foundation for Innovation Leaders Opportunity Fund, and NSERC Discovery grants awarded to S. Short.

References

Brussaard, C. P. D., 2004. Viral control of phytoplankton populations - a review. Journal of Eukaryotic Microbiology 51(2), 125-138.

Chen, F., Suttle, C. A., 1995. Amplification of DNA-polymerase gene fragments from viruses infecting microalgae. Applied and Environmental Microbiology 61(4), 1274-1278.

Clasen, J. L., Suttle, C. A., 2009. Identification of freshwater *Phycodnaviridae* and their potential phytoplankton hosts, using DNA pol sequence fragments and a genetic distance analysis. Applied and Environmental Microbiology 75(4), 991-997.

Fuhrman, J. A., 1999. Marine viruses and their biogeochemical and ecological effects. Nature 399(6736), 541-548.

Gouvea, S. P., Melendez, C., Carberry, M. J., Bullerjahn, G. S., Wilhelm, S. W., Langen, T. A., Twiss, M. R., 2006. Assessment of phosphorus-microbe interactions in Lake Ontario by multiple techniques. Journal Of Great Lakes Research 32(3), 455-470.

Hewson, I., Barbosa, J. G., Brown, J. M., Donelan, R. P., Eaglesham, J. B., Eggleston, E. M., LaBarre, B. A., 2012. Temporal dynamics and decay of putatively allochthonous and autochthonous viral genotypes in contrasting freshwater lakes. Applied and Environmental Microbiology *78*(18), 6583-6591.

Huisman, J., Weissing, F. J., 1999. Biodiversity of plankton by species oscillations and chaos. *Nature* 402(6760), 407-410.

Huisman, J., Weissing, F. J., 2001. Fundamental unpredictability in multispecies competition. American Naturalist 157(5), 488-494.

Jacquet, S., Miki, T., Noble, R., Peduzzi, P., Wilhelm, S., 2010. Viruses in aquatic ecosystems: Important advancements of the last 20 years and prospects for the future in the field of microbial oceanography and limnology. Advances in Oceanography and Limnology 1(1), 97-141.

Middelboe, M., Jacquet, S., Weinbauer, M., 2008. Viruses in freshwater ecosystems: An introduction to the exploration of viruses in new aquatic habitats. Freshwater Biology 53(6), 1069-1075.

Minns, C. K., Munawar, M., Koops, M. A., Millard, E. S., 2011. Long-term ecosystem studies in the Bay of Quinte, Lake Ontario, 1972-2008: A prospectus. Aquatic Ecosystem Health & Management 14(1), 3-8.

Munawar, M., Fitzpatrick, M., Niblock, H., Lorimer, J., 2011. The relative importance of autotrophic and heterotrophic microbial communities in the planktonic food web of the Bay of Quinte, Lake Ontario 2000-2007. Aquatic Ecosystem Health & Management 14(1), 21-32.

Munawar, M., Fitzpatrick, M., Munawar, I. F., Niblock, H., Kane, D., 2012. Assessing ecosystem health impairments using a battery of ecological indicators: Bay of Quinte, Lake Ontario example. Aquatic Ecosystem Health & Management 15(4), 430-441.

Nicholls, K. H., Heintsch, L., Carney, E., 2002. Univariate step-trend and multivariate assessments of the apparent effects of p loading reductions and zebra mussels on the phytoplankton of the Bay of Quinte, Lake Ontario. Journal Of Great Lakes Research 28(1), 15-31.

Nicholls, K. H., Heintsch, L., Carney, E. C., 2004. A multivariate approach for evaluating progress towards phytoplankton community restoration targets: Examples from eutrophication and acidification case histories. Aquatic Ecosystem Health & Management 7(1), 15-30.

Nicholls, K. H., 2011. Detection of regime shifts in multi-species communities: The Bay of Quinte phytoplankton example. Methods in Ecology and Evolution, 2(4), 416-426.

Nicholls, K. H., Carney, E. C., 2011. The phytoplankton of the Bay of Quinte, 1972- 2008: Point-source phosphorus loading control, dreissenid mussel establishment, and a proposed community reference. Aquatic Ecosystem Health Management 14(1), 33-43.

Nicholls, K. H., Hoyle, J. A., Johannsson, O. E., Dermott, R., 2011. A biological regime shift in the Bay of Quinte ecosystem (Lake Ontario) associated with the establishment of invasive dreissenid mussels. Journal of Great Lakes Research 37(2), 310-317.

Nicholls, K. H., 2012. Phosphorus and chlorophyll in the Bay of Quinte: A time-series/ intervention analysis of 1972- 2008 data. Aquatic Ecosystem Health & Management, *15*(4), 421-429.

Rozon, R. M., Short, S. M., 2013. Complex seasonality observed amongst diverse phytoplankton viruses in the Bay of Quinte, an embayment of Lake Ontario. Freshwater Biology 58(12), 2648-2663.

Sandaa, R. A., Short, S. M., Schroeder, D. C., 2010. Fingerprinting aquaticvirus communities. In S. W. Wilhelm, M. G. Weinbauer. C. A. Suttle (Eds.), *Manual of aquatic viral ecology* (pp. 9-18). American Society of Limnology and Oceanography, Waco.

Sekiguchi, H., Tomioka, N., Nakahara, T., Uchiyama, H., 2001. A single band does not always represent single bacterial strains in denaturing gradient gel electrophoresis analysis. Biotechnology Letters 23(15), 1205-1208.

Short, S. M., 2012. The ecology of viruses that infect eukaryotic algae. Environmental Microbiology 14(9), 2253-2271.

Short, S. M., Suttle, C. A., 2002. Sequence analysis of marine virus communities reveals that groups of related algal viruses are widely distributed in nature. Applied and Environmental Microbiology 68(3), 1290-1296.

Short, S. M., Suttle, C. A., 2003. Temporal dynamics of natural communities of marine algal viruses and eukaryotes. Aquatic Microbial Ecology 32(2), 107-119.

Short, S. M., Short, C. M., 2008. Diversity of algal viruses in various North American freshwater environments. Aquatic Microbial Ecology 51(1), 13-21.

Short, S. M., Short, C. M., 2009. Quantitative PCR reveals transient and persistent algal viruses in Lake Ontario, Canada. Environmental Microbiology 11(10), 2639-2648.

Short, S. M., Rusanova, O., Staniewski, M. A., 2011. Novel phycodnavirus genes amplified from Canadian freshwater environments. Aquatic Microbial Ecology, 63(1), 61-67.

Suttle, C. A., 2007. Marine viruses - major players in the global ecosystem. Nature Reviews Microbiology 5(10), 801-812.

Tamura, K., Peterson, D., Peterson, N., Stecher, G., Nei, M., Kumar, S., 2011. Mega5: Molecular evolutionary genetics analysis using maximum likelihood, evolutionary distance, and maximum parsimony methods. Molecular Biology and Evolution 28(10), 2731-2739.

Van Etten, J. L., Lane, L. C., Meints, R. H., 1991. Viruses and virus like particles of eukaryotic algae. Microbiological Reviews, *55*(4), 586-620.

Wilhelm, S. W., Suttle, C. A., 1999. Viruses and nutrient cycles in the sea. BioScience, 49(10), 781-788.

Wommack, K. E., Colwell, R. R., 2000. Virioplankton: Viruses in aquatic ecosystems. Microbiology and Molecular Biology Reviews, 64(1), 69-114.

Zhong, X., Berdjeb, L., Jacquet, S., 2013. Temporal Dynamics and Structure of Picocyanobacteria and Cyanomyoviruses in Two Large and Deep Peri-alpine Lakes. FEMS Microbiology Ecology 86, 312-326.

Zhong, X., Jacquet, S., 2014. Contrasting diversity of phycodnavirus signature genes in two large and deep western European lakes. Environmental Microbiology 16, 759-773.

Phytoplankton ecology in the Bay of Quinte: Spatial distribution, dynamics and heterogeneity during the summer of 2010

M. Munawar[1], M. Fitzpatrick[1,*], H. Niblock[1], H. Kling[2], J. Lorimer[3], R. Rozon[1]

[1]*Fisheries and Oceans Canada, 867 Lakeshore Road, Burlington, ON L7S 1A1*
[2]*Algal Taxonomy and Ecology, Stony Mountain, Manitoba,*
[3]*AEHMS, Burlington, ON*
**Corresponding author: mark.fitzpatrick@dfo-mpo.gc.ca*

Keywords: eutrophication, biodiversity, taxonomy, Area of Concern, algae, photosynthesis

Introduction

Cultural eutrophication has long been a major stressor affecting the health of nearshore areas of the Great Lakes. As the global scientific consensus narrowed in on phosphorus as the leading cause of eutrophication in the 1960s (Vollenweider, 1968), efforts to control the release of phosphorus into watersheds became commonplace. In North America, that led to the governments of Canada and the United States signing a treaty in 1972 – the Great Lakes Water Quality Agreement (GLWQA) – to protect these shared freshwater resources from anthropogenic pollution and to establish specific targets for phosphorus loads (International Joint Commission, 1988). Amid this backdrop, research and monitoring programs were implemented to assess the extent of the problem and establish courses of action.

The Bay of Quinte is a largely riverine embayment, located on the northeastern shores of Lake Ontario, with a long history of eutrophication. Research into algal blooms in the Bay was conducted as early as 1945. Tucker (1948) described blooms of Diatoms (Diatomeae: including species of *Melosira* (syn. *Aulacoseira*), *Tabellaria* and *Fragilaria*), 'blue greens' (Cyanophyta: *Aphanizomenon* spp.) and 'green algae' (Chlorophyta: *Mougeotia* spp.). Although the methods deployed at the time were biased towards sampling larger algae (i.e. > 20 µm in

Ecology of the Bay of Quinte: Health, Management and Global Implications
Editors: C.K. Minns, M. Munawar and M.A. Koops

Aquatic Ecosystem Health & Management Society, Canada

diameter) and the study area limited, the results nonetheless hint at the heterogeneity of phytoplankton communities in the Bay.

A research and monitoring program was established in 1972, coinciding with the signing of the original GLWQA, and has continued to the present (2016) in the process establishing an extensive database of eutrophic algal communities that is unrivalled in the Great Lakes (e.g. Nicholls and Carney, 2011; Minns et al., 2011). This database was amassed by limiting the sampling program to 2 – 3 index stations and by sampling at regular intervals (biweekly from May – October). While this has been a sound strategy for research and monitoring, it does not provide detailed information on the spatial distribution of phytoplankton communities which in turn limits our knowledge of the extent to which eutrophic conditions persist throughout the Bay. Hence the present need to consider the spatial distribution of phytoplankton communities in the Bay of Quinte from both a structural and functional perspective.

In 1975, at a time when the Bay was determined to be in an advanced state of eutrophication (e.g. Johnson and Owen, 1971) and before the introduction of phosphorous controls, Nicholls and Carney (1979) undertook a comprehensive assessment of the phytoplankton communities. Their work included weekly sampling, during May to October, from 8 sites across the Bay. While the bulk of their discussion was devoted to resolving taxonomic identifications among 330 distinct algal taxa using contemporary methodologies, Nicholls and Carney (ibid) did observe that the eutrophic inner bay flora could be described as 'a *Melosira granulata – Aphanizomenon flos-aquae* type plankton'. Similar, at least qualitatively, to Tucker's (1948) findings from 30 years earlier, the results again showed that both Diatomeae and Cyanophyta were important taxa contributing to the algal assemblages in the Bay of Quinte.

While much of the contemporary literature on eutrophication focuses on defining the problem in terms of water quality (e.g. nutrients and chlorophyll-*a*) or in terms of harmful algal blooms, very little attention has been paid to the structure and function of the phytoplankton community.

In this paper, we will consider the spatial distribution of phytoplankton communities (biomass, taxonomy) and primary productivity across the Bay of Quinte during the peak growth period (June – September) alongside standard measurements of water quality. In undertaking this work, our primary objectives were to assess: 1) The spatial extent of eutrophic conditions in the Bay of Quinte; 2) The taxonomy and diversity of phytoplankton assemblages throughout the Bay; and 3) The photosynthetic capacity of algal communities using Size Fractionated Primary Productivity experiments. Our findings will be directly relevant to the continued mitigation of cultural eutrophication in the Bay of Quinte as well as other freshwater ecosystems.

Materials and Methods

The Bay of Quinte is a 'Z' shaped embayment of 254 km^2 located on the northeastern shores of Lake Ontario. Five major tributaries flow into the Bay and it is a largely riverine system. Depth ranges from 4 – 8 m, but approaches 30 m near the outflow into Lake Ontario.

Research surveys were conducted on June 22-24, 2010 (n=12); August 17-19, 2010 (n=12) and September 20-23, 2010 (n=12). Information regarding sampling locations and frequency can be found in Figure 1 and Table S1 (Appendix). All samples were collected from 0800 – 1600 (EDT) and processed or preserved the same day.

Surface temperature was measured using a Hydrolab minisonde 4a surveyor. Integrated epilimnetic water was collected at each site and subsamples were drawn for nutrients, chlorophyll-*a* and phytoplankton. Nutrient analysis, including Total Phosphorus (TP), Soluble Reactive Phosphorus (SRP), Silica, Nitrate + Nitrite, followed the standard protocol of the National Laboratory for Environmental Testing (NLET, 1997). Nutrient data from the September 2010 survey was provided courtesy of S. Watson, Environment Canada (pers. com.). Chlorophyll-*a* was determined by acetone pigment extraction (Strickland and Parsons, 1968).

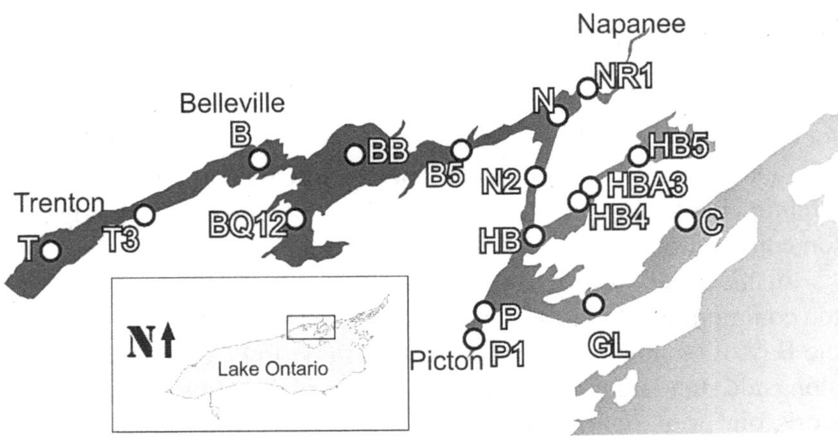

Figure 1. Map of the Bay of Quinte, Lake Ontario and locations sampled during spatial surveys of June, August and September 2010. Coordinates are provided in Table 1.

Phytoplankton samples were fixed immediately with Lugol's iodine. Identification and enumeration were carried out by the Utermöhl (1958) inverted microscope technique as described by Munawar and Munawar (1996). A minimum of 200 units were counted to achieve an acceptable counting efficiency (Lund et al., 1958). Within each sample, cell dimensions were measured directly and the average cell volume for each species was determined by applying the average cell dimensions to a standard geometric shape that most closely resembled the species. In the case of colonial forms, the average number of cells per colony was determined. Cell volume was converted to wet biomass assuming a specific gravity of 1.0 (Strickland, 1960).

Size fractionated primary productivity was estimated for three size categories of phytoplankton (<2 μm, 2-20 μm and >20 μm) by the ^{14}Carbon technique as per the standard protocol of Munawar and Munawar (1996). Whole water samples were spiked with $Na^{14}CO_3$, incubated for 4 hours at surface temperature and exposed to a constant light level of 240 $\mu E\ s^{-1}\ m^{-2}$. Because light and temperature levels were constant in these experiments, the results should be interpreted as potential rather than actual. After incubation, size classes were determined by filtration of the sample through polycarbonate filters. All filters were rinsed with hydrochloric acid (0.5N) in order to remove excess $^{14}C\text{-}CO_2$. Radioactivity was determined by liquid scintillation.

Results

Research cruises on the Bay of Quinte were conducted during June, August and September of 2010. The results presented below include a brief summary of temperature, nutrients, chlorophyll-*a*, phytoplankton biomass, phytoplankton species composition and size fractionated primary productivity. This presentation of results also includes inferences about tributary inputs, hydraulic exchanges and trophic state.

June 22-24 2010 (Early Summer)

Surface temperatures ranged from 16.9 – 23.6 °C in the Bay of Quinte during this early summer period. The warmest temperatures were observed in the upper bay (T, T3, B, BB, B5) and were very much characteristic of 'summer' conditions in temperate climates. Total Phosphorus (TP) ranged from a low of 7.8 $\mu g\ l^{-1}$ near the Lake Ontario interface (C) to a high of 26.4 in the upper bay and below the Remedial Action Plan target of 30 $\mu g\ l^{-1}$. Similarly, the more bioavailable form, SRP, ranged from 1.2 – 4.0 μg l^{-1} with the highest concentrations in the upper and middle bay locations. Silica was also highest in the upper bay (5.4 mg l^{-1}) compared to 0.8 mgl^{-1} in the lower bay. Nitrate+Nitrite varied from below detection to 224.0 μg l^{-1} across the Bay of Quinte with the highest values observed in the lower bay and the smallest values in the upper and middle bay regions. Chlorophyll-*a* also ranged from 2.7 – 10.0 μg l^{-1} with the highest values observed at T3 in the upper and HB4 in the middle bay. These results are summarized in Table S2a (Appendix). In general, the warmer waters and significantly higher concentrations of TP, SRP, SiO_2 and Chl *a*, observed in the upper and middle bay regions compared to the lower bay reflected tributary inputs whereas the lower bay was influenced by the interchange of waters with Lake Ontario. Likewise, the elevated concentrations of NO_3+NO_2 in the lower bay were reflective of conditions in Lake Ontario (Dove and Chapra, 2015).

Phytoplankton biomass was quite variable in June (Table 1a). At sites C and GL in the lower bay, biomass was < 0.3 g m^{-3} and reflective of ultra-oligotrophic conditions. Throughout the Bay of Quinte, biomass was typically ≤ 1.0 g m^{-3}. However, at station B in the upper bay, biomass peaked at ≈2.0 g m^{-3} reflecting a more mesotrophic environment. Station B was also one of the long term monitoring sites, so it should be noted that the biomass observed here was 2 – 10 X higher than at the rest of the sites in the Bay. Algal blooms have been defined as being phytoplankton assemblages where biomass exceeds 3 g m^{-3} (Munawar et al., 2017) and none were observed during this survey.

Phytoplankton assemblages were also quite variable across the bay (Fig 2, Table 1a). Only one Cyanophyta dominated assemblage was observed, at station T in the upper bay, which contained *Aphanocapsa minutissima, A. holsatica, Cyanodictyon planktonica* and *Microcystis botrys*. Diatomeae were most prevalent at station B in the upper bay and stations N, N2 and HB in the middle bay with *Aulacoseira granulata, A. ambigua* and *Fragilaria crotonensis* being the most commonly occurring species. Cryptophyceae dominated assemblages were observed at stations P and GL and included *Cryptomonas reflexa, C. marsonii, C. rostratiformis* and *Plagioselmis nanoplanktonica*.

There were an additional 4 sites that had no dominant algal taxa. In the upper bay, T3 contained a mixture of Diatomeae (*Aulacoseira granulata, A. ambigua*), Cyanophyta (*Aphanothece bachmanii*) and Cryptophyceae (*Cryptomonas reflexa, C. marsonii*). At B5, the main taxa included both Dinophyceae (*Peridinium aciculiferum, P. polinicum*)

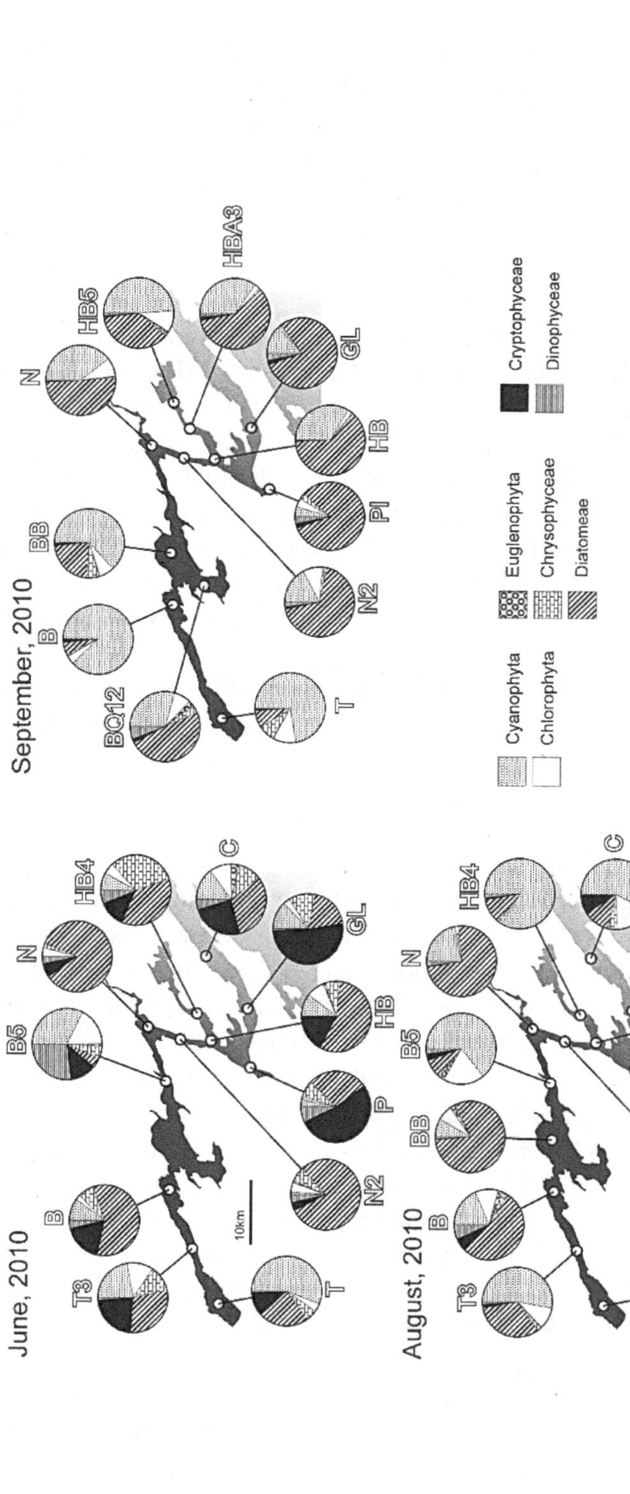

Figure 2. Phytoplankton Composition (% Biomass) in the Bay of Quinte during the June, August and September 2010 surveys.

and Cyanophyta (*Microcystis novacekii*). Station HB4 in the middle bay contained both Diatomeae (*F. crotonensis, A. granulata*) and Chrysophyceae (*Dinobryon divergerns, Chrysochromulina parva*). In the lower bay, Station C contained a mixture of Diatomeae (*F. crotonensis*) and Cryptophyceae (*C. reflexa, C. marsonii, P. nanoplanktica*). See Appendix Table S3 (species contributing > 1% to total biomass) and Appendix Table S4 (species contributing <1% to total biomass) for comprehensive species lists.

Primary productivity showed a 10-fold increase from station C (7.1 mg C m^{-3} h^{-1}) at the interface with Lake Ontario to station B (73.6 mg C m^{-3} h^{-1}) in the upper bay (Fig. 3, Table 1a). These rates were indicative of oligotrophic environments at the low end and hyper-eutrophic at the other extreme. Given the moderately low biomass we observed, the elevated productivity suggests that the phytoplankton were in a growth phase which should be expected at this time of year. Larger net plankton (>20 µm) were the most productive at sites in the upper and middle bay (B, N, N2, HB, HB4) which included primarily filamentous diatoms and colonial blue-greens (Cyanophyta). Nanoplankton (2 – 20 µm) was the most productive fraction at 4 sites in the upper (T, B5), middle (P) and lower (C) bay. At T3 in the upper bay, all three size fractions were equally productive and GL in the lower bay nanoplankton and picoplankton were co-dominant.

August 17-19, 2010 (Mid-Late Summer)

Surface Temperatures showed little variability at this time, ranging from 22.7 – 24.4 °C across the Bay and were at or very near to their annual peaks. Total phosphorus concentrations ranged from 10.6 – 53.0 µg l^{-1} with only 3 sites (T, GL, C) below the Remedial Action Plan target of 30 µg l^{-1} and concentrations at T (29.1) and GL (24.8) were certainly close enough to the target to be of concern. In the upper and middle bay regions, 8 of 9 sites exceeded that target. SRP ranged from 1.6 µg l^{-1} at station C to 11.1 µg l^{-1} in the upper bay (T3). Nitrate + Nitrite was low in the upper and middle bays (<15 µg l^{-1}) and at 6 of the 9 sites was near or below detection limits. Silica ranged from 1.1 – 7.2 mg l^{-1} and showed a distinct gradient from the upper to lower bay. Chlorophyll-*a* was lowest at station C (3.4 µg l^{-1}) and highest at HB4 (32.8 µg l^{-1}) in the middle bay. Complete results are given in Table S2b (Appendix). Taking chlorophyll-*a* as an indicator, 11 of 12 sites showed eutrophic conditions.

Phytoplankton biomass ranged from 0.5 – 7.9 g m^{-3} representing conditions that vary from oligotrophic at C to highly eutrophic at HB

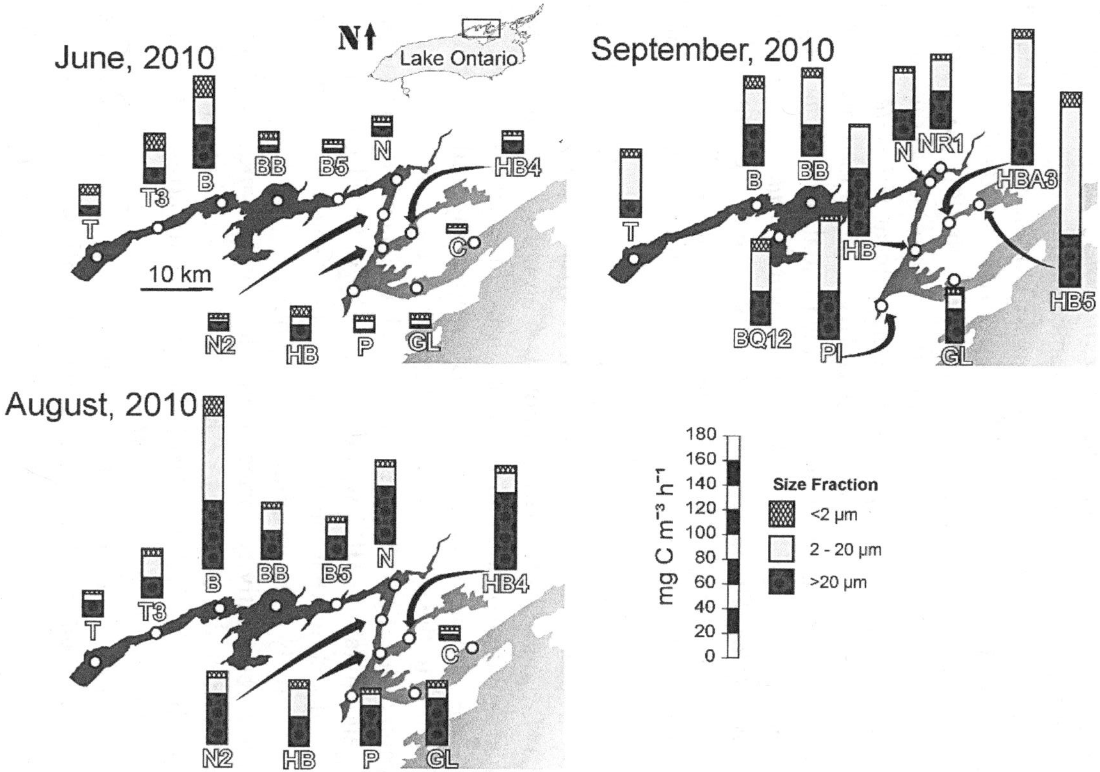

Figure 3. Size Fractionated Primary Productivity (mg C m^{-3} h^{-1}) in the Bay of Quinte during June, August and September 2010.

(Table 1b). With respect to taxonomic composition (Fig. 2, Table 1b), a total of 7 sites were dominated by species of Cyanophyta, 4 contained mostly Diatomeae and 1 was a mixture of diatoms and cyanophytes. In the upper bay, Cyanophyta were the main taxa at stations T (*Microcystis wesenbergii, Rivularia* sp., *Aphanocapsa elachista, A. incerta*) and B5 (*Limnoraphis birgei*) compared to Diatomeae (*Aulacoseira ambigua, A. granulata*) at both B and BB. Station T3 contained almost equal parts Cyanophyta (*Aphanocapsa holsatica, Cyanodictyon reticulatum*) and Diatomeae (*A. ambigua, A. granulata*). In the middle bay, diatoms including *Aulacoseira ambigua, A. granulata* were most common at N and N2 whereas Cyanophyta were dominant at HB (*Gleotrichia echinulate, Microcystis botrys, L. birgei*), HB4 (*Microcystis botrys, M. wesenbergii, L. bergeii*). Station P contained equal parts Cyanophyta (*M. novacekii, M. botrys, M. viridis*) and Diatomeae (*Aulacoseira ambigua, A. granulata*). The taxa in the lower bay were also dominated by species of Cyanophyta including *Dolichospermum* sp, *L. birgeii*, M. *wesenbergii* at GL, and *M. novacekii, L. bergeii* at C. Detailed species lists can be found in Appendix Tables S3 and S4.

Algal blooms (biomass > 3.0 g m^{-3}) were observed at 6 of the 12 sites and were located in the upper (T3, B, BB) and middle bay (N, HB, HB4) regions (Table 1b). As noted, station T3 in the upper bay contained a mixture of colonial Cyanophyta and filamentous Diatomeae, whereas the blooms at stations B and BB were composed of filamentous diatoms. In the middle bay, filamentous diatoms were the largest proportion of the algal bloom at station N, but HB contained filamentous Cyanophyta and HB4 was dominated by colonial Cyanophyta. These results show that algal blooms can be taxonomically diverse, even in relatively close proximity to each other.

Primary productivity varied from a low of 13.0 mg C m^{-3} h^{-1} at station C near Lake Ontario to a high of 161.7 mg C m^{-3} h^{-1} at B in the upper bay (Fig 3, Table 1b) indicative of oligotrophic to hyper eutrophic conditions. Station C, the only site showing oligotrophy, had a fairly even distribution of productivity among algal size fractions including net plankton (39.3%), nanoplankton (30.5%) and picoplankton (30.2%). At sites T3 and B in the upper bay, nanoplankton was the most productive fraction, representing 44.4% and 49.7% of the total, respectively. At the remaining sites, larger net plankton (>20 μm) accounted for 50.4 – 73.6% of the total rate of production and nanoplankton (2-20 μm) accounted for 19.9 – 38.9% of the total. Throughout

Table 1. a) Phytoplankton biomass, primary productivity and common species observed in the Bay of Quinte, June 22-24, 2010. Common phytoplankton specie`s (or highest taxonomic level) include those taxa which contributed > 5% to the total biomass of a given sample.

Station	Phytoplankton Biomass (g m^{-3})	Primary Productivity (mg C m^{-3} h^{-1})	Species
Upper Bay			
T3	1.0	40.2	*Aulacoseira granulata; Cryptomonas reflexa* *Aphanothece bachmanii;* Synechococcales
B	1.9	73.6	*Aulacoseira granulata; Plagioselmis nanoplanktica; Cryptomonas reflexa* *A. ambigua*
BB	n/a	16.1	n/a
B5	0.6	10.0	*Peridinium aciculiferum; Microcystis novacekii;* Synechococcales; *Rhodomonas minuta;* Chlorococcales
Middle Bay			
N	0.9	15.8	*Aulacoseira granulata*
N2	0.9	13.9	*A. granulata; A. ambigua*
HB	0.6	27.4	*A. granulata;* Synechococcales *Cryptomonas reflexa; F. crotonensis* *C. marssonii*
HB4	0.6	17.4	*Dinobryon divergens; F. crotonensis* *C. reflexa; A. granulata*

Table 1a. Continued...

Station	Phytoplankton Biomass (g m^{-3})	Primary Productivity (mg C m^{-3} h^{-1})	Species
Lower Bay			
P	0.7	13.7	*C. reflexa; F. crotonensis* *Gymnodinium helveticum* *C. rostratiformis; C. marssonii*
GL	0.2	11.5	*C. reflexa;* Synechococcales *F. crotonensis; C. rostratiformis* *Rhodomonas minuta; R. lacustris*
C	0.3	7.1	*F. crotonensis; C. reflexa;* Synechococcales Chlorococcales

the bay, except for station C, picoplankton (<2 μm) accounted for 6.4 – 14.8% of the total productivity.

September 20-23, 2010 (Late summer / Early fall)

During this late summer period, surface temperatures ranged from 16.5 – 19.6 °C reflecting the shorter days and cooler air temperatures characteristic of this time of year. Total Phosphorus concentrations ranged from a low of 16.8 μg l^{-1} at GL in the lower bay to a peak of 61.8 at HB5 in the nearshore waters of the middle bay. Most sites however were relatively close to the mean of 36.9 ± 3.1 μg l^{-1}. Soluble Reactive Phosphorus was also lowest at GL (0.9 μg l^{-1}) and highest in the middle bay at N (2.9 μgl^{-1}). Silica ranged from a low of 0.6 mg l^{-1} at P1 in the middle bay to a high of 5.9 mg l^{-1} at B in the upper bay. Nitrogen (Nitrate + Nitrite) was below detection (5 μg l^{-1}) at two sites in the upper bay (B, BB), two in the middle bay (N2, HB) and was near detection limits (6.0 – 15.0) at the remainder. Elevated levels of Nitrate + Nitrite were observed only at sites P1 (53.0) and GL (106.0) and were likely influenced by incursions of water from Lake Ontario. Chlorophyll-*a* ranged from 7.6 – 27.3 μg l^{-1}; the majority of observations were close to the mean of 13.8 ± 1.5 μg l^{-1}. A full summary of these results is included in Table S2c.

Phytoplankton biomass ranged from a low of 1.2 g m^{-3} at T in the upper bay to a peak of 6.9 g m^{-3} at HB5 in the middle bay region (Table 1c) which indicated predominantly eutrophic conditions. Cyanophyta were the dominant taxa at 3 sites in the upper bay (station T: *Cyanodictyon reticulatum, Aphanothece bachmanii, Dolichospermum crassa*; station B: *D. crassa, Microcystis novacekii, M. botrys*; station BB: *C. reticulatum, Dolichospermum* sp.) and 1 site in the middle bay (HB5: *D. crassa, Aphanizomenon gracile*). At the remaining sites, Diatomeae were more common including: *Aulacoseira ambigua, A. granulata* and *Stephanodiscus agassizensis*. These results are summarized in Figure 2 and Table 1c. For detailed phytoplankton species lists refer to Appendix Tables S3 and S4.

Algal blooms (> 3 g m^{-3}) were observed at 9 of the 11 sites spanning the upper, middle and lower reaches of the Bay (Table 1c). One additional site (N) was sufficiently close to this threshold at 2.7 g m^{-3} to be considered in bloom, leaving T in the upper bay as the only site without a bloom. Of these bloom sites, only 2 sites in the upper bay (B, BB) and 1 site in the middle bay (HB 5) were Cyanophyta blooms; the vast majority were Diatomeae blooms. These results showed not just

Table 1. b) Phytoplankton biomass, primary productivity and common species observed in the Bay of Quinte, August 17-19 2010. Common phytoplankton species (or highest taxonomic level) include those taxa which contributed > 5% to the total biomass of a given sample. *Indicates algal bloom conditions (Biomass > 3 g m^{-3}).

Station	Phytoplankton Biomass (g m^{-3})	Primary Productivity (mg C m^{-3} h^{-1})	Species
Upper Bay			
T	1.5	24.6	*Microcystis wesenbergi; Aulacoseira granulata Rivularia* sp.; *Aphanocapsa elachista; A. incerta* Synechococcales
T3*	3.5*	46.0	*Aphanocapsa holsatica; Aulacoseira ambigua A. granulata; Cyanodictyon reticulatum*
B*	4.7*	161.7	*C. reticulatum; Aulacoseira ambigua; A. granulata*
BB*	4.9*	53.7	*A ambigua; A. granulata*
B5	1.1	40.6	*Carteria* sp; *Limnoraphis birgei;* Synechococcales
Middle Bay			
N*	5.3*	79.0	*A. ambigua; A. granulata*
N2	2.7	68.3	*A. ambigua; A. granulata*
HB*	7.9*	120.9	*Gleotrichia echinulata; L. birgei; Microcystis botrys*
HB4*	5.7*	97.6	*M. botrys; L. birgei; M. wesenbergi; A. ambigua Dolichospermum crassa; M. novacekii*
Lower Bay			
P	1.2	53.5	*M. botrys; M. novacekii; M. viridis; A. ambigua; A. granulata*
GL	1.6	60.4	*Dolichospermum* sp; *L. birgei; M. wesenbergi*
C	0.5	13.0	*M. novacekii; L. birgei; Fragilaria crotonensis; Chlorococcales; Synechococcales*

diversity in the composition of algal blooms, but that diatom blooms may be a greater part of the eutrophication problem in the Bay of Quinte.

Primary productivity ranged from 43.3 mg C m^{-3} h^{-1} at GL to 157.7 mg C m^{-3} h^{-1} at HB5 (Fig 3, Table 1c) indicating conditions that ranged from eutrophic to hyper-eutrophic. Net plankton (>20 μm) was the dominant size fraction at HB (60.3%) and HB5 (54.0%) in the middle bay and GL (59.8%) in the lower bay. Nanoplankton (2 – 20 μm) were most productive in the upper bay at T (63.3%), BQ12 (47.3%), BB (54.2%) and in the middle bay at HB5 (65.8%) and P (56.9%). At B in the upper bay and NR1 in the middle bay, both net and nanoplankton were co-dominant. Picoplankton productivity ranged from 2.3%–13.2% of the total productivity across the Bay.

Discussion

Phosphorus abatement, first introduced in the Bay of Quinte in 1978, remains the primary management tool for alleviating eutrophication. While much of the data that has come out of the long term monitoring program has indicated that the Bay is still eutrophic (e.g. Minns et al., 2012; Nicholls and Carney, 2011; Munawar et al., 2012), these conclusions are drawn from a limited number of sites. Our goal for the present study is to reach beyond the prescribed monitoring and assess the spatial extent to which eutrophic conditions may (or may not) persist throughout the Bay during the peak growing season. This includes not only a discussion of water quality, but also the physical variability inherent in a riverine environment. Most importantly, though, we discuss how these factors affect the biomass, composition, diversity and productivity of the algal standing crop.

Tributaries have a major influence on water quality in the Bay of Quinte (e.g. Johnson and Owen, 1971). The upper bay receives inputs from the Trent, Moira and Salmon Rivers; the middle bay is influenced by the Napanee River and to a lesser extent Wilton Creek. Tributary loadings of phosphorus have been estimated at > 600 kg d^{-1} (Minns et al., 2004) which has a significant impact on total phosphorus concentrations in the upper and middle bay (≈30–60 μg l^{-1} in the August and September cruises). These are in addition to Point-sources of phosphorus 15 kg d^{-1} which are managed and other non-Point-sources which are not managed. Furthermore, hydraulic exchanges with Lake Ontario have a major impact on water quality in the lower bay (Freeman and

Table 1. c) Phytoplankton biomass, primary productivity and common species observed in the Bay of Quinte, September 21–23 2010. Common phytoplankton species (or highest taxonomic level) include those taxa which contributed > 5% to the total biomass of a given sample. *Indicates algal bloom conditions (biomass > 3 g m^{-3}).

Station	Total Biomass (g m^{-3})	Primary Productivity (mg C m^{-3} h^{-1})	Species
Upper Bay			
T	1.2	55.3	*Cyanodictyon reticulatum; Aphanothece bachmanii; Dolichospermum crassa;* Synechococcales
B*	4.0*	72.2	*D. crassa; Microcystis novacekii; M. botrys; D. lemmermannii; D. spiroides; C. reticulatum*
BQ 12*	6.5*	69.4	*Aulacoseira ambigua; C. reticulatum* Synechococcales
BB*	4.9*	71.2	*C. reticulatum; Dolichospermum* sp; *A. ambigua; Stephanodiscus agassizensis; M. ichthyoblabe*
Middle Bay			
N	2.7	60.5	*A. granulata; A. ambigua; S. agassizensis Aphanocapsa holsatica; M. novacekii; D. planctonica; C. reticulatum*
N2*	4.4*	58.8	*A. granulata; A. ambigua; S. agassizensis arendsee; M. botrys*
HB*	5.6*	90.6	*Aulacoseira ambigua; S. agassizensis arendsee; M. botrys; A. granulata; M. novacekii*
HBA3*	6.1*	109.6	*A. granulata; A. ambigua; M. wesenbergi Synechococcales*
HB5*	6.9*	157.7	*A. granulata; D. crassa; A. ambigua; Aphanizomenon gracile*
Lower Bay			
P1*	6.3*	99.8	*A. ambigua*
GL*	4.3*	43.3	*A. ambigua; A. granulata; M. wesenbergi*

Prinsenberg, 1986; Oveisy et al., 2015). Lake Ontario water is characterized by elevated levels of Nitrogen (i.e. NO_2+NO_3 can exceed 400 μg l^{-1}) and generally reduced levels of total phosphorus (7 – 8 μg l^{-1}) in the spring (Dove and Chapra, 2015). Using Nitrogen and TP as indicators, incursions of Lake Ontario water are apparent from the mouth of the Bay (C) and extending as far back as Picton Harbour (P). Incursions of Lake Ontario water have also been identified in the upper bay via the Murray Canal near station T, however its impact is less significant (Appendix Table S2, see also Freeman and Prinsenberg, 1986).

Such diverse physical and chemical conditions would be predicted to result in a diverse planktonic flora (e.g. Hutchinson, 1961; Margalef, 1978; Tilman, 1982). This is apparent when we consider the species diversity of the Bay of Quinte (Figure 4; Tables S3, S4). During the June survey, a total of 140 unique species were observed including 19 common species, 28 less common species and 93 rare species. In August there were 209 species observed (19 common, 33 less common, 157 rare) and in September 169 species were identified (19 common, 44 less common and 106 rare). We also note that the largest number of rare species is consistently found among the Chlorophyta (green algae). What these results show is that even eutrophic ecosystems can harbour a diverse algal flora and do not necessarily trend towards monocultures (see also Munawar et al., 2017).

But this top down portrait of species richness does little to explain the observed spatial heterogeneity of phytoplankton communities within each survey. A Mantel test was applied to compare distance between sites with taxonomic group biomass (all 7) for each cruise conducted during 2010. We found no significant correlations in the June (r^2=0.001, P=0.5; 999 repeats) and August (r^2 = 0.04, P= 0.9; 999 repeats) surveys and a significant but weak correlation in the September (r^2 = 0.2, P < 0.05; 999 repeats) survey. In parsing the September results further, we found that only Diatoms had a significant correlation (r^2 = 0.26, P<0.05; 999 repeats) and we can attribute this to the high biomass of filamentous diatoms (1.1 – 5.2 g m^{-3}), notably *Aulacoseira ambigua*, *A. granulata* and *Stephanodiscus agassizensis* at all but two sites (T,B). At the same time, none of the six other taxonomic groups showed any significant correlations. Overall, the Mantel test suggests that the various phytoplankton taxa are distributed randomly throughout the Bay, but that conclusion is obviously superficial.

Underlying this conclusion is the assumption from the Mantel test that distance (in and of itself) is likely to be the sole factor controlling biomass of the various phytoplankton taxa. The reality is that phytoplankton communities are affected by multiple limiting factors (e.g. nutrients, light, turbidity, grazing etc.) and even models that account for 3 – 4 of these factors are still likely to result in random distributions (Spatharis et al., 2009). Likewise, there is ample evidence that the effects of resource limitation on phytoplankton assemblages are not consistent among ecosystems (Grover and Chrzanowski, 2004; Maloufi et al., 2016). More importantly, we have already shown how variable some of these underlying conditions are in the Bay of Quinte given both tributary inputs and hydraulic exchange with Lake Ontario.

In a similar vein, dominant phytoplankton species have been related to ecosystem habitat using a qualitative model (Reynolds et al., 2002; Padisák et al., 2009) wherein each habitat has a unique set of limiting factors that favour particular species. During the June survey, the most commonly occurring species shown in Table 1a reflected: i) shallow lakes or mixed layers with elevated trophic states (*Aulacoseira granulata, Fragilaria crotonensis*); ii) shallow nutrient rich water columns (e.g. *Aphanocapsa minutissima, Cyanodictyon planktonica*); iii) large cryptomonads and small dinoflagellates characteristic of all lentic environments (*Cryptomonas reflexa, C. rostratiformis*), and iv) shallow mesotrophic environments (e.g. *Rhodomonas minuta*). Overall, 11 unique habitats were represented by the dominant phytoplankton although, it should be stressed, most of the habitats reflect elevated trophic states. We repeated this effort for the August and September cruises with similar results (data not shown). The drawback to this approach is that the number of habitats observed is directly proportionate to the number of common species so it will likely under-estimate the impact that rare and less common species have on phytoplankton dynamics (see also Ignatiades and Gotsis-Skretas, 2014; Munawar et al., 2015). Furthermore, top-down effects like grazing are similarly not accounted for (Fox, 2004; Özkan et al., 2014). Nonetheless, it would be challenging to account for all of the *possible* factors that *may* affect phytoplankton diversity.

There is also little practical evidence to support the conclusion that phytoplankton distributions are essentially random. Species of *Aulacoseira* (Diatomeae) and *Dolichospermum* (Cyanophyta) are both historically (Tucker, 1945; Nicholls and Carney, 1979; 2011) and presently among the most common taxa in the Bay of Quinte. That these two genera consistently dominate in this nutrient enriched environment suggests stable co-existence rather than direct competition for limiting

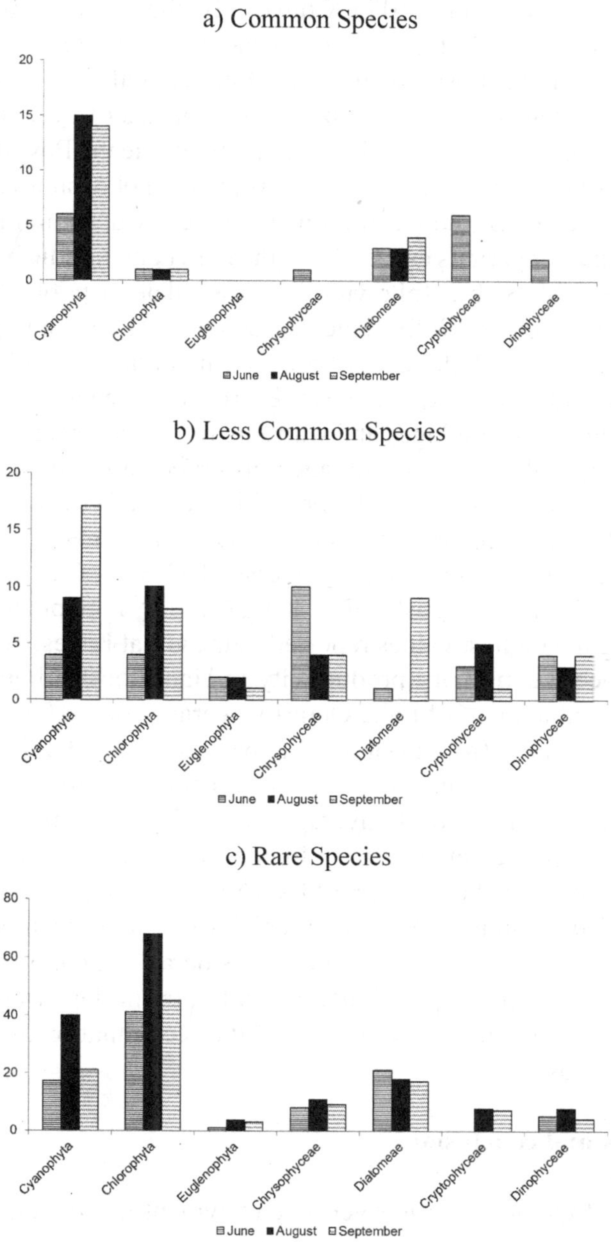

Figure 4. Number of phytoplankton species observed in the Bay of Quinte during the 2010 spatial surveys including: a) Common species (contribute >5% to total biomass), b) Less Common species (contribute 1 – 5% to total biomass), and c) Rare species (contribute < 1% to total biomass).

resources (Brauer et al., 2012). Conversely, there is also evidence of symbiosis between Diatomeae and Cyanophyta reported in the oceanographic literature (Foster et al., 2011; Bonnet et al., 2016); the mechanism being that N_2 fixed by diazotrophic species of cyanobacteria is rapidly utilized by certain species of diatoms. While the Bay of Quinte is often described as nutrient enriched, it is in effect phosphorus enriched. Nitrogen (NO_2+NO_3) concentrations trend below detection in the mid and late summer periods especially in the upper and middle bay reaches (Appendix Table S2 b,c) following the classical description of eutrophication (Vollenweider, 1968). Hence the ecological need for nitrogen-fixing Cyanobacteria. Although speculative, this type of symbiotic relationship would help to explain the long term consistency in Cyanophyta and Diatomeae assemblages and warrants further investigation.

The phytoplankton assemblages observed throughout our study appear to be well adapted to the prevailing conditions. Such an adaptation is further supported by our estimates of primary productivity. Specifically, primary productivity averaged 22.6 mg C m^{-3} h^{-1} in June, 68.3 mg C m^{-3} h^{-1} in August and 82.8 mg C m^{-3} h^{-1} in September. These are among the highest values reported in the Great Lakes. To put these into perspective, primary productivity in Hamilton Harbour, another eutrophic embayment of Lake Ontario, averages 40 – 60 mg C m^{-3} h^{-1} during the summer (Munawar and Fitzpatrick, 2017). Even in western Lake Erie, an ecosystem synonymous with eutrophication in the Great Lakes, primary productivity averages about 30 mg C m^{-3} h^{-1} over the summer (Fitzpatrick et al., 2007). In contrast, primary productivity in the oligotrophic offshore waters of Lake Ontario is approximately 6 mg C m^{-3} h^{-1} in late summer (Munawar et al., 2015). Apart from any broad-based environmental or physiological considerations, there appear to be few stressors (including self-shading and nutrient depletion) limiting photosynthesis in the Bay of Quinte and the accumulation of large algal standing crops.

Summary and conclusions

The Bay of Quinte, a large riverine embayment located on the north-eastern shores of Lake Ontario, has a long history of cultural eutrophication. While the Bay has been the subject of an extensive research and monitoring program that began in 1972, phytoplankton assessments have been limited to 2 — 3 index stations with few exceptions. Our study consisted of three separate surveys conducted during the summer of 2010 (June, August, September) at 12 sites spread evenly throughout the Bay. Among the major findings were that conditions in the Bay ranged from oligo- to mesotrophic in June but were primarily eutrophic in August and September with algal blooms observed at 50 – 75% of the sites; the spatial extent limited only by incursions of oligotrophic water from Lake Ontario. Furthermore, primary productivity in the Bay (e.g. 43.3 – 109.6 mg C m^{-3} h^{-1} in September) was among the highest reported in the Great Lakes, indicative of a phytoplankton community very well adapted to its environment. We also found that the Bay of Quinte supported a diverse algal flora containing 140 unique species in June, 209 in August and 169 in September which we attribute to the variable physical environment. Of the 35 phytoplankton samples assessed, Diatomeae were the most prevalent taxa in 16 of those assemblages followed by Cyanophyta (including many toxigenic species) in 12. The relative importance of diatoms are often overlooked in studies of eutrophication yet two species in particular, *Aulacoseira granulata* and *A. ambigua*, were major contributors to both the algal standing crop and the elevated rates of primary production. We recommend that future research be directed at understanding the ecology, physiology and dynamics of these filamentous diatoms as well as their associations with filamentous and colonial cyanobacteria.

One of the objectives of this study was to track the spatial extent of eutrophication in the Bay of Quinte during the peak growing season (summer). While the June survey indicated oligo to mesotrophic conditions based on standard nutrient and biological measures including phytoplankton biomass of 0.2 – 1.9 g m^{-3}, the phytoplankton assemblages in the upper and middle bay were dominated by eutrophic species (mostly Diatomeae: *Aulacoseira granulata, A. ambigua*) presaging the onset of eutrophic conditions. Furthermore algal blooms were observed at 50% of our sites in August increasing to 75% in September despite the long term reductions in Point-source phosphorus loadings. On that note, only incursions of oligotrophic water from Lake Ontario into the lower bay appeared to limit the spatial extent of eutrophic conditions observed in the August and September surveys.

Another objective was to examine the taxonomic composition of algal assemblages. We observed a high degree of biodiversity with 140 – 209 unique species being identified in each survey, the vast majority of them being rare. Conversely, we also saw a remarkable degree of consistency among the dominant taxa. From a total of 35 observations, we found that nearly half (16) of the algal assemblages were composed largely of

Diatomeae (*A. granulata, A. ambigua*) and 12 sites were dominated by Cyanophyta both filamentous (*Dolichospermum crassa, D. planktonica*) and colonial (*Microcystis* spp., *Gloeotrichia echinulata, Cyanodictyon reticulatum*). It is worth noting that these broad based associations of Diatomeae and Cyanophyta were also reported in historic studies which suggest some degree of adaptation to the prevailing conditions.

Cyanophyta (or Cyanobacteria) often receive considerable attention in discussions regarding eutrophication and algal blooms due to their potential to release toxins like microcystin. However, our study shows that Diatomeae including species of *Aulacoseira* and *Stephanodiscus* are a major part of the algal standing crop in the Bay of Quinte, especially under algal bloom conditions, and hence a major part of the eutrophication problem. Their ecology, physiology and dynamics warrant further research as do their associations with both filamentous and colonial forms of Cyanophyta.

Finally, we considered the photosynthetic capacity of the algal community using radioisotope based measures of primary productivity. We found that primary productivity increased throughout our study from a mean of $\approx$ 20 mg C m^{-3} h^{-1} in June to $\approx$ 75 mg C m^{-3} h^{-1} in September as incipient light and temperature reached their optimum. Given the linear increases observed in both primary productivity and phytoplankton biomass, it would seem that the algal community was physiologically well adapted to this chronically eutrophic environment.

Acknowledgements

The authors would like to thank Robert Bonnell and Ashley Bedford (DFO) for sample collections and technical support during the June and August surveys. The September survey was conducted by Environment Canada and thanks are due to Technical Operations for collecting the samples; we offer special thanks to Dr. S. Watson for inviting us to join the effort and for providing the nutrient data. Finally, we would like to thank Drs. J.H. Leach and C.K. Minns for their constructive reviews.

References

Bonnet, S., Berthelot, H., Turk-Kubo, K., Cornet-Barthaux, V., Fawcett, S., Berman-Frank, I., Barani, A., Grégori, G., Dekaezemacker, J., Benavides, M., Capone, D., 2016. Diazotroph derived nitrogen supports diatom growth in the South West Pacific: A quantitative study using nanoSIMS. Limnol. Oceanogr. 61, 1549-1562.

Brauer, V., Stomp, M., Huisman, J., 2012. The nutrient-load hypothesis: patterns of resource limitation and community structure driven by competition for nutrients and light. Am. Nat. 179(6), 721-740.

Dove, A., Chapra, S. 2015. Long-term trends of nutrients and trophic response variables for the Great Lakes. Limnol. Oceanogr. 60, 696-721.

Fitzpatrick, M.A.J., Munawar, M., Leach, J.H., and Haffner, G.D., 2007. Factors regulating primary production and phytoplankton dynamics in western Lake Erie. Fundam. Appl. Limnol. (Archiv. Hydrobiol.) 169, 137-152.

Foster, R., Kuypers, M., Vagner, T., Paerl, R., Musat, N., Zehr, J. 2011. Nitrogen fixation and transfer in open ocean diatom–cyanobacterial symbioses. ISME 5: 1484-1493.

Fox, J. 2004. Effects of algal and herbivore diversity on the partitioning of biomass within and among trophic levels. Ecology 85(2), 549-559.

Freeman, N., Prinsenberg, S. 1986. Exchange flows in the Adophus Reach / North Channel. In: Minns, C.K.; Hurley, D.A.; Nicholls, K.H. (Eds.), Project Quinte: point-source phosphorus control and ecosystem response in the Bay of Quinte, Lake Ontario Can. Spec. Publ. Fish. Aquat. Sci. 86, 27-39.

Grover, J., Chrzanowski, T. 2004. Limiting resources, disturbance, and diversity in phytoplankton communities. Ecol. Monogr. 74(3), 533-551.

Hutchinson, G.E. 1961. The paradox of the plankton. Am. Nat. 95(882): 137-145.

Ignatiades, L., Gotsis-Skretas, O. 2014. The contribution of rare species to coastal phytoplankton assemblages. Mar. Ecol. 35, 132-145.

International Joint Commission (IJC), 1988. Revised Great Lakes Water Quality Agreement of 1978. Agreement, with Annexes and Terms of Reference, between the United States and Canada signed at Ottawa, November 22, 1978 and Phosphorus Load Reduction Supplement signed October 7, 1983 as amended by protocol signed October 7, 1983, as amended by protocol signed November 18, 1987.

Johnson, M.G. and Owen, G.E. 1971. Nutrients and nutrient budgets in the Bay of Quinte, Lake Ontario. J. Water Pollut. Control Fed. 43, 836-853.

Lund, J.W.G., Kipling, C, and LeCren, E.D., 1958. The inverted microscope method of estimating algal numbers and the statistical basis of estimation by counting. Hydrobiol. 9, 143-170.

Maloufi, S., Catherine, A., Mouillot, D., Louvard, C., Couté, A., Bernard, C., Troussellier, C. 2016. Freshwat. Biol. 61, 633-645.

Margalef, R. 1978. Life-forms of phytoplankton as survival alternatives in an unstable environment. Oceanol. Acta 1(4), 493-509.

Minns, C.K., Owen, G. E., Johnson, M. G., 1986. Nutrient loads and budgets in the Bay of Quinte, Lake Ontario, 1951-81. In: Minns, C.K. Minns, D.A. Hurley, K.H. Nicholls (Eds.), *Project Quinte: point-source phosphorus control and ecosystem response in the Bay of Quinte, Lake Ontario* Can. Spec. Publ. Fish. Aquat. Sci. 86, 59-76.

Minns, C.K., Moore, J.E., Seifried, K.E. 2004. Nutrient loads and budgets in the Bay of Quinte, Lake Ontario, 1972 to 2001. Can. Manuscr. Rep. Fish. Aquat. Sci. 2694: v+49p.

Minns, C.K., Munawar, M., Koops, M.A., Millard, E.S., 2011. Long-term ecosystem studies in the Bay of Quinte, Lake Ontario, 1972–2008: A prospectus. Aquat. Ecosyst. Health Mgmt. 14(1), 3-8.

Munawar, M., Munawar, I.F., 1996. Phytoplankton Dynamics in the North American Great Lakes, Vol. 1 Ontario, Erie and St. Clair. Ecovision World Monograph Series. SPB Academic Publishing, Amsterdam, The Netherlands.

Munawar, M., Munawar, I.F., Fitzpatrick, M., Niblock, H., Lorimer, J. 2015. The phytoplankton community of Lake Ontario in 2008: structure, biodiversity and long term changes. Aquat. Ecosyst. Health Mgmt. 18(1), 28-42.

Munawar, M., Fitzpatrick, M., 2017. Microbial–Planktonic foodweb dynamics of a eutrophic embayment of Lake Ontario: Hamilton Harbour. Aquat. Ecosyst. Health Mgmt. 20(3), 214-229.

Munawar, M., Fitzpatrick, M. Niblock, H., Kling, H., Rozon, R., Lorimer, J., 2017. Phytoplankton ecology of a culturally eutrophic embayment: Hamilton Harbour, Lake Ontario. Aquat. Ecosyst. Health Mgmt. 20(3), 201-213.

National Laboratory for Environmental Testing (NLET), 1997. Manual of Analytical Methods, Major Ions and Nutrients, Volume 1. Environment Canada, Burlington, Ontario.

Nicholls, K.H., Carney, E.C. 1979. The taxonomy of Bay of Quinte phytoplankton and the relative importance of common and rare taxa. Can. J. Bot. 57, 1591-1608.

Nicholls, K.H., Carney, E.C., 2011. The phytoplankton of the Bay of Quinte, 1972-2008: point-source phosphorus loading control, dreissenid mussel establishment, and a proposed community reference. Aquat. Ecosyst. Health Mgmt. 14(1), 33-43.

Oveisy, A., Boegman, L., Rao, Y.R., 2015. A model of the three-dimensional hydrodynamics, transport and flushing in the Bay of Quinte. J. Great Lakes Res. 41(2), 536-548.

Özkan, K., Jeppesen, E., Davidson, T., Søndergaard, M., Lauridsen, T., Bjerring, R., Johansson, L., Svenning, J-C. 2014. Cross-taxon congruence in lake plankton largely independent of environmental gradients. Ecology 95(10), 2778-2788.

Padisák, J., Crossetti, L., Naselli-Flores, L. 2009. Use and misuse in the application of the phytoplankton functional classification: a critical review with updates. Hydobiol. 621, 1-19.

Reynolds, C.S., Huszar, V., Kruk, C., Naselli-Flores, L., Melo, S. 2002. Towards a functional classification of the freshwater phytoplankton. J. Plankt. Res. 24(5), 417-428.

Spatharis, S., Mouillot, D., Chi, T.D., Danielidis, D., Tsirtsis, G. A niche-based modeling approach to phytoplankton community assembly rules. Oecologia 159, 171-180.

Strickland, J.D.H. 1960. Measuring the primary production of marine phytoplankton. Bull. Fish. Res. Board Can. 122.

Strickland, J.D.H., Parsons T.R., 1968. A practical handbook of seawater analysis. Bull. Fish. Res. Bd Can. 167, 311.

Tilman, D. 1982. Resource competition and community structure. Princeton University Press. Princeton, NJ, USA.

Tucker, A. 1948. The phytoplankton of the Bay of Quinte. Trans. Am. Microsc. Soc. 67(4), 365-383.

Utermöhl, H. 1958. Zur vervolkommnung der quantitativen phytoplankton-methodik. Mitt. Internat. Verein. Limnol. 9, 1-38.

Vollenweider, R.A., 1968. Scientific fundamentals of the eutrophication of lakes and flowing waters, with particular reference to nitrogen and phosphorus as factors in eutrophication. Organization for Economic Co-operation and Development, Paris, France.

The above article is a reprinted version of the following previously published article:

Phytoplankton ecology in the Bay of Quinte, Lake Ontario: Spatial distribution, dynamics and heterogeneity
Munawar; M., Fitzpatrick; M., Niblock; H., Kling; H., Lorimer; J., Rozon, R., 2018.
Aquatic Ecosystem Health & Management (2018) 21 (2): 213–226.
https://doi.org/10.1080/14634988.2018.14740588

Supplementary material

Table S1. Sampling locations in the Bay of Quinte. "Y" indicates which sites were sampled during each survey.

Station	Depth	Latitude	Longitude	June	Aug	Sept
Upper Bay						
T	2.9	44.0813	-77.5572	Y	Y	Y
T3	4.7	44.1197	-77.4567	Y	Y	
B	4.7	44.1536	-77.3456	Y	Y	Y
BQ12	2.9	44.1089	-77.3078			Y
BB	5.4	44.1543	-77.2505	Y	Y	Y
B5	6.5	44.1562	-77.1450	Y	Y	
Middle Bay						
NR1	2.9	44.1978	-77.0160			Y
N	5.2	44.1803	-77.0397	Y	Y	Y
N2	7.2	44.1348	77.0634	Y	Y	Y
HB	11.3	44.0933	-77.0717	Y	Y	Y
HBA3	5.2	44.1300	-77.0133			Y
HB4	5.4	44.1146	-77.0191	Y	Y	
HB5	4.4	44.1496	-76.9685			Y
Lower Bay						
P1	4.8	44.0191	-77.1298			Y
P	10.7	44.0450	-77.1167	Y	Y	
GL	20.2	44.0464	-77.0222	Y	Y	Y
C	29.9	44.1089	-76.9089	Y	Y	

Table S2. a) Physical, Chemical and Biological Properties of the Bay of Quinte, June 22–24 2010. Units as indicated. Temp = Surface Temperature, Chl a = Chlorophyll-*a*, TP = Total Phosphorus, NO_3+NO_2 = Nitrate + Nitrite; NH_3 = Ammonia; SiO_2 = Silica. "Det." Indicates values that are at or below detection.

Station	Date	Temp °C	Chl *a* µg l^{-1}	TP µg l^{-1}	SRP µg l^{-1}	NO_3+NO_2 µg l^{-1}	NH_3 µg l^{-1}	SIO_2 mg l^{-1}
Upper Bay								
T	22-Jun	23.6	4.3	16.6	1.2	11.0	Det.	2.3
T3	22-Jun	23.0	9.9	26.4	1.3	15.0	12.0	4.5
B	22-Jun	22.6	8.0	24.7	2	17.0	36.0	5.4
BB	22-Jun	22.5	3.9	18	1.2	Det.	7.0	2.9
B5	23-Jun	22.8	3.6	23.6	3.4	13.0	13.0	2.1
Middle Bay								
N	23-Jun	22.6	6.7	23.1	4	14.0	16.0	2.2
N2	23-Jun	22.2	4.8	19	1.5	15.0	11.0	2.2
HB	23-Jun	21.4	5.7	21.2	1.9	35.0	19.0	2.0
HB4	23-Jun	22.3	10.0	25.8	3.8	30.0	27.0	2.1
Lower Bay								
P	24-Jun	16.9	6.9	14.1	1.4	175.0	18.0	1.6
GL	24-Jun	18.5	4.3	11.3	1.3	173.0	16.0	1.1
C	24-Jun	17.4	2.7	7.8	1.5	224.0	12.0	0.8

Table S2. b) Physical, Chemical and Biological Properties of the Bay of Quinte, August 17–19 2010.

Station	Date	Temp °C	Chl *a* µg l^{-1}	TP µg l^{-1}	SRP µg l^{-1}	NO_3+NO_2 µg l^{-1}	NH_3 µg l^{-1}	SIO_2 mg l^{-1}
Upper Bay								
T	18-Aug	24.1	11.3	29.1	3.0	Det.	Det.	4.5
T3	17-Aug	24.4	12.1	50.4	11.1	Det.	Det.	6.4
B	17-Aug	24.4	29.9	52.4	3.7	9.0	8.0	7.2
BB	17-Aug	23.9	18.2	46.1	2.2	Det.	Det.	6.1
B5	17-Aug	24.1	12.6	33.4	6.7	14.0	10.0	4.5
Middle Bay								
N	18-Aug	23.7	21.4	46.0	5.4	Det.	Det.	3.5
N2	18-Aug	23.9	30.2	50.3	4.9	Det.	Det.	3.7
HB	18-Aug	23.4	25.1	53.0	4.9	9.0	7.0	3.3
HB4	18-Aug	23.6	32.8	46.4	2.4	Det.	Det.	3.1
Lower Bay								
P	19-Aug	23.0	18.4	38.1	4.9	29.0	Det.	3.2
GL	19-Aug	22.7	16.2	24.8	3.4	91.0	6.0	2.3
C	19-Aug	22.7	3.4	10.6	1.6	141.0	11.0	1.1

Table S2. c) Physical, Chemical and Biological Properties of the Bay of Quinte, September 21 – 23 2010.

Station	Date	Temp °C	Chl *a* µg l^{-1}	TP µg l^{-1}	SRP µg l^{-1}	NO_3+NO_2 µg l^{-1}	NH_3 µg l^{-1}	SIO_2 mg l^{-1}
Upper Bay								
T	22-Sep	17.8	14.0	48.2	1.0	8.0	Det.	4.7
B	22-Sep	18.0	13.2	31.7	1.1	Det.	7.0	5.9
BB	22-Sep	17.5	12.8	32.4	1.4	Det.	Det.	4.7
BQ12	22-Sep	17.5	13.9	36.0	1.2	6.0	Det.	4.7
Middle Bay								
NR1	23-Sep	17.5	10.8	11.5	1.5	43.0	Det.	1.1
N	23-Sep	17.5	11.0	33.3	2.8	15.0	6.0	4.2
N2	23-Sep	18.0	12.5	38.1	2.9	Det.	Det.	3.6
HB	21-Sep	17.5	12.8	38.2	2.0	Det.	Det.	2.4
HBA3	21-Sep	17.5	17.3	38.3	1.0	7.0	Det.	0.7
HB5	21-Sep	16.5	27.3	61.8	1.3	8.0	18.0	0.8
Lower Bay								
P1	21-Sep	17.5	9.3	30.9	0.9	53.0	17.0	0.6
GL	21-Sep	19.6	7.6	16.8	0.9	106.0	Det.	0.8

Table S3. List of common phytoplankton species observed in the Bay of Quinte during the 2010 surveys. Species that contributed at least 1% to total biomass in any sample are denoted by an "x".

i. Cyanophyta	June	Aug	Sept
Aphanizomenon sp.			x
A. gracile (Lemmermann) Lemmermann		x	x
Aphanocapsa sp.	x		x
A. delicatissima West & West			x
A. elachista West & West		x	
A. minutissima Thur.	x		
A. holsatica (Lemmermann) G. Cronberg & Komárek	x	x	x
A. incerta (Lemmermann) G. Cronberg & Komárek		x	
Aphanotheca sp.	x	x	
A. bachmanii Kom.-Legner. et Cronb.	x		x
Chroococcus aphanocapsoides Skuja		x	
C. dispersus (Keissl.) Lemmermann		x	
C. minimus (Keissler) Lemmermann			x
C. minutus (Kütz.) Naegeli			x
Coelosphaerium Kuetzingianum Naegeli			x
Cuspidothrix issatschenkoi (Usachev) P. Rajaniemi, Komárek, R. Willame, P. Hrouzek, K. Kastovská, L. Hoffmann & K. Sivonen			x
Cyanodictyon planctonicum B.A. Mayer	x		x
C. reticulatum (Lemmermann) Geitler		x	x
Cylindrospermopsis sp.			x
Dolichospermum sp.		x	x
D. crassa (Lemmermann) Komárková-Legnová & Cronberg		x	x
D. lemmermannii P.G.Richter	x		x
D. macrospora Klebahn		x	
D. mucosa Komárková-Legnerová & Eloranta			x
D. planctonica Brunnthaler		x	x
D. spiroides Klebahn			x

Table S3 Continued…

i. Cyanophyta	**June**	**Aug**	**Sept**
Gloeotrichia echinulata P.G. Richter		x	
Limnoraphis birgei (G.M.Smith) J. Komárek, E. Zapomelová, J. Smarda, J. Kopecky, E. Rejmánková, J. Woodhouse, B.A. Neilan & J. Komárková		x	x
Limnothrix sp.			x
Lyngbya sp.		x	
L. limnetica Lemm.			x
Microcystis sp.	x	x	x
M. aeruginosa Kütz.		x	x
M. botrys Teiling	x	x	x
M. flos-aquae (Wittrock) Kirchner			
M. ichthyoblabe (G.Kunze) Kützing		x	
M. novacekii (Komárek) Compère		x	x
M. viridis (A. Brebison in Rabenhorst) Lemmerman		x	x
M. wesenbergii (Komarek) Starmach		x	x
Pseudanabaena sp.	x		x
Pseudanabaena limnetica (Lemmermann) Komárek			x
Radiocystis geminata Skuja		x	
Rivularia sp.		x	
Woronichinia naegeliana (Unger) Elenkin			x

Table S3 Continued…

ii. Chlorophyta	**June**	**Aug**	**Sept**
Botryococcus braunii Kütz.		x	x
Carteria sp.		x	x
Chlamydomonas sp.	x	x	
Chlorella sp.	x		
Coelastrum reticulatum (Dang.) Senn	x	x	
Collodictyon triciliatum H. Carter		x	x
Crucigeniella apiculata (Lemmermann) Komárek		x	
Dictyosphaerium pulchellum Wood		x	
Franceia sp.		x	
Goniochloris sp.		x	
Hindakia tetrachotoma (Printz) C. Bock, Pröschold & Krienitz			x
Lagerheimia citriformis (Snow) G.M. Smith			x
Mougeotia sp.			x
Oocystis lacustris Chodat			x
P. duplex Meyen			x
Pediastrum simplex (Meyen) Lemm.		x	
Planktosphaeria gelatinosa G.M. Smith		x	
Pseudostaurastrum sp.			x
Tetraëdron minimum (A. Braun) Hansgirg	x		
Tetraselmis sp.	x		
iii. Euglenophyta	**June**	**Aug**	**Sept**
Euglena sp.	x	x	
E. viridis (O.F. Müller) Ehrenberg	x		
Phacus sp.		x	
iv. Chrysophyceae	**June**	**Aug**	**Sept**
Chrysochromulina parva Lackey	x		x
Dinobryon sp.	x		
D. bavaricum Imhof	x		
D. divergens Imhof	x		
Mallomonas sp.	x	x	x
M. acaroides Perty		x	
M. akrokomos Ruttner	x		
M. pseudocoronata Prescott	x		
Ochromonas sp.	x	x	x
Pseudokephyrion sp.	x		
Pseudopedinella sp.	x	x	
Rhizochrysis limnetica G.M. Smith			x
Salpingoeca frequentissima (Zacharias) Lemmermann	x		

Table S3 Continued…

v. Diatomeae	June	Aug	Sept
Asterionella formosa Hassall	x		
Aulacoseira ambigua (Grunow) Simonsen	x	x	x
A. granulata (Ehrenberg) Simonsen	x	x	x
Cocconeis pediculus Ehrenberg			x
Cyclostephanos sp.			x
Cyclotella Meneghiniana Kütz.			x
Fragilaria capucina Desmazières			x
F. crotonensis Kitton	x	x	
Nitzschia sp.			x
N. fonticola Grunow			x
Skeletonema sp.			x
S. agassizensis Håk. & Kling			x
S. binderanus (Kützing) Krieger			x
S. niagarae Ehrenberg			x
vi. Cryptophyceae	**June**	**Aug**	**Sept**
Cryptomonas sp.	x		
C. erosa Ehrenberg		x	
C. Marssonii Skuja	x	x	
C.reflexa Skuja	x	x	
C. rostratiformis Skuja	x		
Gonyostomum semen (Ehrenberg) Diesing		x	
Katablepharis ovalis Skuja	x		
Plagioselmis nannoplanctica Skuja	x		
Rhodomonas lacustris Pascher & Ruttner	x		
R. lens Pascher & Ruttner	x		
R. minuta Skuja	x	x	x
vii. Dinophyceae	**June**	**Aug**	**Sept**
Ceratium furcoides (Levander) Langhans			x
Entzia acuta (Apstein) Lebour	x	x	x
Glenodinium sp.	x	x	
Gymnodinium sp.	x		x
Gymnodinium helveticum Penard	x		
Peridiniopsis polonica (Woloszynska) Bourrelly	x		x
Peridinium aciculiferum Lemmermann	x		
P. inconspicuum Lemmermann		x	

Table S4. List of rare phytoplankton species observed in the Bay of Quinte during the 2010 surveys. Species that contributed less than 1% to total biomass in any sample are denoted by an "x".

i. Cyanophyta	June	Aug	Sept
Anabaenopsis elenkenii Miller		x	
Aphanizomenon flos-aquae (L.) Ralfs		x	
A. aphanizomenoides (Forti) Hortobágyi & Komárek		x	x
A. skujae Komárková-Legnerová & Cronberg			x
Aphanothece clathrata v. *brevis* Bachmann			x
Chroococcus sp.		x	
C. limneticus Lemmermann	x	x	
Coelomoron sp.		x	
Coelosphaerium pallidum Lemmermann			x
Cylindrospermopsis raciborskii (Wolosz.) Seena. & Subbar.	x		x
Cylindrotheca closterium (Ehrenberg) Reimann & J.C. Lewin		x	
Dolichospermum compacta Kuetz (Trivisan)		x	
D. aphanizomenoides Forti		x	
D. ellipsoidea (Compère) Komárek			x
Eucapsis sp.		x	x
Merismopedia sp.	x	x	
M. glauca (Ehrenberg) Kützing		x	x
M. punctata Meyen		x	x
Merismopedia tenuissima Lemmermann		x	x
M. warmingiana (Lagerheim) Forti		x	x
Microcystis smithii Komárek & Anagnostidis			x
Myxobactron sp.		x	
Planktolyngbya sp.			x
Planktothrix suspensa (Pringsheim) Anagnostidis & Komárek		x	x
Pseudanabaena mucicola (Naumann & Huber-Pestalozzi) Schwabe		x	x
P. woronichinii Anagnostidis		x	x
Romeria sp.	x		
Snowella lacustris (Chodat) Komárek & Hindák		x	
Sphaerospermopsis sp.		x	
Synechococcus sp.	x	x	x
Woronichinia sp.		x	

Table S4 Continued…

ii. Chlorophyta (Part 1)	June	Aug	Sept
Actinastrum Hantzschii Lagerheim			x
Ankyra juday (G.M. Smith) Fott	x	x	
A. lanceloata (Kors) Fott	x		
Botryococcus pila J. Komárek & P. Marvan			x
Closteriopsis longissima Lemmermann	x	x	x
Closterium limneticum Lemmermann	x	x	x
C. aciculare T. West	x		x
C. acutum Brébisson	x	x	x
Coelastrum asteroideum De-Notaris		x	
C. cambricum Archer		x	x
C. pseudmicroporum Kors	x	x	x
Coenochloris sp.		x	
Coenococcus planctonicus Korshikov	x	x	x
Cosmarium sp.		x	
C. botyrtis Ehr.	x		
C. depressum (Naeg.) Lund		x	
C. subgranatum (Nordstedt) Lütkemüller		x	
Crucigenia quadrata Morren	x	x	
C. tetrapedia (Kirch.) West & West		x	x
Cryothecomonas scybalophora Thomsen et al.		x	
Dictyochlorella sp.		x	
Dictyosphaerium sp.			x
D. primarium Skuja	x	x	
Didymocystis sp.		x	x
Elakatothrix sp.		x	
E. biplex (Nygaard) Hindák		x	
E. genevensis (Reverd.) Hindak			x
Epithemia sp.		x	
Eudorina elegans Ehrenberg	x	x	x
Gloeotila pelagica (Nyegaard) Skuja		x	
Golenkinia radiata (Chod.) Wille		x	x
Gyromitus cordiformis Skuja			x

Table S4 Continued…

ii. Chlorophyta (Part 1)	June	Aug	Sept
Kirchneriella sp.	x	x	
K. contorta (Schmidle) Bohlin		x	
K. dianae (Bohlin) Comas Gonzalez		x	
K. lunaris (Kirch.) Moebius		x	
K. obesa (W. West) Schmidle		x	
Koliella sp.		x	
K. longiseta (Vischer) Hindák		x	x
K. planctonica Hindák	x		
K. tatrae (Kol) Hindák	x		
Lagerheimia ciliata (Lag.) Chodat		x	x
L. subsalsa Lemmermann	x	x	x
Monoraphidium contortum (Thurs.) Kom	x	x	x
M. minutum (Nägeli) Komárková-legnerová	x	x	x
M. griffithii (Berkeley) Komárková-Legnerová			x
M. komarkovae Nygaard			x
M. tortile (West & G.S. West) Komárková-Legnerová		x	
Nephrochlamy sp.	x		
Nephrochlamys subsolitaria (G.S. West) Korshikov	x	x	x
Nephrocytium agardhianum Nägeli			x

Table S4 Continued…

ii. Chlorophyta (Part 2)	June	Aug	Sept
Oocystis sp.	x	x	x
O. apiculata West		x	
O. parva West & West		x	x
O. pusilla Hansgirg		x	
O. solitaria Wittrock		x	
O. submarina Lagerheim	x		
O. submarina var. *variabilis* Skuja		x	
Pandorina morum (O.F.Müller) Bory de Saint-Vincent	x	x	
Parapediastrum biradiatum var. *longecornutum* (Gutwinski) Tsarenko		x	
Pediastrum kawraiskyi Schmidle		x	x
P. tetras (Ehr.) Ralfs	x		x
Pseudosphaerocystis lacustris (Lemmermann) Nováková	x	x	
Pyramimonas sp.	x	x	
Raphidocelis sp.	x	x	x
Scenedesmus sp.	x	x	x
Schroederia setigera (Shroed.) Lemmermann			x
S. spiralis (Printz) Korshikov	x	x	
Spermatozopsis sp.	x	x	x
Staurastrum sp.	x	x	x
S. avicula Brébisson		x	
S. gracile Ralfs ex Ralfs	x	x	
S. pingue Teiling	x		
S. tetracerum Ralfs ex Ralfs			x
Tetrachlorella alternans (G.M. Smith) Korshikov	x	x	
Tetraëdron sp.		x	
T. caudatum (Corda) Hansgirg	x	x	x
Tetrastrum staurogeniaeforme (Schroeder) Lemmermann			x
T. triappendiculata C. Bernard		x	x
Willea apiculata (Lemmermann) D.M. John, M.J. Wynne & P.M. Tsarenko		x	

Table S4 Continued…

iii. Euglenophyta	**June**	**Aug**	**Sept**
Lepocinclis ovum (Ehrenberg) Lemmermann		X	
Phacus tortus (Lemm.) Skvortzov		x	x
Strobomonas sp.		X	
Trachelomonas hispida (Perty) F. Stein	x		

iv. Chrysophyceae	**June**	**Aug**	**Sept**
Chrysamoeba sp.			x
Chrysostephanosphaera globulifera Scherffel	x	x	x
Dinobryon sertularia Ehrenberg	x		
D. sociale Ehrenberg	x		
D. sociale var. *stipitatum* (Stein) Lemmermann	x	x	
Gloeobotys sp.	x		
G. limneticus (G. M. Smith) Pasher		x	
Mallomonas caudata Ivonof			x
Salpingoeca sp.	x	x	x
Spiniferomonas sp.	x	x	x
Uroglena sp.	x	x	x

Table S4 Continued…

v. Diatomeae	**June**	**Aug**	**Sept**
Achnanthidium minutissimum (Kützing) Czarnecki	x	x	x
Actinocyclus normanii (Gregory) Hustedt			
Atheya zachariasi Brun			x
Cyclostephanos dubius (Hustedt) Round		x	x
Cyclotella sp.	x		
C. tripartita Håkansson			x
Cymbella sp.		x	
Diatoma tenue var. *elongatum* Lyngb.	x		
D. vulgare Bory	x		
Epithemia sp.		x	
Fragilaria sp.	x	x	x
F. nanana Lange-Bertalot	x		x
Melosira varians C.A. Agardh.	x		x
Navicula sp.		x	
Nitzschia acicularis (Kütz) W. Smith	x	x	x
Stephanodiscus sp.	x		x
S. Hantzschii Grunow			x
Synedra sp.	x	x	x
S. acus Kütz.			x
S. acus var *angustissima* Grunow	x		
S. acus var. *radians* (Kütz.) Hustedt	x		
S. ulna (Nitzsch) Ehrenberg	x		x
Tabellaria fenestrata (Lyngb.) Kuetzing	x		
Urosolenia sp.	x		
U. eriensis (H.L. Smith) Round & R.M. Crawford	x	x	x
U. longiseta (O. Zacharias) Edlund & Stoermer		x	x

Table S4 Continued…

vi. Cryptophyceae	**June**	**Aug**	**Sept**
Cryptomonas ovata Ehrenberg		x	
C. pusilla Bachmann		x	x

vii. Dinophyceae	**June**	**Aug**	**Sept**
Amphidinium sp.	x		
Peridiniopsis penardiformis (Lindemann) Bourrelly	x		
P. penardii (Lemmermann) Bourrelly			x
P. quadridens (Stein) Bourrelly		x	
Peridinium sp.		x	x
P. pusillum (Penard) Lemm.	x	x	
P. Willei Huitfeldt-Kaas		x	

A chronicle of nonindigenous species in the Bay of Quinte, Lake Ontario

Sarah A. Bailey[1,*], James A. Hoyle[2], Nicholas E. Mandrak[3] and Ora E. Johannsson[1]

[1]Fisheries and Oceans Canada, 867 Lakeshore Road, Burlington, ON L7S 1A1
[2]Ontario Ministry of Natural Resources, 41 Hatchery Lane, Picton, ON K0K 2T0
[3]Department of Biological Sciences, University of Toronto Scarborough, 1265 Military Trail, Toronto, ON M1C 1A4

[*]Corresponding author: sarah.bailey@dfo-mpo.gc.ca

Introduction

The Bay of Quinte, Lake Ontario, has been the site of intensive fish community sampling since the late 1950s and intensive ecosystem-based research and monitoring activities since 1972 (Project Quinte). Initially focused on measuring ecosystem responses to reductions in phosphorous loading, and later concentrating on setting and evaluating remediation criteria for this Area of Concern, this long-term dataset also documented dramatic changes resulting from biological invasions of nonindigenous species (NIS) (Minns et al., 2011).

In the 1970s, the Bay of Quinte was hyper-eutrophic to the point where excessive algal production limited growth of submerged aquatic vegetation and the fish community was dominated by small-bodied tolerant species including two NIS, White Perch (*Morone americana*) and Alewife (*Alosa pseudoharengus*) (Hoyle et al., 2012). It is not clear if ecosystem-level impacts of these NIS were fully appreciated, particularly since Alewife was good prey for desirable Walleye (*Sander vitreus*), and White Perch had become an important alternative commercial fishery species. It was recognized early on that re-establishment of Bay of Quinte top predators was likely necessary to regulate the abundance of these NIS (Hurley and Christie, 1977). Alewife is an extremely efficient planktivore that consumes a broad size spectrum of zooplankton, including smaller individuals than taken by native fishes (Brooks and Dodson

Ecology of the Bay of Quinte: Health, Management and Global Implications
Editors: C.K. Minns, M. Munawar and M.A. Koops

Aquatic Ecosystem Health & Management Society, Canada

1965). Alewife predation likely altered the composition, biomass, and production of the zooplankton community (Cooley et al., 1986).

Both White Perch and Alewife experienced massive winter die-offs in the late 1970s that, coupled with a 50% reduction in phosphorus loading at local sewage treatment plants, resulted in recovery of native percid fishes, notably Walleye, which effectively controlled the formerly superabundant NIS through predation (Hoyle et al., 2012). These changes in the fish community in the lower Bay revealed that intense predation by overabundant White Perch had been responsible for the near extinction of native amphipods *Diporeia* spp. (formerly *Pontoporeia hoyi*) in this region (Johnson and McNeil, 1986; Hurley, 1992). Also, Nicholls and Hurley (1989) provided evidence that benthic invertebrate consumption by hyper-abundant White Perch could have cascading impacts on lower trophic levels. Although the impacts of these NIS were not fully recognized at the time, they undoubtedly affected the zooplankton community (Johannsson and Bowen, 2012), and likely exerted indirect impacts on nutrient cycling and the phytoplankton community (Nichols and Hurley, 1989). After the corrections within the fish community, these NIS were considered naturalized and of little concern during the 1980s (Minns et al., 2011).

It was the arrival and immediate impact of Dreissenid Mussels in the Great Lakes that raised awareness of the potential for ecosystem level impacts by NIS. Zebra Mussel (*Dreissena polymorpha*) was reported from the Bay of Quinte in 1990 and quickly reached densities of 65,000 individuals per square meter in nearshore waters (1-5 m depth) with an estimated consumption efficiency greater than 100% in the middle and upper sections of the Bay (Bailey et al., 1999). Water transparency rapidly increased, with corresponding increases in submerged aquatic vegetation (Leisti et al., 2012) and phytophilic native fishes (Hoyle et al. 2012). Whole ecosystem changes occurred, causing a regime shift in the phytoplankton, benthic and fish communities (Nicholls et al., 2011). Nonindigenous White Perch and Alewife declined, but so did native Walleye (Hoyle et al., 2012). Because the Bay's popular recreational fishery was focused on Walleye (Hoyle, this monograph), the negative effects of NIS rapidly became evident by the mid-1990s. Although water quality had been improving with decreased phosphorus loading to the Bay, the establishment of Dreissenid Mussels led to shifts in phytoplankton composition towards species preferring more eutrophic conditions (Munawar et al., 2012), particularly *Microcystis* spp., which form blooms and can release toxins (microcystins). A study conducted in 2011 confirmed that microcystin concentrations in raw Bay of Quinte water can exceed World Health Organization guidelines for drinking water (Poste and Ozersky, 2013). These toxins can also accumulate in Dreissenid Mussels and pass into the food chain both through species that eat the mussels and through faecal biodeposits that are consumed by benthos (Poste and Ozerski, 2013).

The presence of dreissenids also facilitated the invasion of co-habitating NIS, the Round Goby (*Neogobius melanostomus*) and the amphipod, *Echinogammarus ischnus*. Round Goby has become an important component of fish biomass in the Bay of Quinte (Taraborelli et al., 2010), and may serve as prey, predator, and competitor with resident fish species. The long-term impact on the Bay of Quinte fish community has not yet been fully played-out but the potential for considerable ecosystem impact is high (Hoyle et al., 2012). Round Goby consume dreissenids and have thereby established a pathway linking Dreissenid Mussels to higher trophic levels (Dietrich et al. 2006); many Bay of Quinte predators were including Round Goby in their diets by the 2003-2005 time-period (Bowlby et al., 2010; Taraborelli et al., 2010). Native Johnny Darter (*Etheostoma nigrum*), like Round Goby a cool water, benthic invertivore, has virtually disappeared in annual Bay of Quinte bottom trawl catches coincident with Round Goby invasion (Hoyle et al., 2012; unpublished data). Although *Echinogammarus ischnus* displaces the native amphipod *Gammarus fasciatus*, no food-web effects are expected (or have been seen) as the two species have similar foodweb roles (Dermott et al., 1998).

Cercopagis pengoi, a predatory cladoceran, arrived in the lower Bay shortly after the establishment of Dreissenid Mussels. It became an obvious nuisance attaching to fishing lines with its hooked tail. Less noticeably, *C. pengoi* altered the composition and biomass of the zooplankton community through selective feeding on copepods (Benoit et al., 2002, Johannsson and Bowen, 2012). The impact of this change on the feeding and survival of both young and planktivorous fishes is not known. In this paper, we chronicle the taxonomic composition of NIS reported in the Bay of Quinte, and the initial vectors and documented impacts associated with these invasions, with a view to better understand the invasion process and improve management activities.

Methods

We compiled a list of NIS recorded from the Bay of Quinte based on literature review and results of the long-term monitoring program (Project Quinte). As a starting point, we searched the scientific literature and unpublished reports from Project Quinte (available at http://www.bqrap.ca/publications/documentlibrary) for occurrences in Quinte of NIS known from the Great Lakes basin as of June, 2014. (Mills et al., 1993; Duggan et al., 2003; Kelly et al., 2009). We recorded the year of first report for each NIS within the Bay of Quinte, while also noting first reports for Lake Ontario proper and the Great Lakes basin. We also recorded the most probable vector(s) of initial introduction to the Great Lakes basin and the type of impact documented for each species. We acknowledge that unknown time lags likely exist between the true date of arrival and the date of first report. If we could not find reports of Great Lakes' NIS in the Bay of Quinte, we assumed they are not established there. We did not consider marginal marsh plants and shoreline trees that are not obligate aquatic species. We also excluded species of cryptogenic origin (uncertain if NIS or native). As a result, our estimate of established NIS in Quinte is undoubtedly conservative.

Chronology of nonindigenous species in the Bay of Quinte

Researchers began documenting NIS in the Bay of Quinte in the mid-1900s, but the dates of establishment for these species are unreliable as these studies generally comprise the first survey of a particular taxonomic group and the NIS may have established, undetected, many years earlier. Timelines of invasion become more reliable with heightened awareness of NIS after the observation of Zebra Mussel in 1990, but only for the best-studied taxa (fishes, molluscs, and pelagic crustaceans). Even with awareness and significant sampling effort, time lags between the actual date of discovery and the date of first record of NIS can occur if the initial introduced population remains small, local and innocuous (Simberloff, 2009). Between 1990 and 2008, the rate of discovery of NIS in the Bay of Quinte is approximately linear with an average of one new species recorded every two years (Figure 1); however, no new species have been recorded since 2008. Three NIS reported from the Bay of Quinte were first reports for species new to the Great Lakes basin, including the European Frogbit (*Hydrocharis morsus-ranae*), a microsporidian (*Heterosporis* sp.) and viral hemorrhagic septicemia virus (*Novirhabdovirus* sp.). All remaining NIS were discovered elsewhere in the Great Lakes prior to observation in the Bay of Quinte, indicating that the majority of NIS

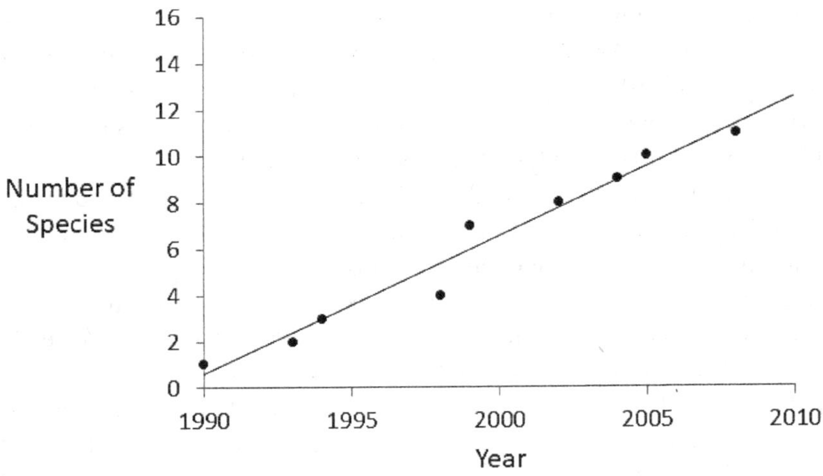

Figure 1. Cumulative number of nonindigenous species recorded in the Bay of Quinte since 1990. Linear regression was used to fit the line (NIS=-1191.001+05987888*post- 1990 years; r^2=0.966616).

probably arrived as a result of secondary introductions (spread) within the Great Lakes basin. We tabulated 36 established NIS in the Bay of Quinte comprised mainly of fishes (n=13), molluscs (n=8), and pelagic crustaceans (n=6), but these are also the best-studied taxa (Table 1). Other taxa are rarely studied (e.g. annelids, microorganisms), or rarely identified to the species level (e.g. algae, harpacticoid copepods).

Vectors of nonindigenous species introductions

A variety of vectors are presumed responsible for the initial introduction to the Great Lakes basin of NIS reported from the Bay of Quinte, including ballast water, canals, deliberate releases, solid ballast and unintentional releases associated with hatchery operations, baitfish use, and the aquarium trade (Table 1). The relative importance of individual vectors has changed over time; initially, NIS introductions to the Great Lakes basin resulted from deliberate stocking and spread through canals, then were associated with solid ballast by the late 1800s, and largely ballast water and unintentional releases since then (Figure 2). However, it is difficult to determine if the vectors associated with initial NIS introductions to the Great Lakes basin have the same relative importance for introductions into the Bay of Quinte.

Table 1. List of nonindigenous species established in the Bay of Quinte, with date of first report in the Bay (if known), Lake Ontario proper, and the Great Lakes basin. Initial vectors of initial introduction to the Great Lakes basin are: BW-ballast water, C-canal, D-deliberate release, SB-solid ballast, UR-unintentional release, U-unknown. Impacts: C–competition; D–disease; G–genetic; H–habitat alteration; P–predation, T–trophic disruption, U-unknown.

Taxon	Species	Initial Vector	First Report Quinte	First Report L Ontario	First Report Great Lakes	Impacts
Fishes	*Petromyzon marinus*	C	unknown	1830	1830	P
Fishes	*Salmo trutta*	D	unknown	1883	1833	C,G
Fishes	*Oncorhynchus tshawytscha*	D	unknown	1873	1873	C
Fishes	*Alosa pseudoharengus*	C	unknown	1873	1873	D,T
Fishes	*Oncorhynchus mykiss*	D	unknown	1900	1876	C
Fishes	*Carassius auratus*	D	unknown	1874	1874	H
Fishes	*Cyprinus carpio*	D	unknown	1896	1879	H
Fishes	*Oncorhynchus kisutch*	D	unknown	1970	1933	C
Fishes	*Oncorhynchus gorbuscha*	UR	unknown	1979	1956	C
Fishes	*Osmerus mordax*	D	1940	1929	1912	T
Molluscs	*Bithynia tentaculata*	SB	1944	1879	1871	C
Fishes	*Morone americana*	C	1952	1950	1950	C,P
Molluscs	*Pisidium amnicum*	SB	1966	1897	1897	U

Table 1. Continued...

Taxon	Species	Initial Vector	First Report Quinte	First Report L Ontario	First Report Great Lakes	Impacts
Molluscs	*Valvata piscinalis*	SB	1966	1897	1897	U
Molluscs	*Pisidium henslowanum*	SB	1966	1916	1916	U
Molluscs	*Sphaerium corneum*	U	1966	1924	1924	U
Annelida	*Potamothrix bedoti*	U	1966	1950	1950	U
Annelida	*Potamothrix moldaviensis*	U	1966	1952	1952	U
Plant	*Potamogeton crispus*	D	1972	1900	1879	C,H
Plant	*Myriophyllum spicatum*	UR	1972	1960	1952	C,H
Plant	*Hydrocharis morsus-ranae*	UR	1972	1972	1972	C,H
Algae	*Stephanodiscus binderanus*	BW	1975	1952	1938	U
Algae	*Cyclotella pseudostelligera*	BW	1975	1975	1946	U
Crustacea	*Eurytemora affinis*	BW	1975	1958	1958	U
Crustacea	*Eubosmina coregoni*	BW	1975	1968	<1959	U
Molluscs	*Dreissena polymorpha*	BW	1990	1990	1988	C,H,T
Molluscs	*Dreissena rostriformis bugensis*	BW	1993	1991	1991	C,H,T
Molluscs	*Potamopyrgus antipodarum*	BW	1994	1991	1991	U
Crustacea	*Bythotrephes longimanus*	BW	1998	1982	1982	C,P

Table 1. Continued...

Taxon	Species	Initial Vector	First Report Quinte	First Report L Ontario	First Report Great Lakes	Impacts
Fishes	*Neogobius melanostomus*	BW	1999	1998	1990	C,D,T
Crustacea	*Cercopagis pengoi*	BW	1999	1998	1998	C,P
Protozoa	*Heterosporis sp.*	UR	1999	1999	1999	D
Fishes	*Scardinius erythrophthalmus*	UR	2002	1989	1989	C
Crustacea	*Echinogammarus ischnus*	BW	2004	1997	1994	C,P
Virus	*Novirhabdovirus* sp.	U	2005	2005	2005	D
Crustacea	*Hemimysis anomala*	BW	2008	2006	2006	U
Fishes, ,	*Tinca tinca*	D	2018	2018	2018	T

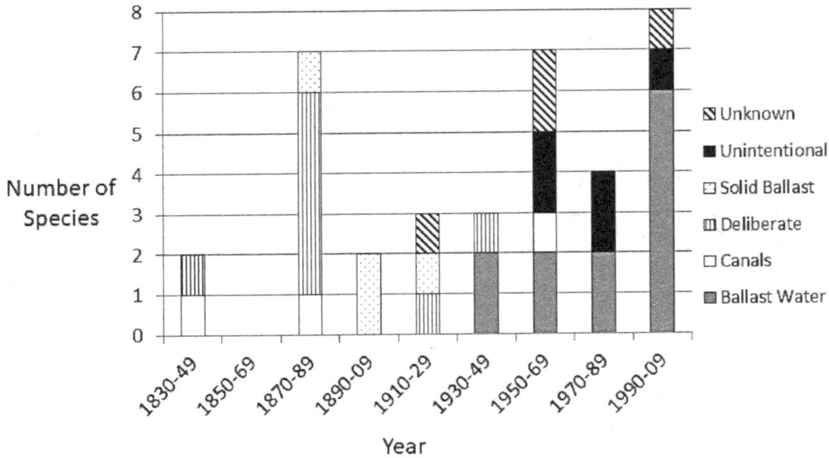

Figure 2. Chronological timeline of nonindigenous species (NIS) established in the Bay of Quinte, arranged by initial vector of introduction (to the Great Lakes basin). Dates and vectors follow Mills et al. (1993) and Ricciardi (2006).

Deliberate releases include authorized stocking programs undertaken by government agencies and unauthorized private releases (Mills et al. 1993). Development of a commercial or recreational fishery is the reason for most authorized introductions in North America. Six nonindigenous fishes reported from the Bay of Quinte are thought to have been initially introduced by stocking in the 1800s (Brown Trout, *Salmo trutta*; Chinook Salmon, *Oncorhynchus tshawytscha*; Rainbow Trout, *Oncorhynchus mykiss*; Common Carp, *Cyprinus carpio*; Coho Salmon, *Oncorhynchus kisutch*; and Rainbow Smelt, *Osmerus mordax*), although two of them (Chinook Salmon, Coho Salmon) did not become established until further stocking in the 1960s (Crossman and Van Meter, 1979). *Potamogeton crispus*, a submergent aquatic plant, is known to have been deliberately introduced to parts of the Great Lakes basin as food for waterfowl (Mills et al. 1993). Deliberate, but unauthorized, release is also the initial vector for at least one fish species, Goldfish (*Carassius auratus*), in the Bay of Quinte. Goldfish is established in Lake Ontario (Crossman and Van Meter, 1979) and it is not known if those caught in the Bay of Quinte were introduced directly into the Bay or secondarily spread after being introduced elsewhere – both are possible. In 2018, Tench (*Tinca tinca*) was captured in the Bay of Quinte, the first record of the species in the Great Lakes. It likely originated from a population that was established in a farm pond in the

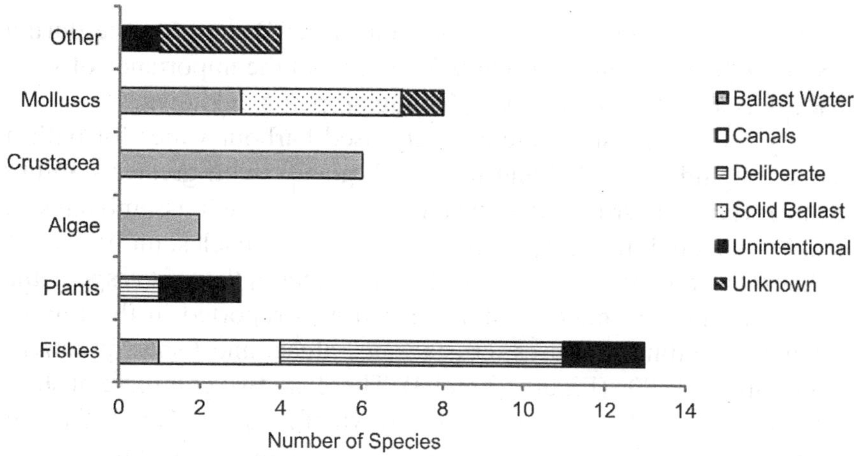

Figure 3. Taxonomic composition of aquatic nonindigenous species established in the Bay of Quinte, arranged by initial vector of introduction (to the Great Lakes' Basin). Initial vector data follow Mills et al. (1993) and Ricciardi (2006).

Richelieu River drainage of Quebec in 1980s (Avlijas et al., 2018). It subsequently became established in the wild in adjacent watershed and has since spread upstream to Lake Champlain and downstream to the St. Lawrence River, where it has spread downstream to the Gulf of St. Lawrence and upstream to the Bay of Quinte. Cumulatively, deliberate releases are presumed responsible for the initial introduction of eight (22%) NIS reported from the Bay of Quinte.

Navigational canals and other water diversions can also facilitate introduction of aquatic NIS (most notably mobile aquatic taxa such as fishes) to previously isolated locations (Figure 3). The Erie Barge Canal links the Hudson and Mohawk rivers, which drain to the Atlantic Ocean, to Oneida Lake, which drains to southeastern Lake Ontario, through the Oneida and Oswego rivers. The canal vector is reported as responsible for the initial introduction of three (8%) NIS reported from the Bay of Quinte: Sea Lamprey (*Petromyzon marinus*), Alewife, and White Perch.

Ballast is defined as any solid or liquid placed into a ship to increase the draft, change the trim, regulate the stability, or maintain stress loads within acceptable limits (NRC 1996). Historically, ships carried rocks, sand, beach debris or other heavy materials as solid ballast which was dumped onto ballast grounds or thrown overboard when the ship arrived in port to load cargo, serving as a vector for plants (seeds) and invertebrates (Mills et al. 1993). Solid ballast was an important initial vector for four (11%) NIS established in the Bay of Quinte during 1870-1920, all molluscs (Figures 2 and 3), but the importance of solid ballast as a vector to the Bay of Quinte proper is unknown.

By the 1800s, ships increasingly used harbour water for ballast since it could be loaded and unloaded quickly with greater stability control (NRC 1996). Ballast water may contain rich assemblages of plankton, including crustaceans, polychaetes, barnacles, molluscs and fishes (Carlton and Geller 1993). Ballast water is thought responsible for the initial introduction of 12 (33%) NIS reported in the Bay of Quinte, including molluscs, crustaceans, algae, and fishes, predominantly in the 1990s (Figures 2 and 3). There are two commercial shipping ports in the Bay of Quinte at the Town of Picton and the Village of Bath. Coastal or foreign ballast water is rarely discharged at these ports (four discharges during 2005-2007; Bailey et al. 2012), rather these ports are primarily used for domestic cargo movements. Picton receives an estimated 500,000 metric tonnes per year of domestic ballast water sourced almost exclusively from Hamilton, Toronto, Rochester, or Cleveland (based on 2005-2007 data summarized in Rup et al. 2010). Shipping activity at Bath was variable across 2005-2007, ranging from 75,000 to 250,000 metric tonnes of ballast water discharged per year, originating predominantly from Toronto, Oswego, Cleveland, Buffalo, and Detroit (Rup et al., 2010). Domestic ballast water originating from other Great Lakes' ports can transfer species with restricted distribution to ports or lakes beyond natural dispersal ability, and is likely a potent mechanism for the rapid and broad distribution of NIS initially introduced by foreign ballast water or other vectors (Rup et al., 2010; Briski et al., 2012). Given that the cement and gypsum trade through these ports has been relatively stable for many years, domestic ballast water may explain why many NIS were reported in the Bay of Quinte soon after their discovery in Lake Ontario; thus, domestic ballast water has likely been an important vector of NIS to the Bay of Quinte proper.

Unintentional releases, defined as release of organisms without intention to create established populations, can occur through 'humane' release of unwanted aquarium species and baitfishes, or accidental release of hatchery fishes (Mills et al., 1993). Pink Salmon (*Oncorhynchus gorbuscha*) is thought to have established from disposal of excess stock by hatchery managers, while the initial introduction of Rudd (*Scardinius erythrophthalmus*) and a protozoan (*Heterosporis* sp.) are related to release of baitfishes (Mills et al., 1993). Two plants (*Myriophyllum spicatum* and *Hydrocharis morsus-ranae*) reported from the Bay of Quinte are thought to be releases associated with the aquarium trade (Mills et al., 1993). In 2018, Tench (*Tinca tinca*) was captured in the Bay of Quinte, the first record of the species in the Great Lakes. It likely originated from a population that was established in a farm pond in the Richelieu River drainage of Quebec in 1980s (Avlijas et al., 2018). It subsequently became established in the wild in adjacent watershed and has since spread upstream to Lake Champlain and downstream to the St. Lawrence River, where it has spread downstream to the Gulf of St. Lawrence and upstream to the Bay of Quinte. Cumulatively, unintentional releases are the initial vector of introduction for five (14%) NIS reported from the Bay of Quinte.

Finally, the initial vector of introduction for four (11%) NIS in the Bay of Quinte is unknown (mollusc *Sphaerium corneum*, annelids *Potamothrix bedoti* and *Potamothrix moldaviensis*, and Viral Hemorrhagic Septicemia (VHS) virus *Novirhabdovirus* sp.).

Impacts

Biological invasions are a pervasive driver of global change that can, but do not always, cause significant ecological, socio-economic and evolutionary impacts affecting recipient ecosystems at the species, community and/or ecosystem levels (Grosholz, 2002; Ricciardi and Cohen, 2007; Strayer, 2010). While invasions are generally described as having negative consequences less than 20% of the time (Williamson, 1996), impacts in freshwater ecosystems appear to be more prevalent (Strayer, 2010). Two thirds of NIS reported from the Bay of Quinte have documented impacts there, or in the Great Lakes, though the severity of those impacts are variable (Table 1). Impacts have not been documented for 12 species, though we caution that these are all smaller taxa receiving very little research effort. All of the fishes and plants, and roughly half of all other taxa, reported from Bay of Quinte have documented impacts, except Algae (Figure 4).

The most common mechanism of impact of NIS in Bay of Quinte is competition with native species for food or physical space (Figure 5). Habitat alteration is the second-most common type of impact, followed by predation and trophic disruption, then disease.

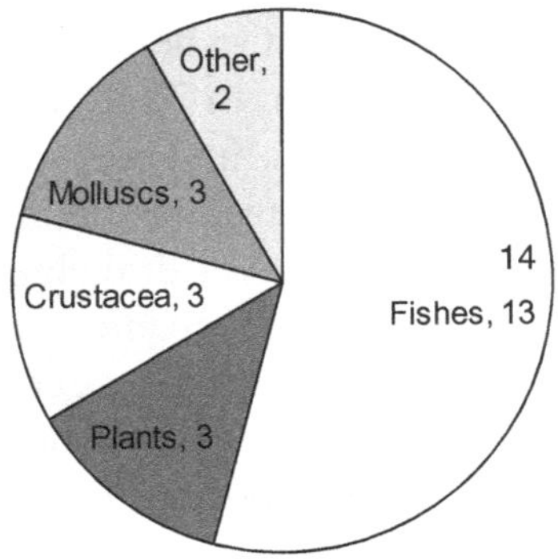

Figure 4. Taxonomic composition, and number, of nonindigenous species in the Bay of Quinte with documented impacts in the Great Lakes basin.

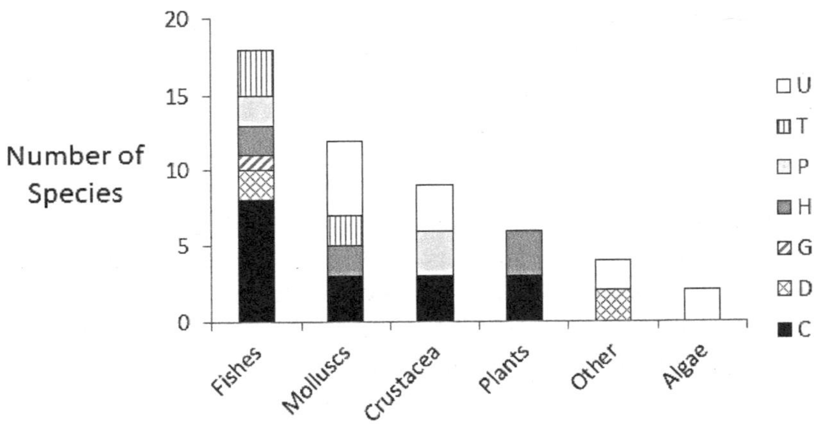

Figure 5. Documented impacts of aquatic nonindigenous species established in the Bay of Quinte, arranged by taxon. Impacts are C–competition; D–disease; G–genetic; H–habitat alteration; P–predation, T–trophic disruption, U-unknown.

Future outlook

It is a daunting and difficult task to accurately predict which species may be introduced next, but meaningful risk assessments can be conducted based on analyses of vector patterns and associated species in major donor regions, environmental tolerances and invasion history (Ricciardi and Rasmussen 1998). The relative importance of initial vectors to the Great Lakes has changed over time and may be expected to change in the future. We discuss here the potential role of historically important vectors as mechanisms for future invasions in order to guide research and management activities in the Bay of Quinte.

Ballast water has been the most likely vector for approximately 55% of NIS introduced to the Great Lakes basin since 1959 (Kelly et al. 2009). Using trait-based models, Kolar and Lodge (2002) predicted that 22 of the 66 Ponto-Caspian fishes examined could successfully invade the Great Lakes if introduced by ballast water, and that five of the 22 species would become nuisance species. However, with the implementation of stronger ballast water regulations for the Great Lakes, the probability of the introduction of these predicted species and, hence, their overall risk may not be as high as originally evaluated. In fact, the rate of ballast-mediated invasions began to decline in 1995, shortly after requirements to exchange ballast water for open ocean water were enacted (Bailey et al., 2011). Ballast water exchange is not a foolproof strategy since species with broad salinity tolerance or dormant life stages might survive the osmotic shock induced by ballast water exchange if not purged from ballast tanks into the ocean. Risk assessments identified 13 high risk, salinity-tolerant species, including the recently established Bloody Red Shrimp (*Hemimysis anomala*), that might be future invaders despite ballast water exchange requirements (Ricciardi and Rasmussen, 1998; Grigorovich et al., 2003). Yet, encouragingly, no new ballast-mediated NIS were reported in the Great Lakes during 2007-2013 (Bailey et al., 2011), thus it is unregulated domestic ballast water that is now thought to pose the higher risk for ballast-mediated invasions in the Great Lakes (Casas-Monroy et al., 2014). Kelly et al. (2009) list 26 ballast-mediated NIS in the Great Lakes that have not been reported from the Bay of Quinte that arguably might be transferred to new locations in domestic ballast. Given the long history of some of these NIS, it is possible that some species are already established in the Bay of Quinte but have not been detected, or that some species pose little risk due to current domestic trade patterns or environmental conditions; however, the amelioration of water quality in the Bay of Quinte coupled with climate change may facilitate future establishment of these species in new locations (see also Carlton 1985; Crossman and Cudmore 1999). Canada has signed on to an international convention that requires ships to meet a ballast water discharge performance standard, by September 2024 most likely achieved through shipboard water treatment systems using processes such as filtration and chlorination. Such treatment systems may prove a viable option for management of domestic ballast water. Further, a combination strategy combining both exchange and treatment may provide the best available protection against new NIS in foreign ballast water in coming years (Briski et al., 2013). Alternative shipping vectors like solid ballast appear to be largely obsolete, while biofouling of exterior ship surfaces does not appear to be an important mechanism for freshwater introductions (Sylvester and MacIsaac, 2010).

Deliberate releases have been the second-most historically important vector of NIS to the Great Lakes. An unintentional "side effect" of deliberate stocking is the release of non-target species when authorized stocks are contaminated by other species, diseases, or pathogens. In part because of these and other impacts, fisheries management is generally moving away from the authorized stocking of non-native species (Helfman, 2007), although *Oncorhynchus* spp. continue to be introduced into the Great Lakes for recreational purposes. Deliberate, but unauthorized, introductions by private individuals have occurred for a variety of reasons including release of unwanted bait or pets, or release for religious or animal rights reasons (Kerr et al., 2005). Recent, recurring reports of nonindigenous sub-tropical macrophytes Water Lettuce (*Pistia stratiotes*) and Water Hyacinth (*Eichhornia crassipes*) in the Lake St. Clair basin indicate that deliberate introductions by private individuals are an extant vector (MacIsaac et al., 2014). These two NIS are currently considered low risk for establishment due to thermal requirements (Rixon et al. 2005), however, future risk may be higher under climate change (MacIsaac et al., 2014).

Unintentional releases associated with live trade of aquatic plants, molluscs and fishes also appear to pose a leading threat in the future (Kerr et al., 2005; Mandrak and Cudmore, 2010). Nonindigenous species are often imported into the Great Lakes basin and kept alive until the point of sale for use in aquaria and ornamental ponds, for school curricula or research purposes, or as human food (Mandrak and Cudmore, 2010). At least 308 species of freshwater fishes and 138 species of aquatic plants are sold live in the aquarium trade in Ontario and Quebec, including a number of high risk, cold-tolerant freshwater NIS (Rixon et al., 2005; Cohen et al., 2007). It has been estimated that more than three thousand aquatic plants associated with the aquarium trade are released annually to the St. Lawrence River from Montreal alone (Cohen et al., 2007). Surveys of aquarium and water garden owners and bait fish users in Ontario indicated that they released unwanted fishes into the wild at rates of 2% (Marson et al., 2009a), 3% (Marson et al., 2009b), and 51%

(Ward et al., 2011) of respondents, respectively. These numbers may seem small but, when extrapolated over the hundreds of thousands of individuals involved in these activities, the risk is significant. For example, there are more than 4.2 million fishing trips involving live baitfish per year in Ontario, including 227,686 trips (median estimate) to the Lake Ontario basin each year (Drake and Mandrak, 2014). Screening-level risk assessments concluded that 11 of 825 freshwater fish species in Canadian live trade (Mandrak et al., 2013), 51 of 149 freshwater plant species (Gantz et al., 2013), and 15 of 73 mollusc species posed a high ecological risk if released into the wild. As the human population is highest and the species availability is greatest in southern Ontario, it, including the Bay of Quinte, is the region of Canada at greatest risk of release and impact of NIS from live trade. The risk posed by deliberate and unintentional releases associated with live trade could be reduced through a variety of efforts ranging from public outreach programs to legislative regulations. Federal and provincial governments and non-government organizations have done much NIS outreach in recent years, the province has banned some NIS such as Asian Carps and Snakeheads, and the federal Fisheries Act has recently been revised to allow the regulation of NIS through prohibition lists. Such efforts, if implemented and enforced, may reduce the risk posed by live trade.

Canals have not led to the establishment of new NIS in the Great Lakes since the 1950s; however, fishes continue to spread through them. Of greatest concern is the connection between the Mississippi and Great Lakes basins through the Chicago Shipping and Sanitary Canal. A series of electrical barriers have been placed in the canal to minimize the probability of species spreading from one basin to the other in either direction. The US Army Corps of Engineers published a study assessing the risk of NIS spreading into the Great Lakes from the Mississippi basin through another 18 artificial connections, concluding that there was a high probability of inter-basin transfer in one connection, medium probability in seven, and a low probability in 10 (USACE, 2013). However, probability of spread between the Great Lakes or between Lake Ontario and the Atlantic Coastal basin through several artificial connections has not been formally evaluated. Two species collected in the Bay of Quinte in recent years, Chain Pickerel (*Esox niger*) (Hoyle and Lake, 2011) and Blueback Herring (*Alosa aestevalis*) (Owens et al., 1998), may have dispersed through the St. Lawrence River with the aid of the Seaway locks, or through the Oswego arm of the Erie Canal.

Pagnucco et al. (2015) indicated that climate change will have an increasingly important influence on future invaders of the Great Lakes, as the lakes become more suitable for a broader range potential invaders, particularly those in trade and occurring south of the basin (Mandrak et al., 1989). However, the threat of the northern expansion of southern species may be reduced by active controls (e.g. electrical barriers) to prevent the northward movement of nonindigenous such as Asian Carps. Although several individuals of one of the Asian Carps, Grass Carp (*Ctenopharyngodon idella*), have been found in Lake Ontario and the St. Lawrence River (Chapman et al., 2021), none have yet been found in the Bay of Quinte).

Summary and conclusions

Nonindigenous species have likely been a part of the Bay of Quinte aquatic community since the early 1800s, although documented reports of new species do not occur until the mid-1900s and the role that NIS may play in terms of ecosystem level impacts was not fully recognized until the arrival of Zebra Mussels in the 1990s. Deliberate releases for stocking purposes were the primary vector of introduction of NIS in the 1800s, followed by ships' ballast water from the mid-1990s. Ballast water regulations have reduced the number of ship-mediated invasions since the late 1990s, with no new NIS reported from the Bay of Quinte during 2009-2013. However, unintentional releases associated with live trade of aquatic plants, molluscs and fishes, and introductions via canals, appear to pose a leading threat for new introductions in the future. Two thirds of NIS reported from the Bay of Quinte have had significant ecological impacts in the Great Lakes, ranging from competition and habitat alteration to trophic disruption and disease. The future colonization of the Bay of Quinte by NIS will undoubtedly be influenced by changes in the ecosystem of the Bay. On one hand, changes such as improved water and habitat quality may increase ecological resistance to invasion, but may in turn also be beneficial for NIS. Similarly, warmer temperatures and lower water levels related to climate change may facilitate invasion by thermophilic NIS, but repel thermophobic species. Risk assessments considering future climatic and habitat changes will improve predictions of future invaders to the Bay of Quinte.

Addendum

In 2018, Tench (*Tinca tinca*) was captured in the Bay of Quinte, the first record of the species in the Great Lakes. It likely originated from a population that was established in a farm pond in the Richelieu River drainage of Quebec in 1980s (Avlijas et al. 2018). It subsequently became established in the wild in the adjacent watershed and has since spread upstream to Lake Champlain and downstream to the St. Lawrence River, where it has spread downstream to the Gulf of St. Lawrence and upstream to the Bay of Quinte. This record has not been included in the preceding chapter, but is included here since it was known by the authors at the time of printing. This demonstrates the difficulty of maintaining a current inventory of NIS while vectors of introduction remain active.

Acknowledgements

We thank Mohi Munawar for encouraging this chapter into existence and Mark Fitzpatrick, Kelly Bowen, Kathy Leisti and Susan Doka for providing data and insight on NIS in the Bay of Quinte. Financial support from Fisheries and Oceans Canada and a NSERC Discovery Grant (SAB) are gratefully acknowledged.

References

Avlijas, S., Ricciardi, A., Mandrak, N.E., 2018. Eurasian Tench (*Tinca tinca*) – the Great Lakes' next invader. Can. J. Fish. Aquat. Sci. 75, 169-179.

Bailey, R.C., Grapentine, L., Stewart, T.J., Schaner, T., Chase, M.E., Mitchell, J.S., Coulas, R.A., 1999. Dreissenidae in Lake Ontario: impact assessment at the whole lake and Bay of Quinte spatial scales. J. Great Lakes Res. 25, 482-491.

Bailey, S.A., Deneau, M.G., Jean, L., Wiley, C.J., Leung, B., MacIsaac, H.J., 2011. Evaluating efficacy of an environmental policy to prevent biological invasions. Environ. Sci. Technol. 45, 2554-2561.

Bailey, S.A., Chan, F., Ellis, S.M., Bronnenhuber, J.E., Bradie, J.N., Simard, N., 2012. Risk assessment for ship-mediated introductions of aquatic nonindigenous species to the Great Lakes and freshwater St. Lawrence River. DFO Can. Sci. Advis. Sec. Res. Doc. 2011/104.

Benoit, H.P., Johannsson, O.E., Warner, D.M., Sprules, W.G., Rudstam, L.G., 2002. Assessing the impact of a recent predatory invader: the population dynamics, vertical distribution, and potential prey of *Cercopagis pengoi* in Lake Ontario. Limnol. Oceanogr. 47, 626-635.

Bowlby, J. N., Hoyle, J.A., Lantry, J.R., Morrison, B.J., 2010. The status of walleye in Lake Ontario 1988-2006. In: Status of walleye in the Great Lakes: proceedings of the 2006 Symposium. Great Lakes Fish. Comm. Tech. Rep. 69. pp. 165-188.

Briski, E., Allinger, L.E., Balcer, M., Cangelosi, A., Fanberg, L., Markee, T.P., Mays, N., Polkinghorne, C.N., Prihoda, K.R., Reavie, E.D., Regan, D.H., Reid, D.M., Saillard, H.J., Schwerdt, T., Schaefer, H., TenEyck, M., Wiley, C.J., Bailey S.A., 2013. A multi-dimensional approach to invasive species prevention. Environ. Sci. Technol. 47, 1216-1221.

Briski, E., Wiley, C.J., Bailey, S.A., 2012. Role of domestic shipping in the introduction or secondary spread of nonindigenous species: biological invasions within the Laurentian Great Lakes. J. Appl. Ecol. 49, 1124-1130.

Brooks, J.L., Dodson, S.I., 1965. Predation, body size and composition of plankton. Science 150, 28-35.

Carlton, J.T., 1985. Transoceanic and interoceanic dispersal of coastal marine organisms: the biology of ballast water. Oceanogr. Mar. Biol. Ann. Rev. 23, 313-371.

Carlton, J.T., Geller, J.B., 1993. Ecological roulette: the global transport of nonindigenous marine organisms. Science, 261, 78-82.

Casas-Monroy, O., Linley, R.D., Adams, J.K., Chan, F.T., Drake, D.A.R., Bailey, S.A., 2014. National Risk Assessment for Introduction of Aquatic Nonindigenous Species to Canada by Ballast Water. DFO Can. Sci. Advis. Sec. Res. Doc. 2013/128.

Chapman, D.C., Benson, A.J., Embke, H.S., King, N.R., Kočovský, P.M., Lewis, T.D., Mandrak, N.E. 2021. Status of the major aquaculture carps of China in the Laurentian Great Lakes basin. J. Great Lakes Res. 47, 3-13.

Cohen, J., Mirotchnick, N., Leung, B., 2007. Thousands introduced annually: the aquarium pathway for non-indigenous plants to the St. Lawrence Seaway. Front. Ecol. Environ. 5, 528-532.

Cooley, J.M., Moore, J.E., Geiling, W.T., 1986. Population dynamics, biomass, and production of the macrozooplankton in the Bay of Quinte during changes in phosphorus loadings. In: C.K. Minns, D.A. Hurley, K.H. Nicholls (Eds.), Project Quinte: point-source phosphorus control and ecosystem response in the Bay of Quinte, Lake Ontario, p. 166-176. Can. Spec. Publ. Fish. Aquat. Sci. 86.

Crossman, E.J., Cudmore, B.C., 1999. Invasive habits of fishes, global warming, and resulting range extensions. In: R. Claudi, J. Leach (Eds.), Nonindigenous freshwater organisms: vectors, biology, and impacts, p. 129-133. Lewis Publishers, Boca Raton, U.S.A.

Crossman, E.J., Van Meter, H.D., 1979. Annotated list of the fishes of the Lake Ontario watershed. Great Lakes Fish. Comm. Tech. Rept 36. 25 p.

Dermott, R., Witt, J., Um, E., Gonzalez, M., 1998. Distribution of the Ponto-Caspian amphipod *Echinogammarus ischnus* in the Great Lakes and replacement of native *Gammarus fasciatus*. J. Great Lakes Res. 24, 442-452.

Dietrich, J.P., Morrison, B.J., Hoyle, J.A., 2006. Alternative ecological pathways in the eastern Lake Ontario foodweb – Round Goby in the diet of Lake Trout. J. Great Lakes Res. 32, 395-400.

Drake, D.A.R. and Mandrak, N.E., 2014. Bycatch, bait, anglers, and roads: quantifying vector activity and propagule introduction risk across lake ecosystems. Ecol. Appl. 24, 877-894.

Duggan, I.C., Bailey, S.A., Colautti, R.I., Gray, D.K., Makarewicz, J.C., MacIsaac, H.J., 2003. Biological invasions in Lake Ontario: past, present and future. In: M. Munawar (Ed.), *State of Lake Ontario (SOLO) – Past, Present and Future*, pp. 541-557. Ecovision World Monograph Series, Schweizerbart Science Publishers, Stuttgart, Germany.

Gantz, C., Mandrak, N.E., Keller, R.P., 2013. Application of an aquatic plant risk assessment to non-indigenous freshwater plants in trade in Canada. DFO Can. Sci. Advis. Sec. Res. Doc. 2013/nnn. v + 24 p.

Grigorovich, I.A., Colautti, R.I., Mills, E.L., Holeck, K., Ballert, A.G., MacIsaac, H.J., 2003. Ballast-mediated animal introductions in the Laurentian Great Lakes: retrospective and prospective analyses. Can. J. Fish. Aquat. Sci. 60, 740-756.

Grosholz, E., 2002. Ecological and evolutionary consequences of coastal invasions. Trends Ecol. Evol. 17, 22-27.

Helfman, G.S., 2007. Fish Conservation: The Degradation and Restoration of Biodiversity, Island Press.

Hoyle, J.A., this monograph. Bay of quinte commercial and recreational fisheries: harvest trends and current status.

Hoyle, J.A., Lake, C., 2011. First occurrence of Chain Pickerel (*Esox niger*) in Ontario: possible range expansion from New York waters of Eastern Lake Ontario. Can. Field Nat. 125, 16-21.

Hoyle, J.A., Bowlby, J.N., Brousseau, C.M., Johnson, T.B., Morrison, B.J., Randall, R.G., 2012: Fish community structure in the Bay of Quinte, Lake Ontario: The influence of nutrient levels and invasive species. Aquatic Ecosyst. Health Mgmt. 15(3), 370-384.

Hurley, D.A., 1992. Feeding and trophic interactions of White Perch (Morone Americana) in the Bay of Quinte, Lake Ontario. Can. J. Fish. Aquat. Sci. 49, 2249-2259.

Hurley, D.A., Christie, W.J., 1977. Depreciation of the warm water fish community in the Bay of Quinte, Lake Ontario. J. Fish. Res. Bd. Can. 34, 1849–1860.

Johannsson, O.E., Bowen, K.L., 2012. Zooplankton production in the Bay of Quinte 1975-2008: relationships with primary production, habitat, planktivory, and aquatic invasive species (Dreissena spp. and Cercopagis pengoi). Can. J. Fish. Aquat. Sci. 69, 2046-2063.

Johnson, M.G., McNeil, O.C., 1986. Changes in abundance and species composition in benthic macroinvertebrate communities of the Bay of Quinte, 1966-84. In: C.K. Minns, D.A. Hurley, K.H. Nicholls (Eds.), Project Quinte: point-source phosphorus control and ecosystem response in the Bay of Quinte, Lake Ontario, p. 177-189. Can. Spec. Publ. Fish. Aquat. Sci. 86.

Kelly, D.W., Lamberti, G.A., MacIsaac, H.J., 2009. The Laurentian Great Lakes as a case study of biological invasion. In: R.P. Keller, D.M. Lodge, M.A. Lewis, and J.F. Shogren (Eds.), *Bioeconomics of Invasive Species: Integrating Ecology, Economics, Policy, and Management*, pp. 205-225. Oxford University Press, New York, U.S.A.

Kerr, S.J., Brousseau, C.S., Muschett, M., 2005. Invasive aquatic species in Ontario: a review and analysis of potential pathways for introduction. Fisheries 30, 21-30.

Kolar, C. S., Lodge D. M., 2002. Ecological predictions and risk assessment for alien fishes in North America. Science 298:1233-1236.

Leisti, K.E., Doka, S.E., Minns, C.K., 2012. Submerged aquatic vegetation in the Bay of Quinte: Response to decreased phosphorus loading and Zebra Mussel invasion. Aquat. Ecosyst. Health Mgmt 15(4), 442–452.

MacIsaac, H.J., Eyraud, A.P., Beric, B., Ghabooli, S., 2016. Can tropical macrophytes establish in the Laurentian Great Lakes? Hydrobiol. 767, 165-174.

Mandrak, N. E., Cudmore, B., 2010. The fall of native fishes and the rise of non-native fishes in the Great Lakes Basin. Aquatic Ecosyst. Health Mgmt. 13(3), 255-268.

Mandrak and Cudmore, 2010.

Mandrak, N.E., 1989. Potential invasion of the Great Lakes by fish species associated with climatic warming. J. Great Lakes Res. 15, 306-316.

Mandrak, N.E., Gantz, C., Jones, L.A., Marson, D., and Cudmore, B., 2013. Evaluation of Five Freshwater Fish Risk Assessment Protocols and Application to Non-Indigenous Organisms in Trade in Canada. DFO Can. Sci. Advis. Sec. Res. Doc. 2013/nnn.

Marson, D., Cudmore, B., Drake, D.A.R., Mandrak, N.E., 2009a. Summary of a survey of aquarium owners in Canada. Can. Manuscr. Rpt. Fish. Aquat. Sci. 2905.

Marson, D., B. Cudmore, D.A.R. Drake, and N.E. Mandrak. 2009b. Summary of a survey of water garden owners in Canada. Can. Manuscr. Rep. Fish. Aquat. Sci. 2906.

Mills, E.L., Leach, J.H., Carlton, J.T., Secor, C.L., 1993. Exotic species in the Great Lakes: a history of biotic crises and anthropogenic introductions. J. Great Lakes Res. 19, 1-54.

Minns, C.K., Munawar, M., Koops, M.A., Millard, E.S., 2011. Long-term ecosystem studies in the Bay of Quinte, Lake Ontario, 1972-2008: a prospectus. Aquatic Ecosyst. Health Mgmt. 14(1), 3-8.

Munawar, M., Fitzpatrick, M., Munawar, I.F., Niblock, H., Kane, D., 2012. Assessing ecosystem health impairments using a battery of ecological indicators: Bay of Quinte, Lake Ontario example. Aquatic Ecosyst. Health Mgmt. 15, 430-441.

NRC [National Research Council], 1996. Stemming the tide: controlling introductions on nonindigenous species by ships' ballast water. National Academy Press, Washington D.C., U.S.A.

Nicholls, K.H., Hoyle, J.A. Johannsson, O.E., Dermott, R., 2011 A biological regime shift in the upper Bay of Quinte ecosystem (Lake Ontario) associated with the establishment of invasive dreissenid mussels. J. Great Lakes Res. 37, 310-317.

Nicholls, K.H., Hurley, D.A., 1989. Recent changes in the phytoplankton of the Bay of Quinte, Lake Ontario: the relative importance of fish, nutrients, and other factors. Can. J. Fish. Aquat. Sci. 46, 770-779.

Owens, R.W., O'Gorman, R., Mills, E.L., Rudstam, L.G., Hasse, J.J., Kulik, B.H., MacNeill, D.B., 1998. Blueback herring (*Alosa aestivalis*) in Lake Ontario: first record, entry route, and colonization potential. J. Great Lakes Res. 24, 723-730.

Pagnucco, K.S., Maynard, G.A., Fera, S.A., Yan, N.D., Nalepa, T.F., Ricciardi, A., 2015. The future of species invasions in the Great Lakes-St. Lawrence River basin. J. Great Lakes Res. 41, 96-107.

Poste, A.E., Ozersky, T., 2013. Invasive dreissenid mussels and round gobies: a benthic pathway for the trophic transfer of microcystin. Environ. Toxicol. Chem. 32, 2159-2164.

Ricciardi, A., 2006. Patterns of invasion in the Laurentian Great Lakes in relation to changes in vector activity. Divers. Distrib. 12, 425-433.

Ricciardi, A., Cohen, J., 2007. The invasiveness of an introduced species does not predict its impact. Biol. Invasions 9, 309-315.

Ricciardi, A., Rasmussen, J.B., 1998. Predicting the identity and impact of future biological invaders: a priority for aquatic resource management. Can. J. Fish. Aquat. Sci. 55, 1759-1765.

Rixon, C.A.M., Duggan, I.C., Bergeron, N.M.N., Ricciard, A., MacIsaac, H.J., 2005. Invasion risks posed by the aquarium trade and live fish markets on the Laurentian Great Lakes. Biodivers. Conserv. 14, 1365-1381.

Rup, M.P., Bailey, S.A., Wiley, C.J., Minton, M.S., Miller, A.W., Ruiz, G.M., MacIsaac, H.J., 2010. Domestic ballast operations on the Great Lakes: potential importance of Lakers as a vector for introduction and spread of nonindigenous species. Can. J. Fish. Aquat. Sci. 67, 256-268.

Simberloff, D., 2009. The role of propagule pressure in biological invasions. Annu. Rev. Ecol. Evol. Syst. 40, 81-102.

Strayer, D., 2010. Alien species in fresh waters: ecological effects, interactions with other stressors, and prospects for the future. Freshwat. Biol. 55, 152-174.

Sylvester, F., MacIsaac, H.J., 2010. Is vessel hull fouling an invasion threat to the Great Lakes? Diversity Distrib. 16, 132-143.

Taraborelli, A.C., Fox, M.G., Johnson, T.B., Schaner, T., 2010. Round goby (Neogobius melanostomus) population structure, biomass, prey consumption and mortality from predation in the Bay of Quinte, Lake Ontario. J. Great Lakes Res. 36:625–632.

USACE [US Army Corps of Engineers], 2013. Great Lakes and Mississippi interbasin study: focus area 2 - aquatic pathways assessment summary report. U.S. Army Corps of Engineers. 108 p.

Ward, J.M., Cudmore, B., Drake, D.A.R., Mandrak, N.E., 2011. Summary of a survey of baitfish users in Canada. Can. Manuscr. Rep. Fish. Aquat. Sci. 2972.

Fish species at risk in the Bay of Quinte and vicinity

Nicholas E. Mandrak

Department of Biological Sciences, University of Toronto Scarborough, Toronto, ON M1C 1A4
nicholas.mandrak@utoronto.ca

Introduction

The highest number of freshwater fish species at risk in Canada is found in southern Ontario. This is the result of historical, environmental, and anthropogenic factors. Southern Ontario has the highest freshwater fish richness as a result of its close proximity to the Mississippian refugium from which the greatest number of Canadian fishes dispersed following the last ice age and its mild climate (Mandrak and Crossman, 1992, Staton and Mandrak, 2006). However, this high richness is found in an area of high anthropogenic stress due to the large human population size and highly modified urban and agricultural landscapes (Chu et al., 2003), and the presence of many invasive species (Ricciardi, 2006). The primary threats to Canadian freshwater fishes are habitat alteration, invasive species, and overexploitation (Dextrase and Mandrak, 2006), all prominent threats in southern Ontario.

Freshwater fish and species at risk richness in Bay of Quinte is not the highest in southern Ontario (Staton and Mandrak, As of 2014), but is exposed to the same threats ubiquitous throughout the region. As of 2014, the 129 species known from the Lake Ontario basin (Crossman and Van Meter, 1979), 65 species have been collected in the Bay itself (Hoyle et al., 2012), and 12 fish species at risk are found in the Bay and its vicinity, including tributaries to the Bay and waters adjacent to Prince Edward County. Fish species richness in southeastern Ontario is lower than in southwestern Ontario as it was more distant from the speciose Mississippian refugium (Mandrak and Crossman, 1992) and has a cooler climate. However, it was closer to the less speciose Atlantic Coastal refugium and, as a result, has several eastern species not found in southwestern Ontario (Mandrak and Crossman, 1992). The Bay itself has undergone substantial habitat degradation, largely as a result of agricultural land use, cultural eutrophication and associated effects (Minns et al., 1986), and many of its tributaries been modified to regulate stream

Ecology of the Bay of Quinte: Health, Management and Global Implications
Editors: C.K. Minns, M. Munawar and M.A. Koops

Aquatic Ecosystem Health & Management Society, Canada

flow for power generation, flood and ice control, and navigation purposes (e.g. Trent-Severn Waterway). As it is contiguous with Lake Ontario, the Bay also has a long history of invasion by nonindigenous species, which have had a variety of impacts including habitat degradation and trophic alteration (Bailey et al., 2014).

This paper provides an overview of the distribution and prospects of, and threats to, fish species at risk in the Bay of Quinte and vicinity including tributaries to the Bay and waters adjacent to Prince Edward County, as of 2014.

Methods

Fish species at risk in the study area were identified in conservation lists published by the federal Committee on the Status of Endangered Wildlife in Canada (COSEWIC) and provincial Committee on the Status of Species at Risk in Ontario (COSSARO). The distributions of the species were determined using COSEWIC conservation status reports and an unpublished Fisheries and Oceans Canada (DFO) fish species at risk database. To examine patterns in threats to, and habitats occupied by, fish species at risk, primary and secondary threats and adult and spawning habitats were compiled from the COSEWIC reports. After the assessment of the conservation status of a species by COSEWIC and prior to the Species at Risk Act (SARA) listing decision by the federal Environment Minister, DFO conducts a recovery potential assessment (RPA) that summarizes all known biological information on the species and determines if recovery is feasible. The COSEWIC and RPA reports were used to provide a summary of the current status of each of the 12 fish species at risk in the Bay and its vicinity.

Results and discussion

Conservation status of 12 species found in the Bay of Quinte and its vicinity have been assessed by COSEWIC and COSSARO (Table 1). Species were generally assessed the same conservation level by COSEWIC and COSSARO and most are listed under the federal Species at Risk Act and provincial Endangered Species Act (ESA) (Table 1). However, the American Eel is assessed by COSEWIC as Threatened and by COSSARO as Endangered (Table 1). The

Table 1. Conservation status of fish species at risk in the Bay of Quinte and vicinity.

Common Name	Designatable Unit	Scientific Name	COSEWIC	SARA	COSSARO/ESA
American Eel		*Anguilla rostrata*	Threatened	Not Listed	Endangered
Atlantic Salmon	Lake Ontario	*Salmo salar*	Extinct	Not Listed	Extinct
Bridle Shiner		*Notropis bifrenatus*	Special Concern	Special Concern	Special Concern
Channel Darter		*Percina copelandi*	Threatened	Threatened	Threatened
Deepwater Sculpin	Great Lakes-Upper St. Lawrence	*Myoxocephalus thompsoni*	Special Concern	Threatened	Special Concern
Eastern Sand Darter	Ontario	*Ammocrypta pellucida*	Threatened	Threatened	Threatened
Grass Pickerel		*Esox americanus vermiculatus*	Special Concern	Special Concern	Special Concern
Lake Sturgeon	Great Lakes-Upper St. Lawrence	*Acipenser fulvescens*	Threatened	Not Listed	Threatened
Pugnose Shiner		*Notropis anogenus*	Threatened	Endangered	Endangered
River Redhorse		*Moxostoma carinatum*	Special Concern	Special Concern	Special Concern
Silver Lamprey	Great Lakes-Upper St. Lawrence	*Ichthyomyzon unicuspis*	Special Concern	Special Concern	Special Concern
Spotted Gar		*Lepisosteus oculatus*	Threatened	Threatened	Threatened

imperilment of American Eel varies across its Canadian range with the species having declined by over 99% in Ontario (COSEWIC, 2006a), explaining the difference in the assessments. Pugnose Shiner was also recently downlisted by COSEWIC to Threatened from Endangered, which it is listed as under COSSARO.

The 12 fish species at risk in the Bay of Quinte and its vicinity have unique distributions for the most part (Figure 1). Only the American Eel, Atlantic Salmon, Lake Sturgeon, and Silver Lamprey had similar distributions within the Bay, being historically present in the Bay of Quinte and using most of its tributaries for spawning. The Bridle Shiner, Pugnose Shiner, and Grass Pickerel prefer wetland habitats and are found in the Bay of Quinte, Prince Edward County, and Lake Ontario wetlands as far west as Presquile Bay. The Pugnose Shiner was also recently (2010) discovered in the Trent River. The Channel Darter prefers riverine habitats in eastern Ontario and is found only in tributaries of the Bay of Quinte. The River Redhorse is found in the Trent River and Bay of

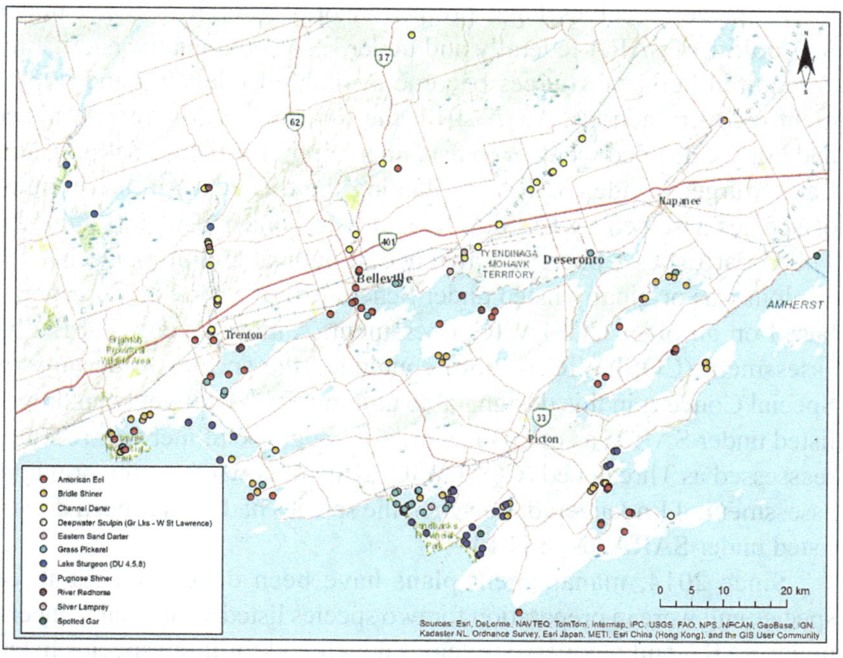

Figure 1. Distribution of fish species at risk in the Bay of Quinte and vicinity based on actual collection records from DFO, OMNR, and the Royal Ontario Museum. Note that no collection records exist for Atlantic Salmon in the area.

Quinte. The Deepwater Sculpin is marginal to the area and found only in the adjacent deeper waters of Lake Ontario. A single Spotted Gar was caught in East Lake in 2007 and there is uncertainty as to whether this represents an established population as extensive sampling has failed to collect additional specimens (COSEWIC 2015). If an established population does exist, whether it should be considered native or introduced as a result of aquarium release is a further debate as it would be the only known Lake Ontario population with the nearest known population being in Long Point Bay in Lake Erie (Holm et al., 2009). Interestingly, a single Spotted Gar was confirmed from upper St. Lawrence River near Kingston in the 1986, but subsequent sampling failed to collect additional specimens (COSEWIC, 2005). The first record of Eastern Sand Darter in eastern Ontario was collected in West Lake in 2013 (Reid and Dextrase, 2014). The numbers subsequently collected (DFO, unpubl. data), suggest an established population. The origin of this population is unknown, but it is found between disjunct native populations in southwestern Ontario and southwestern Quebec (COSEWIC, 2009).

A species receives legal protection only after it is listed under Schedule 1 of SARA federally and under ESA provincially, and listing is required before resources become available to develop and implement recovery actions. As of 2014, the federal Environment Minister had not yet decided whether to list American Eel, Atlantic Salmon, and Lake Sturgeon under SARA, explaining the difference in the number of species assessed by COSEWIC and listed under SARA (Figure 2). The Great Lakes-Upper St. Lawrence designatable unit of Deepwater Sculpin was originally listed under Schedule 2 of SARA as Threatened based on an older COSEWIC assessment. A more recent COSEWIC assessment (COSEWIC, 2006b) concluded the species to be only of Special Concern in this designatable unit, but it had not yet been down-listed under SARA, as of 2014. Similarly, Pugnose Shiner was recently reassessed as Threatened (COSEWIC, 2013b), down from the previous assessment of Endangered; however, the species had not yet been down-listed under SARA, as of 2014.

Since 2014, management plans have been developed for three species and were in preparation for two species listed as Special Concern under SARA and recovery strategies developed for three species listed and in preparation for one species as Threatened or Endangered (Table 2). These management plans and strategies typically identify three types of actions: research and monitoring; management and coordination; and, stewardship, outreach, and awareness (e.g. DFO 2012a,b).

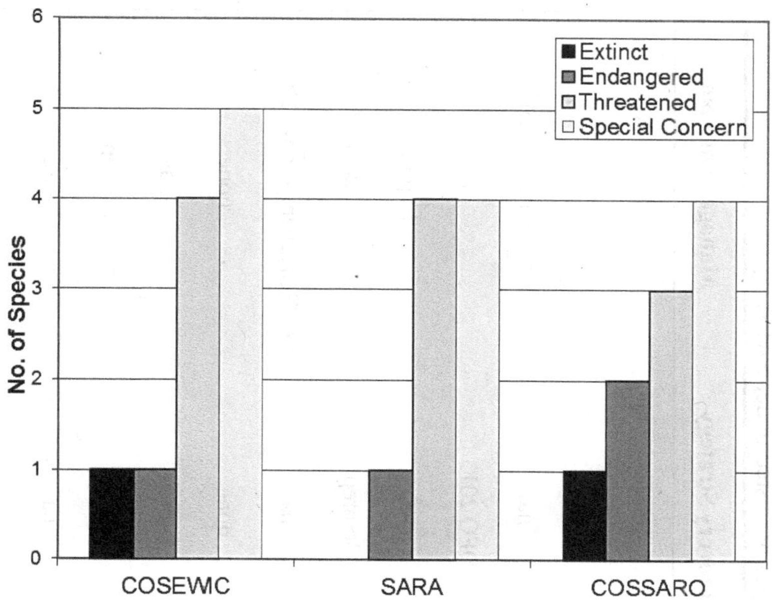

Figue 2. Conservation status of fish species at risk in the Bay of Quinte and vicinity as of 2014.

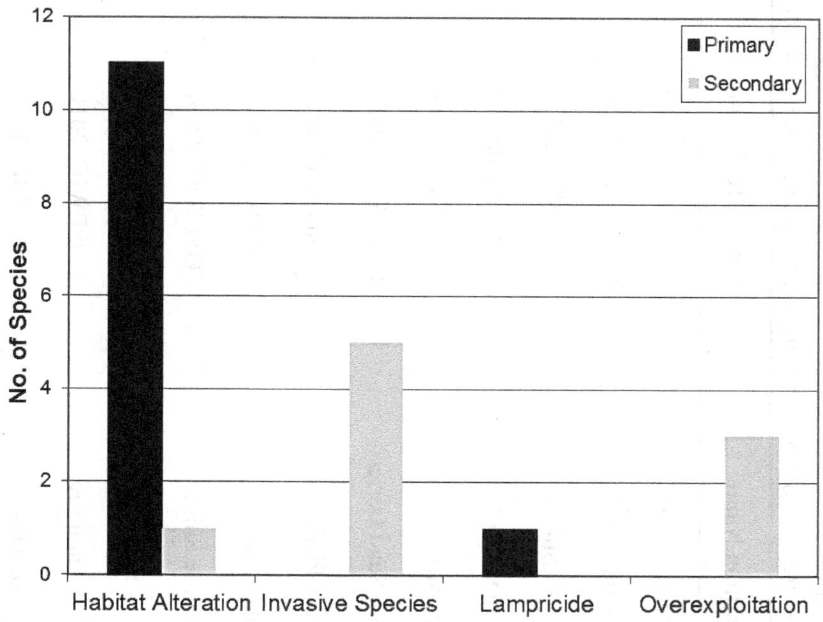

Figure 3. Threats to fish species at risk in the Bay of Quinte and vicinity, as of 2014.

Table 2. COSEWIC reports, recovery strategies, and management plans for fish species at risk in the Bay of Quinte and vicinity. In prep – in preparation; na – not applicable; none – none initiated as species not listed under SARA.

Common Name	COSEWIC Report	Recovery Strategy	Management Plan
American Eel	COSEWIC, 2006a	none	none
Atlantic Salmon	COSEWIC, 2010	none	na
Bridle Shiner	COSEWIC, 2013a	na	Boucher et al., 2011
Channel Darter	Phelps and Francis, 2002	DFO 2013	na
Deepwater Sculpin	COSEWIC, 2006b	na	In prep
Eastern Sand Darter	COSEWIC, 2009	In prep	na
Grass Pickerel	Crossman and Holm, 2005	na	Beauchamp et al., 2012
Lake Sturgeon	COSEWIC, 2006c	none	none
Pugnose Shiner	COSEWIC, 2013b	In prep	na
River Redhorse	COSEWIC, 2006d	na	In prep
Silver Lamprey	COSEWIC, 2011	na	In prep
Spotted Gar	COSEWIC, 2005	In prep	na

Habitat alteration, invasive species, overexploitation, and lampricide were identified as the primary and secondary threats to the 12 species (Figure 3). Habitat alteration, typically wetland loss and stream flow alteration, is the primary threat of 11 of the 12 species (Figure 3). The Bay of Quinte went through extensive habitat degradation largely related initially to deforestation of the watershed for agricultural purposes (1800s) and, more recently, to cultural eutrophication mainly associated with urban areas (1900s), which led to significant declines in water quality, including clarity, resulting in extensive wetland loss (Minns et al., 1986). This led to the Bay of Quinte being listed as a Great Lakes Area of Concern. Remedial activities, primarily related to phosphorus control, and the invasion of Dreissenid Mussels have led to recent improvements in water quality (Nicholls 2012), resulting in increasing aquatic macrophyte cover (Leisti et al., 2012). Many of the tributaries to the Bay of Quinte have been modified for power generating or navigation purposes including the Trent River, modified for both purposes, which is fragmented by a series of locks and dam. Other tributaries, such as the Salmon, Napanee, and Moira rivers have been dammed for power generation and flow regulation, or for control of the invasive Sea Lamprey *(Petromyzon marinus)*. The use of lampricide to assess and control the Sea Lamprey in the Great Lakes is the primary threat to the Great Lakes-Upper St. Lawrence designatable unit of the Silver Lamprey (COSEWIC 2011). In the last 10 years, lampricide has been used in small quantities to assess, but not in large quantities to control, Sea Lamprey in Bay of Quinte tributaries (www.glfc.org/sealamp). Invasive species are the secondary threat to five species. As of 2014, twelve fish species and 23 other species were considered nonindigenous but established in the Bay of Quinte (Bailey et al., this book). The fish species at risk are likely most impacted by Common Carp as a habitat degrader and Round Goby as a competitor. The Common Carp has likely been in the Bay of Quinte since the late1800s (Scott and Crossman 1973) and likely contributed to the decline in wetlands directly through the uprooting of vegetation and indirectly through increasing turbidity by uprooting vegetation and aggressive spawning behaviour, leading to shallower sunlight penetration and resulting in loss of vegetation from deeper littoral areas (Lougheed et al., 1998). Overexploitation was a secondary threat that historically contributed to the decline of American Eel, Atlantic Salmon, and Lake Sturgeon, but is no longer a concern in Lake Ontario as this threat diminished with the extinction of the Atlantic Salmon by the 1890s

Table 3. Threats and adult and spawning habitats of fish species at risk in the Bay of Quinte and its vicinity.

Common Name	Threat		Adult Habitat	Spawning Habitat
	Primary	Secondary		
American Eel	Habitat Alteration	Overexploitation	Lacustrine, Riverine	Sargasso Sea
Atlantic Salmon	Habitat Alteration	Overexploitation	Lacustrine	Riverine
Bridle Shiner	Habitat Alteration	Invasive Species	Wetland	Wetland
Channel Darter	Habitat Alteration	Invasive Species	Riverine	Riverine
Deepwater Sculpin	Habitat Alteration	Invasive Species	Lacustrine	Lacustrine
Eastern Sand Darter	Habitat Alteration	Invasive Species	Lacustrine, Riverine	Lacustrine, Riverine
Grass Pickerel	Habitat Alteration		Wetland	Wetland
Lake Sturgeon	Habitat Alteration	Overexploitation	Lacustrine, Riverine	Riverine
Pugnose Shiner	Habitat Alteration	Invasive Species	Wetland	Wetland
River Redhorse	Habitat Alteration		Riverine	Riverine
Silver Lamprey	Lampricide	Habitat Alteration	Lacustrine, Riverine	Riverine
Spotted Gar	Habitat Alteration		Wetland	Wetland

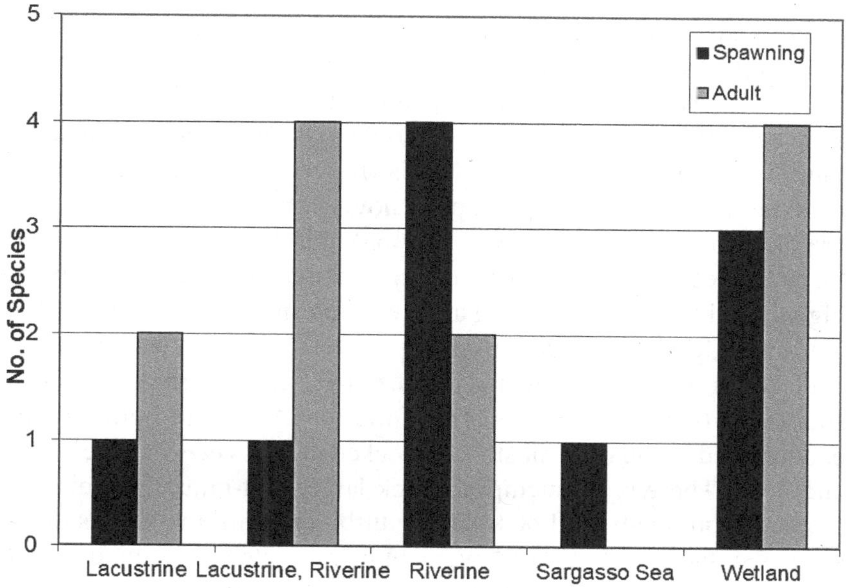

Figure 4. Habitat preferences of fish species at risk in the Bay of Quinte and vicinity.

and a dramatic decline in stocks of the other two species; commercial fishing of these species is now prohibited in Ontario.

The 12 fish species at risk are found in a diversity of habitats in the Bay of Quinte and its vicinity (Table 3). The preferred adult habitat is riverine for seven species, lacustrine for five species, and wetlands for four species, with some species preferring more than one type of habitat (Figure 4). The spawning habitat of five species is riverine, four species is wetlands, two species is lacustrine, and the American Eel spawns in the Sargasso Sea (Holm et al., 2009) (Figure 3). Therefore, it appears that threats to these fish species at risk are ecosystem wide and not restricted to specific habitats; however, species preferring wetland and tributary habitats are more likely to be exposed to local threats. This is consistent with the general degradation of the Bay of Quinte ecosystem that led to its listing as an Area of Concern.

Native distribution of American Eel in Canada extended widely in salt and fresh waters from the Maritimes as far north as Newfoundland, westward up the St. Lawrence and its tributaries and into Lake Ontario and its tributaries to Niagara Falls, which historically prevented farther migration. As the individuals in Lake Ontario were historically all females and the largest and most fecund, it is thought that they contributed significantly to the overall recruitment of this panmictic species (COSEWIC, 2006a). If this is true, then the overall health of the species as a whole may be dependent upon the Lake Ontario eels that successfully return to the Sargasso Sea to spawn. Based on indexed sampling in the Bay of Quinte, abundances in Ontario started dropping precipitously in the 1970s to the point now where they are now less than 1% of historic levels (COSEWIC, 2006a). It is thought that dams in the St. Lawrence River block upstream migration of young eels and cause a high mortality of out-migrating adult eels returning to the Sargasso Sea to spawn (COSEWIC, 2006a). Attempts to mitigate these impacts using eel ladders, truck transport, and turbine modification have not yet been successful (Chaput et al., 2013). Of the few American Eel currently encountered in the Bay, most were stocked as glass eels between 2008 and 2010. There is an attempt to truck larger, out-migrating eels past the dams, but many will be killed by turbines. Until the threats to the species in the St. Lawrence River can be mitigated, the abundance of naturally occurring American Eel is likely to remain very low.

The Atlantic Salmon is a typically anadromous species found in the north Atlantic Ocean and adjacent fresh waters of both North America and Europe; however, a potamodromous population occurred in Lake Ontario (Scott and Crossman, 1973). Given its potamodromy, early age of smoltification, and occurrence of spring and fall spawning runs, the Lake Ontario population was considered a "designatable unit" by COSEWIC (COSEWIC 2010). Historically, in the vicinity of the Bay of Quinte, the Moira, Salmon, and Trent rivers were known spawning and smolt habitat for Atlantic Salmon. Since European settlement in the Lake Ontario basin, the species was heavily exploited commercially and its access to spawning habitat limited by mill dams and associated habitat degradation (Scott and Crossman 1973). The last known capture of the original strain of Atlantic Salmon in the Lake Ontario basin was 1898 and that strain is now considered extinct (COSEWIC, 2010). Recent attempts to introduce non-native strains of Atlantic Salmon have been met with limit success, likely as the result of ongoing habitat degradation in the tributaries and competition with, and predation by, introduced Pacific salmons (*Oncorhynchus* spp.)

Bridle Shiner and Pugnose Shiner were only recently discovered (1979 and 2010, respectively; COSEWIC, 2013a, COSEWIC, 2013b) in the Bay of Quinte area. Both species have a strong preference for low turbidity (Gray et al., 2014) and abundant submergent vegetation. The Bridle Shiner survived the last ice age in an Atlantic Coastal refugium

(Mandrak and Crossman, 1992) and reaches its western range limit in Presquile Bay, just west of Bay of Quinte; whereas, the Pugnose Shiner survived in the Mississippian refugium and reaches its eastern range limit in the upper St. Lawrence River (Mandrak and Crossman, 1992). Both species were likely always present in the Bay of Quinte but went undetected as a result of either misidentification (they are very similar to other 'blackline' shiners; Holm et al., 2009) or low abundance when their preferred submergent aquatic vegetation was historically suppressed due to poor water quality (Leisti et al., 2012). Most small-fish sampling has occurred since the 1970s during the time when water quality was at its poorest, submerged aquatic vegetation was likely sparsest (Leisti et al., 2012), and Bridle Shiner and Pugnose Shiner were likely at their lowest abundance. With the improving water quality in the Bay of Quinte and the consequent increase in submerged aquatic macrophytes, (Leisti et al., 2012), it is expected that these species should increase in abundance.

Grass Pickerel is found sporadically throughout southern Ontario and extreme southwestern Quebec, including the Bay of Quinte and its vicinity. It has a strong preference for wetlands with an abundance of floating and emergent vegetation and low turbidity (Scott and Crossman, 1973). Although this type of habitat has been degraded in the study area (Leisti et al., 2012), there have been no strong trends in its abundance over time, likely as a result of limited sampling of its preferred habitat. Recent improvements in its preferred habitat (Leisti et al., 2012) should lead to healthier Grass Pickerel populations. However, this may be negated by the recent appearance of a congeneric competitor, the Chain Pickerel *(Esox niger),* which is dispersing both eastward and westward from its native range in Canada (Hoyle and Lake, 2011). In its native range, the Chain Pickerel co-occurs with the Redfin Pickerel *(Esox americanus americanus)* (Scott and Crossman 1973), the closest relative of the Grass Pickerel; however, the Chain Pickerel has been shown to impact other fish species where it has been introduced in eastern Canada (Mitchell et al., unpubl., 2010).

Eastern Sand Darter and Channel Darter have disjunct distributions in Canada, with the only populations of both species in the Lake Ontario drainage occurring in the vicinity of Bay of Quinte. The 2013 collection of Eastern Sand Darter in West Lake represents the first known records of this species in the Lake Ontario drainage despite significant prior sampling (Reid and Dextrase, 2014). However, given its disjunct Canadian distribution in the Lake Erie drainage and the

St. Lawrence drainage, it is not totally unexpected that a population has been confirmed in the Lake Ontario drainage. Such a distribution is similar to the Channel Darter distribution in Canada (Phelps and Francis 2002) and is hypothesized to be the result of the contraction of an original postglacial range following the end of the Hypsithermal warm period (ca. 6,000 years ago) (Mandrak, 1990). Given the strong preference of Eastern Sand Darter for sand substrates (Dextrase, 2013), the potential distribution of Eastern Sand Darter will be limited by the extent of surficial sand deposits. Therefore, other areas in the vicinity of the Bay of Quinte with surficial sand deposits should be targeted for sampling using appropriate gears and efforts (i.e. seines; Dextrase et al., 2014). Conversely, the Channel Darter prefers areas of coarser substrates kept clear of finer sediments by flow. Tributaries to the Bay of Quinte with such areas have been dammed, likely resulting in reservoir effects of unsuitable slow-moving, fine-sediment habitats above the dams. As a result, suitable Channel Darter habitat has likely been fragmented in these areas. Given the purposes of the dams, they are unlikely to be removed; therefore, habitat will likely to continue to be limited for Channel Darter in these tributaries.

Lake Sturgeon was once abundant in all of the Great Lakes and its larger tributaries until overexploitation in the late 1800s decimated this late-maturing species (COSEWIC, 2006c). Damming of tributaries, including those to the Bay of Quinte, contributed to the decline of this species by preventing migration to spawning grounds (COSEWIC, 2006c). This large-bodied species can disperse long distances and, as a result, high gene flow has been documented in Great Lakes populations (Welsh et al., 2008). This suggests that the status of Lake Sturgeon in the Bay of Quinte is not strictly influenced by its tributaries itself, but by the availability of spawning habitat in the lower Great Lakes, primarily Lake Ontario, as a whole. Therefore, the future of the Lake Sturgeon in Lake Ontario, including the Bay of Quinte, is dependent upon efforts to rehabilitate spawning habitat through dam removal and substrate enhancement. In the Lake Ontario basin, such efforts have been limited to date.

The River Redhorse has a highly disjunct distribution in southern Ontario and southwestern Quebec, with the only Lake Ontario population occurring in the Bay of Quinte watershed. In this watershed, it has been found in the Trent River and the Bay itself. The Trent River is likely the spawning site for the Bay of Quinte population and adults reside both in the river and the bay. The fragmentation of the Trent

River by dams and locks appears to have a limited effect on River Redhorse, as gene flow appears to be maintained by movement through those structures (Reid et al., 2008). The habitat is unlikely to change in the Trent River and improved wetland quality in Bay (Leisti et al., 2012) is not preferred by the species (Scott and Crossman, 1973); therefore, population abundance is unlikely to change.

Silver Lamprey has a highly disjunct distribution in Canada, occurring in the Great Lakes-St. Lawrence watershed, including Bay of Quinte, as well as the Prairie provinces (COSEWIC, 2011). Our understanding of the distribution and abundance of this species is limited as it is not susceptible to non-targeted sampling and much of what we know about it in the Great Lakes, including Bay of Quinte, is the result of assessment and control efforts to reduce invasive Sea Lamprey abundance. It is also occasionally reported by commercial and recreational anglers when found attached to their catch. Silver Lamprey abundance is dependent upon access to riverine spawning habitat and prey species. Although sufficient prey exist in the Bay of Quinte, access to spawning habitat is limited by dams on its tributaries. Abundance may be further limited by assessment and control treatments of Sea Lamprey nursery habitats using lampricide. The dams are unlikely to be removed and Sea Lamprey treatments are likely to continue; therefore, population abundance is unlikely to change.

Deepwater Sculpin declined so dramatically in Lake Ontario that it was not seen 1964-1996 (COSEWIC, 2006b; Lantry et al., 2007). The cause of its disappearance is unknown but has been hypothesized to be related to contaminants (Scott and Crossman, 1973), invasive species (Casselman and Scott, 2003; Madenjian et al., 2002; Madenjian et al., 2005), and\or the loss of the keystone predators Lake Trout (*Salvelinus namaycush*) and Burbot (*Lota lota*) leading to increased predation from other native sculpins (Brandt, 1986). Since 1996, increasing numbers of specimens have been collected from the main lake (Lantry et al., 2007). Although populations may be increasing in Lake Ontario, the species will likely to continue to be a marginal species in the vicinity of the Bay of Quinte given its preference for very deep (> 100m) waters (Scott and Crossman, 1973).

Several other fish species native to Lake Ontario, not listed by COSEWIC, have been caught only rarely in the Bay of Quinte and its vicinity (Christie 1973, Crossman and Van Meter 1979, Hoyle et al., 2012). Although the Bay and its vicinity appear to provide suitable habitat, some species appear to have been always rare such as Mooneye

(*Hiodon tergisus*), Quillback (*Carpiodes cyprinus*), Bigmouth Buffalo (*Ictiobus cyprinellus*), Ninespine Stickleback (*Pungitius pungitius*), and Sauger (*Sander canadense*). Some coldwater species, Cisco (*Coregonus artedi*) and Lake Whitefish (*C. clupeaformis*), that migrate into the Bay to spawn, declined dramatically in the Bay by the mid-1900s (Christie 1973); others disappeared from Lake Ontario entirely including Kiyi (*C. kiyi*) by 1964, Shortnose Cisco (*C. reighardi*) by 1964 and Bloater (*C. hoyi*) by 1972 (Crossman and Van Meter 1979). Lake Whitefish abundance increased in the 1980s, but declined again in the 1990s (Hoyle, 2005). Another coldwater species, Round Whitefish (*Prosopium cylindraceum*), historically known in the vicinity of the Bay (Christie 1973), was not reported in multi-gear sampling in the Bay between 1969 and 2009 (Hoyle et al., 2012), but are occasionally caught to the west of the Bay off of Presquile Provincial Park (Hoyle, 2012). The historical collection of small fishes (e.g. cyprinids) has been insufficient to examine trends over time.

Summary and conclusions

The highest number of freshwater fish species at risk in Canada occurs in southern Ontario. As of 2014, the 129 species known from the Lake Ontario basin, 65 species have been collected in the Bay of Quinte itself, and 12 fish species at risk are found in the Bay and its vicinity, including tributaries to the Bay and waters adjacent to Prince Edward County. The Bay has undergone substantial habitat degradation, largely as a result of agricultural land use, cultural eutrophication and associated effects, and many of its tributaries been fragmented as a result of power generation and navigation. The Bay also has a long history of invasion by nonindigenous species. The 12 fish species at risk are found in a diversity of habitats, indicating that threats to these species are ecosystem wide and not restricted only to specific habitats. Habitat alteration, typically wetland loss and tributary flow alteration, is the primary threat to 11 of the fish species at risk with invasive species being a secondary threat to most and overexploitation to some. Although ongoing efforts to improve water quality and habitat have had positive effects on some fish species at risk in the Bay and its vicinity, some threats are unlikely to be mitigated (e.g. dams) and, ultimately, the conservation and recovery of fish species at risk will depend upon activities that occur at the scale of the Bay, Lake Ontario, Great Lakes basin, and beyond.

The future of the fish species at risk in the Bay and its vicinity is uncertain. Recent improvements in water quality in the Bay (Nicholls, 2012), and associated improvement in aquatic macrophyte cover (Leisti et al., 2012), are likely to improve populations of fish species at risk that prefer clear, vegetated habitats. However, the extent of macrophyte cover in the Bay is influenced by water levels (Gertzen et al., 2012), which may vary in response to manipulation of Lake Ontario levels at its outflow and potentially by climate change (Mortsch and Quinn, 1996). The impacts of habitat fragmentation in the tributaries and invasive species is ongoing and may be further exacerbated under climate change through decreasing water quantities leading to lower flows (Mortsch and Quinn, 1996), and to increasing water temperatures providing suitable thermal habitat to a greater number of invasive species (e.g. Mandrak, 1989). Ongoing efforts to improve water quality and habitat in the Bay will positively affect some fish species at risk, but not all. Additional efforts are required to reduce fragmentation and minimize the effects of water-level and climate change. Ultimately, the conservation and recovery of fish species at risk in the Bay and its vicinity will be dependent upon activities that occur at the scale of the Bay, Lake Ontario, Great Lakes basin, and beyond.

References

Bailey, S.A., Hoyle, J.A., Mandrak, N.E., Johannsson, O.E., this issue. A chronicle of nonindigenous species in the Bay of Quinte, Lake Ontario.

Beauchamp, J., Boyko, A.L., Dunn, S., Hardy, D., Jarvis, P.L., Staton, S.K., 2012. Management plan for the Grass Pickerel (*Esox americanus vermiculatus*) in Canada. Species at Risk Act Management Plan Series. Fisheries and Oceans Canada, Ottawa.

Boucher, J., Berubé, M., Boyko, A., Bourgeois. M., 2011. Management plan for the Bridle Shiner (*Notropis bifrenatus*) in Canada (Final version). Species at Risk Act Management Plan Series, Fisheries and Oceans Canada, Ottawa.

Bouvier, L.D., Mandrak, N.E., 2010a. Information in support of a Recovery Potentia Assessment of Channel Darter (*Percina copelandi*) in Ontario. DFO Can. Sci. Advis. Sec. Res. Doc. 2010/029.

Bouvier, L.D., Mandrak, N.E., 2010b. Information in support of a Recovery Potential Assessment of Eastern Sand Darter (*Ammocrypta pellucida*) in Ontario. DFO Can. Sci. Advis. Sec. Res. Doc. 2010/093.

Bouvier, L.D., Mandrak, N.E., 2010c. Information in support of a Recovery Potential Assessment of Spotted Gar (*Lepisosteus oculatus*) in Canada. DFO Can. Sci. Advis. Sec. Res. Doc. 2010/079.

Bouvier, L.D., Boyko, A.L., Mandrak N.E., 2010. Information in support of a Recovery Potential Assessment of Pugnose Shiner (*Notropis anogenus*) in Canada. DFO Can. Sci. Advis. Sec. Res. Doc. 2010/009.

Brandt, S.B. 1986. Disappearance of the Deepwater Sculpin (*Myoxocephalus thompsoni*) from Lake Ontario: the keystone predator hypothesis. Journal of Great Lakes Research 12, 18–24.

Casselman, J.M., Scott, K.A., 2003. Fish-community dynamics of Lake Ontario: long-term trends in the fish populations of eastern Lake Ontario and the Bay of Quinte. In: M. Munawar (Ed.), *State of Lake Ontario: Past, present and Future*, pp. 349-383. Ecovision World Monograph Series, Aquatic Ecosystem Health Management Society, Burlington, ON.

Chaput, G., Cairns, D.K., Bastien-Daigle, S., LeBlanc, C., Robichaud, L., Turple, J., Girard, C., 2014. Recovery Potential Assessment for the American Eel (*Anguilla rostrata*) for eastern Canada: mitigation options. DFO Can. Sci. Advis. Sec. Res. Doc. 2013/133.

Christie, W.J., 1973. A review of the changes in the fish species composition of Lake Ontario. Great Lakes Fish. Comm. Tech. Rept. 23.

Chu, C., Minns, C.K., Mandrak, N.E., 2003. Comparative regional assessment of factors impacting freshwater fish resources in Canada. Can. J. Fish. Aquat. Sci. 60, 624-634.

COSEWIC, 2005. COSEWIC assessment and update status report on the spotted gar *Lepisosteus oculatus* in Canada. Committee on the Status of Endangered Wildlife in Canada. Ottawa.

COSEWIC, 2006a. COSEWIC assessment and status report on the American eel Anguilla rostrata in Canada. Committee on the Status of Endangered Wildlife in Canada. Ottawa.

COSEWIC, 2006b. COSEWIC assessment and update status report on the deepwater sculpin *Myoxocephalus thompsonii* (Western and Great Lakes-Western St. Lawrence populations) in Canada. Committee on the Status of Endangered Wildlife in Canada. Ottawa.

COSEWIC, 2006c. COSEWIC assessment and update status report on the lake sturgeon *Acipenser fulvescens* in Canada. Committee on the Status of Endangered Wildlife in Canada. Ottawa.

COSEWIC, 2006d. COSEWIC assessment and update status report on the river redhorse *Moxostoma carinatum* in Canada. Committee on the Status of Endangered Wildlife in Canada. Ottawa.

COSEWIC, 2009, COSEWIC assessment and status report on the Eastern Sand Darter Ammocrypta pellucida, Ontario populations and Quebec populations, in Canada. Committee on the Status of Endangered Wildlife in Canada. Ottawa.

COSEWIC, 2010. COSEWIC assessment and status report on the Atlantic Salmon *Salmo salar* (Nunavik population, Labrador population, Northeast Newfoundland population, South Newfoundland population, Southwest Newfoundland population, Northwest Newfoundland population, Quebec Eastern North Shore population, Quebec Western North Shore population, Anticosti Island population, Inner St. Lawrence population, Lake Ontario population, Gaspé-Southern Gulf of St. Lawrence population, Eastern Cape Breton population, Nova Scotia Southern Upland population, Inner Bay of

Fundy population, Outer Bay of Fundy population) in Canada. Committee on the Status of Endangered Wildlife in Canada. Ottawa.

COSEWIC, 2011. COSEWIC assessment and status report on the Silver Lamprey, Great Lakes-Upper St. Lawrence populations and Saskatchewan-Nelson Rivers populations, *Ichthyomyzon unicuspis* in Canada. Committee on the Status of Endangered Wildlife in Canada. Ottawa.

COSEWIC, 2013a. COSEWIC assessment and status report on the Bridle Shiner *Notropis bifrenatus* in Canada. Committee on the Status of Endangered Wildlife in Canada. Ottawa.

COSEWIC, 2013b. COSEWIC assessment and status report on the Pugnose Shiner *Notropis anogenus* in Canada. Committee on the Status of Endangered Wildlife in Canada. Ottawa.

COSEWIC, 2015. COSEWIC assessment and update status report on the spotted gar *Lepisosteus oculatus* in Canada. Committee on the Status of Endangered Wildlife in Canada. Ottawa.

Crossman, E.J., Holm, E., 2005. COSEWIC status report on the grass pickerel *Esox americanus vermiculatus* in Canada. Committee on the Status of Endangered Wildlife in Canada. Ottawa.

Crossman, E.J., Van Meter, H.D., 1979. Annotated list of the fishes of the Lake Ontario watershed. Great Lakes Fish. Comm. Tech. Rept. 23.

Dextrase, A.J., 2013. Modelling occupancy and abundance of Eastern Sand Darter (*Ammocrypta pellucida*) while accounting for imperfect detection. Doctoral Thesis, Trent University, Peterborough, ON.

Dextrase, A.J., Mandrak N.E., 2006. Impacts of invasive alien species on freshwater fauna at risk in Canada. Biological Invasions 8, 13–24.

Dextrase, A.J., Mandrak, N.E., Barnucz, J., Bouvier, L.D., Gaspardy, R., Reid, S.M., 2014. Sampling effort required to detect fishes at risk in Ontario. Can. Manuscr. Rep. Fish. Aquat. Sci. 3024.

DFO, 2008, Recovery Potential Assessment of Great Lakes and St. Lawrence River watersheds (Designatable Unit 8) Lake Sturgeon (*Acipenser fulvescens*) population. DFO Can. Sci. Advis. Sec. Sci. Advis. Rep. 2008/042.

DFO, 2009, Zonal Science Peer Review of the American eel (*Anguilla rostrata*) prior to assessment by COSEWIC; 11-12 October, 2005. DFO Can. Sci. Advis. Sec. Proceed. Ser. 2009/028.

DFO, 2012a, Recovery strategy for the Eastern Sand Darter (*Ammocrypta pellucida*) in Canada: Ontario populations. Species at Risk Act Recovery Strategy Series, Fisheries and Oceans Canada, Ottawa.

DFO, 2012b, Recovery strategy for the Pugnose Shiner (*Notropis anogenus*) in Canada (Proposed). Species at Risk Act Recovery Strategy Series. Fisheries and Oceans Canada, Ottawa ON.

DFO, 2013. Recovery Strategy for the Channel Darter (*Percina copelandi*) in Canada. Species at Risk Act Recovery Strategy Series. Fisheries and Oceans Canada, Ottawa.

Gertzen, E.L., Doka, S.E., Minns, C.K., Moore, J.E., Bakelaar, C., 2012. Effects of water levels and water level regulation on the supply of suitable spawning habitat for eight fish guilds in bays of prince Edward County, Lake Ontario. Aquat. Ecosyst. Health Mgmt 15(4), 397-409.

Gray, S.M., Bieber, F.M.E., McDonnell, L.H., Chapman, L.J., Mandrak. N.E, 2014. Experimental evidence for species-specific response to turbidity in imperilled fishes. Aquatic Conservation: Marine and Freshwater Ecosystems 24, 546-560.

Holm, E., Mandrak, N.E., Burridge, M., 2009. The field guide to freshwater fishes of Ontario. Royal Ontario Museum, Toronto, ON.

Hoyle, J.A. 2005. Status of lake whitefish (*Coregonus clupeaformis*) in Lake Ontario and the response to the disappearance of *Diporeia* spp. L.C. Mohr, T.F. Nalepa (Eds.),. Proceedings of a workshop on the dynamics of lake whitefish (*Coregonus clupeaformis*) and the amphipod *Diporeia* spp. in the Great Lakes. Great Lakes Fish. Comm. Tech. Rep. 66, 47-66.

Hoyle, J.A., 2015. Fish species composition, distribution, and abundance trends in the open-coastal waters of northeastern Lake Ontario, 1992-2012. Aquat. Ecosyst. Health Mgmt. 18(1), 89-100.

Hoyle, J.A., Lake, C., 2011. First occurrence of Chain Pickerel (*Esox niger*) in Ontario: possible range expansion from New York waters of eastern Lake Ontario. Can. Field Nat.125, 16-21.

Hoyle, J.A., Bowlby, J.N., Brousseau, C.M., Johnson, T.B., Morrison, B.L., Randall, R.G., 2012. Fish community structure in the Bay of Quinte, Lake Ontario: the influence of nutrient levels and invasive species. Aquat. Ecosyst. Health Mgmt. 15(4), 370-384.

Lantry, B.F., O'Gorman, R., Walsh, M.G., Casselman, J.M., Hoyle, J.A., Keir, M.J., Lantry, J.R., 2007. Reappearance of Deepwater Sculpin in Lake Ontario: resurgence or last gasp of a doomed population? Journal of Great Lakes Research 33(sp1), 34-45.

Leisti, K.E., Doka, S.E., Minns, C.K., 2012. Submerged aquatic vegetation in the Bay of Quinte: Response to decreased phosphorus loading and Zebra Mussel invasion. Aquat. Ecosyst. Health Mgmt 15(4), 442–452.

Lougheed V.L., Crosbie B., Chow-Fraser P.,1998. Predictions on the effect of common carp (*Cyprinus carpio*) exclusion on water quality, zooplankton, and submergent macrophytes in a Great Lakes wetland. Canadian Journal of Fisheries and Aquatic Sciences, 55, 1189–1197.

Madenjian, C.R., Fahnenstiel, G.L., Johengen, T.H., Nalepa, T.F., Vanderploeg, H.A., Fleischer, G.W., Schneeberger, P.J., Benjamin, D.M., Smith, E.B., Bence, J.R., Rutherford, E.S., Lavis, D.S., Robertson, D.M., Jude, D.J., Ebener. M.P., 2002. Dynamics of the Lake Michigan food web, 1970-2000. Canadian Journal of Fisheries and Aquatic Sciences 59,736-753.

Madenjian, C.R., Hondorp, D.W., Desorcie, T.J., Holuszko, J.D., 2005. Sculpin community dynamics in Lake Michigan. Journal of Great Lakes Research 31, 267-276.

Mandrak, N.E., 1989. Potential invasion of the Great Lakes by fish species associated with climatic warming. J. Great Lakes Res. 15, 306-316.

Mandrak, N.E., 1990. The zoogeography of Ontario freshwater fishes. MSc. Thesis. Department of Zoology, University of Toronto. Toronto, ON.

Mandrak, N.E., Crossman, E.J., 1992. Postglacial dispersal of freshwater fishes into Ontario. Can. J. Zool. 70, 2247-2259.

Minns, C.K., Hurley, D.A., Nicholls, K.H., 1986. Project Quinte: point-source phosphorus control and ecosystem response in the Bay of Quinte, Lake Ontario. Can. Spec. Publ. Fish. Aquat. Sci. 86.

Mortsch, L.D., Quinn, F.H., 1996. Climate change scenarios for Great Lakes Basin ecosystem studies. Limn. Oceano. 41, 903-911.

Nicholls, K.H., 2012. Phosphorus and chlorophyll in the Bay of Quinte: a time-series/intervention analysis of 1972-2008 data. Aquat. Ecosyst. Health Mgmt 15(4), 421-429.

Phelps, A., Francis, A., 2002. Update COSEWIC status report on the channel darter *Percina copelandi* in Canada. Committee on the Status of Endangered Wildlife in Canada. Ottawa.

Reid, S., Dextrase, A., 2014. First record of *Ammocrypta pellucida* (Agassiz, 1863) (Actinopterygii: Perciformes) from the Lake Ontario drainage basin. Check List 10(5), 1201–1203.

Reid, S.M., Wilson, C.C., Carl, L.M., Zorn, T., 2008. Species traits influence the genetic consequences of river fragmentation on two co-occurring redhorse (*Moxostoma*) sucker species. Canadian Journal of Fisheries and Aquatic Sciences 65, 1892-1904.

Ricciardi, A., 2006. Patterns of invasion in the Laurentian Great Lakes in relation to changes in vector activity. Divers. Distrib. 12, 425-433.

Scott, W.B., Crossman, E.J., 1973. Freshwater fishes of Canada. Fisheries Research Board of Canada Bulletin 184. Ottawa, ON.

Staton, S.K., Mandrak N.E., 2006. Focusing conservation efforts for freshwater biodiversity. pp. 197-204 In: G. Nelson et al., (eds). Protected areas and species and ecosystems at risk: research and planning challenges. Proceedings of the Parks Research Forum of Ontario Annual Meeting 2005. Parks Research Forum of Ontario, University of Waterloo, Waterloo, ON.

Welsh, A., Hill, T., Quinlan, H., Robinson, C., May, B., 2008. Genetic assessment of Lake Sturgeon population structure in the Laurentian Great Lakes. N. Amer. J. Fish. Manag. 28, 572-591.

The Ecology of the Bay of Quinte: Health, Management and Global Implications

C.K. Minns[1], M. Munawar, J. Hoyle[3], O.E. Johannsson, M. Koops, E.S. Millard[1] and B.J. Morrison

[1]*Department of Ecology and Evolutionary Biology, University of Toronto, Toronto, ON M5S 3B2*
[2]*Fisheries and Oceans Canada*
867 Lakeshore Road, Burlington, ON L7S 1A1
[3]*Ontario Ministry of Natural Resources, 41 Hatchery Lane, Picton, ON K0K 2T0*

Keywords: ecosystem management, development, lessons, recommendations

Introduction

The long-term partnership between Project Quinte (PQ), a group of agencies and scientists studying the ecology and responses of the Bay of Quinte to management actions, and the group implementing the Bay of Quinte Remedial Action Plan (BQRAP) has provided lessons on the management of ecosystems. A review of the evolution of ecosystem management in the Bay from the pre-PQ Past (Pre-1972) through to the BQRAP stages to the Present (2012-2014) provides the context for assessing future prospects. Reflection on that evolution allowed identification of valuable lessons learned since the early 1970s. Finally, these considerations form the basis for a set of recommendations for other groups seeking to tackle ecosystem management challenges elsewhere in the future.

This paper was developed through a combination of (i) reviews of the papers comprising this monograph as well as those published earlier; (ii) the collective experience and deliberations of the authors; and (iii) results of a brief informal survey conducted among past and present contributors to Project Quinte.

Evolution of ecosystem management in the Bay of Quinte

The evolution of PQ/BQRAP activities in the Bay may be grouped into three periods – Early (1972-1985), Middle (1986-2000), and Late (2001-2011), bracketed by the Past and Present (Table 1). In the Past (pre-1972)

Ecology of the Bay of Quinte: Health, Management and Global Implications
Editors: C.K. Minns, M. Munawar and M.A. Koops

Aquatic Ecosystem Health & Management Society, Canada

Table 1. A summary overview of key characteristics of the periods in the evolution of ecosystem-based management in the Bay of Quinte with the early, middle and later periods of Project Quinte/BQRAP bracketed by the past and present.

Period	Timing	Institutions	Management issues	Science issues	Science paradigms	Societal context
Past	pre-1972	OWRC/OMNR	Water quality, Fisheries, Macrophytes, Declining Walleye	Top down/bottom up, Fishing vs P, Energy flow	Jurisdictionally divided assessment and management	Rise of environmentalism, IJC
Early	1972-1985	Project Quinte (multi-agency)	Eutrophication, Walleye	Effects of P control, Productivity	Nutrient budgets, Energy flow,	GLWQA, Clean-up, GLFC
Middle	1986-2000	BQRAP Phases I&II (multi-agency)	Assessment of ecological problems Development of possible solutions	Eutrophication Fisheries response	Ecosystem approach	Restoration, Tightening budgets
Late	2001-2011	BQRAP Phase III (multi-agency)	Public engagement, Cost effectiveness, Post-ZM Walleye decline, First Nations fisheries	Toxic chemicals, Invasive species, Habitat, Microbial foodweb	Ecosystem modelling, Uncertainty	Economic limitations
Present	2012-2014	EC/Local Authorities	BUI Status, Imminent Delisting, Loss of institutional memory, Economics	Invasive species, Cyanobacteria blooms, Species at risk	Adaptive management	Declining budgets and governance, Jurisdictional fatigue

limited environmental management was jurisdictionally divided in an era of algal blooms, decreased macrophyte cover, potential anoxia and lower Walleye harvests. The Ontario Water Resources Commission and Ontario Ministry of Natural Resources were the primary management institutions. The end of the Past saw the rise of environmentalism with the International Joint Commission (IJC) as a major platform for enlarged assessments and expression of rising public concern.

In the Early (1972-1985) period, Project Quinte's empirical studies of nutrient dynamics and energy flow in relation to productivity began as a response to the first Great Lakes Water Quality Agreement and its focus on reducing Point-source phosphorus loadings as justified by Johnson and Owen (1971) after a success pilot in Gravenhurst Bay, Lake Huron. The Middle (1986-2000) period was triggered by the IJC's identification of Great Lakes Areas of Concern (AOCs) and an urgent call for Remedial Action Plans (RAP) that included the Bay of Quinte. The Great Lakes Fishery Commission and its partners began to expand its scope of interest in this period cross-connecting with the IJC and its partner water quality agencies. The BQRAP process focussed in Stage 1 on the status of the ecosystem and problem definition in relation to the impaired uses and then in Stage 2 on analyses of potential remedies and how to implement them. After Stage 2 a remedial action plan was set in motion. Project Quinte members were drawn into the RAP as the core of its technical advisory committee. The renewed GLWQA 1978 also called for a holistic ecosystem approach. In the Bay this expanded the scope of issues to include toxic and bioaccumulating chemicals, and habitat for fish and wildlife. As the assessment of remedial options matured, new invasive species, particularly the Driessenid Mussels, emerged as a major disruptive force. In the Late (2001-2011) period as the RAP focus shifted to implementation of preferred options, the role of the technical advisory group/Project Quinte evolved to help more with ecological assessments, enhanced monitoring and ecosystem modelling (Koops and Blukacz-Richards this volume). The science became increasingly focused on modelling the growing monitoring time-series and gauging uncertainties in complex ecosystem responses to an array of changes both natural and man-induced. Given the large expenditures already undertaken to reduce Point-source nutrient inputs, management agencies and local governments were increasingly concerned with the potential costs of additional measures identified for RAP implementation. Local conservation authorities assumed responsibility for RAP implementation with assistance from federal and provincial agencies as the BQRAP Restoration Council. Further analysis by the technical committee (Project Quinte) indicated greater control of nutrient inputs and reduced internal flux would be needed to reach and secure a sustainable healthy state in the upper and middle Bay. The Restoration Council has been wrestling with this issue ever since as STP needs increase. In this period the role of the IJC was diminished as Environment Canada (EC) and the US Environmental Protection Agency managed activities and public involvement through the Bi-National Committee.

Post-2000, declining agency budgets and the loss of institutional memory through changing representation, have led to a decreased influence of scientific advice in BQRAP deliberations. Recently the Bay became a candidate for Delisting under the RAP process whereby targets set to overcome designated beneficial use impairments (BUIs) are deemed to have been met. Under delisting the Bay may be designated an "Area in Recovery" though it is unclear if this will ensure sustained interest and funding. Many Project Quinte members do not favour delisting as many of the eutrophication related BUIs have not been adequately addressed and problems related to eutrophication and invasive species are persistent. Phosphorus loading reductions stands out as the most significant remedial action although the Driessenid invasion undoubtedly both assisted (increased water clarity) and hindered (increased algal blooms) its effectiveness. The main issues of concern in the Bay at present are eutrophication, unmeasured diffuse watershed inputs, algal blooms and invasive species. Environment Canada-funded local government engagement is the main source of governance now with EC administratively committed to delisting a set quota of AOCs across the Great Lakes. Attention is shifting from the AOCs to the near-shore zone as a whole with less-focused activities and commitments.

The future period

Much about the future in the Bay of Quinte is uncertain, particularly after the delisting of the AOC which is expected to follow acceptance of a Stage 3 RAP report detailing the achievements of the implementation plan in the next year or so. The trajectories for some key factors such as population growth and climate change are fairly clear though the exact timing is uncertain. Some of the uncertainties relate to key slow-changing features of the ecosystem, such as accumulated internal phosphorus reflux, macrophyte cover in the upper Bay and oxygen depletion in the lower Bay hypolimnion.

While much progress has been achieved with Point-source phosphorus control (Arhonditsis et al, this volume), full realization of the benefits awaits the dissipation of long-accumulated internal phosphorus reflux and, probably, long accumulated terrestrial phosphorus loadings both coupled with continuing commitments to limit external loadings (e.g. high concentration-low flow stormwater inputs) and rethink agricultural land use practices (i.e. periodic tilling). Ontario's total population is projected to grow from 13.63 to 17.37 million in the 2013-2036 period (27.4%; OMF 2013) mainly in the larger urban areas and although little growth is projected for the three census districts adjoining the Bay of Quinte (Hastings, Lennox and Addington, and Prince Edward): 203,710 to 205,120 (0.7%) the Bay and surround lands are popular with those living in the nearby urban centres particularly the Greater Toronto Area, Ottawa, Hull and Gatineau. Hence local population growth should not exert upward pressure on loadings and existing facilities should have sufficient capacity (for higher flows and lower final effluent concentrations); however, increased use by visitors to the area will require services to meet their needs requiring well planned local storm water control and rural sewage management. That said, increased costs for enhanced loading controls for storm-water and un-sewered sources in developed areas may prove burdensome to those local communities.

Despite the phosphorus loading reductions, blooms of toxin-producing cyanobacteria have become a severe problem since the Driessenid invasion. These blooms were identified as the least addressed issue in the survey. The algal blooms present both a human health risk via untreated drinking water and bathing and structural and toxic stresses within the ecosystem with dense blooms cloaking macrophyte beds and restricting light penetration in offshore areas. The mechanisms are poorly understood but (1) selective rejection of bloom-forming Microcystis by filter-feeding Driessenids, (2) shifts in nutrient cycling due to Driessenid filtration, (3) excess internal P loading, (4) high inorganic nitrogen levels, and (5) imbalance in N:P ratios are contributory.

The outlook for invasive species remains fraught despite enhanced ballast water legislation and basin-wide efforts to prevent further invasions as live-trade movements of fish, molluscs, and plants are increasingly a problem (Bailey et al, this volume). The Round Goby has become abundant in recent years and the ramifications of their rise for the ecosystem are unrealized to date. Other invasive species such as several Thalassiosiraceae diatom species; Trapa natans, European

Water Chestnut; two Potamothrix species of tubificid worms; Gammaris tigrinus, an amphipod; Hemimysis anomala, a mysid; Myriophyllum spicatum, Eurasian milfoil; and bacterial kidney disease (BKD) which affects fish, are recent arrivals in the basin and are an invasion threat for the Bay (http://www.glerl.noaa.gov/res/Programs/glansis/glansis.html).

Climate change was seen as a major addition to the stressors affecting the Bay and its biodiversity later in the 21st C. Currently a dozen fish species are at risk in the Bay area, mainly threatened by wetland loss and stream flow alteration (Mandrak, this volume). Much climate warming is projected for the Bay of Quinte area (Observations and projections for the Quinte West municipality from Minns et al., 2014). The observed 30-year mean climate norms showed little change during the 20th Century. In the 2041-2070 and 2071-2100 periods, spring is projected to advance by 9-23 and 15-65 days and fall to come later by 9-17 and 14-37 days making winter shorter relative to the 20th C norms. Given prevailing rates of increase greenhouse gas emissions, air temperatures are projected to rise by 2.6-3.3 and 4.0-5.8°C in the 2041-2070 and 2071-2100 periods respectively. Projected changes in precipitation straddle past observed norms though with a bias toward increases likely with more extremes. Increased evaporation levels due to higher temperatures are likely to offset any possible increases in precipitation, and hence, lower river runoff and lower water levels should be expected. Increased precipitation is expected in the form of less frequent but more intense storms, hence local run-off will be more episodic and more acutely severe. Reduced dilution effects from tributary flows may exacerbate any shortcomings in nutrient load management with respect to sustaining improved water quality in the Bay.

Climate warming can be expected to influence the species composition of the fisheries as warmwater species are favoured over coolwater ones. Hoyle (in this Volume) showed centrarchids increased after the Driessenid invasion as facilitated by the recovery of extensive macrophyte beds, particularly in the upper Bay, Walleye stabilized at levels well below past peaks. Currently-valued species like Walleye and Northern Pike prefer coolwater and can be expected to decline in the upper Bay as increasingly popular sport fish species like Largemouth Bass and commercially important sunfish species further increase. Walleye which prefer areas with less light and intermediate temperatures may adapt by shifting deeper where possible and into the cooler lower Bay and eastern Lake Ontario although spawning and early life history requirements may be limiting. Decreased ice cover will likely reduce the size of the winter recreational fishery for Walleye and, in summer, Walleye may be less accessible to anglers. Climate change will also have wider-reaching implications for the biodiversity of the ecosystem as shifts in realizable niche cascade through the various assemblages. The successful colonization of the Bay by both 'native' species, like the Chain Pickerel (Hoyle and Lake, 20xx), and invasive species may influence this cascade.

The prospects outlined for the Bay of Quinte later in this century, particularly, as the climate warms demonstrate that the need for active adaptive ecosystem management will continue. Over the 40+ years, despite sustained monitoring and assessment, unexpected changes such as the massive cold winter White Perch die-off in 1977-78 as P loadings dropped (Hoyle et al., this volume), the Driessenid invasion in 1994 (Dermott et al, this volume), and the emergence of cyanobacteria blooms in the last decade (Nicholls and Carney, this volume), have precipitated large changes in the ecosystem, in our scientific paradigms, and management agendas.

If the expected delisting of the AOC after the Stage 3 RAP report reduces attention to and investments in maintenance of a healthy sustainable ecosystem, the RAP process begun in the mid-1980s will have been only partially successful because the lessons learned were not used to advance our management of the system.

Global lessons and recommended actions

Many environmental problems like those encountered in the Bay of Quinte are present across the Great Lakes and in water bodies throughout the world. After all the time and effort that have been applied to the Bay of Quinte since 1972 (40+ years) it is possible to identify several useful lessons and recommendations that can be shared with those seeking to restore or maintain healthy aquatic ecosystems. In the Great Lakes Basin, the "Ecosystem Approach" concept that has been in place since the mid-1980s, increasingly guided environmental scientists since the mid-1980s, while management agencies have remained jurisdictionally divided in part along an eutrophication control versus fisheries production axis. The latest GLWQA 2012 moved the partner agencies closer to establishing an ecosystem management framework though without the necessary dedicated resources.

In parallel, the custodians of the many large marine ecosystem fisheries around the world have pursued the goal of "Ecosystem-based Fisheries Management". More recently, the phrase "Ecosystem Goods and Services" has gained much currency. All these guiding concepts are tied to the philosophy of "Adaptive Learning and Management" which has been widely advocated (Minns,; Randall et al, this volume). Adaptive management (AM) depends on persistent cycles of problem identification, assessment of solution options, implementation of management actions, and sustained monitoring of outcomes. Successful implementation of AM requires persistent cycling of the entire process, and for the Bay of Quinte the 40+ years of AM previously described is a remarkable achievement and provided a great deal of information and lessons learned. Lessons gleaned from the studies in the Bay of Quinte can be summarized under four headings: target, people, evidence, and solution; each with recommended actions:

Target

Lesson - Take a Holistic View: With the advent of the ecosystem approach much more attention was paid to the connections and interactions among all the elements of an ecosystem and among the intervention actions (Koops and Blukacz-Richards, this volume). More recently there has been some regression toward narrow jurisdictionally divided management centered on phosphorus control as economic constraints and reduced government resources limit institutional effectiveness. Sustainable ecosystem management requires a whole ecosystem approach balancing multiple demands and responses.

Recommended Action: Focus on an ecosystem rather than an issue as it will be easier to draw local support. The Bay of Quinte was the focus not eutrophication. The ecosystem will change over time but the need for sustainable management will continue. Issues can come and go as specific problems are solved and as society's priorities shift.

People

Lesson - Work outside-the-box: Events in the Bay of Quinte provide a strong local example of the potential power of groups that reach beyond the confines of single institutions. Originally, Project Quinte was a pre-existing informal grouping drawn from several agencies and universities with a shared interest in learning from the unique opportunity of coordinated P control in a single basin. The RAP process was successful because it took the responsibility for finding solutions beyond the purview of single agencies into a multi-agency context with the added feature of increased public involvement. This worked at the local RAP level. Once the process began shifting from assessment, diagnosis, and identification of options to implementation economic and administrative considerations took centre-stage with less input and reduced levels of engagement of scientists. Evidence shows that this process can also work at larger scales of governance. Within the Great Lakes basin as a whole, management success relied on insight and direction from successive Great Lakes Water Quality Agreements and the co-operation and commitment of two countries and nine state and provincial governments represented by numerous agencies. Success depended on the commitment of all players to common goals, and common goals depend on the political will of the governing agencies. The activities of the European Union with respect to the natural environment, e.g. the Water, Habitat and Environmental Liability Directives (Kallis and Butler 2001; Rundcrantz and Skärbäck, 2003), have demonstrated on an even larger scale how such collaborations under common political direction can allow the quality of decision-making to rise to the challenges.

Lesson - Collaborate and Communicate: Effective sustainable ecosystem management continually requires the energy, experience and expertise of many different participants. Interactions among, within and between groups of scientists, managers, politicians, and the public at large played a big part in the progress made in the Bay. From the beginning PQ was nourished and sustained by two activities: an annual report and an annual meeting. The first facilitated the sharing of data and analyses while the second encouraged ongoing interpretation of results and refinement of questions and monitoring activities. The BQRAP committees met regularly and maintained an active advisory group drawn from the public at large. BQRAP maintained a visibility in the area and today continues with monthly newsletters and annual reports. In recent years both federal and provincial agencies have committed much effort to ensuring the many data sets are properly archived for use by current and future users.

Lesson - Recruit and Renew: As sustainable, adaptive and successful ecosystem management is a very long-term process, it's important to renew and adjust group membership slowly but continuously, maintaining the engagement of all participant groups (scientists, managers, and public) as the years pass. Much institutional memory was lost as long-serving Project Quinte and BQRAP participants retired or moved on. Rapidly altered committee memberships often create new agendas and priorities without sufficient attention to the lessons learned and the historical course of the enterprise.

Recommended Action: Establish an informal technical working group drawing from a range of agencies and institutions to develop the assessment and diagnosis of the ecosystem problem of concern. Individuals often have pre-existing involvements in the ecosystem under the auspices of their home institutions making it easier to justify involvement in the technical group. Once that group has established a basis for collaboration and a proven record it can evolve into an advisory group. Then renew the group gradually and add new specialists as needed.

Evidence

Lesson - Use science-informed decision-making: Decisions on the management of natural ecosystems cannot be based solely on scientific evidence about ecological processes as factors in the social and economic domains must to be considered. Social factors may include community/cultural acceptance of the proposed solutions and health and safety concerns. Economic factors include how solutions might affect segments of the economy and the ability to pay for the scientifically preferred solutions. Nonetheless, the ecological requirement must be at the core of any ecosystem decision-process if the solutions are to be sustainable in the long-term and provide ecological goods and services essential to society.

Lesson - Sustain monitoring and research: Data and understanding are the essential ingredients for sound ecosystem management just as in medicine, agriculture, financial markets, and government. Beyond the confirmatory assessments following AOC rehabilitation, monitoring programs will be necessary to detect and diagnose unexpected changes and expected responses to interventions assessed. This need not mean large expensive measurement programs but rather continually assessed and adaptive programs that carefully maintain core elements for continuity and modular elements that target new or changing conditions.

Lesson - Paradigms shift: Ecosystem management is a long-term changing process framed by a continuous evolution in ecological, social and economic thinking. The adaptive management approach is well-suited to this continual flux.

Lesson - Expect surprises: Ecosystem management is not a precise laboratory science where all factors extraneous to proposed management

actions can be controlled. Surprises happen as has been seen several times in the Bay of Quinte, altering ecosystem structures and function, and thus our perception of the state of eutrophication. Some facets appeared to have improved over expectations based on phosphorus levels (water transparency) and Driessenid (macrophyte cover) while other facets had worsened (algal blooms). Management now has to occur under the new regime. Also in the future climate change can be expected to trigger further large-scale changes in the ecosystem as different species dominance patterns are induced by changes in thermal and hydrological dynamics.
Recommended Action: Establish a sustainable monitoring and research program directed to providing intelligence on the state of the ecosystem and capable of detecting change. Ensure that the data collected are reviewed, shared and archived. Distribute regular status reports to agencies, institutions, and the public.

Solutions

Lesson - Only human activities and expectations can be managed: There are many features of ecosystems that operate ecologically and evolutionarily beyond the control of humans. Adaptive management provides a framework for probing ecosystem responses to unexpected changes and interventions with the goal of securing sustainable goods and services.
Recommended Action: Identify all the issues that need to be addressed to move the target ecosystem toward ecological sustainability. Evaluate alternative sets of ecologically coherent solutions to provide the basis for evidence-based decision-making. Make decisions and start the cycle again.

Summary

Ecosystem management activities in the Bay of Quinte developed from the long-term partnership between the Project Quinte group of agencies and scientists studying ecosystem ecology and responses to management, and the agencies implementing the Bay of Quinte Remedial Action Plan. The evolution of ecosystem management in the Bay from the pre-Project Quinte Past (pre-1972), through the Bay of Quinte Remedial Action Plan stages to the Present (2012-2014), provides the context for assessing Future prospects for the Bay and for identifying globally useful lessons and recommendations. A survey of scientists involved in Quinte indicated: 1) Eutrophication remains the dominant concern although invasive species and climate change compound the restoration process, and 2) Broad-based collaboration founded on a sustained program of monitoring and research was the key to successful ecosystem management. The lessons learned were grouped: Target – Take a holistic view; People – Work outside-the-box, Collaborate and communicate, and Recruit and renew; Evidence – Use science-informed decision-making, Sustain monitoring and research, Paradigms shift, and Expect surprises; and Solutions – Only human activities and expectations can be managed. The recommendations flowed from the lessons.

Acknowledgements

The authors wish to thank the many participants in Project Quinte and the Bay of Quinte Remedial Action Plan who have contributed papers, experience and ideas to this synthesis. Thanks to Mark Fitzpatrick for his contributions as the papers came in for this volume. The feedback from the two anonymous referees on the penultimate version helped clarify key points.

References

Arhonditsis, G.B., Kim, D.-K., Shimoda, Y., Zhang, W., Watson, S., Mugalingam, S., Dittrich, M., Geater, K., McClure, C., Keene, B., Morley, A., Richards, A., Long, T., Rao, Y.R., Kalinauskas, R., *This volume*. Integration of best management practices in the Bay of Quinte watershed with Phosphorus dynamics in the receiving water body: What does the model predict?

Bailey, S., Mandrak, N.E., Hoyle, J., Johannsson, O., *This volume*. Aquatic Invasive Species in the Bay of Quinte.

Dermott, R., Martchenko, D., Johnson, M., *This volume*. Changes in the benthic community of the Bay of Quinte, Lake Ontario over a 40 year period.

Hoyle, J.A., *This Volume*. Bay of Quinte Commercial and Recreational Fisheries: Harvest Trends and Current Status.

Hoyle, J.A., Bowlby, J.N., Brousseau, C.M., Johnson, T.B., Morrison, B.J., and Randall, R.G., *This volume*. Fish community structure in the Bay of Quinte, Lake Ontario: The influence of nutrient levels and invasive species.

Johnson, M.G. and Owen, G.E., 1971. Nutrients and nutrient budgets in the Bay of Quinte, Lake Ontario. J. Water Pollution Control Fed. 43(5), 836-853.

Kallis, G and Butler, D., 2001. The EU water framework directive: measures and implications. Water Policy 3, 125-143.

Koops, M., Blukacz-Richards, A. This volume. An integrated approach to identifying ecosystem recovery targets: Application to the Bay of Quinte.

Mandrak, N.E., *This volume*. Fish Species at Risk in the Bay of Quinte Watershed.

Minns, C.K., 2014. Canadian fish habitat management: symptoms and remedies. Amer. Fish. Soc. Symp. 78.

Minns, C.K., Shuter, B.J., and Fung, S., 2014. Further Regional Projections of Climate Change Effects on Ice Cover and Open-Water Duration for Ontario Lakes Using Updated Ice-Date Models. Ontario Ministry of Natural Resources, Climate Change Research Report.

Nicholls, K.H., Carney, E.C., this volume. The phytoplankton of the Bay of Quinte, 1972-2008: point-source phosphorus loading control, Dreissenid Mussel establishment, and a proposed community reference.

Ontario Ministry of Finance (OMF), 2013. Ontario Population Projections Update, 2012–2036. ISBN 978–1–4606–2022–9 (PDF).

Randall, R.G., Koops, M.A., and Minns, C.K., *This volume*. A comparison of approaches for integrated management in coastal marine areas of Canada with the historical approach used in the Great Lakes (Bay of Quinte).

Rundcrantz, K and Skärbäck, E., 2003. Environmental compensation in planning: a review of five different countries with major emphasis on the German system. European Environment 13, 204–226.

Shuter, B. J., Minns, C. K., Fung, S. R., 2013. Empirical models for forecasting changes in the phenology of ice cover for Canadian lakes. Can. J. Fish. Aquat. Sci. 70(7), 982-991.

Figure 1. A collage of images from research on the Bay of Quinte, Ontario. From top to bottom, left to right (including photo credits): the mouth of the Bay of Quinte flows in Lake Ontario (R. Rozon, DFO); the Leslie J research vessel (DFO); a variety of phytoplankton species from the Bay; M. Frank (M. Burley) processing a water sample from the Bay (DFO); an algal bloom in 2005 (DFO); collecting fish samples (J. Hoyle, OMNR); R. Bonnell and J. Gerlofsma collecting data (DFO); R. Rozon persevering samples for later processing (DFO); H. Niblock deploys the Schindler to collect samples (DFO); a Pumpkinseed sunfish (*Lepomis gibbosus*) from the Bay of Quinte (J. Hoyle, OMNR); boating on the Bay (DFO).

Acknowledgements

Project Quinte (PQ) has been the umbrella term for aquatic ecosystem studies in the Bay of Quinte (BoQ), a Great Lakes Area of Concern (AOC), initiated by scientists in the 1970s with the decision to reduce phosphorus loadings. Its core activities have focused on monitoring and assessment of many features of the ecosystem. With the designation of BoQ as an Area of Concern, Project Quinte would go on to cover related work on all aspects of ecosystem health and management, e.g. drinking water quality and fisheries in support of the Remedial Action Plan process.

Agencies and departments: 1) Department of Fisheries and Oceans (DFO) in particular the Great Lakes Laboratory for Fisheries and Aquatic Sciences (GLLFAS) in Burlington at the Canada Centre for Inland Waters (CCIW); Environment and Climate Change Canada units at CCIW: National Laboratory for Environmental Testing (NLET) and National Water Research Institute (NWRI); Ontario Ministry of the Environment, Conservation and Parks (OMECP) units at Environmental Monitoring and Reporting Branch (EMRB) at Rexdale, ON and the regional office in Kingston, ON; Ontario Ministry of Natural Resources and Forestry (OMNRF) unit at Lake Ontario Management Unit (LOMU) at the Glenora Fisheries Station; and Quinte Conservation Authority (QCA) at Belleville, ON and Lower Trent Conservation, Trenton, ON (See www.bqrap.ca for on-going activities and a cumulative library of reports and publications).

Original contributors: DFO- Murray G. Johnson (Lead PQ instigator), Scott Millard, Ken Minns; OMECP- Ken Nicholls (Algae), Murray German (RAP Chair) Fred Stride (RAP Secretary), Glen Owen, Michael Michalski (PQ instigator); OMNRF- Jack Christie (PQ instigator), Donald Hurley, Jim Hoyle.

Primary contributors: DFO- John Cooley (Zooplankton), Mohi Munawar (phytoplankton, primary productivity, microbial loop), Ron Dermott(Benthos), Ora Johannsson(Benthos), John Leslie (Larval fish), Jim Moore(Modelling), Bud Timmins, May Frank (M. Burley), Robert Bonnell, Kelly Bowen and Ashley Bedford (Fieldwork), Deborah Myles, Heather Niblock, Mark Fitzpatrick, Jocelyn Gerlofsma, Julie Dahl, Warren Currie, Warren Norwood, J. Lorimer, Carl Wanghofer; Marten Koops; NWRI- Susan Watson (Algae, Microcystis); OMECP- Gord Hopkins, Lucy Heintsch, Elaine Carney, Glenn Robbins ; OMNRF- Tim

Johnson, John Casselman, Ron MacIntosh and Cecil Lobb (Skippers of the *Namaycush*); Coast Guard who provided the boats and personnel used to conduct field programs; Queens Univ.-Mike Bristow, Adele Crowder(Macrophytes); Bill and Dee Geiling (Zooplankton).

Data management and archiving: Ken Scott (OMNRF), May Frank (M. Burley), Melissa Risto (DFO-GLLFAS).

Special Thanks: The many hundreds of summer students and part-time workers over nearly 50 years who have assisted with running all aspects of PQ: monitoring cruises, analyzing samples, entering data, identifying biota, etc., etc.; the participants who make projects like this feasible and successful.

Funding agencies: Under successive bilateral Great Lakes Water Quality Agreements PQ was supported federally by the Great Lakes Action Plan (GLAP) and provincially by Canada Ontario Agreements (COA) principally with operating budgets and supporting staff and contracts. Further departmental budgets at DFO, ECC, OMECP, OMNRF and elsewhere provided for salaries and infrastructure. Partial funding for typesetting and publishing of the monograph was provided by the Great Lakes Laboratory for Fisheries and Aquatic Sciences, Fisheries and Oceans Canada, Burlington, ON.

Sponsors and endorsers: International Joint Commission (IJC), Great Lakes Fishery Commission (GLFC).

Appendix

GOALS AND OBJECTIVES OF PROJECT QUINTE

The **1980 Annual Report of Project Quinte** included a consensus statement of the project's goals and objectives. That statement was qualified by this accompanying footnote statement: "Project Quinte participants consider goals (and sub goals) to be the ideal and broadest definition of purpose for the project, while objectives specifically describe the main scientific and technical questions, which are amenable to scheduling, budgeting and defining of work required." This statement provides an historical benchmark against which the long term evolution of the project may be assessed.

Overall Goal of Project Quinte

"To understand the relationship among nutrient loading, ecological characteristics and processes, and societal benefits of the Bay of Quinte ecosystem."

Subgoal A

To determine the relationship among Point-source Phosphorus control, runoff, hydrodynamics, sediment nutrient dynamics, and biota, in regulating Phosphorus concentrations in the bay.

Objectives

1) Determine annual and seasonal nutrient budgets for the bay and parts thereof, to measure the response to reduced Point-source P inputs and to evaluate components of the P-control strategy.

2) Determine the relationship between external P loading and internal P regeneration, particularly from the sediments and assess the degree to which internal loading will retard or prevent full recovery.

3) Determine the relationship between P loading and hypolimnetic oxygen depletion, short-term in upper bay during brief thermal stratification and over the season in the lower bay.

4) Determine the role of macrophytes in reducing re-suspension and sequestering of nutrients, in sediments, especially in the upper bay.

Subgoal B

To examine the productivity of aquatic communities and of identified trophic levels, and the efficiencies of material and energy flow, in relation to trophic state.

Objectives

1) Determine the factors controlling dynamics of phytoplankton (and selected components e.g. bloom-forming and non-bloom-forming blue green algae, diatoms etc) and to examine the role of both internal and external phosphorus loadings relative to other factors (water, temperature and mixing, light, nitrogen, silica).

2) Assess how presence or absence of the macrophyte community and associated macro invertebrates affects the foodweb and allocation of production among the biota.

3) Examine rates of production of benthic macro invertebrates and efficiencies of use of imported energy in relation to quality and quantity of imported food, temperature, type of macro invertebrate community, especially in the context of P loading and trophic state.

4) Determine the productivity of the important zooplankton communities in the bay in the pre- and post- P control periods in relation to primary production, algal species composition, presence/absence of macrophytes.

5) Examine the efficiency of use of benthos and zooplankton by fish species and communities.

6) Determine production of fish in the bay, of planktivores, bottom feeders and carnivores, in relation to trophic state and qualitative and quantitative aspects of food resources.

7) Determine the materials transported in and out of the bay in fish biomass, e.g. White Perch and Alewife seasonal migrations.

Subgoal C

To examine the effect of trophic state, in a nutrient loading context, on the structure, composition and diversity of aquatic communities.

Objectives

1) Determine seasonal and spatial similarities and differences in species composition, biomass and foodweb efficiency of phytoplankton in relation to nutrient loading and concentrations and other basin characteristics under both pre and post-phosphorus control regimes.

2) Determine whether there are significant changes in the zooplankton communities (i.e. structure, composition etc) resulting from P control.

3) Determine the impact of p1anktivorous fish on the structure of zooplankton communities.

4) Examine structure of benthic macro invertebrate communities (i.e. species composition, diversity, variability) in relation to trophic state in both space and time (pre- and post- P control periods) and assess potential implications for fish communities and production of fish.

5) Compare the structure, diversity and production of fish among sites and in the pre- and post- P control periods, especially to examine the influence of habitat diversity macrophytes, substrate, turbidity on fish community composition.

Subgoal D

To predict and determine successional changes in biota and describe them in terms of material and energy flow.

Objectives

1) Test the hypothesis that macrophyte development in earlier stages of eutrophication may "buffer" the impact of rising P inputs before a swift decline in macrophytes allows a transition to a hypereutrophic state. (This "buffering" may be necessary for recovery of the bay).

2) Determine if there has been any important successional changes in plankton and benthos communities resulting from P control.

3) Investigate concepts and models to allow prediction of structure and production of plankton and benthos communities with changing trophic state, particularly re implications for fish communities.

4) Determine if succession can be shown to exist in a fish community that has been exposed to changing nutrient loads, particularly to examine the possibility that shifts in food base influence relative abundance of fish species. (Possible significance of competitive exclusion in White Perch and Yellow Perch will be examined).

5) To describe the sequence of changes in lake chemistry and biota that have occurred since 1900 as a result of eutrophication using the sediment record and historical records.

Subgoal E

To examine the significance of natural forces and man induced stresses affecting variability in ecosystem behaviour, especially when confounded with responses to nutrient management.

Objectives

1) Assess year-to-year variability in phytoplankton composition, biomass and primary production in relation to fluctuations in "natural" factors like weather, hydrodynamics.

2) Determine apparent changes between 1945 and 1980 in net phytoplankton at several water depths in the lower bay near Indian Point.

3) Examine year-to-year variability in composition and production of benthic macro invertebrates attributable to sediment type, temperature, food import and other "natural" factors.

4) Determine to what extent natural forces may impact zooplankton communities of the bay.

5) Examine the role of climatic factors, especially temperature and water level, in determining year-class strength of several fish species (e.g. are Yellow Perch and Walleye year-class strengths determined by spring water temperature; is White Perch year-class strength determined by over winter conditions?).

Subgoal F

To examine the natural limits of variation in the biotic community components, particularly as related to mechanisms compensating for natural stressing factors.

Objectives

1) Assess importance of compensatory mechanisms which contribute toward stability in populations of selected annual species (e.g. *Pontoporeia affinis, Chironomus)* especially interactions among growth, predation, survival, etc.

2) Examine effect of variation in year-class strength of one species of fish on year-class strength of other species (e.g. are Walleye and Yellow Perch year-class strength influenced by that of White Perch?).

Subgoal G

To examine social benefits of the Bay of Quinte ecosystem in the context of trophic state and management policies and programs.

Objectives

1) Develop a strategy for the management of P loading and concentration which allows for the effects of future population growth in the system.

2) Assess how expected variation in runoff, both secular and seasonal, should be dealt with in selection of P control strategies.

3) If the macrophyte community "overshoots" during recovery of the bay, what societal problems can be expected and what can be done, especially to avoid premature intervention (e.g. chemical weed control).

4) Identify benefits that accrue solely from the prescribed reduction in P load and those that would require additional restorative measures (including further reductions as well as different measures).

5) Examine whether fishery management might be used to improve water quality i.e. to manipulate fish-zooplankton-algae interactions.

6) Determine the extent to which biological activity, by direct and indirect measures of production, biomass and biological community characteristics, is related to water quality diagnosed by chemical methods and compare both biological and chemical measures to direct assessment of benefits.

Subgoal H

To be aware of lessons in methodology that might be usefully transferred to the planning of other management programs.

Objectives

1) Develop a low-cost efficient and effective scheme for monitoring which would serve to indicate improvements and as an early-warning system should degradation occur in the future.

Index

D

E

H

I

Q

R

S

About the Editors

Dr. C.K. Minns has been a senior scientist of DFO since 1974 and played a key role in implementing ecosystem approach in the management of project Quinte. He has published 103 papers and 16 book chapters. He has served as guest editor of two special issues of AEHM *on Bay of Quinte and also as the head of the Fish Habitat section and lead advisor of the Project Quinte.*

Dr. Minns continues to be highly productive even after his retirement in 2005, as an adjunct professor at the University of Toronto working with students and colleagues at DFO and other agencies. He is also an editorial advisor of the AEHMS and serves as its executive committee member. This book is an evidence of his deep dedication to Great Lakes science. He was awarded the Queen's Golden Jubilee Medal in 2002 for public service and the DFO Prix d'Excellence in 2008 for his contribution to IPCC 4.

Since 1969, Dr. Mohiuddin Munawar has been actively involved in Great Lakes research as a senior research scientist at Fisheries & Oceans Canada, Burlington, Ontario. Dr. Munawar has published over 280 peer-reviewed papers and delivered over 400 scientific presentations, including several keynotes, plenaries and invited presentations. He has published two original treaties dealing with the phytoplankton dynamics of the North American Great Lakes. His field of specialization is phytoplankton structure and ecology, microbial loop, planktonic foodweb, ecosystem health and management of the Laurentian Great Lakes. In addition, Dr. Munawar has convened and chaired many symposia and conferences across the world.

He has been professionally active on a global basis as the founder-president of the Aquatic Ecosystem Health & Management Society (AEHMS), Canada; founder/chief editor of Aquatic Ecosystem Health & Management (AEHM) *and also founder/editor of the Ecovision World Monograph Series (EWMS) with worldwide distribution by Michigan State University Press. He has organized and edited more than 50 special issues of the* AEHM *on various topics and in addition, has edited more than 20 books on a variety of ecological, environmental and ecosystem science topics.*

Dr. Munawar has received several awards and distinctions such as:

- *The Chandler-Misener Award for best scientific paper of the year, IAGLR, 1978*
- *The Anderson-Everett Award for life-time contribution to Great Lakes science by IAGLR, 1998*
- *Honorary doctorate (Doctor Honoris Causa) from Göteborg University, Sweden, 2007*
- *Assistant Deputy Minister's Distinction Award by Fisheries & Oceans Canada, 2007*
- *Fellowship of the Academy of Science, Engineering & Technology (ASET), India, 2008*
- *Life Time Achievement Award (LTAA) for Great Lakes Science by IAGLR, 2012*
- *Honorary Fellowship by the Inland Fisheries Society of India (IFSI), Kolkota, 2013*
- *Jack Vallentyne Outreach award by IAGLR, 2019*

Marten A. Koops is a Research Scientist with the Great Lakes Laboratory for Fisheries and Aquatic Sciences, Fisheries and Oceans Canada, where he served as the lead for the Bay of Quinte foodweb monitoring program for a ten-year period (2007-2017). His research aims to better understand the factors that determine or limit the productivity of freshwater fishes, focusing on anthropogenic stressors, environmental conditions, life history traits, foodweb function, and the interactions among these factors. This research is applied through ecological models and risk assessments to inform management. His research and science advice has been recognized by Fisheries and Oceans Canada through a Prix d'Excellence and an ADM Science Distinction Award.